AF344531

In Vitro Digestibility in Animal Nutritional Studies

In Vitro Digestibility in Animal Nutritional Studies

Special Issue Editor

Pier Giorgio Peiretti

MDPI • Basel • Beijing • Wuhan • Barcelona • Belgrade • Manchester • Tokyo • Cluj • Tianjin

Special Issue Editor
Pier Giorgio Peiretti
Institute of Sciences of Food Production
National Research Council
Italy

Editorial Office
MDPI
St. Alban-Anlage 66
4052 Basel, Switzerland

This is a reprint of articles from the Special Issue published online in the open access journal *Animals* (ISSN 2076-2615) (available at: https://www.mdpi.com/journal/animals/special_issues/In_vitro_digestibility).

For citation purposes, cite each article independently as indicated on the article page online and as indicated below:

LastName, A.A.; LastName, B.B.; LastName, C.C. Article Title. *Journal Name* **Year**, *Article Number*, Page Range.

ISBN 978-3-03936-459-6 (Hbk)
ISBN 978-3-03936-460-2 (PDF)

Contents

About the Special Issue Editor

Pier Giorgio Peiretti, Doctor in Veterinary Medicine and Post-Graduate specialization at the school Inspection of Products of Animal Origin, is a Senior Researcher with the Italian National Research Council. He has a Full Professorship in Animal Nutrition with the Italian Ministry of Universities and Research. He is an expert peer reviewer for *Italian Scientific Evaluation* (REPRISE), an expert in evaluation of VQR 2011–2014 with the National Agency for the Evaluation of Universities and Research Institutes, an expert in evaluation for the Third Mission at National Agency for the Evaluation of Universities and Research Institutes (ANVUR), a principal investigator or scientific person in charge for workpackage in several projects, and Guest Editor of: Animals and the Journal of Food Quality. He is a reviewer for: *Agriculture; Agronomy Journal; Animal, Animals, Animal Feed Science and Technology; Animal Production Science; Annals of Animal Science; Antioxidants; Aquaculture; BMC Veterinary Research; British Journal of Nutrition; CyTA—Journal of Food, Dyes and Pigments; Food Chemistry; Food & Function, Food Research International, Foods, Industrial Crops and Products, International Food Research Journal, International Journal of Livestock Production; Italian Journal of Food Science; Journal of Animal Physiology and Animal Nutrition; Journal of Animal Science; Journal of Applied Animal Research; Journal of Equine Veterinary Science; Journal of Food Biochemistry; Journal of Functional Foods; Journal of the American Oil Chemists' Society; Journal of the Science of Food and Agriculture; Large Animal Review; Livestock Research for Rural Development; Livestock Science; Meat Science; Metabolites; Molecules; Natural Product Research; Plant Biosystems; Plant Foods for Human Nutrition; and World Rabbit Science* (see https://publons.com/researcher/1329844/pier-giorgio-peiretti/). His research activity is well documented by 262 papers about studies on animal nutrition and food science (see: http://www.cnr.it/people/piergiorgio.peiretti).

Editorial

Introduction to the Special Issue: In Vitro Digestibility in Animal Nutritional Studies

Pier Giorgio Peiretti

Institute of Sciences of Food Production, National Research Council, 10095 Grugliasco, Italy;
piergiorgio.peiretti@ispa.cnr.it

Received: 22 May 2020; Accepted: 25 May 2020; Published: 27 May 2020

The use of animals in research elicits a diverse range of attitudes and emotions, with some people demanding the abolition of research on animals and others expressing strong support. Typically, opponents of animal research cite animal welfare and suffering, as well as the uselessness of digestibility trials. In vitro techniques for feed evaluation are important methodologies for studying the physiology of certain segments of the digestive tract and the fermentative and digestive characteristics of feed.

Given the above, the trend is to prefer the use of in vitro enzymatic analysis over in vivo studies, which are more costly, laborious, and require animals anyway. Different closed-system fermentation apparatuses have been used in digestibility studies in ruminants and monogastric and companion animals. The incubators, which have been developed for multiple analyses of feeds, have reduced labor demands and improved precision compared to traditional in vitro methods, which are time-consuming and imprecise. Furthermore, they could offer an alternative system to traditional in vivo methods, but further studies are needed to fully assess the potential of different methods and apparatuses in animal nutritional studies.

This book is a Reprint of the papers published in the Special Issue: "In Vitro Digestibility in Animal Nutritional Studies".

Chapter 1—In vitro gas production systems are regularly utilized to screen feed ingredients for inclusion in ruminant diets. However, not all in vitro systems are set up to measure methane (CH_4) production, nor do all papers report in vitro CH_4. Therefore, the objective of this study was to develop models to predict in vitro production of CH_4, a greenhouse gas produced by ruminants, from in vitro gas and volatile fatty acid (VFA) production data, and to identify the major drivers of CH_4 production in these systems. Meta-analysis and machine learning (ML) methodologies were applied to predict CH_4 production from in vitro gas parameters. The meta-analysis results indicate that equations, containing apparent dry matter (DM) digestibility, total VFA production, propionate, valerate and feed type (forage vs. concentrate) resulted in best prediction of CH_4. The ML models far exceeded the predictability achieved using meta-analysis, but further evaluation of an external database would be required to assess their generalization capacity. The models developed can be utilized to estimate CH_4 emissions in vitro.

Chapter 2—In the last years, there has been increasing interest in the use of forages containing condensed tannins (CT) in ruminant nutrition. Condensed tannins can reduce the methane emissions and the ruminal degradation of protein, improving the animal performances to different extents depending on the source and dose of CT. In vitro fermentation of sainfoin has not been studied in fresh forage. The effect of CT can be studied in comparison with a similar CT-free forage or using polyethylene glycol (PEG), which is a tannin-blocking agent. The maturity stage influences the chemical composition to a different degree depending on the legume species, and can affect the content and fractions of CT. This trial aimed to compare the fermentation parameters of sainfoin with or without PEG, to detect the differences arising from CT, at different stages of maturity (vegetative, start-flowering, and end-flowering) and compare them with the fermentation parameters of alfalfa. The main results were that sainfoin had greater in vitro organic matter degradability (IVOMD) and lower ammonia and acetic:propionic ratio than alfalfa. Sainfoin CT affected the ammonia and individual fatty acid proportions. In conclusion, fermentation end-products were affected both by the chemical composition and CT contents.

Chapter 3—Dietary methane mitigation strategies do not necessarily make food production from ruminants more energy-efficient, but reducing methane (CH_4) in the atmosphere immediately slows down global warming, helping to keep it within 2 °C above the pre-industrial baseline. There is no single most efficient strategy for mitigating enteric CH_4 production from domestic ruminants on forage-based diets. This study assessed a wide variety of dietary CH_4 mitigation strategies in the laboratory, to provide background for future studies with live animals on the efficiency and feasibility of dietary manipulation strategies to reduce CH_4 production. Among different chemical and plant-derived inhibitors and potential CH_4-reducing diets assessed, inclusion of the natural antimethanogenic macroalga Asparagopsis taxiformis showed the strongest, and dose-dependent, CH_4 mitigating effect, with the least impact on rumen fermentation parameters. Therefore, applying Asparagopsis taxiformis at a low daily dose was the best potential dietary mitigation strategy tested, with promising long-term effects, and should be further studied in diets for lactating dairy cows.

Chapter 4 - The present study comparatively investigates the inhibitory difference of nitroethane (NE), 2-nitroethanol (NEOH), and 2-nitro-1-propanol (NPOH) on in vitro rumen fermentation, microbial populations, and coenzyme activities associated with methanogenesis. The results showed that both NE and NEOH were more effective in reducing ruminal methane (CH_4) production than NPOH. This work provides evidence that NE, NEOH, and NPOH were able to inhibit methanogen population and dramatically decrease methyl-coenzyme M reductase gene expression and the content of coenzymes F_{420} and F_{430} with different magnitudes in order to reduce ruminal CH_4 production.

Chapter 5—Inoculum from different feeding types of the ruminant species host has unequal tolerance and effects to condensed tannin (CT) due to their respective feeding strategies behavior producing different ruminal microbiota profiles. This paper describes that in long-term incubation, CT plant extract addition affects in vitro fermentation kinetics more severely in grazing ruminant than browsing ruminants.

Chapter 6—A sudden change from a milk/forage diet to a high concentrate diet in young ruminants increases the rate and extent of rumen microbial fermentation, leading to digestive problems, such as acidosis. The magnitude of this effect depends on the nature of the ingredients. Six carbohydrate sources were tested: Three cereal grains (barley, maize and brown sorghum), as high starch sources of different availability, and three byproducts (sugarbeet pulp, citrus pulp and wheat bran), as sources of either insoluble or soluble fibre. An in vitro semi-continuous incubation system was used to compare the fermentation pattern of substrates incubated with inocula-simulating concentrate or forage diets, under the pH and liquid outflow rate conditions of intensive feeding systems. The magnitude of microbial fermentation was higher with the concentrate than the forage inoculum, and the drop in pH in the first part of incubation was more profound. Among the substrates, citrus pulp had a greater acidification potential and was fermented at a higher extent, followed by wheat bran and barley. In conclusion, the acidification capacity of substrates plays an important role in environmental conditions, depending on the type of diet given to the ruminant. This in vitro system allows us to compare the substrates under conditions simulating high-concentrate feeding.

Chapter 7—Essential oils (EO) can be used as natural alternatives to in-feed antibiotics. Most EO products in the market are based on a combination of EO or their active molecules but prove that additivity or synergy is lacking. The effect of six EO (tea tree oil—TeTr, oregano oil—Ore, clove bud oil—Clo, thyme oil—Thy, rosemary oil—Ros and sage oil—Sag) and different mixes on in vitro microbial fermentation profile of a feedlot beef cattle type fermentation were evaluated for their additive, synergistic or antagonistic effects. Mixing TeTr with Thy, Ore or Thy + Ore modified rumen microbial fermentation profile, but the size of the effect was similar to that obtained with TeTr alone, suggesting that the effects were not additive. When Thy, Ore or Thy + Ore were mixed with Clo, most effects on rumen fermentation profile disappeared, even when TeTr was part of the mix, suggesting an antagonistic interaction of Clo with Thy and Ore. The results do not support the hypothesis of additivity among the EO tested, and antagonistic effects may occur among some of them, at least in a low pH, beef-type fermentation conditions.

Chapter 8—Forages are an essential portion of ruminant rations to maintain rumen function. Exploring how orchardgrass and alfalfa interact in the rumen is necessary to better understand their feed use potential as both hay and silage. This study evaluated in vitro rumen degradation, fermentation characteristics, and methane production responses to different forage ratios of alfalfa and orchardgrass. The results indicate that dry matter and organic matter degradability and methane production were greater for mixed silages, compared to mixed hays. A forage ratio of 50:50 for orchardgrass and alfalfa favor the growth of rumen microorganisms without compromising nutrient digestion and rumen fermentation.

Chapter 9—Currently, vegetable protein sources such as soybean meal and rapeseed meal are expensive and with volatile prices. These economic circumstances are driving the research of potential new protein resources for beef cattle diets that can reduce the ration cost without compromising animal productive yields. As possible candidates, camelina meal and camelina expeller have been studied; they are co-products with a high protein percentage, obtained after oil extraction from the oil seeds of Camelina sativa. The objectives of this study were to characterize these camelina co-products and ascertain if they could be useful ingredients for beef cattle diets. The results indicate that the diets, formulated with camelina meal and camelina expeller, do not show differences in the efficiency of microbial protein synthesis, compared to the current reference proteins, camelina meal diet being the most similar to soybean meal and rapeseed meal diets, and camelina expeller the diet with the highest fermentation potential. The results of soybean meal as an individual ingredient reveal more differences with camelina co-products. In vivo studies are necessary to draw conclusions, but in vitro results obtained suggest that camelina meal and camelina expeller are potential substitutes for rapeseed meal in beef cattle diets.

Chapter 10—The objective of this study was to determine the relationships between milk odd- and branched-chain fatty acids (OBCFAs) and ruminal fermentation parameters, microbial populations, and base contents. Significant relationships existed between the concentrations of C11:0, iso-C15:0, anteiso-C15:0, C15:0, and anteiso-C17:0 in rumen and milk. The total OBCFA content in milk was positively related to the acetate molar proportion but negatively correlated with isoacid levels. The adenine/N ratio was negatively related to milk OBCFA content but positively associated with the iso-C15:0/iso-C17:0 ratio.

Chapter 11—The evaluation of fibre digestibility is very important for the formulation of ruminant diets. Fibre digestibility is usually determined in a laboratory, with rumen inoculum obtained from cannulated cows. The research of alternative and less invasive inoculum sources is a critical issue that should be addressed. The present study evaluated the potential of faecal inocula, obtained from cows fed different diets, to assess fibre digestibility of different substrates at different incubation times (48, 240 and 360 h). At short incubation times, fibre digestibility obtained with rumen fluid was always higher than those obtained with faecal inocula, confirming a lower activity of the faecal inocula, compared with rumen fluid. However, the type of diets fed to the donor animals had a significant effect on fibre digestibility, with a more active faecal inoculum for cows fed a diet based on maize silage. Despite the differences obtained at the short incubation time, the digestibility values at longer intervals showed that faecal inoculum could replace rumen inoculum. As a consequence, faeces may replace rumen fluid as inoculum for end-point measures, avoiding the use of cannulated animals and decreasing the analytical costs.

Chapter 12—The utilization of animal donors of rumen fluid for laboratory experiments can raise ethical concerns due to invasive methods of collection (rumen cannulated or intubated animals). Societies are strongly oriented to support cruelty free experiments and alternatives to the collection of rumen fluids from live animals are urgently requested from the scientific community. Thus, in order to attenuate the dependence of laboratories on animal donors, this study compared the rumen inoculum, collected at slaughter, with the fermentation liquid from a rumen continuous fermenter and both rumen inoculum were used fresh or preserved (by refrigeration, chilling and freeze-drying). The results support the theory for using continuous fermenters to generate inoculum for in vitro purposes, and short-term refrigeration is confirmed to be a valuable storage system to facilitate transfer inoculum from the collection sites. These findings should attenuate the need for laboratories' frequent collections from animals while continuing research in ruminant nutrition.

Chapter 13—The use of seaweeds as ingredients of ruminant diets can be an alternative to conventional feedstuffs, but it is necessary to assess their nutritive value. The aim of this study was to analyze the chemical composition and in vitro rumen fermentation of eight brown, red and green seaweed species collected in Norway during both, spring and autumn. The in vitro ruminal fermentation characteristics of 17 diets composed of oat hay:concentrate in a 1:1 ratio, with the concentrate, containing no seaweed or including one of the 16 seaweed samples, was also studied. Species and season determined differences in chemical composition and in vitro fermentation of seaweeds. Most of the tested seaweeds can be included in the diet (up to 200 g/kg concentrate) without negative effects on in vitro ruminal fermentation.

Chapter 14—Winter brassica crops such as kales and swedes are used to supply feed in times of seasonal shortage. However, to the best of our knowledge, there is little information about the fermentation characteristics of these forages in the rumen. This study assessed the nutrient concentration, in vitro fermentation and in situ rumen degradation characteristics of Brassica oleracea (L.) ssp. acephala (kales) and Brassica napus (L.) ssp. napobrassica (swedes). The kales and swedes both showed different nutrient concentrations and fermented fast and extensively in the rumen. However, in vitro fermentation of swedes resulted in lower acetate and greater proportions of butyrate and propionate. Varieties of swedes showed more differences in terms of degradation and fermentation in the rumen compared to kale varieties.

Chapter 15—The common vetch (Vicia sativa L.) is an important legume crop of mixed crop-livestock systems that provides high-quality grains used as food/feed and straw used as ruminant feed. The objective of this study was to determine the variability in grain yield, straw yield, straw chemical composition, carbohydrate and protein fractions, in vitro gas production, and in situ ruminal degradability of four different varieties of common vetch grown on the Qinghai-Tibetan Plateau. The results showed that grain yield, straw yield, and straw nutrient value varied significantly among the four varieties. Overall, the findings indicated that in terms of straw yield and nutritive quality, variety Lanjian No. 1 has the greatest potential as a crop for supplementing ruminant diets in the smallholder mixed crop–livestock systems on the Qinghai-Tibetan Plateau.

Chapter 16—Common vetch (Vicia sativa L.) grain is an important source of protein in rations for ruminants, but little information is available on the protein value of common vetch grains, both in terms of chemical composition and protein degradability, and regarding variation between intra-species and year. The objective of this study was to evaluate grain chemical composition, ruminal protein degradability in vivo, and intestinal protein digestibility in vitro of four common vetch varieties over two cropping years on the Tibetan Plateau. This study was also conducted to establish correlations of grain chemical composition with ruminal degradability parameters of grain protein and with intestinal digestibility of grain protein. The results of this study demonstrated that grain quality characteristics varied significantly among varieties and years. The relationship between grain chemical composition and intestinally absorbable digestible protein (IADP) was best described by a linear regression equation, and coefficients of determination remained very high ($R^2 = 0.891$). Overall, the results indicated that in terms of effective crude protein degradability and IADP of grain, common vetch varieties Lanjian Number 2 and Lanjian Number 3 have the greatest potential among varieties examined for supplementing ruminant diets when grown on the Tibetan Plateau.

Chapter 17—Pea grains may partially replace soybean or rapeseed meals and cereals in ruminant diets, but this is limited by high solubility of pea protein in the rumen. Hydro-thermic treatments, such as toasting may stabilize the protein and shift digestion from the rumen to the small intestine.

The effect of toasting of ensiled pea grains on rumen-undegraded protein was tested in vitro and on apparent digestibility of organic matter, gross energy, and proximate nutrients in a digestion trial with sheep. Ensiling plus toasting increased rumen-undegraded protein from 20 to 62% of crude protein, but it also increased acid detergent insoluble protein, which is unavailable for digestive enzymes in the small intestine from 0.5 to 2.6% of crude protein. Ensiling plus toasting did not, however, affect total tract apparent digestibility of organic matter, energy, crude protein, or any other nutrient fraction, nor did it alter the concentration of metabolizable energy or net energy lactation in the peas. The technique can be implemented on farms and might have a positive impact on field pea production.

Chapter 18—Feedstuff evaluation through animal trials is time consuming and expensive. An alternative, the gas production method, measures the amount of fermentation gas produced from incubating feedstuffs with microbes from ruminal fluid or faecal samples. The models can be applied to gas production profiles to determine extent of feedstuff degradation, either in the rumen or in the hindgut. Typical gas production profiles show a monotonically increasing monophasic pattern. However, atypical gas production profiles exist whereby at least two consecutive phases of gas production are present; these profiles are much less well described. Two models are proposed to fit these biphasic profiles, a sum of two Mitscherlich equations, and sum of Mitscherlich + linear equations. Additionally, two models that describe typical monophasic gas production curves, the simple Mitscherlich and the generalised Mitscherlich (root-t) model, were assessed for comparison. The models were fitted to 25 gas production profiles, arising from incubating feedstuffs with faecal inocula from equines. Of these 25 profiles, 17 displayed atypical biphasic patterns, and 8 displayed typical monophasic patterns. The two biphasic models were found to describe both the atypical and typical gas production profiles accurately. These models allow for the evaluation of feedstuffs using cost- and time-efficient methods.

Chapter 19—Various in vitro methodologies have been developed and used to estimate the digestibility of feed ingredients, such as corn distillers dried grains with solubles (cDDGS) and soybean hulls (SBH) which contain high concentrations of dietary fiber. This study evaluated two in vitro gas production recording systems (manual versus automated) and two initial fecal inoculum volumes (30 versus. 75 mL) on the parameters of in vitro fermentation of cDDGS and SBH. The results showed that the use of 75-mL inoculum volume with 0.5 g substrate tended to reduce the variation of measurements compared to the 30-mL inoculum volume with 0.2 g substrate regardless of the gas production recording system. These findings suggest that using larger inoculum volume with more substrate increases the precision of measurements. Furthermore, the automated system decreases labor for conducting the assay.

Chapter 20—Over the years, broiler chickens have been selected for rapid growth which makes them very efficient at depositing body protein in a short period of time. This is important since the broiler sector is expected to contribute to the growing global demand for poultry meat. In light of this, the quality of proteins fed to poultry is becoming more important. The concept of protein nutrition is based on the sequential process through which proteins are digested, and the amino acids are absorbed and become available for metabolic processes. The nutritional quality of protein ingredients for poultry is based on their amino acid bioavailability. Animal and plant ingredients are the main sources of protein used in poultry diets and they vary in digestibility and amino acid composition. Although, in vivo digestibility assays for poultry are available, they are expensive and time consuming to conduct. In vivo digestibility assays are the optimum tools for characterizing protein sources to be used in commercial production. However, it is not always practical to conduct these assays in commercial settings. Commercial production, therefore, relies on the use of other assays such as in vitro assays to evaluate the quality of protein sources.

Chapter 21—The Ankom Daisy[II] incubator (AD[II]; Ankom Technology Corporation Fairport, NY, USA) has gained acceptance as an alternative to traditional in vitro procedures. It reduces the labour requirement and increases the number of determinations that can be completed by a single operator. The apparatus allows for the simultaneous incubation of several feedstuffs in sealed polyester bags in the same incubation vessel, which is rotated continuously at 39.5 °C. With this method, the material that disappears from the bag during incubation is considered digestible. The method, which was first developed to predict the digestibility of feedstuffs for ruminants, has been modified and adapted to improve its accuracy and prediction capacity. Modifications used by various researchers include the use of different inocula, buffer solutions, and sample weights. Recently, attempts have been made to adapt the method to determine nutrient digestibility of feedstuff in non-ruminant animals, including pets.

Article

Application of Meta-Analysis and Machine Learning Methods to the Prediction of Methane Production from In Vitro Mixed Ruminal Micro-Organism Fermentation

Jennifer L. Ellis [1],*, Héctor Alaiz-Moretón [2], Alberto Navarro-Villa [3,4], Emma J. McGeough [3,5], Peter Purcell [3], Christopher D. Powell [1], Padraig O'Kiely [3], James France [1] and Secundino López [6],*

[1] Centre for Nutrition Modelling, Department of Animal Biosciences, University of Guelph, 50 Stone Road East, Guelph, ON N1G 2W1, Canada; cpowell@uoguelph.ca (C.D.P.); jfrance@uoguelph.ca (J.F.)
[2] Departamento de Ingeniería Eléctrica de Sistemas y Automática, Escuela de Ingeniería Industrial e Informática, Universidad de León, Campus Universitario de Vegazana, 24071 León, Spain; hector.moreton@unileon.es
[3] Animal & Grassland Research and Innovation Centre, Teagasc, Grange, Dunsany, Co. Meath C15 PW93, Ireland; alberto.navarro.villa@trouwnutrition.com (A.N.-V.); emma.mcgeough@umanitoba.ca (E.J.M.); peter.purcell86@gmail.com (P.P.); Padraig.OKiely@teagasc.ie (P.O.)
[4] Trouw Nutrition R&D, Ctra. CM-4004 km 10.5, 45950 El Viso de San Juan, Spain
[5] Department of Animal Science, University of Manitoba, Winnipeg, MB R3T 2N2, Canada
[6] Instituto de Ganadería de Montaña (IGM), CSIC-Universidad de León, Departamento de Producción Animal, Universidad de León, 24007 León, Spain
* Correspondence: jellis@uoguelph.ca (J.L.E.); s.lopez@unileon.es (S.L.); Tel.: +1-519-824-4120 (ext. 56522) (J.L.E.); +34-987-291-291 (S.L.)

Received: 10 April 2020; Accepted: 16 April 2020; Published: 21 April 2020

Simple Summary: In vitro gas production systems are regularly utilized to screen feed ingredients for inclusion in ruminant diets. However, not all in vitro systems are set up to measure methane (CH_4) production, nor do all papers report in vitro CH_4. Therefore, the objective of this study was to develop models to predict in vitro production of CH_4, a greenhouse gas produced by ruminants, from in vitro gas and volatile fatty acid (VFA) production data, and to identify the major drivers of CH_4 production in these systems. Meta-analysis and machine learning (ML) methodologies were applied to predict CH_4 production from in vitro gas parameters. Meta-analysis results indicate that equations containing apparent dry matter (DM) digestibility, total VFA production, propionate, valerate and feed type (forage vs. concentrate) resulted in best prediction of CH_4. The ML models far exceeded the predictability achieved using meta-analysis, but further evaluation on an external database would be required to assess their generalization capacity. The models developed can be utilized to estimate CH_4 emissions in vitro.

Abstract: In vitro gas production systems are utilized to screen feed ingredients for inclusion in ruminant diets. However, not all in vitro systems are set up to measure methane (CH_4) production, nor do all publications report in vitro CH_4. Therefore, the objective of this study was to develop models to predict in vitro CH_4 production from total gas and volatile fatty acid (VFA) production data and to identify the major drivers of CH_4 production in these systems. Meta-analysis and machine learning (ML) methodologies were applied to a database of 354 data points from 11 studies to predict CH_4 production from total gas production, apparent DM digestibility (DMD), final pH, feed type (forage or concentrate), and acetate, propionate, butyrate and valerate production. Model evaluation was performed on an internal dataset of 107 data points. Meta-analysis results indicate that equations containing DMD, total VFA production, propionate, feed type and valerate resulted in best predictability of CH_4 on the internal evaluation dataset. The ML models far exceeded the predictability achieved using meta-analysis, but further evaluation on an external database would be required to assess generalization ability on unrelated data. Between the ML methodologies assessed, artificial neural networks and support vector regression resulted in very similar predictability, but differed in fitting, as assessed by behaviour analysis. The models developed can be utilized to estimate CH_4 emissions in vitro.

Keywords: in vitro gas production; methane; rumen; feed; meta-analysis; machine learning; neural network

---

1. Introduction

Globally, greenhouse gas (GHG) emissions from the agriculture, forestry and other land use (AFOL) sector account for ~23% of the global anthropogenic GHG total emissions [1], with enteric methane (CH_4) from fermentation in the forestomach of ruminants representing 32%–40% of that total [1] (thereby 7.4%–9.2% of the global anthropogenic total). From the farmer's perspective, CH_4 also represents an energy loss and an inefficiency of production, ranging from approximately 3.0 (feedlot cattle) to 7.0 (forage fed cattle) percent of gross energy intake, with a ±20% uncertainty [2]. As a result, and to meet public expectation for sustainably produced food products, the agriculture sector has mobilized to examine a large array of potential CH_4 (as well as N and P excretion) mitigation strategies [3–5], to reduce the environmental impact of livestock and food production.

At the animal level, CH_4 is produced as a byproduct of anaerobic fermentation in the rumen and hindgut of ruminants, whereby methanogens utilize H_2 to obtain ATP by reducing CO_2 to CH_4 [6]. The removal of H_2 through methanogenesis, the main H-sink in the rumen [6], prevents the inhibitory effect of H_2 on ruminal fermentation and allows for the degradation and fermentation of feed to proceed. When methanogenesis is reduced, other pathways must be promoted to utilize H_2 or otherwise fermentation, digestibility and intake may be negatively affected [6].

As animal experiments to evaluate feedstuffs and feed additives are costly, time consuming and do not guarantee conclusive outcomes, the in vitro gas production technique represents a viable option for prescreening or screening of feedstuffs/additives for potential inclusion in the ration of modern dairy cows, beef cattle and other ruminants. However, CH_4 is often, but not always, included in the gases measured during in vitro incubation (particularly in developing countries where equipment may be unaffordable, unavailable or limited, for example). A reliable measure of CH_4 from in vitro cultures of mixed ruminal micro-organisms would be a useful tool to assess the potential dietary effects on methanogenesis. Estimation of CH_4 from the output of other fermentation end-products commonly measured in vitro could be a suitable alternative, and Jayanegara [7] proposed the use of the stoichiometric equations of Hegarty and Nolan [8] and of Moss et al. [9] to predict in vitro CH_4. However, using this approach CH_4 was generally overpredicted, presumably because in vitro H_2 recovery observed in practice was substantially less than that assumed by the stoichiometric models. The objectives of this study were therefore to: (1) to develop empirical models to predict in vitro CH_4 production from in vitro gas production measures—via meta-analysis (multiple linear regression) and machine learning (ML) methods (artificial neural networks, ANN, and support vector regression, SVR), and (2) to identify the fermentation parameters most closely related to CH_4 production in vitro.

2. Materials and Methods

2.1. Database

The database compiled for this study consisted of 397 in vitro rumen fermentation bottle means (each the average of 3–5 replicate measurements), taken after 24 h of incubation, from 13 experiments reported in 10 publications [10–19] (experiments 1–3 were from publication [10]), plus 1 unpublished study [20]. As a result, experimental animals were not directly employed in this study. In accordance with the National Centre for the Replacement Refinement and Reduction of Animals in Research (NC3Rs), per Directive 2010/63/EU, all study data used were publicly available (with the exception of the one unpublished study) as reported in the aforementioned articles. Studies evaluated the in vitro gas and CH_4 production from oven-dried feedstuffs, including ryegrass, forbs, grass silages, clover, maize silage and other whole-crop cereal silages and concentrate feeds (no feed additives or rumen modifiers were included in the database). Feed type (FT) was categorized as either forage (FT = 1) or concentrate (FT = 2). The database included in vitro measurements of CH_4 gas production (CH_4i, mL/g DM incubated, and CH_4d, mL/g DM apparently digested), total gas production (TGP, mL/g DM incubated), apparent DM digestibility (DMD, g/g), volatile fatty acid production (VFA, mmol/g DM incubated), molar proportions of acetic acid (AC, mmol/mol VFA), propionic acid (PR, mmol/mol VFA), butyric acid (BT, mmol/mol VFA) and valeric acid (VL, mmol/mol VFA), the acetate to propionate ratio (C2C3) and the incubation medium final pH (pH). Daily production of each volatile fatty acid (ACp, PRp, BTp or VLp for mmol AC, PR, BT or VL produced per g DM incubated, respectively) was calculated from total VFA and the corresponding molar proportions. Variable abbreviations, units and descriptions are also summarized in Appendix A (Table A1). When digestibility is measured during in vitro batch cultures of mixed ruminal micro-organisms, it is assumed that DM disappearance after the incubation time (in this particular case 24 h) is an acceptable metric of apparent DM digestibility. Due to missing data, two studies [11,12], were removed from the database, leaving 354 observations from 11 experiments.

For model development and evaluation purposes, the dataset (n = 354) was divided into two subsets, the first one for training and model development purposes (70% of data, n = 247, with 4 outlier data points removed for meta-analysis), and the second one for model testing and evaluation (internal evaluation) purposes (30% of data, n = 107). Aside from 4 data points which were removed for the meta-analysis (statistical outliers), the 'two' developmental datasets were identical. Division of data points into the training or evaluation datasets was via random assignment, but each contained a proportional number of observations relative to the FT variable. Descriptive statistics for the training and evaluation datasets are provided in Table 1.

Table 1. Summary of the training (n = 247, 243) and internal evaluation (n = 107) datasets.

Variable [1]	pH	DMD	TGP	CH_4i	CH_4d	VFA	AC	PR	BT	VL	C2C3
Training Dataset (Machine learning, n = 247)											
Mean	6.59	0.67	163	26.6	40.7	5.48	632.3	238.5	94.8	31.8	2.7
Median	6.64	0.68	160	25.3	38.0	5.33	632.2	237.7	94.4	31.8	2.7
Minimum	5.45	0.20	51	6.9	19.5	2.12	477.6	117.5	43.4	5.0	1.4
Maximum	6.78	0.91	276	50.8	71.5	9.79	812.1	346.9	181.3	79.3	7.0
Training Dataset (Meta-analysis, n = 243)											
Mean	6.59	0.67	164	26.7	40.6	5.50	631.7	239.2	94.9	31.9	2.7
Median	6.64	0.68	161	25.3	38.0	5.35	632.1	237.9	94.4	31.8	2.7
Minimum	5.45	0.22	71	11.2	19.5	2.12	477.6	117.5	60.0	10.3	1.4
Maximum	6.78	0.91	276	50.8	71.5	9.79	812.1	346.9	181.3	79.3	7.0
Evaluation Dataset (Machine learning/Meta-Analysis, n = 107)											
Mean	6.60	0.66	162	26.0	40.2	5.51	630.6	241.5	94.5	31.8	2.7
Median	6.64	0.68	162	25.4	37.9	5.25	629.0	238.5	96.4	32.5	2.7
Minimum	5.49	0.21	61	8.7	22.0	4.02	503.7	152.1	53.8	6.9	1.7
Maximum	6.84	0.87	246	43.0	70.9	9.81	787.5	333.1	178.9	60.8	5.1

[1] Variables (units): pH = final pH in the incubation medium. DMD = apparent dry matter (DM) digestibility (g DM disappeared/g DM incubated). TGP = total gas production (mL gas/g DM incubated). CH_4i = methane production (mL CH_4/g DM incubated). CH_4d = methane production (mL CH_4/g DM apparently digested). VFA production (mmol total VFA/g DM incubated). AC = acetic acid (mmol AC/mol VFA). PR = propionic acid (mmol PR/mol VFA). BT = butyric acid (mmol BT/mol VFA). VL = valeric acid (mmol VL/mol VFA). C2C3 = acetate to propionate ratio.

Independent testing of the model gives a measure of the model's 'generalization ability' ('test error'), or the ability to make predictions on unseen data. This is particularly important for some ML approaches, which may achieve very accurate predictions, but essentially model the noise in the data.

2.2. Model Fitting—Meta-Analysis

The main effects of in vitro fermentation variables (TGP, DMD, VFA, AC, ACp, PR, PRp, BT, BTp, VL, VLp, C2C3 and final pH) were analyzed for inclusion in predictive models using the PROC MIXED procedure of SAS [21] to predict CH_4i (mL/g incubated DM), or CH_4d (mL/g DM apparently digested). Equations were fitted to the training dataset (Table 1—meta-analysis).

The mixed model analysis was chosen because the data were compiled from multiple studies, and thus the experiment was considered as a random effect [22]. If, when running the model, the random covariance or the random slope was not significant, they were removed from the model or simplified [22], though the random intercept term was always retained. The dual quasi-Newton technique was used for optimization with an adaptive Gaussian quadrature as the integration method. Normal distribution of the random study effect was assessed via Q–Q distribution plot, and normality of residuals via examination of the residual plots (PROC MIXED).

Three approaches were taken to fitting mixed models to this dataset: (1) univariate analysis of each dependent–independent variable combination (explanatory variable in linear, quadratic or cubic form); (2) multivariate analysis, preceded by examination in PROC REG (MaxR) and assessment for collinearity between driving variables in PROC CORR/visual plotting; and (3) multivariate analysis based on known biological principles. Approaches (2) and (3) are not distinguished/presented separately in the results, as both are considered 'multivariate'. A fourth approach was also included for comparison with the ML models (described below): (4) where all driving variables were included, irrespective of significance or collinearity (linear equation form). With the exception of approach (4), only equations with significant slope parameters ($p < 0.05$) and normally distributed residuals/random effects were retained and evaluated.

2.3. Model Fitting—Machine Learning

The in vitro fermentation variables (TGP, DMD, VFA, AC, PR, BT, VL, C2C3 and final pH) were retained as potential driving variables for development of ML-based predictive models for CH_4i and CH_4d. Predictive models were fitted on the training dataset (Table 1—Machine learning). The raw dataset was subjected to a preprocessing normalization process (standard scalar) [23] according to:

$$Z = (X - u)/S \tag{1}$$

where Z is the normalized value, X is the raw value, u is the mean of the training samples and S is the standard deviation of the training samples. The objective of this normalization step was to improve the convergence of the training process in the regression methods utilized [24]. Subsequently, two ML techniques (support vector regression and artificial neural network) were implemented using the Scikit-learn software library [25] for the Python programming language [26]. For both ML approaches, a 10-fold cross-validation procedure was used to fit the predictive models to the training dataset ($n = 247$; the evaluation database, $n = 107$, was therefore not included in this analysis). The training dataset was subsequently randomly split into 10 equal subgroups, and the model was trained using nine of the subsets and validated on the remaining one part of the data to compute a performance measure. This holdout process was repeated for each of the 10-folds, such that each subset was utilized for validation, whereas the other nine subsets were pooled for the training, in turn. The error estimation was averaged over the 10 iterations to assess the fit performance.

2.3.1. Support Vector Regression

Support Vector Machine (SVM) is a ML technique based on supervised learning with a modality oriented for regression problems, namely Support Vector Regression (SVR), able to forecast continuous variables [27] (in this case, CH_4 from in vitro cultures of mixed ruminal micro-organisms). The SVR method transforms the input data (previously normalized) into a multidimensional space by using nonlinear mapping, and a linear regression procedure is applied to each hyperplane obtained to calculate the desired output. The SVR method is developed by changing the kernel function and tuning the parameters C (the regularization parameter), γ (the kernel coefficient), Tol (tolerance for stopping criterion) and degree of the polynomial. Three 'kernel' functions were considered—linear, radial basis function and polynomial. The ranges of values used for the parameter optimization were $C \in \{1, 10, 100, 1000\}$, Tol $\in \{0.1, 0.01\}$, $\gamma \in \{1, 0.1, 0.01, 0.001\}$ and degree (only for the polynomial function) $\in \{2, 3, 4, 5, 6\}$. Grid search combined with cross-validation [27] were used to achieve the best combination of parameters resulting in the optimal and most robust SVR model solution, on the basis of the ε-insensitive loss function. The best SVR models for both variables to be predicted (CH_4i or CH_4d) were obtained using the radial basis function as kernel, with the parameter values $C = 1000$, Tol $= 0.1$ and $\gamma = 1$.

2.3.2. Artificial Neural Network—Multilayer Perceptron

A multilayer perceptron (MLP) is a machine learning method based on supervised learning, and is a specific topology of a feedforward artificial neural network (ANN) [28]. The MLP network used for the current study was composed of three layers of nodes: the input layer, one hidden layer and an output layer. Achieving the optimal MLP architecture can require tuning a number of hyperparameters such as the number of hidden layers, neurons or iterations. For the current study, one hidden layer was applied and the rectified linear unit (ReLU) nonlinear activation function was implemented in each node (neuron) of this hidden layer (except the input nodes). On the other hand, the single neuron of the output layer utilized the linear activation function [28]. The training procedure was based on the backpropagation technique, using grid search combined with cross-validation [28] to derive the best combination of parameters resulting in the optimal and most robust MLP model solution. The square-error loss function and the limited-memory Broyden–Fletcher–Goldfarb–Shanno (L-BFGS) numerical method were used for optimization. The number of hidden neurons in the best MLP models was 26 and 27 for CH_4i and CH_4d predictions, respectively, (the range tested was from 12–30 neurons in the hidden layer). Other hyperparameters were included in the grid for tuning aiming to optimize the training of the ANN. The best results were observed when early stopping was activated (to prevent overfitting), any prior attributes stored on the estimator were cleared ("warm start" disabled), initial learning rate was set at 10^{-7} and kept constant, and the batch size for each iteration was equal to 1.

2.4. Model Evaluation

Model predictions developed in the current study (via meta-analysis, SVR or ANN) were evaluated using an independent data subset (internal evaluation, as the data are independent but related to the training dataset), described in Section 2.1 and in Table 1. Models were evaluated for their predictability using mean square prediction error (MSPE), calculated as:

$$\text{MSPE} = \sum_{i=1}^{n} \frac{(O_i - P_i)^2}{n} \tag{2}$$

where O_i is the observed value, P_i is the predicted value and n is the number of observations. Square root of the MSPE (RMSPE), expressed as a proportion of the observed mean (RMSPE, %), gives an estimate of the overall prediction error. The RMSPE was decomposed into random (disturbance) error

(ED), error due to deviation of the regression slope from unity (ER), and error in central tendency due to overall bias (EB) [29]. The EB, ER and ED fractions of MSPE were calculated as:

$$EB = \left(\overline{P} - \overline{O}\right)^2 \tag{3}$$

$$ER = \left(S_p - R \times S_o\right)^2 \tag{4}$$

$$ED = \left(1 - R^2\right) \times S_o{}^2 \tag{5}$$

where $\overline{P}$ and $\overline{O}$ are the predicted and observed means, S_p is the standard deviation of predicted values, S_o is the standard deviation of observed values and R is the Pearson correlation coefficient.

Correspondence between predicted and observed values was also assessed by the concordance correlation coefficient (CCC) [30], which was calculated as:

$$CCC = R \times C_b \tag{6}$$

where C_b is a bias correction factor (a measure of accuracy), and R is the Pearson correlation coefficient (a measure of precision). The C_b variable is calculated as:

$$C_b = \frac{2}{\left(v + \frac{1}{v} + \mu^2\right)} \tag{7}$$

where

$$v = \frac{S_o}{S_p} \tag{8}$$

$$\mu = \frac{\overline{O} - \overline{P}}{\sqrt{\left(S_o - S_p\right)}} \tag{9}$$

so that v provides a measure of scale shift, while μ provides a measure of location shift. The v value indicates the change in standard deviation, if any, between predicted and observed values. A positive μ value indicates underprediction, while a negative μ indicates overprediction. Predictions were further evaluated visually against observations (via predicted vs. observed plots) as well as against residuals (residual vs. predicted, not shown).

As one criticism of many ML methodologies remains their lack of transparency (i.e., no predictive equation is produced), models developed via ANN and SVR were further evaluated using behaviour analysis, where model inputs were systematically altered ±10% (in isolation) and the model's 'behavioural' response (% change in output prediction) was assessed (direction and magnitude).

3. Results

3.1. Correlation Matrix Analysis

Potential X variables were evaluated against each other via correlation matrix analysis, to determine the extent of collinearity between X variables (Table 2). X variables that were highly collinear with each other (correlation >0.500) are highlighted in grey (Table 2). X variables that were highly collinear included TGP and DMD, TGP and VFA/ACp/PRp, VFA and ACp/PRp/BTp, AC and PR/PRp, ACp and PRp/BTp, PR and PRp, PRp and BTp, BT and BTp, pH and BTp. These combinations were therefore avoided in multivariate meta-analysis equation development.

Correlation analysis was also used to examine potential correlations between X and Y variables (Table 2). The X variables moderately correlated (>0.300) with CH_4d included DMD (−0.408), AC (0.405), ACp (0.350) and PR (−0.396), while X variables moderately correlated (>0.300) with CH_4i included DMD (0.399), TGP (0.755), VFA (0.472), ACp (0.472), PRp (0.326) and BTp (0.315).

Table 2. Correlation matrix (Pearson correlation coefficients, R) of dependent and independent variables, meta-analysis development dataset ($n = 243$) [1,2].

Variable	CH_4d	CH_4i	pH	DMD	TGP	VFA	AC	ACp	PR	PRp	BT	BTp	VL	VLp
pH	0.027	0.007												
DMD	−0.408	0.399	−0.060											
TGP	0.146	0.755	−0.206	0.707										
VFA	0.223	0.472	−0.437	0.282	0.648									
AC	0.405	−0.014	0.208	−0.474	−0.341	−0.159								
ACp	0.350	0.472	−0.340	0.136	0.544	0.944	0.168							
PR	−0.396	−0.045	−0.125	0.426	0.316	0.094	−0.863	−0.184						
PRp	−0.050	0.326	−0.431	0.439	0.663	0.817	−0.622	0.605	0.640					
BT	−0.035	0.045	−0.360	0.065	0.115	0.250	−0.279	0.136	−0.150	0.143				
BTp	0.103	0.315	−0.566	0.234	0.482	0.798	−0.294	0.679	−0.001	0.633	0.762			
VL	−0.113	0.223	0.273	0.369	0.096	0.007	−0.246	−0.071	0.023	0.016	0.145	0.071		
VLp	0.059	0.458	−0.021	0.439	0.435	0.555	−0.279	0.458	0.069	0.465	0.236	0.491	0.820	
C2C3	0.420	−0.038	0.070	−0.516	−0.373	−0.103	0.921	0.194	−0.924	−0.606	−0.078	−0.133	−0.186	−0.198

[1] Statistical significance of R values ($n = 243$): $p < 0.05$ if $|R| > 0.126$, $p < 0.01$ if $|R| > 0.165$, $p < 0.001$ if $|R| > 0.210$. Light grey boxes (for pairwise correlations between X variables) show $|R| > 0.500$, light blue boxes (for correlations between X and Y variables) show $|R| > 0.300$, [2] Variables (units): CH_4i = methane production (mL CH_4/g DM incubated). CH_4d = methane production (mL CH_4/g DM apparently digested). pH = final pH in the incubation medium. DMD = apparent dry matter (DM) digestibility (g DM disappeared/g DM incubated). TGP = total gas production (mL gas/g DM incubated). VFA production (mmol total VFA/g DM incubated). AC = acetic acid (mmol AC/mol VFA). ACp = AC production (mmol AC/g DM incubated). PR = propionic acid (mmol PR/mol VFA). PRp = PR production (mmol PR/g DM incubated). BT = butyric acid (mmol BT/mol VFA). BTp = BT production (mmol BT/g DM incubated). VL = valeric acid (mmol VL/mol VFA). V_p = VL production (mmol VL/g DM incubated). C2C3 = acetate to propionate ratio.

3.2. Univariate Meta-Analysis Models

Seventy-eight univariate equations to predict CH_4d or CH_4i were developed and evaluated with the variables presented in Table 2, in linear, quadratic or cubic form. Those with nonsignificant slope parameters or model fitting (fixed or random) problems were discarded, and the remaining equations ($n = 22$) were assessed on the evaluation dataset. On average, the CH_4i outcome was predicted with higher CCC and lower RMSPE values compared to the CH_4d outcome (Table 3). The best six performing equations have their model evaluation results presented in Table 3. The best performing univariate equations included the X variables ACp, PRp, DMD, VLp, VFA and TGP. The best performing univariate equations were those predicting CH_4i with TGP as a driving X variable, with a CCC on the evaluation database of 0.644 (quadratic) and 0.650 (linear) (Table 3).

Table 3. Univariate equations—meta-analysis model evaluation via root mean square prediction error (RMSPE) and concordance correlation coefficient (CCC) analysis on the internal evaluation ($n = 107$) database.

Equation [1]	Y	X	Form	Mean [2]	SEM	RMSPE, %	EB, %	ER, %	ED, %	CCC	R	C_b
U1	CH_4d	C2C3	Linear	44.7	0.37	25.7	19	2	79	0.113	0.194	0.579
U2	CH_4d	PR	Quad	44.7	0.35	25.9	18	3	78	0.116	0.185	0.623
U3	CH_4d	AC	Linear	44.8	0.29	25.6	20	1	79	0.124	0.228	0.541
U4	CH_4d	PR	Linear	44.8	0.31	26.1	19	4	77	0.128	0.198	0.649
U5	CH_4d	DMD	Quad	44.3	0.23	24.0	18	2	81	0.182	0.391	0.466
U6	CH_4d	DMD	Linear	44.4	0.24	23.9	19	3	78	0.196	0.432	0.454
U7	CH_4i	PRp	Quad	27.7	0.29	21.0	11	3	87	0.303	0.377	0.803
U8	CH_4i	DMD	Cubic	28.2	0.19	21.8	15	4	80	0.305	0.375	0.813
U9	CH_4i	VLp	Quad	28.0	0.26	20.7	14	0	85	0.314	0.420	0.747
U10	CH_4i	VFA	Linear	27.4	0.16	20.9	7	7	86	0.346	0.390	0.889
U11	CH_4i	TGP	Quad	27.2	0.33	15.5	10	1	90	0.644	0.717	0.898
U12	CH_4i	TGP	Linear	27.3	0.31	15.5	10	0	89	0.650	0.717	0.906

[1] Equation ID corresponds to equations presented in subsequent tables, with Y as the response (predicted) variable [either CH_4i (observed mean 26.0 ± 0.53 mL CH_4/g DM incubated) or CH_4d (observed mean 40.2 ± 0.91 mL CH_4/g DM apparently digested) and X as the explanatory variable (see Tables 1 and 2 for abbreviations and units of each variable), [2] Mean = mean of predicted values; SEM = standard error of the mean of predicted values; RMSPE = root mean square prediction error expressed as a percentage of the observed mean; EB, ER and ED = error due to bias, regression and disturbance, respectively (all as % of total MSPE); CCC = concordance correlation coefficient; R = Pearson correlation coefficient (measure of precision); C_b = bias correction factor (measure of accuracy).

The results of univariate equation development (Table 3) agreed roughly with the correlation analysis (Table 2), where the variables most highly correlated with CH_4i (TGP, VFA, ACp, VLp, DMD) and CH_4d (DMD, AC, C2C3, PR) (Table 2) appeared in the best performing univariate equations (Table 3). Some differences were evident, for example in the R-values, which may be explained by the difference in approach (correlation across all data points vs. correlation within study). The best performing univariate equations (U6, U12) were as follows:

$$CH_4d \text{ (mL } CH_4\text{/g DM digested)} = 58.52 \ (\pm 3.210) - 21.24 \ (\pm 3.045) \times DMD \text{ (DM digestibility) (U6)} \quad (10)$$

$$CH_4i \text{ (mL } CH_4\text{/g DM incubated)} = 3.00 \ (\pm 1.546) + 0.149 \ (\pm 0.005) \times TGP \text{ (mL/g DM incubated) (U12)} \quad (11)$$

3.3. Multivariate Meta-Analysis Models

Seventy-two multivariate equations, to predict CH_4d or CH_4i, were developed and evaluated with the variables presented in Table 2, in linear combinations. Those with nonsignificant slope parameters, model fitting problems (fixed or random) or had multiple X variables which were previously deemed to be collinear (Table 2) were discarded, and the remaining equations ($n = 29$) were evaluated on the evaluation dataset. Evaluation of the top six performing multivariate equations (for each of CH_4d and CH_4i) is reported in Table 4.

Best performing equations for CH_4d included (1) equation M5 (CCC = 0.419) with DMD, VFA, PR, FT and VL as X variables, as well as (2) M6 (CCC = 0.425) with DMD and VFA as X variables (Table 4). Best performing equations for CH_4i included (1) equation M11 (CCC = 0.438) with VFA and FT as X variables, and (2) equation M12 (CCC = 0.703) with PR, VL and TGP as X variables (Table 4).

The overall best performing equations (from univariate or multivariate origin, CH_4d, CH_4i) are presented in Table 5, and their predicted vs. observed plots are illustrated in Figure 1.

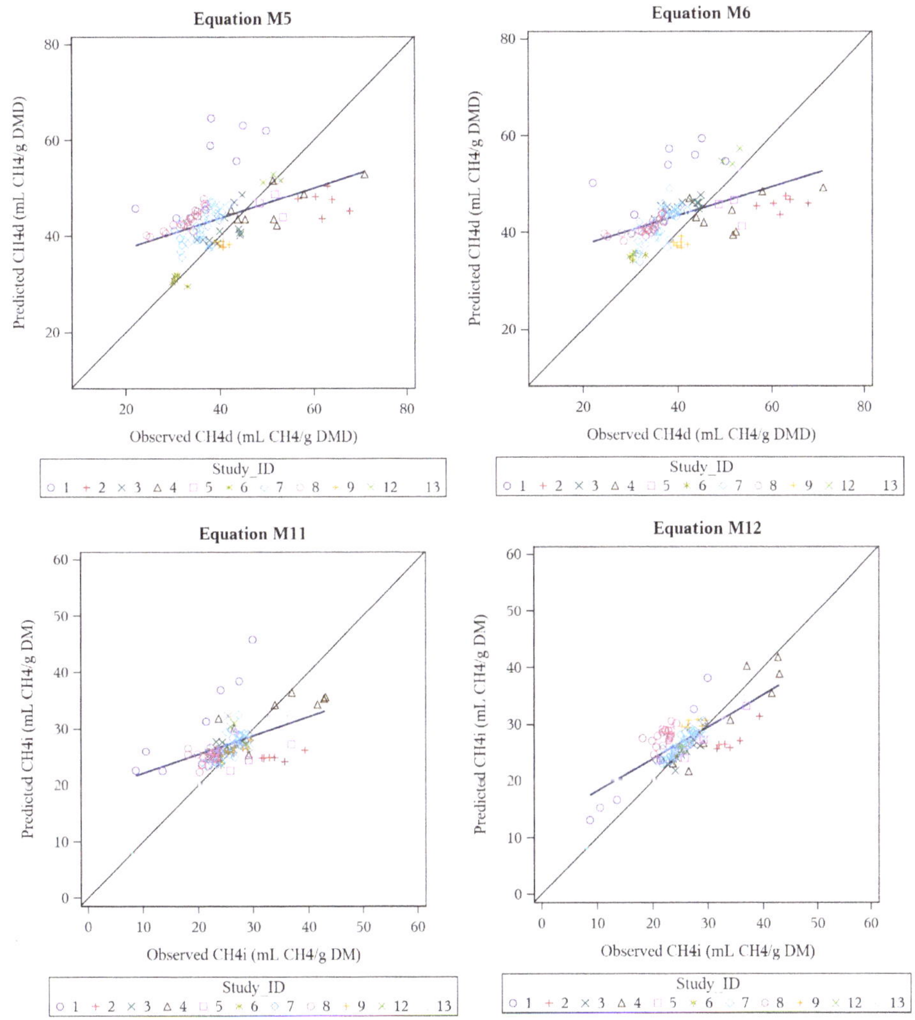

Figure 1. Predicted vs. observed plots for the top four performing meta-analysis equations M5 (CH_4d), M6 (CH_4d), M11 (CH_4i) and M12 (CH_4i), as evaluated on the evaluation dataset ($n = 107$).

Table 4. Linear multivariate equations—meta-analysis model evaluation via root mean square prediction error (RMSPE) and concordance correlation coefficient (CCC) analysis on the internal evaluation ($n = 107$) database.

Equation [1]	Y	X	Mean [2]	SEM	RMSPE, %	EB, %	ER, %	ED, %	CCC	R	C_b
M1	CH_4d	BTp, DMD	44.0	0.36	22.7	17	1	82	0.306	0.476	0.643
M2	CH_4d	PRp, VLp, DMD	43.9	0.51	22.6	16	1	83	0.379	0.473	0.800
M3	CH_4d	PR, VL, VFA, DMD	43.8	0.55	22.9	16	2	82	0.383	0.461	0.830
M4	CH_4d	DMD, VFA, pH, PR	43.5	0.65	23.3	12	7	81	0.401	0.448	0.896
M5	CH_4d	DMD, VFA, PR, FT, VL	43.8	0.58	22.5	16	2	82	0.419	0.492	0.853
M6	CH_4d	DMD, VFA	43.5	0.53	21.7	14	1	85	0.425	0.516	0.823
M7	CH_4i	pH, DMD, VLp, FT, BTp	28.5	0.34	21.1	21	3	76	0.407	0.496	0.833
M8	CH_4i	pH, DMD, PRp, VLp, FT	28.4	0.33	20.8	21	2	78	0.410	0.520	0.826
M9	CH_4i	pH, DMD, BTp, FT	28.3	0.32	20.2	20	1	79	0.428	0.520	0.823
M10	CH_4i	DMD, VFA, FT	27.7	0.32	19.4	12	1	87	0.434	0.514	0.844
M11	CH_4i	VFA, FT	27.4	0.37	19.8	8	5	87	0.438	0.484	0.905
M12	CH_4i	PR, VL, TGP	27.2	0.40	14.8	11	0	89	0.703	0.752	0.936

[1] Equation ID corresponds to equations presented in subsequent tables, with Y as the response (predicted) variable [either CH_4i (observed mean 26.0 ± 0.53 mL CH_4/g DM incubated) or CH_4d (observed mean 40.2 ± 0.91 mL CH_4/g DM apparently digested)] and X as the explanatory variables (see Tables 1 and 2 for abbreviations and units of each variable; FT is feed type either forage or concentrate), [2] Mean = mean of predicted values; SEM = standard error of the mean of predicted values; RMSPE = root mean square prediction error expressed as a percentage of the observed mean; EB, ER and ED = error due to bias, regression and disturbance, respectively (all as % of total MSPE); CCC = concordance correlation coefficient; R = Pearson correlation coefficient (measure of precision); C_b = bias correction factor (measure of accuracy).

Table 5. Best performing equations from the meta-analysis [1].

Equation ID	Y	Intercept	X1	X2	X3	X4	X5
M5	CH_4d	76.35 ($\pm$ 4.511)	-31.03 ($\pm$ 3.922) $\times$ DMD	3.21 ($\pm$ 0.352) $\times$ VFA	-0.094 ($\pm$ 0.01202) $\times$ PR	-3.017 ($\pm$ 1.460) (if FT = 1)	-0.133 ($\pm$ 0.0380) $\times$ VL
M6	CH_4d	51.35 ($\pm$ 3.086)	-41.98 ($\pm$ 3.429) $\times$ DMD	3.65 ($\pm$ 0.390) $\times$ VFA			
M11	CH_4i	15.8 ($\pm$ 2.614)	3.06 ($\pm$ 0.241) $\times$ VFA	-5.70 ($\pm$ 1.028) (if FT = 1)			
M12	CH_4i	11.58 ($\pm$ 1.627)	-0.0633 ($\pm$ 0.0057) $\times$ PR	0.0947 ($\pm$ 0.01728) $\times$ VL	0.172 ($\pm$ 0.0046) $\times$ TGP		

[1] Variables (units): CH_4i = methane production (mL CH_4/g DM incubated). CH_4d = methane production (mL CH_4/g DM apparently digested). DMD = apparent dry matter (DM) digestibility (g DM disappeared/g DM incubated). TGP = total gas production (mL gas/g DM incubated). VFA production (mmol total VFA/g DM incubated). PR = propionic acid (mmol PR/mol VFA). VL = valeric acid (mmol VL/mol VFA). FT = feed type either forage or concentrate.

3.4. Support Vector Regression and Artificial Neural Network Models

Evaluation of SVR and ANN models developed are presented in Table 6. Both SVR and ANN models demonstrated high predictability on the test dataset, with CCC values >0.90 for both CH_4d and CH_4i. For comparison purposes, meta-analysis equations METd and METi were also developed, via meta-analysis, but included all X variables (in linear form, regardless of significance). The CCC values for these equations were 0.645 and 0.734, respectively (Table 6), indicating that the SVR and MLP models must consider a complex multiple-nonlinear response surface between the X variables and Y variables, in order to achieve substantially higher CCC values. The predicted vs. observed plots for these models are illustrated in Figure 2.

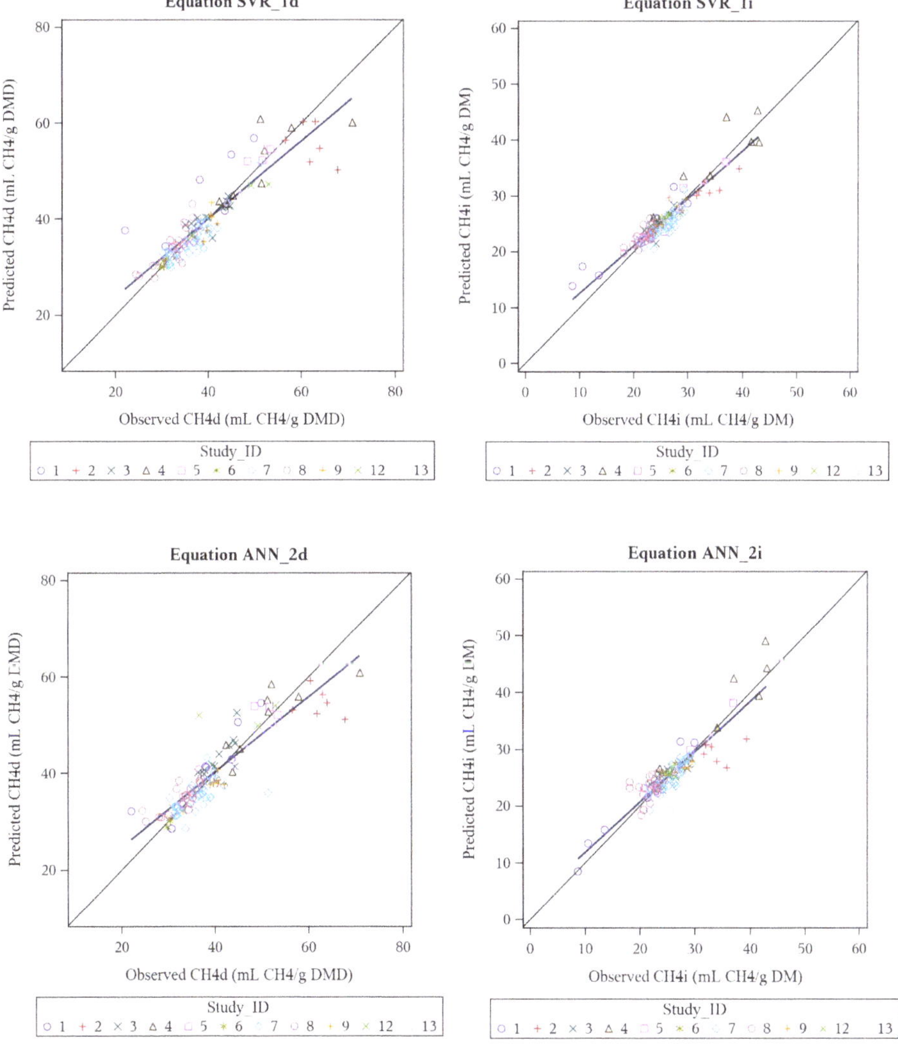

Figure 2. Predicted vs. observed plots for the machine learning equations SVR_1d (CH_4d), SVR_1i (CH_4i), ANN_2d (CH_4d) and ANN_2i (CH_4i), as evaluated on the evaluation dataset (n = 107).

Table 6. Machine learning and meta-analysis (including all the X variables) model evaluation via root mean square prediction error (RMSPE) and concordance correlation coefficient (CCC) analysis on the internal evaluation (n = 107) database.

Equation [1]	Y	X	Mean [2]	SEM	RMSPE, %	EB, %	ER, %	ED, %	CCC	R	C_b
SVR_1d	CH_4d	all, nonlinear	40.2	0.82	9.9	0.5	0	99.5	0.899	0.905	0.994
SVR_1i	CH_4i	all, nonlinear	26.1	0.49	8.3	0.6	0.1	99.3	0.917	0.920	0.997
ANN_2d	CH_4d	all, nonlinear	40.5	0.80	9.5	0.5	0.8	98.7	0.907	0.915	0.991
ANN_2i	CH_4i	all, nonlinear	26.0	0.52	9.1	0	2.9	97.1	0.906	0.906	1.000
METd	CH_4d	all, linear	42.9	0.54	17.1	16	6	79	0.643	0.762	0.844
METi	CH_4i	all, linear	27.2	0.40	14.0	12	0	88	0.734	0.782	0.939

[1] Equation IDs with 'd' refer to CH_4d, and 'i' refer to CH_4i equations. METd and METi are meta-analysis equations. Y is the response (predicted) variable [either CH_4i (observed mean 26.0 ± 0.53 mL CH_4/g DM incubated) or CH_4d (observed mean 40.2 ± 0.91 mL CH_4/g DM apparently digested) and X are the explanatory variables (all variables included in this analysis for comparison purposes), [2] Mean = mean of predicted values; SEM = standard error of the mean of predicted values; RMSPE = root mean square prediction error expressed as a percentage of the observed mean; EB, ER and ED = error due to bias, regression and disturbance, respectively (all as % of total MSPE); CCC = concordance correlation coefficient; R = Pearson correlation coefficient (measure of precision); C_b = bias correction factor (measure of accuracy).

3.5. Behaviour Analysis—Machine Learning Models

Unlike the meta-analysis method that results in a predictive equation, the ML methods SVR and ANN do not have the same degree of transparency. To understand the causal pathways to obtain the predictive result, behaviour analysis was performed (Table 7) by systematically changing the inputs in isolation and determining the degree of change in the output prediction. This was performed at +10% and −10% to determine direction of change in the response variable.

Results show (Table 7) that the models ANN_2i and SVR_1i (predicting CH_4i) were highly sensitive to the X variables pH and TGP, to varying extents (dependent on the model and FT). Secondary to these variables, the CH_4i predictions were sensitive to AC, PR, BT and DMD. Each model (ANN, SVR) demonstrated different sensitivity to these driving variables, and the sensitivity differed between the FT 1 (forage) and FT 2 (concentrate) substrates (Table 7).

For the models ANN_2d and SVR_1d (predicting CH_4d), these were shown to be highly sensitive to the X variables pH, DMD, TGP, AC and BT (Table 7), again dependent on the method (ANN, SVR) and the FT (forage vs. concentrate). Driving variables that differed greatly in sensitivity between models (ANN vs. SVR) included pH (14% and 36% vs. −6% and 5% change with ±10%, FT = 1, CH_4i), DMD (0% vs. −5% and 9% change with ±10%, FT = 1, CH_4i) and AC (11% and −11% vs. 2% and 0% change with ±10%, FT = 1, CH_4d) (Table 7), indicating that each approach fit the data slightly differently.

Some responses had different directional effects in the different models. For example, increasing pH increased CH_4i by 14% in ANN_2i, but decreased it by 6% in SVR_1i (FT = 1) (and similarly for FT = 2, pH increased CH_4i in ANN_2i by 4% and by 30% in SVR_1i); increasing BT did not change CH_4i in ANN_2i, but decreased CH_4i by 6% in SVR1i (FT = 1); and increasing AC reduced CH_4i by 3% in ANN_2i but increased CH_4i by 7% in SVR_1i (FT = 2). Similar results were found for CH_4d predictions, where, for example, increasing pH decreased CH_4d with ANN_2d by 14%, but increased it by 9% with SVR_1d (FT = 1), and for FT = 2, raising pH increased CH_4d by 11% with ANN_2d, and by 37% with SVR_1d.

Table 7. Behaviour analysis [1] of the artificial neural network (ANN) and support vector regression (SVR) models.

X-Variable	CH_4i (on Average 25.5 and 36.3 mL CH_4/g Dry Matter Incubated, for Forage and Concentrate, Respectively)				CH_4d (on Average 40.0 and 46.4 mL CH_4/g Dry Matter Apparently Digested for Forage and Concentrate, Respectively)			
	ANN (ANN_2i)		SVR (SVR_1i)		ANN (ANN_2d)		SVR (SVR_1d)	
Change in X-variable	+10% [2]	−10% [3]	+10% [2]	−10% [3]	+10% [2]	−10% [3]	+10% [2]	−10% [3]
Feed type = forage (FT = 1)								
pH	14%	36%	−6%	5%	−14%	7%	9%	35%
DMD	0%	0%	−5%	9%	−7%	18%	−13%	18%
TGP	12%	−10%	20%	−16%	15%	−8%	20%	−16%
Total VFA	−1%	1%	−1%	0%	−1%	1%	−1%	0%
Acetate (AC)	5%	11%	4%	3%	11%	−11%	2%	0%
Propionate (PR)	0%	0%	−2%	2%	−1%	1%	−1%	2%
Butyrate (BT)	0%	0%	−6%	6%	−1%	1%	−6%	8%
Valerate (VL)	−2%	2%	0%	−1%	−1%	1%	1%	−1%
C2C3	−1%	1%	1%	0%	−4%	4%	1%	0%
Feed type = concentrate (FT = 2)								
pH	4%	−24%	30%	−39%	11%	−23%	37%	−37%
DMD	−2%	−1%	5%	−2%	−8%	8%	−3%	5%
TGP	11%	−11%	12%	−10%	10%	−10%	9%	−8%
Total VFA	−4%	2%	−3%	2%	−2%	2%	−2%	1%
Acetate (AC)	−3%	3%	7%	−6%	3%	−3%	6%	−6%
Propionate (PR)	−3%	3%	−4%	4%	−3%	0%	−3%	3%
Butyrate (BT)	2%	−2%	−1%	1%	2%	−3%	−2%	2%
Valerate (VL)	1%	−1%	0%	0%	1%	−1%	0%	0%
C2C3	0%	−2%	1%	−1%	0%	0%	1%	−1%

[1] Expected change (in %) in the predicted variable (either CH_4i or CH_4d), with a ±10% change in the driving variable. Changes in the predicted variable exceeding |10%| are highlighted in grey, [2] Increase of 10% in the explanatory variable, [3] Decrease of 10% in the explanatory variable.

Some behaviour responses within the ML methods were also directionally different between FT. For example, CH_4i (ANN_2i) increased as pH was increased (+14%, FT = 1), but also increased when pH was decreased (+36%), indicating a nonlinear/polynomial response surface. This is in contrast to when FT = 2, where increasing pH increased CH_4i by 4%, and decreasing pH decreased CH_4i by 24% (Table 7). For CH_4i (SVR_1i, FT = 1), increasing pH decreased CH_4i by 6%, while increasing pH increased CH_4i by 5%. This is in contrast to when FT = 2, where increasing pH increased CH_4i by 30%, and decreasing it decreased CH_4i by 39%. Similar directional differences were observed for CH_4d (FT = 1 vs. 2).

4. Discussion

To predict CH_4 (in vivo or in vitro) based on stoichiometry principles alone and considering a H recovery of 100%, Hegarty and Nolan [8] proposed the equation: CH_4 (mmol/L) = 0.5AC + 0.5BT − 0.25PR − 0.25VL (where all VFAs are expressed in mmol/L). Similarly, Moss et al. [9], considering a H recovery of 90%, proposed the equation: CH_4 (mmol/L) = 0.45AC − 0.275PR + 0.40BT (where all VFAs are expressed in mmol/L). These equations reflect the net production of H as a result of AC and BT synthesis by rumen microbes and the net utilization of H as a result of PR and VL synthesis by rumen microbes during fermentation of feed. The resulting H is utilized by methanogens to reduce CO_2 to H_2O ($CO_2 + 8H \rightarrow CH_4 + 2H_2O$). However, predicting CH_4 from the above stoichiometric equations is only valid if (1) these VFAs are the only end-products of fermentation, (2) no free H_2 accumulates or escapes, (3) the microbial digestion process is strictly anaerobic, and (4) H_2 is not used in other reactions (e.g., reduction of sulphates to sulphides, or saturation of double bonds in fatty acids) [8]. In practice, the production of CH_4 will be less than the stoichiometry prediction given by the above equations, because these assumptions are generally not held. Jayanegara et al. [7] used both stoichiometric equations to predict CH_4 from VFA concentrations in vitro, and found that indeed, the equations overpredicted CH_4, likely due to a much lower observed H recovery (observed range of 28.9% to 56.2%) compared to the recoveries assumed by the models (100% and 90%). In agreement with Jayanegara et al. [7], when the stoichiometric equations [8,9] were applied to the current test dataset, CH_4i was overpredicted (observed CH_4i (mmol/L) = 11.6 ± 2.44, using [8], predicted CH_4i = 16.2 ± 3.07; using [9], predicted CH_4i = 14.0 ± 2.68) and had poor CCC evaluation statistics (0.135 and 0.227 for [8,9], respectively). For the test dataset, the average H_2 recovery, calculated according to [31], was 80%, a value that is substantially lower than the theoretical recovery rates [8,9], and also different from those observed by Jayanegara et al. [7], indicating the potential value of an empirical approach, such as those developed in our work.

The objective of the current study was to utilize meta-analysis and ML methodologies to predict CH_4 emissions from in vitro gas and VFA production data. Results of this work found that via meta-analysis, the best predictive equations of in vitro CH_4d included the variables − DMD + VFA, or − DMD + VFA − PR − FT − VL (Equations M6 and M5 respectively, Table 5), while the best predictive equations of in vitro CH_4i included the variables + VFA − FT, or − PR + VL + TGP (Equations M11 and M12, respectively, Table 5). The significant positive sign on VL in Equation M12 is concerning, as stoichiometrically, the production of VL utilizes H and therefore is associated with a lower CH_4 emission. This illustrates a limitation of empirical modelling (whether it be meta-analysis or a ML), that the resulting equations strive to find the best statistical relationship to the data, regardless of biological principles. It is possible this could be related to the relatively small contribution to total VFA made by VL, or correlation with specific feed ingredient properties.

The best performing univariate equations (U6 (CH_4d), U12 (CH_4i)) were based on DMD and TGP, respectively. The correlation between CH_4d and TGP was low (Table 2), indicating that the DMD correction to CH_4i (CH_4d) accounted for much of the strong relationship between CH_4i and TGP. Such simple regressions may be used when VFA data are not reported, but would miss considerable variance explained by defining the type of VFA being produced (see multivariate equations).

The models produced via ML methodologies, ANN and SVR, have much higher predictability (CCC, RMSPE analysis) of the CH_4i and CH_4d outcomes compared to the meta-analysis models. This was a result of the meta-analysis models being limited to including only significant X variables ($p < 0.05$), while the ML methodologies have no such limitation. As well, the ML methodologies mapped more complex response surfaces between multiple X variables and the Y variable, based on linear, radial or polynomial shapes. While this resulted in a greatly improved prediction on related (internal evaluation) data (Table 6), it may end up fitting noise or other unrelated data characteristics in the training dataset, resulting in a diminished predictability on unrelated data (external evaluation). Such an external evaluation would be a required next step to test the globalization ability of such ML models—in particular considering the relatively small size of the training dataset and the data hungry nature of ML models.

Unsurprisingly, in both meta-analysis and ML models TGP was a significant driving variable, as an indicator of the overall extent of fermentation occurring in vitro. Directionally, the meta-analysis and ML methods agree, whereby increasing TGP increases CH_4i and CH_4d. The variable DMD was particularly relevant with the CH_4d models, where increasing DMD resulted in a lower CH_4d (Tables 5 and 7).

Interestingly, while pH did not appear in many highly significant meta-analysis equations (Tables 3 and 4), it did appear to have a strong presence in the ML models, as illustrated by the behaviour analysis (Table 7). According to the pH dependent VFA stoichiometry of [32], an increase in ruminal pH causes a shift in soluble carbohydrate fermentation towards AC and away from PR and BT, and a shift in starch fermentation towards AC and BT and away from PR. For FT = 2 (concentrates), the ANN_2d and SVR_1d equations to predict CH_4d show a tendency for CH_4 to increase as pH increases (by 11% and 37%, respectively) (Table 7). In line, when pH is decreased, CH_4d also decreased (Table 7). However, when FT = 1 (forages), and pH increases the ANN_2d prediction decreases (−14%), while the SVR_1d prediction increases (9%). For the ANN_2d equation, it is difficult to conceptualize where the −14% in CH_4d comes from, aside from a nuance in the database.

5. Conclusions

The current study successfully delivered models (using both meta-analysis and ML methodologies) which can be used to estimate CH_4 production from in vitro fermentation systems. Meta-analysis results indicate that equations containing DMD, VFA, PR, FT and VL resulted in the best prediction of CH_4 on an internal evaluation dataset of in vitro data. The ML models by far exceed the predictability achieved using meta-analysis methods, but should be evaluated on an external database to assess predictability and generalization potential on unrelated data, in particular given the limited database size and the data hungry nature of such ML methodologies. Between the ML methodologies assessed, ANN and SVR resulted in very similar predictive performance, but differences in fitting, as assessed by behaviour analysis, were evident. The models developed may be utilized to estimate CH_4 emissions in vitro, in instances where total gas and VFA production, but not CH_4, are measured.

Author Contributions: Conceptualization, S.L. and J.F.; methodology, J.L.E. and H.A.-M.; software, J.L.E. and H.A.-M.; validation, J.L.E., H.A.-M. and S.L.; formal analysis, J.L.E., H.A.-M.; investigation, J.L.E.; resources, S.L.; data curation, S.L., A.N.-V., E.J.M., P.P. and P.O.; writing—original draft preparation, J.L.E.; writing—review and editing, J.L.E., S.L., H.A.-M., J.F., C.D.P., A.N.-V., E.J.M., P.P. and P.O.; visualization, J.L.E.; supervision, S.L. and J.F.; project administration, S.L.; funding acquisition, S.L. All authors have read and agreed to the published version of the manuscript.

Funding: The in vitro digestion study research was funded by the National Development Plan through the Research Stimulus Fund administered by the Department of Agriculture, Fisheries and Food, Ireland, grant number RSF 05 224, RSF 06 361 and RSF 07 517.

Appendix A

Table A1. Summary of variable abbreviations and units/descriptions.

Variable Abbreviation	Unit	Description
CH$_4$i	mL CH$_4$/g DM incubated	In vitro methane production
CH$_4$d	mL CH$_4$/g DM apparently digested	In vitro methane production
pH	-	Final pH in the incubation medium
DMD	g DM disappeared/g DM incubated	Apparent dry matter (DM) digestibility
TGP	mL gas/g DM incubated	Total gas production
VFA	mmol total VFA/g DM incubated	Total VFA production
AC	mmol AC/mol VFA	Acetic acid, proportion of total VFA
PR	mmol PR/mol VFA	Propionic acid, proportion of total VFA
BT	mmol BT/mol VFA	Butyric acid, proportion of total VFA
VL	mmol VL/mol VFA	Valeric acid, proportion of total VFA
ACp	mmol AC/g DM incubated	Acetic acid production
PRp	mmol PR/g DM incubated	Propionic acid production
BTp	mmol BT/g DM incubated	Butyric acid production
VLp	mmol VL/g DM incubated	Valeric acid production
C2C3	AC/PR	Acetate to propionate ratio

References

1. IPCC. *Climate Change 2014 Synthesis Report. Contribution of Working Groups I, II and III to the Fifth Assessment Report of the Intergovernmental Panel on Climate Change*; Core Writing Team, Pachauri, R.K., Meyer, L.A., Eds.; IPCC: Geneva, Switzerland, 2014; pp. 1–151.
2. IPCC. *2019 Refinement to the 2006 IPCC Guidelines for National Greenhouse Gas Inventories*; Volume 4: Agriculture, Forestry and Other Land Use - Chapter 10: Emissions from livestock and manure management; Calvo Buendia, E., Tanabe, K., Kranjc, A., Baasansuren, J., Fukuda, M., Ngarize, S., Osako, A., Pyrozhenko, Y., Shermanau, P., Federici, S., Eds.; IPCC: Geneva, Switzerland, 2019; pp. 10.1–10.209.
3. Martin, C.; Morgavi, D.P.; Doreau, M. Methane mitigation in ruminants: From microbe to the farm scale. *Animal* **2010**, *4*, 351–365. [CrossRef] [PubMed]
4. Kumar, S.; Choudhury, P.K.; Carro, M.D.; Griffith, G.W.; Dagar, S.S.; Puniya, M.; Calabro, S.; Ravella, S.R.; Dhewa, T.; Upadhyay, R.C.; et al. New aspects and strategies for methane mitigation from ruminants. *Appl. Microbiol. Biotechnol.* **2014**, *98*, 31–44. [CrossRef] [PubMed]
5. Hristov, A.N.; Oh, J.; Firkins, J.L.; Dijkstra, J.; Kebreab, E.; Waghorn, G.; Makkar, H.P.S.; Adesogan, A.T.; Yang, W.; Lee, C.; et al. Special Topics—Mitigation of methane and nitrous oxide emissions from animal operations: I. A review of enteric methane mitigation options. *J. Anim. Sci.* **2013**, *91*, 5045–5069. [CrossRef] [PubMed]
6. Ellis, L.L.; Dijkstra, J.; Kebreab, E.; Bannink, A.; Odongo, N.E.; McBride, B.W.; France, J. Aspects of rumen microbiology central to mechanistic modelling of methane production in cattle. *J. Agric. Sci.* **2008**, *146*, 213–233. [CrossRef]
7. Jayanegara, A.; Ikhsan, I.; Toharmat, T. Assessment of methane estimation from volatile fatty acid stoichiometry in the rumen in vitro. *J. Ind. Trop. Anim. Agric.* **2013**, *38*, 103–108. [CrossRef]
8. Hegarty, R.S.; Nolan, J.V. Estimation of ruminal methane production from measurement of volatile fatty acid production. In *Measuring Methane Production from Ruminants*; Makkar, H.P.S., Vercoe, P.E., Eds.; Springer: Dordrecht, The Netherlands, 2007; pp. 69–92.
9. Moss, A.; Jouany, J.-P.; Newbold, J. Methane production by ruminants: Its contribution to global warming. *Ann. Zootech.* **2000**, *49*, 231–253. [CrossRef]
10. McGeough, E.J.; O'Kiely, P.; O'Brien, M.; Kenny, D.A. An evaluation of the methane output associated with high-moisture grains and silages using the in vitro total gas production technique. *Anim. Prod. Sci.* **2011**, *51*, 627–634. [CrossRef]
11. Navarro-Villa, A.; O'Brien, M.; López, S.; Boland, T.M.; O'Kiely, P. Modifications of a gas production technique for assessing in vitro rumen methane production from feedstuffs. *Anim. Feed Sci. Technol.* **2011**, *166*, 163–174. [CrossRef]

12. Navarro-Villa, A.; O'Brien, M.; López, S.; Boland, T.M.; O'Kiely, P. In vitro rumen methane output of red clover and perennial ryegrass assayed using the gas production technique (GPT). *Anim. Feed Sci. Technol.* **2011**, *168*, 152–164. [CrossRef]

13. Navarro-Villa, A.; O'Brien, M.; López, S.; Boland, T.M.; O'Kiely, P. In vitro rumen methane output of grasses and grass silages differing in fermentation characteristics using the gas-production technique (GPT). *Grass Forage Sci.* **2013**, *68*, 228–244. [CrossRef]

14. Purcell, P.J.; O'Brien, M.; Boland, T.M.; O'Kiely, P. In vitro rumen methane output of perennial ryegrass samples prepared by freeze drying or thermal drying (40 °C). *Anim. Feed Sci. Technol.* **2011**, *166*, 175–182. [CrossRef]

15. Purcell, P.J.; O'Brien, M.; Boland, T.M.; O'Donovan, M.; O'Kiely, P. Impacts of herbage mass and sward allowance of perennial ryegrass sampled throughout the growing season on in vitro rumen methane production. *Anim. Feed Sci. Technol.* **2011**, *166*, 405–411. [CrossRef]

16. Purcell, P.J.; Boland, T.M.; O'Brien, M.; O'Kiely, P. In vitro rumen methane output of forb species sampled in spring and summer. *Agric. Food Sci.* **2012**, *21*, 83–90. [CrossRef]

17. Purcell, P.J.; O'Brien, M.; Navarro-Villa, A.; Boland, T.M.; McEvoy, M.; Grogan, D.; O'Kiely, P. In vitro rumen methane output of perennial ryegrass varieties and perennial grass species harvested throughout the growing season: In vitro rumen methane output of perennial grasses. *Grass Forage Sci.* **2012**, *67*, 280–298. [CrossRef]

18. Purcell, P.J.; Grant, J.; Boland, T.M.; Grogan, D.; O'Kiely, P. The in vitro rumen methane output of perennial grass species and white clover varieties, and associative effects for their binary mixtures, evaluated using a batch-culture technique. *Anim. Prod. Sci.* **2012**, *52*, 1077. [CrossRef]

19. Purcell, P.J.; Boland, T.M.; O'Kiely, P. The effect of water-soluble carbohydrate concentration and type on in vitro rumen methane output of perennial ryegrass determined using a 24-hour batch-culture gas production technique. *Irish J. Food Agric. Res.* **2014**, *53*, 21–36.

20. Navarro-Villa, A.; O'Brien, M.; López, S.; Boland, T.M.; O'Kiely, P. Determination of the in vitro rumen methane output of contrasting feeds using the gas production technique (GPT). (unpublished).

21. SAS Institute Inc. *SAS/STAT® 14.1 User's Guide*; SAS Institute Inc.: Cary, NC, USA, 2015.

22. St-Pierre, N.R. Invited review: Integrating quantitative findings from multiple studies using mixed model methodology. *J. Dairy Sci.* **2001**, *84*, 741–755. [CrossRef]

23. Coelho, L.P.; Richert, W.; Brucher, M. *Building Machine Learning Systems with Python: Explore Machine Learning and Deep Learning Techniques for Building Intelligent Systems Using Scikit-Learn and TensorFlow*; Packt Publishing: Birmingham, UK, 2018.

24. Buitinck, L.; Louppe, G.; Blondel, M.; Pedregosa, F.; Mueller, A.; Grisel, O.; Niculae, V.; Prettenhofer, P.; Gramfort, A.; Grobler, J.; et al. API design for machine learning software: Experiences from the scikit-learn project. In Proceedings of the European Conference on Machine Learning and Principles and Practice of Knowledge Discovery in Databases, Workshop on Languages for Data Mining and Machine Learning, Prague, Czech Republic, 23–27 September 2013; pp. 108–122. Available online: https://arxiv.org/pdf/1309.0238.pdf (accessed on 15 April 2020).

25. Pedregosa, F.; Varoquaux, G.; Gramfort, A.; Michel, V.; Thirion, B.; Grisel, O.; Blondel, M.; Prettenhofer, P.; Weiss, R.; Dubourg, V.; et al. Scikit-learn: Machine learning in Python. *J. Mach. Learn. Res.* **2011**, *12*, 2825–2830.

26. Rossum, G.V. *Python Tutorial Release 3.6.4*; Python Software Foundation: Wilmington, DE, USA, 2018.

27. Drucker, H.; Burges, C.J.C.; Kaufman, L.; Smola, A.J.; Vapnik, V. Support vector regression machines. In *Advances in Neural Information Processing Systems 9*; Mozer, M.C., Jordan, M.I., Petsche, T., Eds.; MIT Press: Cambridge, MA, USA, 1997; pp. 155–161.

28. Paliwal, M.; Kumar, U.A. Neural networks and statistical techniques: A review of applications. *Expert Syst. Appl.* **2009**, *36*, 2–17. [CrossRef]

29. Bibby, J.; Toutenburg, T. *Prediction and Improved Estimation in Linear Models*; John Wiley & Sons: Chichester, UK, 1977.

30. Lin, L.I. A concordance correlation coefficient to evaluate reproducibility. *Biometrics* **1989**, *45*, 255–268. [CrossRef]

31. Demeyer, D.; Van Nevel, C. Protein fermentation and growth by rumen microbes. *Ann. Rech. Vet.* **1979**, *10*, 277–279.

32. Bannink, A.; France, J.; López, S.; Gerrits, W.J.J.; Kebreab, E.; Tamminga, S.; Dijkstra, J. Modelling the implications of feeding strategy on rumen fermentation and functioning of the rumen wall. *Anim. Feed Sci. Technol.* **2008**, *143*, 3–26. [CrossRef]

 *animals*

Article

Methane Production of Fresh Sainfoin, with or without PEG, and Fresh Alfalfa at Different Stages of Maturity is Similar but the Fermentation End Products Vary

Pablo José Rufino-Moya, Mireia Blanco, Juan Ramón Bertolín and Margalida Joy *

Centro de Investigación y Tecnología Agroalimentaria de Aragón (CITA), Instituto Agroalimentario de Aragón–IA2 (CITA-Universidad de Zaragoza), Avda, Montañana 930, 50059 Zaragoza, Spain; pjrufino@cita-aragon.es (P.J.R.-M.); mblanco@aragon.es (M.B.); jrbertolin@cita-aragon.es (J.R.B.)
* Correspondence: mjoy@aragon.es; Tel.: +34-976-716-442

Received: 14 March 2019; Accepted: 18 April 2019; Published: 26 April 2019

Simple Summary: In the last years, there has been increasing interest in the use of forages containing condensed tannins (CT) in ruminant nutrition. Condensed tannins can reduce the methane emissions and the ruminal degradation of protein, improving the animal performances to different extents depending on the source and dose of CT. In vitro fermentation of sainfoin has not been studied in fresh forage. The effect of CT can be studied in comparison with a similar CT-free forage or using polyethylene glycol (PEG), which is a tannin-blocking agent. The maturity stage influences the chemical composition to a different degree depending on the legume species, and can affect the content and fractions of CT. The aims of this trial were to compare the fermentation parameters of sainfoin with or without PEG, to detect the differences due to CT, at different maturity stages (vegetative, start-flowering, and end-flowering) and compare them with the fermentation parameters of alfalfa. The main results were that sainfoin had greater in vitro organic matter degradability (IVOMD) and lower ammonia and acetic:propionic ratio than alfalfa. Sainfoin CT had effect on ammonia and individual fatty acid proportions. In conclusion, fermentation end products were affected both by the chemical composition and CT contents.

Abstract: Alfalfa and sainfoin are high-quality forages with different condensed tannins (CT) content, which can be affected by the stage of maturity. To study the effects of CT on fermentation parameters, three substrates (alfalfa, sainfoin, and sainfoin+PEG) at three stages of maturity were in vitro incubated for 72 h. Sainfoin had greater total polyphenol and CT contents than alfalfa. As maturity advanced, CT contents in sainfoin decreased ($p < 0.05$), except for the protein-bound CT fraction ($p > 0.05$). The total gas and methane production was affected neither by the substrate nor by the stage of maturity ($p > 0.05$). Overall, sainfoin and sainfoin+PEG had greater in vitro organic matter degradability (IVOMD) than alfalfa ($p < 0.05$). Alfalfa and sainfoin+PEG presented higher ammonia content than sainfoin ($p < 0.001$). Total volatile fatty acid (VFA) production was only affected by the stage of maturity ($p < 0.05$), and the individual VFA proportions were affected by the substrate and the stage of maturity ($p < 0.001$). In conclusion, alfalfa and sainfoin only differed in the IVOMD and the fermentation end products. Moreover, CT reduced ammonia production and the ratio methane: VFA, but the IVOMD was reduced only in the vegetative stage.

Keywords: polyethylene glycol; gas production; in vitro organic matter degradability; condensed tannins; ammonia; volatile fatty acid; in vitro assay

---

1. Introduction

There is increasing interest in legume-based forage production systems because of their low reliance on fertilizer nitrogen, potential environmental benefits, and high protein content that contribute to low-input and sustainable livestock production systems [1]. Alfalfa (*Medicago sativa L.*) and sainfoin (*Onobrychis viciifolia Scop.*) are two pluriannual legumes widely grown in the Mediterranean area, presenting high forage productive capacity, high nutritional value, and restorative action for soil fertility [2]. However, alfalfa has a low content of polyphenols and is considered virtually free of condensed tannins (CTs), whereas sainfoin presents a high content of polyphenols and a medium to high content of CTs [3–5].

Alfalfa is usually fed as hay to ruminants mainly to avoid bloat, although continuous grazing is possible without bloat incidence both in sheep [6] and growing cattle [7]. Thus, the majority of studies that compared the ruminal fermentation of alfalfa and sainfoin have been performed using hays [3,4,8]. The differences between alfalfa and sainfoin have been ascribed to differences in the chemical compositions [4], but also to the presence of CTs [8]. To clarify whether the differences between alfalfa and sainfoin are only due to CTs, polyethylene glycol (PEG) must be added as a tannin-blocking agent [8]. To the best of our knowledge, the ruminal fermentation of both species was compared in fresh forages only by McMahon et al. [9] and Chung et al. [10]. Depending on their content, characteristics, and properties [11], CTs from sainfoin alter both the breakdown of protein in the rumen to ammonia and methane, gas and the production of total volatile fatty acids (VFAs) [3,4,12]. These, in turn, are associated with the variety, stage of maturity, and environmental factors [5,13,14].

The objectives of this trial were to compare the in vitro fermentation of alfalfa and sainfoin at three stages of maturity and to clarify whether the differences between both legumes were due to the CTs of sainfoin using PEG.

2. Materials and Methods

2.1. Experimental Design

Three substrates (alfalfa, sainfoin, and sainfoin+PEG) at three stages of maturity (vegetative, start-flowering, and end-flowering) were used to evaluate the effect of sainfoin CTs on in vitro fermentation. Alfalfa was used as a tannin-free legume.

2.2. Animal and Diets

2.2.1. Forages, Crop Management, and Harvest

The experiment was performed at the CITA Research Institute at Zaragoza (41°42′ N, 0°49′ W), altitude 216 m a.s.l., located in Ebro Valley, north-eastern Spain. The silt loam soil at the site had pH 8.1 and 1.81% organic matter and contained 16% clay, 53.5% silt, and 30.5% sand. Alfalfa and sainfoin were cultivated and managed as described by Lobón et al. [15]. Samples of forage were collected fortnightly from 14 April to 22 September 2015. The stage of maturity of the sainfoin and alfalfa was classified into vegetative, start-flowering, and end-flowering according to Borreani et al. [16] and Kalu and Fick [17], respectively. In each sampling, 10 samples per legume were obtained from 0.25 m^2 areas randomly allocated in the plot. These samples were mixed homogenously, and a part of the sample was separated manually into stems, leaves, and flowers to study their respective percentages. Another part of the samples was dried at 60 °C for 48 h for chemical analyses, and the rest of the sample was freeze-dried in a Genesis Freeze Dryer 25 (Hucoa Erlöss, SA/Thermo Fisher Scientific, Madrid, Spain) for polyphenol and CT analyses and in vitro fermentation assays. Samples for the chemical analyses were ground and sieved through a 1-mm screen (Rotary Mill, ZM200 Retsch, Hann, Germany), and a part of the samples was ground and sieved through a 0.2-mm screen for crude protein (CP), polyphenol, and CT determination. All the samples were stored at −20 °C in total darkness.

The number of samples of vegetative, start-flowering and end-flowering stages of each forage to study the in vitro fermentation were chosen in concordance with the representativeness of the plant development. The vegetative stage was the most frequent stage (55%), followed by end-flowering (27%) and start-flowering (18%) stages. Three samples of the vegetative stage and two samples of the start- and end-flowering of each legume species were studied.

2.2.2. Animals and Sampling of Ruminal Digesta

The procedures used in the trial followed the Spanish guidelines for experimental animal protection (RD 53/2013) and were approved by the Institutional Animal Care and Use Committee of the Research Centre (Procedure number 2011-05). Rasa Aragonesa wethers (n = 4; Live weight: 65 ± 2.1 kg) were used as donors of ruminal content. The management of the rumen-cannulated wethers and the sampling of the ruminal digesta was made as reported in Rufino-Moya et al. [18]. Briefly, wethers were fed a diet composed by alfalfa hay (70%) and barley grain (30%) at energy maintenance level. Before morning feeding, ruminal digesta from each wether was collected into a prewarmed insulated thermo, individually strained through four layers of cheesecloth and homogenized. Rumen fluid from the four wethers was mixed (pH: 6.76 ± 0.099), and a buffer solution was added in a proportion of 1:2 (*v/v*) based on the protocol of Menke and Steingass [19]

2.2.3. In Vitro Gas Production Technique and Sampling

The production of gas was measured with the Ankom system (Ankom Technology Corporation, Fairport, NY, USA), which had 310 mL capacity bottles fitted with pressure and temperature sensors. The valve open time was one second, the threshold for gas release was 5 PSI and the bottles were not shaken. Five hundred mg of freeze-dried substrate (alfalfa, sainfoin, or sainfoin+PEG) were incubated with 60 mL of buffered solution:rumen fluid (2:1 *v/v*) in a water bath (at 39 °C) for 72 h. To make the sainfoin+PEG samples, PEG-4000 (Merck, Darmstadt, Germany) was added to the buffered rumen fluid at a concentration of 2.3 g/L [12]. Four runs were conducted on four separate days, and each sample was incubated in duplicate in each run. Gas production and corrected with the blanks (two bottles without substrate were included in each run).

After 72 h of incubation, the bottles were placed in ice to stop fermentation (5–10 min), and then tempered (at room temperature for 10–15 min). A sample of the gas produced was transferred (at atmospheric pressure) with a syringe attached to a manometer into a Vacutainer®tube to determine CH_4 and conserved at 4 °C until analysis. At the end of gas sampling, the pH was measured immediately with a microPH 2002 (Crison Instruments S.A., Barcelona, Spain). The sampling to determine ammonia (NH_3-N) content and VFA were carried out as reported in Rufino-Moya et al. [18]. Briefly, to determine the ammonia content, 2.5 mL of liquid was mixed with 2.5 mL HCl 0.1 N. For VFA determination, 0.5 mL of the liquid was added to 0.5 mL of deproteinizing solution and 1 mL of distilled water. Tubes with samples for determination of ammonia and VFAs were stored at −20 °C until future analyses. The entire incubated sample was filtered through a preweighed bag (50 μm; Ankom) to estimate the in vitro organic matter degradability (IVOMD). Briefly, the bags were sealed, washed, dried at 103 °C for 48 h, and finally, placed in a muffle at 550 °C to obtain the ashes. The organic matter of the bag content was obtained as DM-ashes, and the IVOMD was calculated.

2.3. Analytical Methods

2.3.1. Chemical Composition

All the analyses of the chemical composition were analyzed as reported in Rufino-Moya et al. [18] according to official methods. Briefly, AOAC methods were used to determine the contents of dry matter (index no. 934.01), ash (index no. 942.05), and CP (index no. 968.06) [20]. Neutral detergent fiber (NDFom), acid detergent fiber (ADFom), and lignin (sa) contents were determined according to the method described by Van Soest et al. [21] using the Ankom 200/220 fiber analyzer (Ankom Technology

Corporation). The NDFom was assayed with a heat stable amylase. The lignin (sa) was analyzed in ADF residues by the solubilization of cellulose with sulfuric acid. All the values were corrected for ash-free content. The ether extract (EE) was determined following the approved procedure Am 5-04 [22] using an XT10 Ankom extractor (Ankom Technology Corporation). The nonstructural carbohydrates (NSC) were calculated as $NSC = 1000 - CP - EE - NDF - ash$, according to Guglielmelli et al. [3].

The content of total polyphenol (TP) and the fractions of CT were determined as described in Rufino-Moya et al. [18]. Briefly, the TP were extracted using the method of Makkar [23] and were quantified following the method of Julkunen-Tiitto [24]. The extractable CT (ECT), protein-bound CT (PBCT), and fiber-bound CT (FBCT) were extracted and fractioned following the method of Terrill et al. [25] and quantified by the colorimetric HCl-butanol method described by Grabber et al. [26]. The standard used for the quantification was extracted and purified from sainfoin using the method described by Wolfe et al. [27]. An Heλios β spectrophotometer was used to measure the samples and standard calibration at 550 nm.

2.3.2. Determination of Parameters of the In Vitro Gas Production Technique

Gas production, recorded hourly for 72 h, was used to estimate the parameters of the kinetics of fermentation adjusting the gas produced to the model described by France et al. [28]:

$$P = A \times \left(1 - e^{-ct}\right)$$

where P is the cumulative gas production (mL) at time t (h), A is the potential gas production (mL), and c is the rate of gas production (h^{-1}).

An HP-4890 (Agilent, St. Clara, CA, USA) gas chromatograph (GC) equipped with a flame ionization detector (FID) and a TG-BOND Q+ capillary column (30 m × 0.32 mm id × 10 μm film thickness, Thermo Scientific, Waltham, MA, USA) was used to determine CH$_4$. Helium was used as the carrier gas at a flow rate of 1 mL/min. The oven temperature was maintained at 100 °C (isothermal program). The splitless injection volume was 200 μL. Methane identification was based on the retention time compared with the standard. The methane concentration was calculated from the peak concentration:area ratio using the peak area generated from standard gas as the reference (CH$_4$; 99.995% purity [C45], Carburos Metálicos, Barcelona, Spain).

The content of ammonia was measured in Epoch microplate Spectrophotometer (BioTek Instruments, Inc., Winooski, VT, USA) using a colorimetric method described by Chaney and Marbach [29].

A Bruker Scion 460 GC (Bruker, Billerica, MA, USA) equipped with CP-8400 autosampler, FID and a BR-SWax capillary column (30 m × 0.25 mm i.d. × 0.25 μm film thickness, Bruker) was used to determine the concentration of acetic, propionic, iso-butyric, butyric, iso-valeric, and valeric acids. Helium was the carrier gas (flow rate of 1 mL/min). The oven temperature program was 100 °C, followed by a 6 °C/min increase to 160 °C. The injection volume was 1 μL at a split ratio of 1:50. The individual VFAs were identified based on retention time comparisons with commercially available standards of acetic, propionic, iso-butyric, butyric, iso-valeric, valeric, and 4-methyl-valeric acids at ≥99% purity (Sigma-Aldrich, St. Louis, MO, USA).

2.4. Statistical Analyses

Data were analyzed using SAS v. 9.3 (SAS Inst. Inc., Cary, NC, USA). The fermentation kinetic parameters (A and c) were estimated using a nonlinear regression model (NLIN procedure). The contents of secondary compounds were analyzed using the GLM procedure with the substrate, the stage of maturity and its interaction as fixed effects. Total gas and CH$_4$ production, A, c, IVOMD and the fermentation end products were analyzed using mixed models considering the substrate, the stage of maturity and its interaction as fixed effects and the run as random effect. If the interaction was not significant, it was deleted from the model and the analysis was repeated. The least square means, their associated standard errors and the differences were obtained. Pearson correlation coefficients between variables were calculated using the CORR procedure. For the entire test, the effects were considered a significant probability at a value of $p < 0.05$ or a trend at $p = 0.05$.

3. Results

3.1. Chemical Composition

The chemical composition and the percentage of stems, leaves, and flowers of both legume species at the three stages of maturity are shown in Table 1. On average, alfalfa and sainfoin had similar ADFom (231 g/kg DM), CP (198 g/kg DM), EE (15 g/kg DM), and NSC (335 g/kg DM) contents. Alfalfa, however, had higher ash and NDFom contents and a lower lignin (sa) content than sainfoin. For both forages, NDFom and ADFom content increased as the stage of maturity progressed, whereas the CP content decreased. As the forage matured, the lignin (sa) content increased only in alfalfa whereas the contents of EE and NSC decreased only in sainfoin.

Table 1. Chemical composition and plant components of alfalfa and sainfoin at three stages of maturity.

	Alfalfa			Sainfoin		
Item	Vegetative	Start-Flowering	End-Flowering	Vegetative	Start-Flowering	End-Flowering
Chemical composition						
Dry Matter (DM) (g/kg)	249	261	333	241	262	224
Ash (g/kg DM)	103	111	98	83	113	82
CP [1] (g/kg DM)	227	207	166	204	201	181
NDFom [2] (g/kg DM)	336	352	405	313	324	394
ADFom [3] (g/kg DM)	201	230	276	199	213	264
Lignin (sa), (g/kg DM)	45	53	66	78	76	76
Ether extract (g/kg DM)	18	10	16	22	15	11
NSC [4] (g/kg DM)	317	320	316	379	347	332
Plant components (%)						
Leaves	59.3	53.7	42.7	66.3	56.0	38.5
Stems	40.7	43.7	51.3	33.7	40.8	44.6
Flowers	0.0	2.5	5.9	0.0	3.2	16.8

[1] crude protein; [2] Neutral detergent fiber; [3] acid detergent fiber; [4] nonstructural carbohydrates.

Alfalfa and sainfoin had similar proportions of leaves (51.9 vs. 53.6%). However, alfalfa had a greater proportion of stems (45.2 vs. 39.7%) and a lower proportion of flowers (2.8 vs. 6.7%) than sainfoin. Regarding the stage of maturity, the proportion of stems and flowers increased, whereas the proportion of leaves decreased as the stage of maturity advanced.

3.2. Contents of Total Polyphenols and Condensed Tannins

The content of total polyphenols and the total (TCT) and fractions of CT were affected by the interaction between the legume species and the stage of maturity ($p < 0.05$) (Figure 1). Alfalfa presented steady contents of total polyphenols, TCTs, ECTs, PBCTs, and FBCTs, which were lower than those of sainfoin ($p < 0.001$) regardless of the stage of maturity. The contents of polyphenols, TCTs, ECTs, and FBCTs decreased as maturity advanced ($p < 0.05$).

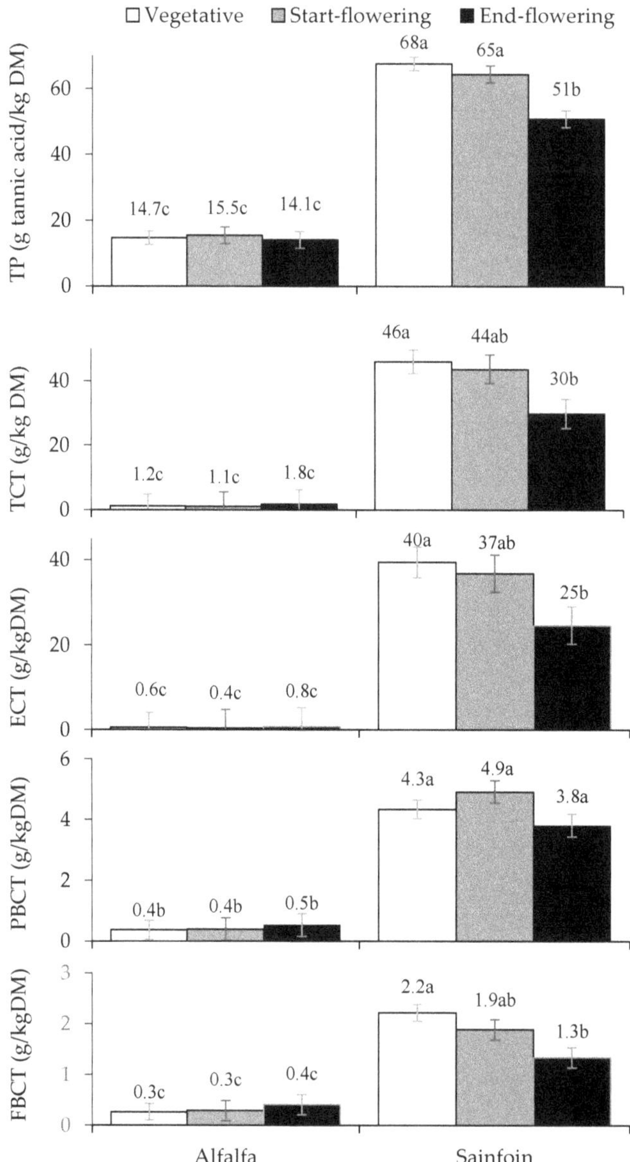

Figure 1. Effect of the species and the stage of maturity on the contents of total polyphenols (TP), total condensed tannins (TCT), extractable CT (ECT), protein-bound CT (PBCT), and fiber-bound (FBCT). Within a parameter, means with different letter differ at $p < 0.05$. (Within each species: n = 3 for the vegetative stage, n = 2 for the start-flowering, and n = 2 for the end-flowering stages.)

3.3. In Vitro Fermentation

The pH was affected by the interaction between the substrate and the stage of maturity ($p < 0.01$; Table 2). Alfalfa had greater pH than sainfoin and sainfoin+PEG ($p < 0.05$). Sainfoin and sainfoin+PEG were affected similarly by the stage of maturity ($p < 0.001$), with the greatest pH value in the vegetative stage ($p < 0.05$) Total gas and CH_4 production were affected neither by the substrate nor by the stage of maturity ($p > 0.05$; Table 2). However, the interaction between the substrate and the stage of maturity affected A ($p < 0.001$) and c ($p = 0.05$). Alfalfa showed lower A values at start-and end-flowering stages ($p < 0.001$) and greater c in the vegetative and start-flowering stages ($p < 0.05$) than sainfoin and sainfoin+PEG ($p < 0.05$). Regarding the effect stage of maturity, sainfoin and sainfoin+PEG had the lowest A values in the vegetative stage ($p < 0.05$). As the stage of maturity progressed, c increased in sainfoin and sainfoin+PEG substrates ($p < 0.05$).

The IVOMD was also affected by the interaction between the substrate and the stage of maturity ($p < 0.001$; Table 2). Alfalfa had lower IVOMD than both sainfoin substrates in the start-flowering and end-flowering stages ($p < 0.05$), whereas sainfoin+PEG had greater IVOMD than alfalfa and sainfoin in the vegetative stage ($p < 0.05$). In alfalfa, the IVOMD decreased as the stage of maturity advanced ($p < 0.05$). The sainfoin and sainfoin+PEG showed the greatest IVOMD in the start-flowering stage ($p < 0.05$). The IVOMD was correlated with A (r = 0.60; $P < 0.01$) and with the total VFA production (r = 0.51; $p < 0.05$).

The NH_3-N content was only affected by the substrate ($p < 0.001$); sainfoin produced a lower NH_3-N concentration than alfalfa and sainfoin+PEG (Table 2). In contrast, total VFA production was only affected by the stage of maturity ($p < 0.05$), the start-flowering stage presented greater VFA production than the rest of the stages (105, 99, and 100 mmol/L for start-flowering, vegetative, and end-flowering, respectively). Regarding the individual VFAs, alfalfa had a lower acetic acid proportion and greater proportions of the rest of the individual VFAs than sainfoin ($p < 0.001$). When comparing both sainfoin substrates, sainfoin had a greater acetic acid proportion and lower proportions of the rest of the VFAs than sainfoin+PEG ($p < 0.001$). Sainfoin presented the greatest C_2:C_3 ratio, followed by sainfoin+PEG and alfalfa, which had the lowest ratio ($p < 0.001$). Regarding the effect of the stage of maturity, the vegetative stage had a lower proportion of acetic acid and greater proportions of propionic, iso-butyric, and iso-valeric acid than the rest of stages of maturity ($p < 0.001$). The vegetative stage had a lower C_2:C_3 ratio than the other stages ($p < 0.001$). The CH_4:VFA ratio was only affected by the substrate ($p = 0.01$); sainfoin+PEG presented a greater CH_4:VFA ratio than the other substrates ($p < 0.05$).

Table 2. Effect of the substrate (S) and the stage of maturity [1] (SM) on gas and methane production (CH_4), potential gas production (A), rate of gas production (c), in vitro organic matter degradability (IVOMD), ammonia (NH_3-N), and volatile fatty acids (VFAs).

Item	Alfalfa			Sainfoin			Sainfoin+PEG			r.s.d. [2]	p-Values		
	VEG	Start-F	End-F	VEG	Start-F	End-F	VEG	Start-F	End-F		S	SM	SxSM
pH	6.42 [a]	6.44 [a]	6.41 [a]	6.35 [bx]	6.32 [by]	6.33 [by]	6.37 [bx]	6.31 [bz]	6.34 [by]	0.032	<0.001	0.002	0.01
Gas production (mL/g dOM [3])	179	184	188	183	163	181	180	199	173	26.5	0.43	0.97	0.12
A (mL)	68 [x]	67 [bxy]	62 [y]	63 [y]	74 [bx]	70 [x]	65 [y]	78 [ax]	67 [y]	5.7	0.01	<0.001	<0.001
c (h^{-1})	0.2 [a]	0.22 [a]	0.19	0.14 [by]	0.16 [bxy]	0.19 [y]	0.15 [bx]	0.15 [bx]	0.19 [y]	0.037	<0.001	0.02	0.05
CH_4 production (mL/g dOM [3])	37	38	39	37	33	38	37	39	36	5.0	0.29	0.73	0.17
CH_4: gas (mL/L)	145	142	141	141 [xy]	134 [y]	148 [x]	145	144	144	8.8	0.35	0.13	0.11
IVOMD (g/kg)	851 [bx]	837 [bx]	770 [by]	850 [by]	926 [ax]	856 [ay]	868 [ay]	936 [ax]	864 [ay]	16.9	<0.001	<0.001	<0.001
NH_3-N (mg/L)	240 [a]	247 [a]	210	206 [b]	184 [b]	201	242 [a]	234 [a]	215	31.7	<0.001	0.06	0.20
Total VFAs (mmol/L)	102 [xy]	105 [x]	97 [y]	96 [y]	103 [x]	101 [xy]	99 [y]	105 [x]	102 [xy]	7.1	0.61	0.02	0.35
Acetic acid (C_2) (mol/100 mol)	63.1 [cy]	63.9 [bx]	64.0 [bx]	65.4 [ay]	66.5 [ax]	66.3 [ax]	63.9 [by]	64.6 [bx]	64.5 [bx]	0.76	<0.001	<0.001	0.95
Propionic acid (C_3) (mol/100 mol)	15.5 [a]	15.3 [a]	15.3 [a]	15 [bx]	14.5 [cy]	14.5 [by]	15.3 [ax]	14.9 [by]	15.2 [ax]	0.28	<0.001	<0.001	0.09
Butyric acid (mol/100 mol)	13.0 [a]	12.6 [a]	12.6 [a]	12.1 [c]	12.2 [b]	12 [b]	12.5 [b]	12.8 [a]	12.4 [a]	0.45	<0.001	0.20	0.16
Iso-butyric acid (mol/100 mol)	1.97 [ax]	1.91 [axy]	1.87 [ay]	1.82 [bx]	1.68 [by]	1.72 [by]	1.97 [ax]	1.83 [ay]	1.84 [ay]	0.100	<0.001	<0.001	0.63
Valeric acid (mol/100 mol)	2.14 [axy]	2.09 [ay]	2.22 [ax]	1.87 [bx]	1.7 [cy]	1.83 [cx]	2.12 [ay]	1.95 [bx]	2.04 [by]	0.107	<0.001	<0.001	0.11
Iso-valeric acid (mol/100 mol)	4.27 [ax]	4.09 [axy]	4.04 [ay]	3.87 [bx]	3.5 [by]	3.65 [by]	4.19 [ax]	3.91 [ay]	3.93 [ay]	0.223	<0.001	<0.001	0.69
C_2:C_3 (mol/mol)	4.1 [cy]	4.18 [cxy]	4.2 [bx]	4.39 [ay]	4.62 [ax]	4.59 [ax]	4.18 [by]	4.36 [bx]	4.25 [by]	0.109	<0.001	<0.001	0.13
CH_4:VFA (mL/mol)	2.41	2.24 [b]	2.35	2.53	2.22 [b]	2.48	2.62	2.73 [a]	2.4	0.313	0.01	0.26	0.10

[1] VEG: vegetative; Start-F: start of flowering; End-F: end of flowering; [2] residual standard deviation; [3] degraded organic matter. Within a parameter, means with different superscript (a,b,c) differ at $p < 0.05$ for the substrate effect in each stage of maturity; with different superscripts (x,y,z) at $p < 0.05$ for the stage of maturity effect in each substrate.

4. Discussion

In Mediterranean areas, there is an increasing interest to reintroduce forage-based systems in ovine production to ensure the viability and sustainability of the farms. Legumes are especially advisable due to their nutritional quality for ruminants and environmental benefits [1]. Moreover, the presence of CT in some legumes may decrease CH_4 production and improve the performance of ovine to different extents depending on the source and dose of CT. As these legumes are usually fed conserved, either as silage or as hay, there is scarce information on the fermentation parameters when alfalfa and sainfoin are offered fresh. Previous experiments showed differences between the fermentation parameters of alfalfa and sainfoin hay and silage [4,30], however, it is not clear if the differences were due to the chemical composition, the presence of CT in sainfoin or both. With the present study, the fermentation parameters of alfalfa and sainfoin, with or without PEG, in fresh at different stages of maturity were compared to try to clarify the origin of the differences in fermentation. The use of gas production technique is a good tool to evaluate the effect of CT on fermentation parameters, but the fermentation is influenced by the time of incubation, species of the animal donor and the diet [31,32]. Furthermore, the effects of sainfoin CT vary according the variety, harvest time, and cultivation site [13,14], making it difficult to compare the results with other studies. In the present study, the content of total CT and their fractions were analyzed, however, the chemical characteristics of CT (molecular weight, degree of polymerization, prodelphinidin/procyanidin ratio, cis/trans ratio, etc.) were not evaluated.

4.1. Effect of the Substrate

References showed noticeable variability in the chemical composition of alfalfa and sainfoin among studies, which is related to the stage at harvest, leaf:stem ratio, soil characteristics, weather conditions, and the cultivars utilized [5,13,14]. In the present study, the similar CP and ADF contents and different NDF contents of alfalfa and sainfoin agree partially with the results reported by Chung et al. [10], who observed similar NDF, ADF and CP contents in fresh alfalfa and sainfoin at the late vegetative stage. However, at the early vegetative stage, the same authors observed greater NDF and ADF contents in alfalfa than sainfoin and similar CP contents.

Alfalfa has a low content of total polyphenols and is considered a CT-free legume [5], although it may present very low CT content in the seed coats. Therefore, the present results related to the presence of CTs agree with previous studies that analyzed both legumes offered fresh [33,34] or as preserved forages [4,30].

The pH did not negatively affect the fermentation environment because the values were within the range of 6.2 to 6.8 and these values ensure a favorable environment for the activity of cellulolytic bacteria [35]. The inclusion of PEG in the current experiment did not affect pH, as reported in fresh sainfoin [12] and sainfoin hay [8]. Regarding gas production, the similar production of alfalfa and sainfoin agrees with the similar gas production observed in alfalfa and sainfoin leaves incubated in Rusitec units [9] and alfalfa and sainfoin silages [30]. However, when alfalfa and sainfoin hays were studied, differences in gas production were reported [3,4,8]. The inconsistency of the results of the type of substrate on gas production might be related to the differences in chemical composition, the different characteristics of CTs and of type and settings of the in vitro assay [14]. In that sense, the similar gas production between sainfoin and sainfoin+PEG was unexpected because previous experiments reported increases in gas production when PEG was added to fresh sainfoin [12,14]. According to Azuhnwi et al. [13], the inclusion of PEG increased gas production by 2.7 to 9.6%, depending on the sainfoin variety, site, and stage at harvest. In the current experiment, the inclusion of PEG slightly increased the gas production, although not statistically significantly.

The similar CH_4 production of alfalfa and sainfoin recorded in the present study agrees with results reported using fresh forages [10] and silages [30]. However, the inclusion of extracts from sainfoin accessions in alfalfa decreased CH_4 production, but the effect was greatly dependent on the accession and the dose of inclusion [11]. Moderate CT content may have beneficial effects reducing rumen CH_4 emission production [36]. The action of CT on methanogenesis can be attributed to indirect effects via reduced hydrogen production (and presumably reduced forage digestibility) and via direct inhibitory effects on methanogens [34]. Regarding the effect of PEG, the inclusion of PEG in sainfoin did not affect CH_4 production in previous studies [12,37] as in the current experiment. The structural features of condensed tannins affect in vitro CH_4 production, which may be linked to the interaction of CTs with dietary substrate or microbial cells [11,38]. Therefore, the type of CT and dose present in the current experiment might not be sufficient to modify CH_4 production.

The reduced *A* in alfalfa compared with sainfoin and sainfoin+PEG in the start- and end-flowering stages can be related to the higher fiber fraction, as reported by Guglielmelli et al. [3]. In the current experiment, the presence of CTs in sainfoin had no effect on *A*, as reported by Calabrò et al. [8]. The higher *c* in alfalfa when compared to sainfoin agrees with the results reported by Hatew et al. [11], although the effect on *c* depends on the types and concentrations of sainfoin CTs.

The lower IVOMD of alfalfa, when compared to sainfoin and sainfoin+PEG, was also reported using fresh forage estimated in situ [10] and in vitro [39] and could be related to the greater fiber fraction. The increased IVOMD in sainfoin+PEG with respect to sainfoin at the vegetative stage could be related to the blockage of CTs by the PEG. However, Theodoridou et al. [12] reported no effect of the inclusion of PEG in sainfoin on IVOMD studied at 24 h, regardless of the stage of maturity. The discrepancy between studies could be related to the content, characteristics and structures of CTs, which depends on the botanical species and variety of the source [13,14].

The NH_3-N contents recorded in the present study are in line with most similar studies that compared alfalfa and sainfoin [3,4,12]. The reduced NH_3-N concentration in sainfoin with respect to alfalfa and sainfoin+PEG confirmed the inhibition elicited by CTs in the ruminal degradation of dietary proteins due to the formation of complexes CT-protein at ruminal pH [40]. In contrast, the effect of CTs on total VFA production is not clear. Some studies reported a lower total VFA production in sainfoin silage than in alfalfa silage [30], and the inclusion of different doses of extracted accession of sainfoin in alfalfa decreased or maintained the total VFA production, depending on the accession [11]. In the current experiment, the total VFA production was not affected by the legume species, as observed by other authors [4,10]. The inclusion of PEG did not affect total VFA production in the current experiment as reported for sainfoin hay [8,37], which is contrary to the increase of total VFA production observed by Hatew et al. [14]. The differences between the studies could be due to the time of incubation, species of the animal donor, chemical structure, and biological activity of CTs [14,31].

Generally, the presence of CT from sainfoin leads to an increase in the propionic acid proportion and a reduction in the C_2:C_3 ratio [10,11,38]. However, the effect on each individual VFA proportion is variable due to the type of substrate, types, and contents of CTs and length of incubation period or the donor animal [11,31]. In the present study, sainfoin had a greater acetic acid proportion than alfalfa, which was similar to results from Guglielmelli et al. [3] using hays and Grosse-Brinkhaus et al. [30] using silages. The C_2:C_3 ratio recorded in the present study was greater in sainfoin than in alfalfa and sainfoin+PEG, which is in contrast with other studies that did not observe effects of the type of substrate or the addition of PEG [3,12,37]. Sainfoin had lower valeric acid and branched-chain VFA proportions than alfalfa and sainfoin+PEG because of the presence of CTs [4,30,38]. Condensed tannins reduce the proportions of branched-chain VFAs due to reduce protein degradation in the rumen because these VFA are products of the breakdown of the carbon skeleton of amino acids during rumen fermentation [41].

4.2. Effect of the Stage of Maturity

The decrease in CP content and the concomitant increase in the cell wall (NDFom and ADFom) content as the stage of maturity progressed in both forages is a result of the decrease of the proportion of leaves to stems and the increase of lignified tissues [10]. The steady lignin (sa) content in sainfoin during the development of plants can be due to some interference between this compound and CTs during analysis, as reported by Guglielmelli et al. [3].

In the current experiment, the TP and TCT contents were affected by the stage of maturity, as reported in previous studies [3,5,42], although the magnitude of the effect varied among the studies. Regarding the CT fractions, there is little information about the influence of the stage of maturity. As in the present study, Chung et al. [10] observed a reduction of the ECT fraction in the end-flowering stage with respect to the vegetative stage in sainfoin. However, Jin et al. [43] observed greater TCT, ECT, and PBCT contents in *Dalea purpurea* at flowering than at the vegetative stage due to the high percentage of flowers, which are very rich in CTs [44]. From a physiological point of view, the reduction in the secondary compound concentrations as maturity advances could be due to a sort of dilution as a consequence of the growth and expansion of plant cells [45] and to the decrease in the leaf: stem ratio as a consequence of the reduction of leaves, which are rich in CTs [44].

The lack of an effect of the stage of maturity on pH values in alfalfa agrees with the results reported by Chung et al. [10], but the reduction of pH in sainfoin at both flowering stages disagrees with the abovementioned study. In this sense, the stage of maturity had no effect on pH when sainfoin hay was incubated [3,8]. More studies considering the stage of maturity, the chemical composition and the presence of secondary compounds in alfalfa and sainfoin must be performed to discern the importance of these factors on the ruminal pH.

Previous studies reported a reduction of in vitro gas and CH_4 production as the stage of maturity advanced, associated with the chemical composition and the CT content [12,46]. However, the stage of maturity had no effect on gas and CH_4 production in the current experiment, which is in agreement with in vitro [3] and in vivo [10] experiments. The similar chemical composition observed between stages of maturity, the low biological activity of CT and the interactions between nutritive components and antinutritional factors could be responsible for the similar gas and CH_4 production [3].

As expected, the IVOMD of alfalfa decreased as the fiber fraction increased with maturity [47]. In contrast, sainfoin, with or without PEG, presented considerably high IVOMD at the start-flowering stage with respect to the rest of the stages in agreement with Theodoridou et al. [12]. These results could be due to the different biological activity of CTs at vegetative and start-flowering stages [5], because the chemical composition and CT contents were similar in both stages.

The advancing of the stage of maturity tended to reduce NH_3-N production as a consequence of the decrease of CP content in concordance with several studies carried out in vitro [3,12] and in vivo [10,42]. Studies concerning the effect of the stage of maturity on the total production of VFAs and their proportions show discrepancies. The chemical composition of the substrates, the length of the incubation period, and the inoculum donor animal are determinant factors that can influence VFA production and proportions [31]. In the current experiment, the start-flowering stage presented the highest total VFA production, in concordance with the highest IVOMD observed. However, Theodoridou et al. [12] studied the effect of the stage of maturity of fresh sainfoin with a similar CT content in a 24 h in vitro assay and did not find an effect on the total VFA production.

In relation to the proportion of individual VFAs, the effect of the stage of maturity on these parameters has been reported in vitro and in vivo in previous studies [3,10], but the results are not consistent. In the current experiment, as maturity advances, there is an increase in acetic acid and a decrease in propionic acid proportions, thus increasing the C_2:C_3 ratio due probably to increase of fiber and reduction of CT content [42]. The reduction of the proportion of iso-butyric and iso-valeric acids in the start- and end-flowering stages and valeric acid at the start-flowering stage in comparison with the vegetative stage might be explained by the decrease in CP content, because they are products of the breakdown of the carbon skeleton of amino acids during rumen fermentation, as the maturity of the forage advanced [10,47].

5. Conclusions

In conclusion, sainfoin might be an alternative to alfalfa due to the high IVOMD and the potential protection against ruminal protein degradation, according to the results of ammonia content, branched-chain VFAs, and valeric acid proportion from sainfoin in vitro. The effect of the stage of maturity was less than expected, probably due to the high quality of the forages. It is required to study the effects of the type of substrate and stage of maturity on animal performance to recommend the best stage of maturity to cut sainfoin and alfalfa.

Author Contributions: Conceptualization, M.B.; Data curation, M.B. and M.J.; Formal analysis, P.J.R.-M. and M.B.; Funding acquisition, M.J.; Investigation, P.J.R.-M. and J.R.B.; Methodology, M.B. and M.J.; Project administration, M.J.; Resources, M.B. and M.J.; Supervision, M.B. and M.J.; Visualization, P.J.R.-M.; Writing—original draft, P.J.R.-M.; Writing—review & editing, M.B. and M.J..

Funding: This research was funded by the Ministry of Economy and Competitiveness of Spain, the European Union Regional Development Funds (INIA RTA2012-080-00, INIA RZP2017-00001) and the Research Group Funds of the Aragón Government (A14_17R). P.J. Rufino's contract is supported by a doctoral grant from the INIA-EFS, M. Blanco has a contract supported by INIA-EFS.

Acknowledgments: Appreciation is expressed to the staff of CITA de Aragón for their help in data collection. Special thanks to the lab team and to the crop team for their assistance.

Conflicts of Interest: The authors declare no conflicts of interest.

References

1. Phelan, P.; Moloney, A.; McGeough, E.; Humphreys, J.; Bertilsson, J.; O'Riordan, E.; O'Kiely, P. Forage legumes for grazing and conserving in ruminant production systems. *Crit. Rev. Plant Sci.* **2015**, *34*, 281–326. [CrossRef]

2. Delgado, I.; Muñoz, F.; Andueza, D. Evaluación comparativa de la alfalfa y la esparceta, en condiciones de secano y regadío de Aragón. *Pastos Y PAC* **2014**, *2020*, 304–310.

3. Guglielmelli, A.; Calabrò, S.; Primi, R.; Carone, F.; Cutrignelli, M.I.; Tudisco, R.; Piccolo, G.; Ronchi, B.; Danieli, P.P. In vitro fermentation patterns and methane production of sainfoin (Onobrychis viciifolia Scop.) hay with different condensed tannin contents. *Grass Forage Sci.* **2011**, *66*, 488–500. [CrossRef]

4. Toral, P.G.; Hervás, G.; Missaoui, H.; Andrés, S.; Giráldez, F.J.; Jellali, S.; Frutos, P. Effects of a tannin-rich legume (Onobrychis viciifolia) on in vitro ruminal biohydrogenation and fermentation. *Span. J. Agric. Res.* **2016**, *14*, 1–9. [CrossRef]

5. Theodoridou, K.; Aufrère, J.; Andueza, D.; Le Morvan, A.; Picard, F.; Stringano, E.; Pourrat, J.; Mueller-Harvey, I.; Baumont, R. Effect of plant development during first and second growth cycle on chemical composition, condensed tannins and nutritive value of three sainfoin (Onobrychis viciifolia) varieties and lucerne. *Grass Forage Sci.* **2011**, *66*, 402–414. [CrossRef]

6. Alvarez-Rodriguez, J.; Sanz, A.; Ripoll-Bosch, R.; Joy, M. Do alfalfa grazing and lactation length affect the digestive tract fill of light lambs? *Small Rumin. Res.* **2010**, *94*, 109–116. [CrossRef]

7. Blanco, M.; Joy, M.; Ripoll, G.; Sauerwein, H.; Casasús, I. Grazing lucerne as fattening management for young bulls: Technical and economic performance and diet authentication. *Animal* **2011**, *5*, 113–122. [CrossRef]

8. Calabrò, S.; Guglielmelli, A.; Iannaccone, F.; Danieli, P.P.; Tudisco, R.; Ruggiero, C.; Piccolo, G.; Cutrignelli, M.I.; Infascelli, F. Fermentation kinetics of sainfoin hay with and without PEG. *J. Anim. Physiol. Anim. Nutr.* **2012**, *96*, 842–849. [CrossRef]

9. McMahon, L.; Majak, W.; McAllister, T.; Hall, J.; Jones, G.; Popp, J.; Cheng, K.-J. Effect of sainfoin on in vitro digestion of fresh alfalfa and bloat in steers. *Can. J. Anim. Sci.* **1999**, *79*, 203–212. [CrossRef]

10. Chung, Y.H.; Mc Geough, E.J.; Acharya, S.; McAllister, T.A.; McGinn, S.M.; Harstad, O.M.; Beauchemin, K.A. Enteric methane emission, diet digestibility, and nitrogen excretion from beef heifers fed sainfoin or alfalfa. *J. Anim. Sci.* **2013**, *91*, 4861–4874. [CrossRef]

11. Hatew, B.; Stringano, E.; Mueller-Harvey, I.; Hendriks, W.H.; Carbonero, C.H.; Smith, L.M.J.; Pellikaan, W.F. Impact of variation in structure of condensed tannins from sainfoin (Onobrychis viciifolia) on in vitro ruminal methane production and fermentation characteristics. *J. Anim. Physiol. Anim. Nutr.* **2016**, *100*, 348–360. [CrossRef]

12. Theodoridou, K.; Aufrère, J.; Niderkorn, V.; Andueza, D.; Le Morvan, A.; Picard, F.; Baumont, R. In vitro study of the effects of condensed tannins in sainfoin on the digestive process in the rumen at two vegetation cycles. *Anim. Feed Sci. Technol.* **2011**, *170*, 147–159. [CrossRef]

13. Azuhnwi, B.; Boller, B.; Martens, M.; Dohme-Meier, F.; Ampuero, S.; Günter, S.; Kreuzer, M.; Hess, H. Morphology, tannin concentration and forage value of 15 Swiss accessions of sainfoin (Onobrychis viciifolia Scop.) as influenced by harvest time and cultivation site. *Grass Forage Sci.* **2011**, *66*, 474–487. [CrossRef]

14. Hatew, B.; Hayot Carbonero, C.; Stringano, E.; Sales, L.F.; Smith, L.M.J.; Mueller-Harvey, I.; Hendriks, W.H.; Pellikaan, W.F. Diversity of condensed tannin structures affects rumen in vitro methane production in sainfoin (Onobrychis viciifolia) accessions. *Grass Forage Sci.* **2015**, *70*, 474–490. [CrossRef]

15. Lobón, S.; Blanco, M.; Sanz, A.; Ripoll, G.; Joy, M. Effects of feeding strategies during lactation and the inclusion of quebracho in the fattening on performance and carcass traits in light lambs. *J. Sci. Food Agric.* **2019**, *99*, 457–463. [CrossRef]

16. Borreani, G.; Peiretti, P.G.; Tabacco, E. Evolution of yield and quality of sainfoin (Onobrychis viciifolia Scop.) in the spring growth cycle. *Agronomie* **2003**, *23*, 193–201. [CrossRef]

17. Kalu, B.A.; Fick, G.W. Quantifying Morphological Development of Alfalfa for Studies of Herbage Quality. *Crop Sci.* **1981**, *21*, 267–271. [CrossRef]

18. Rufino-Moya, P.J.; Blanco, M.; Bertolín, J.R.; Joy, M. Effect of the method of preservation on the chemical composition and in vitro fermentation characteristics in two legumes rich in condensed tannins. *Anim. Feed Sci. Technol.* **2019**, *251*, 12–20. [CrossRef]

19. Menke, K.H.; Steingass, H. Estimation of the energetic feed value obtained from chemical analysis and in vitro gas production using rumen fluid. *Anim. Res. Dev.* **1988**, *28*, 7–55.

20. AOAC. *Official Methods of Analysis*, 17th ed.; Association of Official Analytical Chemist: Arlington, VA, USA, 2000.

21. Van Soest, P.J.; Robertson, J.B.; Lewis, B.A. Methods for dietary fiber, neutral detergent fiber, and nonstarch polysaccharides in relation to animal nutrition. *J. Dairy Sci.* **1991**, *74*, 3583–3597. [CrossRef]

22. AOCS. *Approved Procedure Am 5-04, Rapid Determination of Oil/Fat Utilizing High. Temperature Solvent Extraction*; AOCS Press: Urbana, IL, USA, 2005.

23. Makkar, H.P.S. *Quantification of Tannins in Tree Foliage*; FAO/IAEA Working Document; IAEA: Vienna, Austria, 2000.

24. Julkunen-Tiitto, R. Phenolic Constituents in the Leaves of Northern Willows: Methods for the Analysis of Certain Phenolics. *J. Agric. Food Chem.* **1985**, *33*, 213–217. [CrossRef]

25. Terrill, T.H.; Windham, W.R.; Hoveland, C.S.; Amos, H.E. Forage preservation method influences on tannin concentration, intake and digestibility of Sericea lespedeza by sheep. *Agron. J.* **1989**, *81*, 435–439. [CrossRef]

26. Grabber, J.H.; Zeller, W.E.; Mueller-Harvey, I. Acetone enhances the direct analysis of procyanidin- and prodelphinidin-based condensed tannins in lotus species by the butanol-HCl-iron assay. *J. Agric. Food Chem.* **2013**, *61*, 2669–2678. [CrossRef] [PubMed]

27. Wolfe, R.M.; Terrill, T.H.; Muir, J.P. Drying method and origin of standard affect condensed tannin (CT) concentrations in perennial herbaceous legumes using simplified butanol-HCl CT analysis. *J. Sci. Food Agric.* **2008**, *88*, 1060–1067. [CrossRef]

28. France, J.; Dijkstra, J.; Dhanoa, M.S.; Lopez, S.; Bannink, A. Estimating the extent of degradation of ruminant feeds from a description of their gas production profiles observed in vitro: Derivation of models and other mathematical considerations. *Br. J. Nutr.* **2000**, *83*, 143–150. [CrossRef] [PubMed]

29. Chaney, A.L.; Marbach, E.P. Modified reagents for determination of urea and ammonia. *Clin. Chem.* **1962**, *8*, 130–132. [PubMed]

30. Grosse-Brinkhaus, A.; Wyss, U.; Arrigo, Y.; Girard, M.; Bee, G.; Zeitz, J.O.; Kreuzer, M.; Dohme-Meier, F. In vitro ruminal fermentation characteristics and utilisable CP supply of sainfoin and birdsfoot trefoil silages and their mixtures with other legumes. *Animal* **2017**, *11*, 580–590. [CrossRef]

31. Frutos, P.; Hervás, G.; Giráldez, F.J.; Mantecón, A.R. An in vitro study on the ability of polyethylene glycol to inhibit the effect of quebracho tannins and tannic acid on rumen fermentation in sheep, goats, cows, and deer. *Aust. J. Agric. Res.* **2004**, *55*, 1125–1132. [CrossRef]

32. Martínez, T.F.; McAllister, T.A.; Wang, Y.; Reuter, T. Effects of tannic acid and quebracho tannins on in vitro ruminal fermentation of wheat and corn grain. *J. Sci. Food Agric.* **2006**, *86*, 1244–1256. [CrossRef]

33. Scharenberg, A.; Arrigo, Y.; Gutzwiller, A.; Soliva, C.R.; Wyss, U.; Kreuzer, M.F.D. Palatability in sheep and in vitro nutritional value of dried and ensiled sainfoin (*Onobrychis viciifolia*) birdsfoot trefoil (*Lotus corniculatus*), and chicory (*Cichorium intybus*). *Arch. Anim. Nutr.* **2007**, *61*, 481–496. [CrossRef] [PubMed]

34. Tavendale, M.H.; Meagher, L.P.; Pacheco, D.; Walker, N.; Attwood, G.T.; Sivakumaran, S. Methane production from in vitro rumen incubations with Lotus pedunculatus and Medicago sativa, and effects of extractable condensed tannin fractions on methanogenesis. *Anim. Feed Sci. Technol.* **2005**, *123–124 Pt 1*, 403–419. [CrossRef]

35. Doane, P.H.; Pell, A.N.; Schofield, P. The Effect of Preservation Method on the Neutral Detergent Soluble Fraction of Forages. *J. Anim. Sci.* **1997**, *75*, 1140–1148. [CrossRef] [PubMed]

36. Bodas, R.; Prieto, N.; García-González, R.; Andrés, S.; Giráldez, F.J.; López, S. Manipulation of rumen fermentation and methane production with plant secondary metabolites. *Anim. Feed Sci. Technol.* **2012**, *176*, 78–93. [CrossRef]

37. Niderkorn, V.; Mueller-Harvey, I.; Le Morvan, A.; Aufrère, J. Synergistic effects of mixing cocksfoot and sainfoin on in vitro rumen fermentation. Role of condensed tannins. *Anim. Feed Sci. Technol.* **2012**, *178*, 48–56. [CrossRef]

38. Huyen, N.; Fryganas, C.; Uittenbogaard, G.; Mueller-Harvey, I.; Verstegen, M.; Hendriks, W.; Pellikaan, W. Structural features of condensed tannins affect in vitro ruminal methane production and fermentation characteristics. *J. Agric. Sci.* **2016**, *154*, 1474–1487. [CrossRef]

39. Lobón, S.; Molino, F.; Legua, A.; Eseverri, M.P.; Céspedes, M.A.; Joy, M. Efecto del Forraje y de La Inclusión de Concentrado En la Dieta Sobre La Producción de Gas y Metano En Ovino. 2015. Available online: http://www.uibcongres.org/imgdb/archivo_dpo19392.pdf (accessed on 1 March 2019).

40. Mueller-Harvey, I. Unravelling the conundrum of tannins in animal nutrition and health. *J. Sci. Food Agric.* **2006**, *86*, 2010–2037. [CrossRef]

41. Hassanat, F.; Benchaar, C. Assessment of the effect of condensed (acacia and quebracho) and hydrolysable (chestnut and valonea) tannins on rumen fermentation and methane production in vitro. *J. Sci. Food Agric.* **2013**, *93*, 332–339. [CrossRef]

42. Theodoridou, K.; Aufrére, J.; Andueza, D.; Le Morvan, A.; Picard, F.; Pourrat, J.; Baumont, R. Effects of condensed tannins in wrapped silage bales of sainfoin (Onobrychis viciifolia) on in vivo and in situ digestion in sheep. *Animal* **2012**, *6*, 245–253. [CrossRef] [PubMed]

43. Jin, L.; Wang, Y.; Iwaasa, A.; Xu, Z.; Schellenberg, M.; Zhang, Y.; Liu, X.; McAllister, T. Effect of condensed tannins on ruminal degradability of purple prairie clover (Dalea purpurea Vent.) harvested at two growth stages. *Anim. Feed Sci. Technol.* **2012**, *176*, 17–25. [CrossRef]

44. Piluzza, G.; Bullitta, S. The dynamics of phenolic concentration in some pasture species and implications for animal husbandry. *J. Sci. Food Agric.* **2010**, *90*, 1452–1459. [CrossRef]

45. Joseph, R.; Tanner, G.; Larkin, P. Proanthocyanidin synthesis in the forage legume Onobrychis viciifolia. A study of chalcone synthase, dihydroflavonol 4-reductase and leucoanthocyanidin 4-reductase in developing leaves. *Funct. Plant Biol.* **1998**, *25*, 271–278. [CrossRef]

46. Bal, M.A.; Ozturk, D.; Aydin, R.; Erol, A.; Ozkan, C.O.; Ata, M.; Karakas, E.; Karabay, P. Nutritive value of sainfoin (Onobrychis viciaefolia) harvested at different maturity stages. *Pak. J. Biol. Sci.* **2006**, *9*, 205–209. [CrossRef]

47. Navarro-Villa, A.; O'Brien, M.; López, S.; Boland, T.M.; O'Kiely, P. In vitro rumen methane output of red clover and perennial ryegrass assayed using the gas production technique (GPT). *Anim. Feed Sci. Technol.* **2011**, *168*, 152–164. [CrossRef]

Article

In Vitro Evaluation of Different Dietary Methane Mitigation Strategies

Juana C. Chagas, Mohammad Ramin and Sophie J. Krizsan *

Department of Agricultural Research for Northern Sweden, Swedish University of Agricultural Sciences (SLU), Skogsmarksgränd, 90183 Umeå, Sweden; juana.chagas@slu.se (J.C.C.); mohammad.ramin@slu.se (M.R.)
* Correspondence: sophie.krizsan@slu.se; Tel.: +46-90-7868748

Received: 15 October 2019; Accepted: 5 December 2019; Published: 11 December 2019

Simple Summary: Dietary methane mitigation strategies do not necessarily make food production from ruminants more energy-efficient, but reducing methane (CH_4) in the atmosphere immediately slows down global warming, helping to keep it within 2 °C above the pre-industrial baseline. There is no single most efficient strategy for mitigating enteric CH_4 production from domestic ruminants on forage-based diets. This study assessed a wide variety of dietary CH_4 mitigation strategies in the laboratory, to provide background for future studies with live animals on the efficiency and feasibility of dietary manipulation strategies to reduce CH_4 production. Among different chemical and plant-derived inhibitors and potential CH_4-reducing diets assessed, inclusion of the natural antimethanogenic macroalga *Asparagopsis taxiformis* showed the strongest, and dose-dependent, CH_4 mitigating effect, with the least impact on rumen fermentation parameters. Thus, applying *Asparagopsis taxiformis* at a low daily dose was the best potential dietary mitigation strategy tested, with promising long-term effects, and should be further studied in diets for lactating dairy cows.

Abstract: We assessed and ranked different dietary strategies for mitigating methane (CH_4) emissions and other fermentation parameters, using an automated gas system in two in vitro experiments. In experiment 1, a wide range of dietary CH_4 mitigation strategies was tested. In experiment 2, the two most promising CH_4 inhibitory compounds from experiment 1 were tested in a dose-response study. In experiment 1, the chemical compounds 2-nitroethanol, nitrate, propynoic acid, p-coumaric acid, bromoform, and *Asparagopsis taxiformis* (AT) decreased predicted in vivo CH_4 production (1.30, 21.3, 13.9, 24.2, 2.00, and 0.20 mL/g DM, respectively) compared with the control diet (38.7 mL/g DM). The 2-nitroethanol and AT treatments had lower molar proportions of acetate and higher molar proportions of propionate and butyrate compared with the control diet. In experiment 2, predicted in vivo CH_4 production decreased curvilinearly, molar proportions of acetate decreased, and propionate and butyrate proportions increased curvilinearly with increased levels of AT and 2-nitroethanol. Thus 2-nitroethanol and AT were the most efficient strategies to reduce CH_4 emissions in vitro, and AT inclusion additionally showed a strong dose-dependent CH_4 mitigating effect, with the least impact on rumen fermentation parameters.

Keywords: antimethanogenic; chemical inhibition; global warming; halogenated compound; macroalgae; methane production; methanogenic inhibitor; plant inhibitory compound

---

1. Introduction

The global population is growing and, although there is enough food in the world today, there are major differences in how people live. Meat and milk from ruminants are high-quality foods and a large proportion of their production is based on grass, but production is still resource-intensive. Future intensification of agriculture can reinforce negative effects such as greenhouse gas (GHG) emissions, the main contributor to climate change through global warming [1,2].

Methane (CH_4) is a powerful GHG that plays a key part in global climate change and concentrations have been rising rapidly in the atmosphere over the past decade. Recently published data based on radioactive carbon (C^{14}) content in CH_4 indicate that anthropogenic emissions of CH_4 in recent decades have been higher than previously estimated [3]. Satellite data [4] suggest that the increased global CH_4 emissions in the period 2005–2015 were mostly due to increased extraction of shale gas, and that the natural gas and oil industry contributes twice as much CH_4 emissions as animal agriculture.

Methanogenesis in the rumen is an essential metabolic process required to remove molecular hydrogen generated during fermentation. The production of CH_4 is influenced by animal species, age, management, and diet. The Rumen Census project sequenced a wide variety of rumen and camelid foregut microbial communities in many samples from a wide variety of animal species and countries, to identify factors such as diet, host species, or geography causing the greatest variation in CH_4 emissions [5]. The results showed that rumen archaeal diversity was similar irrespective of host or diet, and a core rumen bacterial population of 67% of the community occurred irrespective of host or diet. The main diversity changes in other bacteria present were caused by diet, and not host genetics [6]. Although dietary strategies for mitigating enteric CH_4 production in ruminants have been intensively studied, no single most efficient dietary strategy has been identified for dairy cows on forage-based diets.

Methane losses from typical dairy cow diets are 6–7% of gross energy intake, but losses are approximately 3% in feedlot situations, indicating that feeding high-concentrate diets can reduce CH_4 production [7]. However, recent data indicate that the effect on CH_4 production of including more grain in dairy cows diets is small [8] and use of this strategy can therefore be questioned. Use of antimethanogenics or plant inhibitory compounds in ruminant diets can also reduce GHG emissions, and has been suggested as an effective and feasible strategy in the livestock sector [9]. Dietary mitigation strategies do not necessarily make food production from ruminants more energy-efficient, but they reduce CH_4 emissions to the atmosphere and thus immediately slow down global warming [10], contributing to keep the planet within 2 °C of the pre-industrial baseline [11]. The use of CH_4 inhibitors might be the most immediate and efficient strategy to reduce CH_4 emissions from dairy cows. The most successful inhibitor suggested in vivo so far is 3-nitroxypropanol that has showed CH_4 reducing effects when provided to dairy cows in a low dose [12]. The tropical macroalgae *Asparagopsis Taxiformis* is a recent and natural supplement that has shown very promising CH_4 inhibitory effects in vitro [13].

In vitro gas production technique has been developed to evaluate factors influencing digestibility and fermentation kinetics from feeds. The technique has been used to estimate CH_4 emission with the advantage of screening large number of samples, providing large amount of data points, and allowing accurate predictions of in vivo CH_4 production [14].

This study assessed and ranked a wide variety of dietary CH_4 mitigation strategies using an automated gas in vitro system, in order to provide background for future in vivo evaluations of dietary manipulation strategies for efficiently reducing CH_4 production from domestic ruminants.

2. Materials and Methods

Two in vitro experiments were conducted to assess different dietary antimethanogenic compounds. In experiment 1, the dietary CH_4 mitigating strategies tested comprised six chemical inhibitory compounds at two levels, three plant-derived inhibitory treatments at two levels, five different potentially CH_4-reducing diets with the active ingredients in two levels except for one of the diets, and two typical grass silage fermentation acids at two levels to mimic different silage fermentation qualities. In experiment 2, the two most promising CH_4 inhibitory treatments from experiment 1 were tested in a dose-response experiment designed to represent a wide range of treatment levels.

2.1. Experimental Treatments

2.1.1. Experiment 1

All experimental diets were composed from a control diet that consisted of timothy grass (*Phleum pratense*), rolled barley (*Hordeum vulgare*), and rapeseed (*Brassica napus*) meal in a ratio of 545:363:92 g/kg diet dry matter (DM). The grass and rolled barley originated from Röbäcksdalens research farm in Umeå (63°45′ N, 20°17′ E), Sweden. The rapeseed meal was a commercial solvent-extracted and heat-moisture-treated protein supplement ExPro-00SF (Aarhus Karlshamn AB, Malmö, Sweden). All potential dietary CH_4 mitigating strategies tested in experiment 1 are listed in Table 1. The chemical compounds 2-nitroethanol (2-NE), propynoic acid, ferulic acid, p-coumaric acid, and bromoform (Sigma-Aldrich Sweden AB, Stockholm, Sweden) were added without replacing any DM of the control diet. Nitrate was added to the control diet to represent one level of 21 g NO_3/kg DM or 0.0890 g $Ca(NO_3)_2 \times 4H_2O$/g DM (Sigma-Aldrich Sweden AB, Stockholm, Sweden). The nitrate treatment was compared with a zero-nitrate treatment in which 0.0350 g urea/g DM and 0.051 g $CaCO_3$/g DM (J.T. Baker BV, Deventer, The Netherland) were added to the control diet to achieve an isonitrogenous and equivalent diet (159 g crude protein (CP)/kg DM). The plant-derived compounds rowan (*Sorbus aucuparia*) berries and the forb fireweed (*Chamerion angustifolium*) were added to replace grass and barley in the control diet, such that the ratio of forage:concentrate was kept constant relative to all other diets. These ingredients were collected in Umeå (63° N, 20° E), Sweden in October and July 2018, respectively. The red seaweed *Asparagopsis taxiformis* (AT) was added in such a small dose in both levels of the treatment that no replacement of control dietary ingredients was made. The seaweed was harvested in the Azores (38.6° N, 28° W), Portugal, in October 2018. Replacements in the potentially CH_4-reducing diets were also made so that the forage:concentrate ratio was kept constant relative to all other diets and to contain 160 g CP/kg diet DM. Rapeseed oil and oats (*Avena sativa*) were added to replace grass and barley on a DM basis. These ingredients were also collected in Umeå in July 2018. Dried distiller's grains (Agrodrank 90, Agroetanol, Östergötland, Sweden) replaced rapeseed meal in the control diet and was added to represent an increment of 20 g/kg DM in CP between the levels (CP concentration 160 g/kg DM and 180 g/kg DM, respectively). The CP concentration was made iso-nitrogenous to the control diet for the lowest level when dried distiller's grain replaced rapeseed meal.

Table 1. Experimental treatments evaluated in vitro in experiment 1 for methane (CH_4) mitigation potential.

Treatments	Levels	
Chemical compounds		
2-nitroethanol	5 mM	10 mM
Nitrate	None [1]	21 g/kg DM [2]
Propynoic acid	2 mM	4 mM
Ferulic acid	10 mM	20 mM
p-Coumaric acid	10 mM	20 mM
Bromoform	1.5 mg/g DM	3 mg/g DM
Plant-derived treatments		
Rowan berries	50 g/kg DM	100 g/kg DM
Fireweed	50 g/kg DM	100 g/kg DM
Asparagopsis taxiformis	10 g/kg OM	20 g/kg OM
Potentially CH_4-reducing treatments		
Rapeseed oil	40 g/kg DM	80 g/kg DM
Dried distiller's grain	90 g/kg DM	180 g/kg DM
Barley:oat	175:175 g/kg	0:350 g/kg
Maize silage:grass	275:275 g/kg [3]	545:0 g/kg [4]
Red clover:grass	275:275 g/kg	None
Lactic acid	60 g/kg DM	120 g/kg DM
Lactic acid + acetic acid	80 + 30 g DM	80 + 60 g DM

DM = dry matter; [1] 0.035 g of urea + 0.051 g of $CaCO_3$ on DM basis included in control diet in comparison with nitrate treatment; [2] 0.089% $Ca(NO_3)_2 \times 4H_2O$ on DM basis; [3] Urea was added to correct CP at 160 g/kg DM; [4] Urea was added to correct CP at 160 g/kg DM.

In the treatments where maize (*Zea mays*) silage replaced grass silage, urea was added to make diets isonitrogenous to the control diet. No correction of CP concentration was made in the diet when red clover (*Trifolium pratense*) replaced grass.

2.1.2. Experiment 2

In experiment 2, AT (0, 0.06, 0.13, 0.25, 0.5, and 1.0 g/kg of diet organic matter (OM)) and 2-NE (0, 0.3, 0.7, 1.3, 2.6, and 5.1 mM) were tested in a dose-response experiment comprising six different treatment levels. The same control diet as in experiment 1, of timothy, rolled barley, and rapeseed meal, was used in experiment 2.

2.2. In Vitro Incubations

The handling of animals in this experiment was approved by the Swedish Ethics Committee on Animal Research (Dnr A 32-16), represented by the Court of Appeal for Northern Norrland in Umeå, and the experiment was carried out in accordance with laws and regulations governing experiments performed with live animals in Sweden.

Two lactating Swedish Red cows, fed ad libitum on a diet of 600 g/kg grass silage and 400 g/kg concentrate on a DM basis (presenting chemical composition as 509 g/kg of DM, 425 g/kg NDF, and 171 g/kg CP), were used as donor animals of rumen inoculum for all incubations. The rumen fluid from each cow was filtered separately using a double layer of cheesecloth into Thermos flasks that were pre-warmed and flushed with carbon dioxide (CO_2) prior to collection. Rumen fluid was transported to the laboratory within 15 min. Equal amounts from each cow were immediately blended, strained through four layers of cheesecloth, and added to buffered mineral solution [15] including Peptone™ (pancreatic digested casein; Merck, Darmstadt, Germany) at 39 °C under constant mixing and CO_2 flushing, to give a buffered rumen fluid solution with a rumen fluid:buffer ratio of 1:4 by volume.

Prior to each in vitro incubation, dietary ingredients were dried at 60 °C for 48 h and milled in a Retsch SM 2000 cutting mill (Retsch GmbH, Haan, Germany) to pass through a 1-mm screen. Then 1003 ± 38 mg of DM substrate were weighed into serum bottles flushed with CO_2, and 60 mL of the previously prepared buffered rumen fluid were added. All bottles were placed in a water bath and gently and continuously agitated at 39 °C during an incubation period of 48 h.

These procedures were repeated for six runs in total and all samples were incubated, with three replicates of each sample. All runs included triplicate bottles with blanks (i.e., bottles with 60 mL of buffered rumen fluid with no sample or treatment in), and samples were randomly allocated to the in vitro incubation bottles and never incubated in the same bottle in more than one run.

2.3. In Vitro Gas Production Measurements and Sampling

Gas production was measured using a fully automated system (Gas Production Recorder, GPR-2, Version 1.0 2015, Wageningen UR), with readings made every 12 min and corrected to the normal air pressure (101.3 kPa) [16].

Measurement of CH_4 in vitro was performed according to Ramin and Huhtanen [14] on gas samples withdrawn during the incubation period (0.2 mL) from each bottle at 2, 4, 8, 24, 32, and 48 h. Concentration of CH_4 was determined with a Varian Star 3400 CX gas chromatograph (Varian Analytical Instruments, Walnut Creek, CA, USA) equipped with a thermal conductivity detector.

Liquid samples of 0.6 mL were collected from the bottles at 8, 24, and 48 h of incubation and immediately stored at −20 °C until analysis of volatile fatty acids (VFA). Liquid samples for ammonia-nitrogen (NH_3-N) analysis were taken at 8 and 24 h of incubation, and also stored at −20 °C before further analysis. Liquid samples from the replicate treatments between runs were pooled before NH_3-N and VFA analysis.

After 48 h of incubation, all bottles were removed from the water bath and placed on ice to stop fermentation. The residue was used for in vitro determination of true organic matter digestibility (TOMD).

2.4. Chemical Analysis

The concentrations of DM and OM in the individual dietary ingredients were quantified by AOAC [17] method 930.15 and method 942.05, respectively. Concentrations of nitrogen were determined by Kjeldahl digestion of 1000 mg sample in 12 M sulfuric acid using Foss Tecator Kjeltabs Cu (Höganäs, Sweden) in a Block Digestion 28 system (SEAL Analytical Ltd., Mequon, WI, USA), followed by determination of total nitrogen by continuous flow analysis using an Auto Analyzer 3 (SEAL Analytical Ltd., Mequon, WI, USA). The samples were analyzed for neutral detergent fiber (NDF) using a heat-stable α-amylase [18] in an ANKOM200 Fiber Analyzer (Ankom Technology Corp., Macedon, NY, USA).

In vitro TOMD was determined for all samples in all runs by analyzing ash-free NDF concentrations in the residues using 07-11/5 Sefar Petex (Sefar AG, Heiden, Switzerland) in situ bags according to Krizsan et al. [19].

Individual VFA concentrations in rumen fluid samples were determined using a Waters Alliance 2795 UPLC system as described by Puhakka et al. [20], and NH_3-N concentration according to the method provided by SEAL Analytical (Method no. G-102-93 multitest MT7) using AutoAnalyzer 3.

Bromoform concentration in AT was analyzed according to Roque et al. [21] using an Agilent 7890B GC applied to Agilent 7000C triple quad Mass Spectrometer equipped with a ZB-5ms column (Agilent Technologies, Inc. Santa Clara, CA, USA).

2.5. Calculations

Mean blank gas production within run was subtracted from sample gas production. In vivo predicted CH_4 production was calculated as described by Ramin and Huhtanen [14] as:

$$CH_4 = 265 \times CH_4 \text{ concentration} + \text{total gas production} \times CH_4 \text{ concentration} \times 0.55$$

where total gas production is in mL/g sample, 265 is the total headspace volume (mL), and 0.55 is the ratio of CH_4 emissions in the outflow gas from the in vitro system. A mean retention time of 50 h (20 h in the first compartment and 30 h in the second compartment), corresponding to the maintenance level of feed intake, was used in model simulations.

Total VFA (TVFA) production was calculated as: the molar proportion of individual VFA were calculated related to TVFA.

$$TVFA \text{ (mmol)} = (\textstyle\sum \text{ individual VFA concentration} - \text{mean of blank VFA}) \times$$
$$0.06 \text{ (amount of buffered rumen fluid)}$$

The molar proportion of individual VFA were calculated related to TVFA.
The in vitro TOMD was calculated as:

$$TOMD \text{ (g/kg)} = \frac{\text{incubated OM (g)} - \text{NDF residue corrected for ash and blank (g)}}{1000 \times \text{incubated OM (g)}}$$

2.6. Statistical Analysis

Data on in vivo predicted CH_4 production and in vitro TOMD from Experiment 1 were analyzed using the MIXED procedure in SAS (SAS Institute Inc., Cary, NC, version 9.4), by a model correcting for random effect of bottle and fixed effect of run and treatment:

$$Y_{ijk} = \mu + T_i + R_j + B_k + e_{ijk}$$

where Y_{ijk} is dependent variable ijk, μ is overall mean, T_i is treatment i, R_j is run j, B_k is bottle k, and $e_{ijk} \sim N(0,\sigma_e^2)$ is the random residual error. Orthogonal contrasts were included for evaluation of control diet vs. treatment and of linear responses to level of treatment.

Data on measured VFA and NH_3-N concentrations from Experiment 1 were evaluated in a repeated measurements model using the Toeplitz function in the MIXED procedure in SAS (SAS Institute Inc., Cary, NC, USA, version 9.4) (level within treatment was used as subject). The model accounted for effects of treatment and time, and interactions between treatment and time:

$$y_{ij} = \mu + T_i + A_j + (TA)_{ij} + e_{ij}$$

where y_{ij} is the dependent variable ij, μ is overall mean, T_i is treatment i, A_j is time j, $(TA)_{ij}$ is interaction between treatment i and time j, and $e_{ij} \sim N(0,\sigma_e^2)$ is the random residual error.

Data on predicted in vivo CH_4 production, in vitro TOMD, total VFA (TVFA), and molar proportions of individual VFA and NH_3-N from experiment 2 were subjected to linear and quadratic regression analysis using the REG procedure in SAS (SAS Institute Inc., Cary, NC, USA, version 9.4). Best fit was judged from lowest root mean square error and highest adjusted R^2.

Effects were considered statistically significant at p-value ≤ 0.05.

3. Results

The chemical composition of the control diet and the potential CH_4 reducing diets is shown in Table 2. The AT bromoform concentration was 6.84 mg/g DM.

Table 2. Chemical composition (g/kg DM) of control and potential methane (CH_4) reducing diets evaluated in vitro in experiment 1.

Treatment	Level	Organic Matter	Crude Protein	Neutral Detergent Fiber
Control diet	—	944	160	387
Rapeseed oil	40 g/kg DM	906	154	372
Rapeseed oil	80 g/kg DM	869	149	356
Dried distiller's grain	90 g/kg DM	946	161	378
Dried distiller's grain	180 g/kg DM	946	181	366
Barley: oat	175:175 g/kg	944	165	385
Barley: oat	0:350 g/kg	944	170	383
Maize silage: grass	275:275 g/kg	954	160	355
Maize silage: grass	545:0 g/kg	963	160	323
Red clover: grass	275:275 g/kg	932	171	345
Lactic acid	60 g/kg DM	887	151	364
Lactic acid	120 g/kg DM	831	143	341
Lactic acid + acetic acid	80 + 30 g/kg DM	840	144	345
Lactic acid + acetic acid	80 + 60 g/kg DM	812	140	333

NDF = neutral detergent fibre.

3.1. Experiment 1

Predicted in vivo CH_4 production derived from analysis of 48 h gas in in vitro incubation of the control diet was 38.7 mL/g DM, in vitro TOMD was 867 g/kg, TVFA was 3.62 mmol, and molar proportion of acetate, butyrate, and propionate was 583, 125, and 237 mmol/mol, respectively. In comparison with the control diet the chemical compounds 2-NE, nitrate, propynoic acid, p-coumaric acid, bromoform, and the plant compound AT, decreased ($p \leq 0.01$) in vivo CH_4 predicted production (Table 3). Addition of 2-NE, bromoform, and AT gave the strongest inhibition ($p < 0.01$) of predicted in vivo CH_4 production among all experimental treatments (97%, 95%, and 99% reduction in the value for the control diet). The reduction in predicted in vivo CH_4 production achieved by the other compounds ranged between 38% and 64% of the value for the control diet. Surprisingly, none of the potential CH_4 reducing diets or lactic acid and acetic acid addition affected CH_4 production in this study ($p \geq 0.20$). In vitro TOMD was negatively affected by the chemical compounds p-coumaric acid and bromoform ($p < 0.01$), while rapeseed oil inclusion in the diet increased in vitro TOMD compared with the control diet ($p = 0.04$). Propynoic acid and bromoform decreased ($p \leq 0.01$) TVFA compared with the control diet. Several of the treatments altered the molar proportions of individual VFA. Acetate decreased ($p \leq 0.03$) on adding 2-NE, propynoic acid, p-coumaric acid, bromoform, AT, or lactic acid to the control diet. For all those treatments except p-coumaric acid and bromoform, there was a concomitant increase ($p \leq 0.05$) in molar proportions of propionic and butyric acid compared with the control diet. Results of nitrate vs. zero nitrate treatment were: TVFA 2.91 vs. 3.01 mol, acetate 597 vs. 604 mmol/mol propionate 250 vs. 227 mmol/mol and butyrate 87 vs. 123 mmol/mol.

The molar proportion of isobutyrate, isovalerate and valerate, and NH_3-N for the control diet and experimental treatments are given in, Table 4. The molar proportions of the branched-chain volatile fatty acids (BCVFA) were altered by many of the CH_4 mitigating strategies tested. Compared to the control diet, isobutyrate increased ($p \leq 0.01$) for p-coumaric acid treatment, while for bromoform treatment the molar proportion decreased ($p \leq 0.01$). The treatments, 2-NE, propynoic, p-coumaric, ferulic acid, AT, lactic acid, and lactic acid + acetic acid, increased ($p \leq 0.04$), and bromoform decreased ($p \leq 0.01$) the molar proportion of isovalerate compared to the control diet. Propynoic acid decreased ($p \leq 0.05$) while bromoform and AT increased ($p \leq 0.05$) molar proportion of valerate. Results of nitrate vs. zero nitrate treatment were: isobutyrate 6.63 vs. 9.96 mmol/mol, isovalerate 4.01 vs. 4.94 mmol/mol, valerate 16.4 vs. 16.3 mmol/mol, and NH_3-N concentration 436 vs. 557 mg/L.

Tests for linear effects between the two levels according to Table 1 and the control diet (0 here) revealed no significant effect on in vitro TOMD ($p = 0.148$) for all treatments (data not presented). However, there was a significant linear decrease ($p < 0.01$) in predicted in vivo CH_4 production for propynoic acid (24 and 0 mL/g DM) and p-coumaric acid (27.1 and 19.8 mL/g DM) when the inclusion level was increased.

Table 3. Effect of experimental treatments on predicted in vivo CH_4 production (mL/g DM), in vitro true organic matter digestibility (TOMD, g/kg), total volatile fatty acid production (TVFA, mmol), and molar proportions of acetate, propionate, and butyrate (mmol/mol of TVFA) measured in 48 h gas from the in vitro incubation in experiment 1.

Treatment	CH_4	TOMD	TVFA	Acetate	Propionate	Butyrate	p-Value [1]					
							C_{CH4}	C_{TOMD}	C_{TVFA}	$C_{Acetate}$	$C_{Propionate}$	$C_{Butyrate}$
Control	38.7	867	3.62	583	237	125	–	–	–	–	–	–
2-nitroethanol	1.30	858	3.01	440	309	211	<0.01	0.30	0.10	<0.01	<0.01	<0.01
Nitrate [2]	21.3	874	2.96	619	250	87	<0.01	0.82	NA	NA	NA	NA
Propynoic acid	13.9	839	2.57	476	297	209	<0.01	0.25	0.01	<0.01	<0.01	<0.01
Ferulic acid	27.5	859	3.54	597	229	109	0.06	0.71	0.82	0.62	0.68	0.32
p-Coumaric acid	24.2	763	3.01	492	176	121	0.01	<0.01	0.10	<0.01	<0.01	0.82
Bromoform	2.00	822	2.30	436	270	261	<0.01	<0.01	<0.01	<0.01	0.08	<0.01
Fireweed	38.1	858	4.01	583	226	140	0.34	0.69	0.24	0.99	0.55	0.34
Rowan berries	28.9	843	3.71	586	241	117	0.96	0.35	0.80	0.93	0.82	0.64
A. taxiformis	0.20	852	3.61	418	327	184	<0.01	0.97	0.98	<0.01	<0.01	<0.01
Rapeseed oil	38.2	896	4.04	600	217	128	0.82	0.04	0.24	0.56	0.28	0.83
Dried distiller's grain	35.3	877	3.74	549	241	152	0.43	0.48	0.73	0.66	0.91	0.33
Barley: oat	37.4	863	3.66	596	225	124	0.73	0.73	0.89	0.65	0.51	0.98
Maize silage: grass	30.7	846	3.71	577	240	136	0.61	0.56	0.80	0.84	0.87	0.48
Red clover: grass	47.9	882	3.16	599	234	129	0.20	0.53	0.27	0.62	0.88	0.85
Lactic acid	34.1	866	3.65	516	271	158	0.25	0.54	0.93	0.03	0.07	0.05
Lactic acid + acetic acid	35.2	885	3.34	598	224	135	0.34	0.11	0.43	0.59	0.49	0.53
SEM	1.75	4.3	0.120	7.2	4.2	5.5	–	–	–	–	–	–

NA = not analyzed; SEM = standard error mean. [1] Orthogonal contrasts of control diet vs. treatment of the different in vitro traits. [2] Nitrate treatment was compared to the zero nitrate diet made by adding urea and $CaCO_3$ to the control diet according to Table 1; numerical differences of TVFA and molar proportions of volatile fatty acids are given in the text.

Table 4. Effect of experimental treatments on molar proportions of isobutyrate, isovalerate, and valerate (mmol/mol of TVFA), and ammonia concentration (NH_3-N, mg/L) measured in 48 h gas from the in vitro incubation in experiment 1.

Treatments	Isobutyrate	Isovalerate	Valerate	NH_3-N	p-Value [1]			
					$C_{Isobutyrate}$	$C_{Isovalerate}$	$C_{Valerate}$	C_{NH3-N}
Control	10.9	0.51	21.4	282	–	–	–	–
2-nitroethanol	5.16	2.77	16.4	436	0.80	<0.01	0.14	0.39
Nitrate [2]	6.63	4.01	16.43	270	NA	NA	NA	NA
Propynoic acid	8.34	2.59	0.77	311	0.91	<0.01	<0.01	0.25
Ferulic acid	14.4	5.23	20.0	320	0.88	0.03	0.67	0.98
p-Coumaric acid	165	4.23	22.6	263	<0.01	<0.01	0.73	0.81
Bromoform	0.00	0.00	28.7	302	0.64	<0.01	0.03	0.17
Fireweed	10.8	5.97	18.5	304	0.99	0.20	0.38	0.78
Rowan berries	10.1	5.85	21.5	289	0.97	0.15	0.98	0.81
A. taxiformis	5.95	5.33	35.4	354	0.83	0.04	0.00	0.52
Rapeseed oil	13.3	8.32	19.1	319	0.92	0.08	0.49	0.25
Dried distiller's grain	11.5	6.87	20.7	301	0.98	0.91	0.83	0.07
Barley: oat	12.5	6.98	18.5	359	0.95	0.98	0.38	0.75
Maize: grass	10.0	6.10	17.8	281	0.97	0.26	0.28	0.20
Red clover: grass	9.04	5.23	14.6	306	0.94	0.06	0.08	0.45
Lactic acid	9.30	5.29	21.3	323	0.94	0.03	0.97	0.86
Lactic acid + acetic acid	8.95	4.61	16.7	282	0.93	<0.01	0.16	0.77
SEM	4.426	0.370	0.87	12.2	–	–	–	–

NA = not analyzed; SEM = standard error mean. [1] Orthogonal contrasts of control diet vs. treatment of the different in vitro traits. [2] Nitrate treatment was compared to the zero nitrate diet made by adding urea and $CaCO_3$ to the control diet according to Table 1; numerical differences molar proportions of branched-chain volatile fatty acids and NH_3-N are given in the text.

3.2. Experiment 2

Predicted in vivo CH_4 production decreased curvilinearly ($p < 0.01$) with increased levels of both 2-NE (Figure 1).

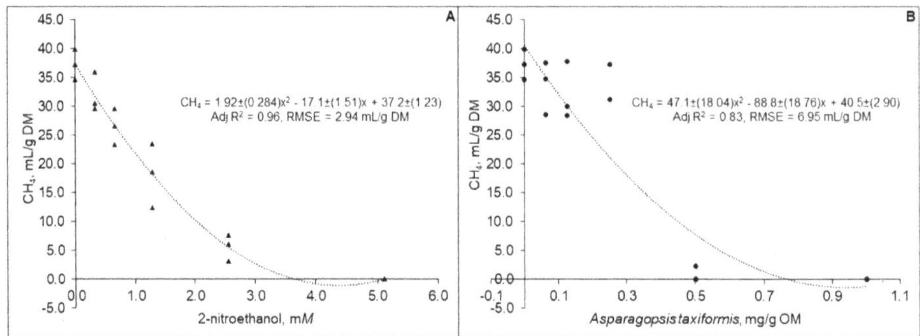

Figure 1. Predicted in vivo methane production based on analysis of 48 h gas from in vitro incubation of a control diet (545:363:92 g/kg of grass silage:barley:rapeseed meal) treated with different levels (three replicates per level) of (**A**) 2-nitroethanol and (**B**) *Asparagopsis taxiformis* in experiment 2.

The TVFA content decreased linearly ($p < 0.01$) from 5.35 to 3.00 mmol at 48 h for 2-NE and from 4.71 to 4.33 mmol at 24 h for AT for the lower to higher level of supplementation (Figure 2). The TVFA content for 2-NE at 8 h ($p < 0.01$; adj $R^2 = 0.38$; RSME = 0.22 mmol) and 24 h ($p < 0.01$; adj $R^2 = 0.55$; RSME = 0.25 mmol) showed curvilinear responses, while for AT the curvilinear pattern was verified at 8 h ($p < 0.01$; $R^2 = 0.55$; RSME = 0.25 mmol) and 48 h ($p = 0.01$; adj $R^2 = 0.33$; RSME = 0.33 mmol).

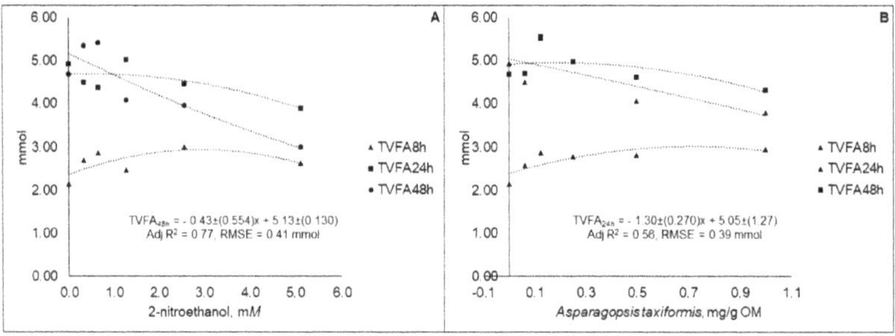

Figure 2. Total volatile fatty acid (TVFA) content in fluid samples taken at different time points during 48 h in vitro incubation of a control diet (545:363:92 g/kg of grass silage:barley:rapeseed meal) treated with different levels (three replicates per level) of (**A**) 2-nitroethanol and (**B**) *Asparagopsis taxiformis* in experiment 2.

Molar proportion of acetate decreased, while propionate and butyrate proportions increased curvilinearly ($p < 0.01$), at all-time points studied for 2-NE and AT (Figure 3). There were no statistical difference ($p > 0.05$) between the coefficients generated for the equations of the different time points. The best fit equations of molar proportions of VFA were generated for both 2-NE and AT from different sampling time points (Figure 3), but the equations generated were not statistically different ($p > 0.05$) from the other sampling time points (results not presented).

There were no linear or curvilinear relationships between TOMD and level of supplementation for 2-NE ($p = 0.152$) or AT ($p = 0.142$) (results not presented).

The equations of the molar proportions of BCVFA (isobutyrate, isovalerate, and valerate) were statistically different ($p < 0.05$) between the different sampling time points. The best fit equations are presented in Figure 4.

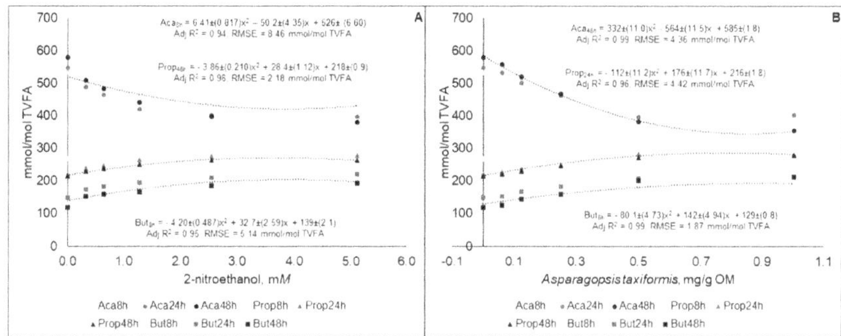

Figure 3. Molar proportions of acetate (Ace), propionate (Prop), and butyrate (But) in fluid samples gas samples taken at different time points during 48 h in vitro incubation of a control diet (545:363:92 g/kg of grass silage:barley:rapeseed meal) treated with different levels (three replicates per level) of (**A**) 2-nitroethanol and (**B**) *Asparagopsis taxiformis* in experiment 2.

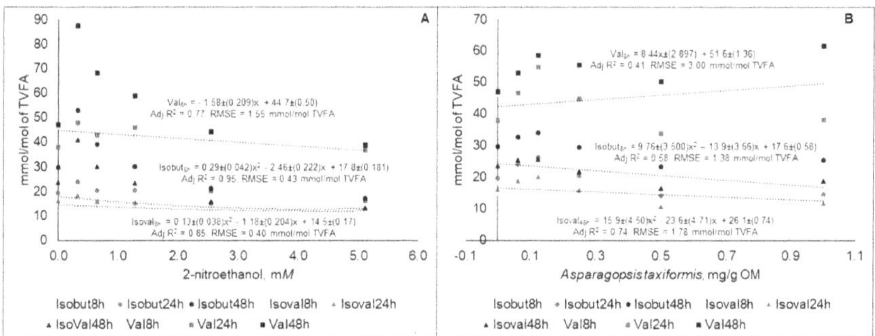

Figure 4. Molar proportions of isobutyrate (Isobut), isovalerate (Isoval), and valerate (Val) in fluid taken at different time points during 48 h in vitro incubation of a control diet (545:363:92 g/kg of grass silage:barley:rapeseed meal) treated with different levels (three replicates per level) of (**A**) 2-nitroethanol and (**B**) *Asparagopsis taxiformis* in experiment 2.

The NH_3-N concentration responses decreased linearly ($p < 0.05$) for both 2-NE and AT at both 8 and 24 h, and the best fit equations are presented in Figure 5.

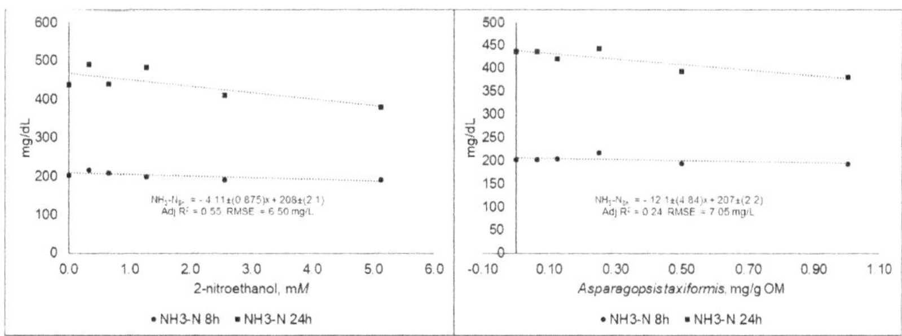

Figure 5. Ammonia concentration (NH_3-N) in fluid samples taken at different time points during 48 h in vitro incubation of a control diet (545:363:92 g/kg of grass silage:barley:rapeseed meal) treated with different levels (three replicates per level) of (**A**) 2-nitroethanol and (**B**) *Asparagopsis taxiformis* in experiment 2.

4. Discussion

The CP concentration of the potential CH_4 reducing diets varied between 140 and 181 g/kg DM, and reflected the characteristics of the dietary ingredient studied in each diets. Regarding that Peptone™ was included in the buffered rumen fluid, none of the diets supplied an insufficient amount of CP in terms of CP available for rumen microbial growth in comparison with in vivo requirements [22].

Ruminants are valuable food producers world-wide, since they are able to utilize fibrous non-human-edible resources (forages and pasture) through microbial fermentation of feed in the rumen. Recent data indicate that domesticated ruminants are not the major contributor to anthropogenic CH_4 emissions. The fermentation of feed and decomposition of manure are the foremost sources of GHG emissions caused by domesticated ruminants [23]. Estimates suggest that livestock are responsible for around 9% and 37% of anthropogenic CO_2 and CH_4 emissions, respectively [24]. Long-term strategies to improve feed efficiency through targeted breeding [25] and improved longevity or lifetime productivity [26] could reduce CH_4 emissions from dairy cows. According to Knapp et al. [9], nutrition and feeding approaches may be able to reduce CH_4 emissions per unit of energy-corrected milk by 2.5–15%, while reductions of 15–30% can be achieved by combined genetic and management approaches.

4.1. In Vitro Measurements of CH_4 Production in Ruminants

This in vitro study evaluated a wide variety of dietary CH_4 inhibitors, which would not have been feasible in an in vivo study. A main advantage of the in vitro gas production system in measuring CH_4 emissions is that it provides a large number of data points, allowing accurate estimates of CH_4 emissions. However, it is a batch culture approach and has some limitations compared with in vivo studies (e.g., no absorption of VFA over time). The in vitro method used here was developed by Ramin and Huhtanen [14] to overcome this problem and involves a modeling approach based on data obtained from the gas in vitro system. They assumed a gross energy concentration of 18.5 MJ/kg DM, while the predicted proportion of CH_4 energy for a sample size of around 1000 mg was calculated to be 0.061. This value is close to observed in vivo values at production levels of intake in dairy cows [27]. Recently, Danielsson et al. [28] evaluated the in vitro technique developed by Ramin and Huhtanen [16] using data (diets) from in vivo studies using a respiration chamber to measure CH_4 emissions. The results showed a high correlation (R^2 = 0.96) between observed (chamber) data and predicted in vivo CH_4 values, confirming that the in vitro system is a useful tool for screening diets and evaluating feed additives.

4.2. Dietary Strategies to Decrease CH_4 Production from Ruminants

In this study, we screened many different dietary strategies with known potential to mitigate CH_4 production from ruminants and also a few new potential inhibitors. It is known that improved forage quality, feeding balanced diets to ensure efficient utilization of nutrients, and optimized microbial protein synthesis in the rumen can decrease CH_4 production in relation to animal productivity [29]. With respect to improved forage quality, the effects on CH_4 production reported in the literature are contradictory. Enteric CH_4 production increases with more digestible substrate available for rumen microbes, but overall emissions of CH_4 in lactating dairy cows can decrease per kg increase in digestible OM [8]. The mechanism behind this effect is likely that better forage quality improves intake, and thereby increases passage rate. Increased passage rate (i.e., decreased feed retention) and larger animals (i.e., greater body mass) have been associated with reduced CH_4 emissions in sheep [30–32].

Contrasted to our results, in measurements in vivo, rapeseed oil added at 50g/kg DM to a grass silage-based diet reduced ruminal CH_4 emissions from lactating cows by 22% [33], with the reduction observed being entirely explained by decreases in DM intake and the dilution effect on fermentable OM. Use of dried distiller's grain to replace soybean meal in diets based on grass silage decreased CH_4 production in an in vitro study by Franco et al. [34]. The effect was explained by a shift in the ruminal fermentation pattern to decreased acetate and butyrate production and increased propionate production. A similar shift in ruminal fermentation pattern was observed when rapeseed meal replaced soybean meal in vitro in that study [35]. In the present study, dried distiller's grain replaced rapeseed meal in the control diet and the suggested similarities in ruminal fermentation pattern of these protein supplements would explain the lack of effect on predicted in vivo CH_4 production. In contrast to our results, Fant et al. [35] observed a significant reduction in predicted in vivo CH_4 production of 2.1 mL/g DM when using oats instead of barley as the concentrate carbohydrate source. Inclusion of maize silage is also reported to promote propionate fermentation in the rumen, and thereby decrease CH_4 production in dairy cows [36–38]. However, we did not observe this effect with inclusion of maize silage in the diets. Greater molar proportions of acetate and lower proportions of propionate in VFAs when replacing grass with red clover have been reported both in vivo [39] and in vitro [40], suggesting that CH_4 production potential is greater when ruminants are fed red clover. Maize silage and red clover diets were only numerically lower respectively higher in in vivo predicted CH_4 production compared with the control diet in this study. On the other hand, grasses are generally more likely to accumulate nitrates than legumes, and nitrate inhibits enteric CH_4 production by replacing reduction of CO_2 to CH_4 as a major sink for disposal of H_2 in the rumen [41]. Interactions between ruminant physiological responses and diet quality affecting CH_4 production might explain the lack of impact on in vivo predicted CH_4 production by the potential CH_4-reducing diets screened in vitro in this study.

The chemical inhibitors 2-NE and bromoform, and the plant-derived inhibitor AT, gave a very large reduction in predicted in vivo CH_4 production in this study. The bromoform concentrations of 1.5 and 3.0 mg/kg DM used in this study were representative of concentrations occurring naturally in AT [42]. Vucko et al. [43] analyzed bromoform concentrations in AT biomass subjected to a wide variety of post-harvesting processes and found a maximum concentration of 4.4 mg/g DM for unrinsed, frozen, and freeze-dried AT. Those authors suggest a bromoform threshold of 1.0 mg/g DM in AT for 100% inhibition of CH_4 production in vitro, which corresponds with our results and the levels used in this study.

Machado et al. [44] tested different dosages of AT in vitro and found that production of CH_4 was decreased by 84.7% at an inclusion level of 1% (OM basis), while at AT doses greater than 2% (OM basis), CH_4 production was decreased by more than 99% compared with the control treatment. In the present study, in vivo predicted CH_4 production was inhibited almost completely by AT already at a level of 0.5% on an OM basis. Li et al. [45] added AT to diets fed to sheep and observed a reduction of CH_4 production at inclusion levels exceeding 1% of OM intake, but with altered rumen fermentation at all inclusion rates, i.e., at inclusions ≥0.5% of OM intake. On the other hand, in a short-term in vivo experiment by Stefenoni et al. [46], inclusion of AT at 0.5% of DM intake decreased CH_4 emission in lactating dairy cows by 80%, with no negative effects on DM intake and milk yield (rumen fermentation parameters were not measured).

An in vitro study by Zhang and Yang [47] indicated high potential of 2-NE to mitigate CH_4 production, as also found in the present study. However, they observed a negative effect on in vitro digestibility already at their lowest dose of 5 mM, which was not observed in this study. Use of 2-NE in an in vivo trial would not be realistic, considering that the concentration we used in vitro would equate to a daily dose of 0.9 L of 2-NE for a dairy cow with a rumen volume of 200 L.

Except for molar proportion of valerate with increased AT supplementation, all of the BCVFA decreased with increased supplementation in the dose response experiment. The BCVFA are mainly a consequence of the degradation of the amino acids valine, isoleucine, leucine and proline and are used for the biosynthesis of those amino acids and higher branched chain volatile fatty acids. The BCVFA are specific nutrients for the ruminal cellulolytic bacteria, and are believed to have a general positive influence on microbial fermentation [48].

Nitrate, propynoic acid, and p-coumaric acid had much lower inhibitory effects on predicted in vivo CH_4 production. Nitrate is reported to be an effective CH_4 production mitigating dietary component [49,50]. For example, a 24.8% reduction in CH_4 production by lactating cows receiving nitrate at 21.1 g NO_3^-/kg of DM was observed by Olijhoek et al. [51]. The dose of nitrate that can be toxic to ruminants' ranges between 198 and 998 mg/kg live weight and is dependent on diet, administration, and consumption [52]. However, the negative effects of nitrate can be reduced through gradual adaptation of animals to consumption of this nitrogen source, which could contribute to reducing CH_4 emissions. The CH_4 inhibitory effect of propynoic acid in this study was lower than that observed by Zhou et al. [53] at a comparable inclusion rate (75.7% reduction compared with a control diet with no inhibitor added). Also the lower amount of TVFA compared with the control diet indicated that propynoic acid can potentially affect digestibility. The inhibitory effect of p-coumaric acid on CH_4 production by ruminants has not been studied previously and there are no in vitro results with which to compare, but the treatment decreased the dietary TOMD. Lactate in the rumen are metabolized to propionate, which could hypothetical induce changes in ruminal fermentation pattern providing an alternative hydrogen sink to reduce methanogenesis. Likely, the lactic acid preservation has to be more extensive than the levels suggested in this study to have an effect on CH_4 production in dairy cows.

5. Conclusions

This study confirmed that natural bioactives produced by the red seaweed *Asparagopsis taxiformis* can act as a strong natural inhibitor of CH_4 production in domesticated ruminants. Use of CH_4 inhibitors with high mitigation potential at a reasonable dietary supplementation level could be an important and effective strategy to mitigate CH_4 emissions by ruminants. However, *Asparagopsis taxiformis* needs to be further evaluated in vivo to ensure it has no negative effects on animal health, productivity, or product quality. It is also important to establish the long-term CH_4 mitigation effect of using this inhibitor.

Author Contributions: S.J.K.: conceived and supervised the study and acquired funding. S.J.K., J.C.C.: inputs to data analysis. S.J.K., M.R.: methodology. J.C.C., S.J.K., M.R.: conducted the experiment and wrote the full paper.

Funding: This research was funded by FACCE ERA-GAS and FORMAS.

Acknowledgments: The authors would like to extend their sincere appreciation to Ann-Sofi Hahlin, Azam Jafari, and Amanda Poppi for their support in laboratory work, and to Carl Tryggers Foundation for providing a postdoc scholarship for J.C.C.

Conflicts of Interest: There is no conflict of interest relevant to this publication.

References

1. FAO. Food and Agriculture Organization of the United Nations. The State of Food Security and Nutrition in the World 2018. Available online: http://www.fao.org/3/i9553en/i9553en.pdf (accessed on 10 August 2019).
2. IPCC. Intergovernmental Panel on Climate Change. Climate Change 2013. Available online: https://www.ipcc.ch/site/assets/uploads/2018/03/WG1AR5_SummaryVolume_FINAL.pdf (accessed on 10 August 2019).
3. Shindell, D. The social cost of atmospheric release. *Clim. Chang.* **2015**, *130*, 313–326. [CrossRef]
4. Howarth, R.W. Ideas and perspectives: Is shale gas a major driver of recent increase in global atmospheric methane? *Biogeosciences* **2019**, *16*, 3033–3046. [CrossRef]
5. Petrenko, V.V.; Smith, A.M.; Schaefer, H.; Riedel, K.; Brook, E.; Baggenstos, D.; Hart, C.; Hua, Q.; Buizert, C.; Schilt, A.; et al. Minimal geological methane emissions during the Younger Dryas-Preboreal abrupt warming event. *Nature* **2017**, *548*, 443–446. [CrossRef] [PubMed]
6. Henderson, G.; Cox, F.; Ganesh, S.; Jonker, A.; Young, W.; Global Rumen Census Collaborators; Janssen, P.H. Rumen microbial community composition varies with diet and host, but a core microbiome is found across a wide geographical range. *Sci. Rep.* **2015**, *5*, 14567. [CrossRef]
7. Johnson, K.A.; Johnson, D.E. Methane emissions from cattle. *J. Anim. Sci.* **1998**, *73*, 2483–2492. [CrossRef]
8. Ramin, M.; Huhtanen, P. Development of equations for predicting methane emissions from ruminants. *J. Dairy Sci.* **2013**, *96*, 2476–2493. [CrossRef]

9. Knapp, J.R.; Laur, G.L.; Vadas, P.A.; Weiss, W.P.; Tricarico, W. Invited review: Enteric methane in dairy cattle production: Quantifying the opportunities and impact of reducing emissions. *J. Dairy Sci.* **2014**, *97*, 3231–3326. [CrossRef]

10. Shindell, D.; Kuylenstierna, J.C.; Vignati, E.; van Dingenen, R.; Amann, M.; Klimont, Z.; Anenberg, S.C.; Muller, N.; Janssens-Maenhout, G.; Raes, F.; et al. Simultaneously mitigating near-term climate change and improving human health and food security. *Science* **2012**, *335*, 183–189. [CrossRef]

11. IPCC. Intergovernmental Panel on Climate Change 2018. Available online: https://report.ipcc.ch/sr15/pdf/sr15_spm_final.pdf (accessed on 10 August 2019).

12. Hristov, A.N.; Oh, J.; Giallongo, F.; Frederick, T.W.; Harper, M.T.; Weeks, H.L.; Brabco, A.F.; Moate, P.J.; Deighton, M.H.; Williams, R.O.; et al. An inhibitor persistently decreased enteric methane emission from dairy cows with no negative effect on milk production. *Proc. Natl. Acad. Sci. USA* **2015**, *112*, 10663–10668. [CrossRef]

13. Machado, L.; Magnusson, M.; Paul, N.A.; de Nys, R.; Tomkins, N. Effects of marine and freshwater macroalgae on in vitro total gas and methane production. *PLoS ONE* **2014**, *9*, e85289. [CrossRef]

14. Ramin, M.; Huhtanen, P. Development of an in vitro method for determination of methane production kinetics using a fully automated in vitro gas system—A modelling approach. *Anim. Feed Sci. Technol.* **2012**, *174*, 190–200. [CrossRef]

15. Menke, K.H.; Steingass, H. Estimation of the energetic feed value obtained from chemical analysis and in vitro gas production using rumen fluid. *Anim. Res. Dev.* **1988**, *28*, 7–25.

16. Cone, J.W.; Van Gelder, A.H.; Visscher, G.J.W.; Oudshoorn, L. Influence of rumen fluid and substrate concentration on fermentation kinetics measured with a fully automated time related gas production apparatus. *Anim. Feed Sci. Technol.* **1996**, *61*, 113–128. [CrossRef]

17. AOAC International. *Official Methods of Analysis of AOAC International*, 18th ed.; AOAC International: Gaithersburg, MD, USA, 2005; pp. 24–56.

18. Mertens, D.R. Gravimetric determination of amylase-treated neutral detergent fiber in feeds using refluxing in beakers or crucibles: Collaborative study. *J. AOAC Int.* **2002**, *85*, 1217–1240.

19. Krizsan, S.J.; Rinne, M.; Nyholm, L.; Huhtanen, P. New recommendations for the ruminal in situ determination of indigestible neutral detergent fibre. *Anim. Feed Sci.* **2015**, *205*, 31–41. [CrossRef]

20. Puhakka, L.; Jaakkola, S.; Simpura, I.; Kokkonen, T.; Vanhatalo, A. Effects of replacing rapeseed meal with fava bean at two concentrate crude protein levels on feed intake, nutrient digestion, and milk production in cows fed grass silage-based diets. *J. Dairy Sci.* **2016**, *99*, 7993–8006. [CrossRef]

21. Roque, B.M.; Salwen, J.K.; Kinley, R.; Kebreab, E. Inclusion of *Asparagopsis armata* in lactating dairy cows' diet reduces enteric methane emission by over 50 percent. *J. Clean. Prod.* **2019**, *234*, 132–138. [CrossRef]

22. Broderick, G.A.; Huhtanen, P.; Ahvenjärvi, S.; Reynal, S.M.; Shingfield, J. Quantifying ruminal nitrogen metabolism using the omasal sampling technique in cattle—A meta-analysis. *J. Dairy Sci.* **2010**, *93*, 3216–3230. [CrossRef]

23. Chang, J.; Peng, S.; Ciais, P.; Saunois, M.; Dangal, S.R.S.; Herrero, M.; Havlík, P.; Tian, H.; Bousquet, P. Revisiting enteric methane emissions from domestic ruminants and their $\delta^{13}C_{CH4}$ source signature. *Nat. Commun.* **2019**, *10*, e3420. [CrossRef]

24. Kingston-Smith, A.H.; Edwards, J.E.; Huws, S.A.; Kim, E.J.; Abberton, M. Plant-based strategies towards minimising 'livestock's long shadow'. *Proc. Nutr. Soc.* **2010**, *69*, 613–620. [CrossRef]

25. Li, B.; Fikse, W.F.; Løvendahl, P.; Lassen, J.; Lidauer, M.H.; Mäntysaari, P.; Berglund, B. Genetic heterogeneity of feed intake, energy-corrected milk and body weight across lactation in Holstein, Nordic Red, and Jersey cows. *J. Dairy Sci.* **2018**, *101*, 10011–10021. [CrossRef] [PubMed]

26. Grandl, F.; Furger, M.; Kreuzer, M.; Zehetmeier, M. Impact of longevity on greenhouse gas emissions and profitability of individual dairy cows analysed with different system boundaries. *Animal* **2019**, *13*, 198–208. [CrossRef] [PubMed]

27. Yan, T.; Agnew, R.E.; Gordon, F.J.; Porter, M.G. Prediction of methane energy output in dairy and beef cattle offered grass silage-based diets. *Livest. Prod. Sci.* **2000**, *64*, 253–263. [CrossRef]

28. Danielsson, R.; Ramin, M.; Bertilsson, J.; Lund, P.; Huhtanen, P. Evaluation of an in vitro system for predicting methane production in vivo. *J. Dairy. Sci.* **2017**, *100*, 8881–8894. [CrossRef] [PubMed]

29. Hristov, N.; Oh, J.; Lee, C.; Meinen, R.; Montes, F.; Ott, T.; Firkins, J.; Rotz, A.; Dell, C.; Adesogan, A.; et al. *Mitigation of Greenhouse Gas Emissions in Livestock Production—A Review of Technical Options for Non-CO$_2$ emissions*; FAO Animal Production and Health: Rome, Italy, 2013; pp. 10–60.

30. Pinares-Patiño, C.S.; Ulyatt, M.J.; Lassey, K.R.; Barry, T.N.; Holmes, C.W. Rumen function and digestion parameters associated with differences between sheep in methane emissions when fed chaffed lucerne hay. *J. Agric. Sci.* **2003**, *140*, 205–214. [CrossRef]

31. Pinares-Patiño, C.S.; Ebrahimi, S.H.; McEwan, J.C.; Clark, H.; Luo, D. Is rumen retention time implicated in sheep differences in methane emission? In Proceedings of the New Zealand Society of Animal Production, Wellington, New Zeland, 24–26 June 2011; Volume 71, pp. 219–222.

32. Goopy, J.P.; Donaldson, A.; Hegarty, R.; Vercoe, P.E.; Haynes, F.; Barnett, M.; Oddy, V.H. Low-methane yield sheep have smaller rumens and shorter rumen retention time. *Br. J. Nutr.* **2014**, *111*, 578–585. [CrossRef]

33. Bayat, A.R.; Tapio, I.; Vilkki, J.; Shingfield, K.J.; Leskinen, H. Plant oil supplements reduce methane emissions and improve milk fatty acid composition in dairy cows fed grass silage-based diets without affecting milk yield. *J. Dairy Sci.* **2018**, *101*, 1136–1151. [CrossRef]

34. Franco, M.O.; Krizsan, S.J.; Ramin, M.; Spörndly, R.; Huhtanen, P. In vitro evaluation of agro-industrial by-products replacing soybean meal in two different basal diets for ruminants. In Proceedings of the 8th Nordic Feed Science Conference, Uppsala, Sweden, 13–14 June 2017; pp. 170613–170614.

35. Fant, P.; Ramin, M.; Jaakkola, S.; Grimberg, Å.; Carlsson, A.S.; Huhtanen, P. Effects of different barley and oat varieties on methane production, digestibility and fermentation pattern in vitro. *J. Dairy. Sci.* **2019**, in press. [CrossRef]

36. O'Mara, F.P.; Fitzgerald, J.J.; Murphy, J.J.; Rath, M. The effect on milk production of replacing grass silage with maize silage in the diet of dairy cows. *Livest. Prod. Sci.* **1998**, *55*, 79–87. [CrossRef]

37. Beauchemin, K.A.; Kreuzer, M.; O'Mara, F.; McAllister, T.A. Nutritional management for enteric methane abatement: A review. *Aust. J. Exp. Agric.* **2008**, *48*, 21–27. [CrossRef]

38. Haque, M.N. Dietary manipulation: A sustainable way to mitigate methane emissions from ruminants. *J. Anim. Sci. Technol.* **2018**, *60*, 15. [CrossRef] [PubMed]

39. Vanhatalo, A.; Kuoppala, K.; Ahvenjärvi, S.; Rinne, M. Effects of feeding grass or red clover silage cut at two maturity stages in dairy cows. Nitrogen metabolism and supply of amino acids. *J. Dairy Sci.* **2009**, *92*, 5620–5633. [CrossRef] [PubMed]

40. Navarro-Villa, A.; O'Brian, M.; López, S.; Boland, T.M.; O'Kiely, P. In vitro rumen methane output of red clover and perennial ryegrass assayed using the gas production technique (GPT). *Anim. Feed Sci. Tech.* **2011**, *168*, 152–164. [CrossRef]

41. Allison, M.J.; Reddy, C.A.; Cook, H.M. The effects of nitrate and nitrite on VFA and CH$_4$ production by rumen microbes. *J. Anim.l Sci.* **1981**, *53*, 391–399.

42. Machado, L.; Magnusson, M.; Paul, N.A.; Kinley, R.; Nys, R.; Tomkins, N. Identification of bioactives from the red seaweed *Asparagopsis taxiformis* that promote antimethanogenic activity in vitro. *J. Appl. Phycol.* **2016**, *28*, 3117–3126. [CrossRef]

43. Vucko, M.J.; Magnusson, M.; Kinley, R.D.; Villart, C.; Nys, R. The effects of processing on the in vitro antimethanogenic capacity and concentration of secondary metabolites of *Asparagopsis taxiformis*. *J. Appl. Phycol.* **2017**, *29*, 1577–1586. [CrossRef]

44. Machado, L.; Magnusson, M.; Paul, N.A.; Kinley, R.; de Nys, R.; Tomkins, N. Dose-response effects of *Asparagopsis taxiformis* and *Oedogonium* sp. on in vitro fermentation and methane production. *J. Appl. Phycol.* **2016**, *28*, 1443–1452. [CrossRef]

45. Li, X.; Norman, H.C.; Kinley, R.D.; Laurence, M.; Wilmot, M.; Bender, H.; Nys, R.; Tomkins, N. *Asparagopsis taxiformis* decreases enteric methane production from sheep. *Anim. Prod. Sci.* **2018**, *58*, 681–688. [CrossRef]

46. Stefenoni, H.; Räisänen, S.; Melgar, A.; Lage, C.; Young, M.; Hristov, A. Dose-response effect of the macroalga *Asparagopsis taxiformis* on enteric methane emission in lactating dairy cows. In Proceedings of the American Dairy Science Association Annual Meeting, Cincinnati, OH, USA, 23–26 June 2019; pp. W163, 378.

47. Zhang, D.F.; Yan, H.J. In vitro ruminal methanogenesis of a hay-rich substrate in response to different combination supplements of nitrocompounds; pyromellitic diimide and 2-bromoethanesulphonate. *Anim. Feed Sci. Technol.* **2011**, *163*, 20–23. [CrossRef]

48. Andries, J.I.; Buysse, F.X.; DeBrabander, D.L.; Cottyn, B.G. Isoacids in ruminant nutrition: Their role in ruminal and intermediary metabolism and possible influences on performances—A review. *Anim. Feed Sci. Technol.* **1987**, *18*, 169–180. [CrossRef]
49. Newbold, J.R.; van Zijderveld, S.M.; Hulshof, R.B.A.; Fokkink, W.B.; Leng, R.A.; Terencio, P.; Powers, W.J.; van Adrichem, P.S.J.; Paton, N.D.; Perdok, H.B. The effect of incremental levels of dietary nitrate on methane emissions in Holstein steers and performance in Nelore bulls. *J. Anim. Sci.* **2014**, *92*, 5032–5040. [CrossRef] [PubMed]
50. Klop, G.; Hatew, B.; Bannink, A.; Dijkstra, J. Feeding nitrate and docosahexaenoic acid affects enteric methane production and milk fatty acid composition in lactating dairy cows. *J. Dairy Sci.* **2016**, *99*, 1161–1172. [CrossRef] [PubMed]
51. Olijhoek, D.W.; Hellwing, A.L.F.; Brask, M.; Weisbjerg, M.R.; Højberg, O.; Larsen, M.K.; Dijkstra, E.J.; Erlandsen, E.J.; Lund, P. Effect of dietary nitrate level on enteric methane production, hydrogen emission, rumen fermentation, and nutrient digestibility in dairy cows. *J. Dairy Sci.* **2016**, *99*, 6191–6205. [CrossRef] [PubMed]
52. Department of Climate Change, Commonwealth Government of Australia. The Potential of Feeding Nitrate to Reduce Enteric Methane Production in Ruminants. Available online: http://www.penambulbooks.com/ (accessed on 14 September 2019).
53. Zhou, Z.; Meng, Q.; Yu, Z. Effects of methanogenic inhibitors on methane production and abundances of methanogens and cellulolytic bacteria in in vitro ruminal cultures. *Appl. Environ. Microbiol.* **2011**, *77*, 2634–2639. [CrossRef]

Article

Rumen Methanogenesis, Rumen Fermentation, and Microbial Community Response to Nitroethane, 2-Nitroethanol, and 2-Nitro-1-Propanol: An In Vitro Study

Zhenwei Zhang, Yanlu Wang, Xuemeng Si, Zhijun Cao, Shengli Li and Hongjian Yang *

State Key Laboratory of Animal Nutrition, College of Animal Science and Technology, China Agricultural University, Beijing 100193, China; qingyibushuo@163.com (Z.Z.); yanluwang@yeah.net (Y.W.); sxmswun@126.com (X.S.); caozhijun@cau.edu.cn (Z.C.); lisheng0677@163.com (S.L.)
* Correspondence: yang_hongjian@sina.com

Received: 7 February 2020; Accepted: 5 March 2020; Published: 13 March 2020

Simple Summary: The present study comparatively investigates the inhibitory difference of nitroethane (NE), 2-nitroethanol (NEOH), and 2-nitro-1-propanol (NPOH) on in vitro rumen fermentation, microbial populations, and coenzyme activities associated with methanogenesis. The results showed that both NE and NEOH were more effective in reducing ruminal methane (CH_4) production than NPOH. This work provides evidence that NE, NEOH, and NPOH were able to inhibit methanogen population and dramatically decrease methyl-coenzyme M reductase gene expression and the content of coenzymes F_{420} and F_{430} with different magnitudes in order to reduce ruminal CH_4 production.

Abstract: Nitroethane (NE), 2-nitroethanol (NEOH), and 2-nitro-1-propanol (NPOH) were comparatively examined to determine their inhibitory actions on rumen fermentation and methanogenesis in vitro. Fermentation characteristics, CH_4 and total gas production, and coenzyme contents were determined at 6, 12, 24, 48, and 72 h incubation time, and the populations of ruminal microbiota were analyzed by real-time PCR at 72 h incubation time. The addition of NE, NEOH, and NPOH slowed down in vitro rumen fermentation and reduced the proportion of molar CH_4 by 96.7%, 96.7%, and 41.7%, respectively ($p < 0.01$). The content of coenzymes F_{420} and F_{430} and the relative expression of the *mcr*A gene declined with the supplementation of NE, NEOH, and NPOH in comparison with the control ($p < 0.01$). The addition of NE, NEOH, and NPOH decreased total volatile fatty acids (VFAs) and acetate ($p < 0.05$), but had no effect on propionate concentration ($p > 0.05$). Real-time PCR results showed that the relative abundance of total methanogens, *Methanobacteriales*, *Methanococcales*, and *Fibrobacter succinogenes* were reduced by NE, NEOH, and NPOH ($p < 0.05$). In addition, the nitro-degradation rates in culture fluids were ranked as NEOH (-0.088) > NE (-0.069) > NPOH (-0.054). In brief, the results firstly provided evidence that NE, NEOH, and NPOH were able to decrease methanogen abundance and dramatically decrease *mcr*A gene expression and coenzyme F_{420} and F_{430} contents with different magnitudes to reduce ruminal CH_4 production.

Keywords: nitrocompounds; methanogenesis; rumen fermentation; microbial community; coenzyme

---

1. Introduction

Nitrocompounds are classified into aromatic compounds containing nitro groups in the aromatic ring and aliphatic–aromatic compounds containing nitro groups only in the aliphatic side chain. Among the naturally occurring aliphatic nitrocompounds, 3-nitro-1-propanol (3-NPOH) and 3-nitro-1-propionic acid (NPA) are regarded as nitrotoxins that can accumulate to toxic levels in certain forages (e.g., *Astragalus*); their toxic and metabolic properties were well investigated in ruminants and monogastric animals nearly a century ago [1]. Over the last decades, certain commercially available short-chain aliphatic nitrocompounds have been chemically synthesized and are known to be potent methane-inhibiting compounds, since methane (CH_4) from livestock is increasingly considered as a significant greenhouse gas [2].

Among these commercially available nitrocompounds, Latham et al. [2] summarized the supplemental effects of nitroethane (NE), 2-nitroethanol (NEOH), and 2-nitro-1-propanol (NPOH) in comparison with 3-NPOH and NPA on CH_4 and volatile fatty acid (VFAs) production during in vitro rumen incubations, finding that the yield of VFAs was almost not affected, whereas CH_4 production consistently decreased by up to 58%–99%. Regarding in vivo studies in sheep [3] and feedlot steer [4–6], CH_4-reducing activity varied depending on dose levels and depletion of nitrocompounds. More recently, in the study of Hristov et al. [7], as much as a 64-fold increase in hydrogen (H_2) production was observed in 3-nitrooxypropanol-treated cattle, which was still only about 3% of the H_2 spared from ruminal production. Despite the fact that CH_4 emission is negatively associated with energy retention and greenhouse gas production, rumen archaea play an important ecological role in methanogenesis, however, few studies have examined the microbial response to nitrocompounds. In many of the aforementioned studies, the inhibited archaea populations or methanogenesis-associated enzymes were not characterized though methyl-coenzyme M reductase (*mcr*). The coenzymes F_{420} and F_{430} are known in nature as key enzymes involved in CH_4 formation from H_2 and CO_2 by archaea [8–10]. Although the naturally occurring nitrocompounds 3-NPOH and NPA are recognized to be metabolized by rumen microorganisms to aminopropanol and β-alanine, a nonessential amino acid may be utilized by ruminant animals and potentially used as a source of carbon, nitrogen, and energy, making it an attractive candidate [1,2]. To avoid unknown risks of toxic intermediates in feeding practice trials regarding commercially available NE, NEOH, and NPOH, the present study comparatively investigated the inhibitory difference of each nitrocompound on in vitro rumen fermentation, microbial populations, and coenzyme activities associated with methanogenesis.

2. Materials and Methods

The donor animals and experimental procedures were approved by the requirements of Beijing Municipal Council on Animal Care according to the protocol of CAU20171014-1.

2.1. Nitrocompounds

Nitroethane (NE), 2-nitroethanol (NEOH), and 2-nitro-1-propanol (NPOH) were purchased commercially (Sigma Aldrich, Inc., LA, USA). Structures of these nitrocompounds are shown in Figure 1. Their analytical grades were 99%, 90%, and 98%, respectively.

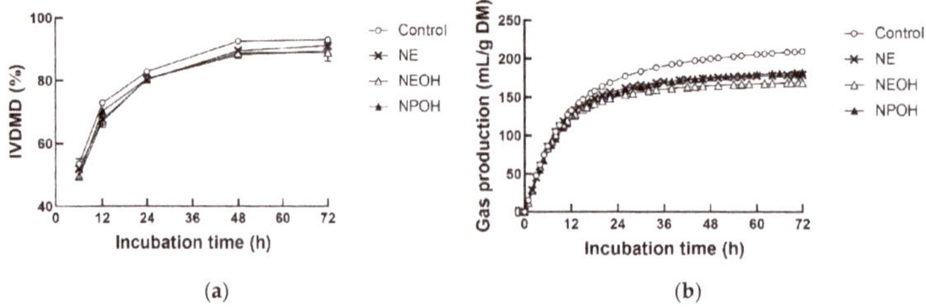

Figure 1. Effect of NE, NEOH, and NPOH addition to culture fluids on in vitro dry matter disappearance (IVDMD (**a**) and cumulative gas production (**b**) of grain-rich feed incubated with rumen fluids obtained from lactating dairy cows. NE: nitroethane; NEOH: 2-nitroalcohol; NPOH: 2-nitro-1-propanol. Statistical analyses showed that the effects of nitrocompounds on IVDMD and gas production were significant at $p < 0.01$.

2.2. Animals

Five multiparous and rumen-cannulated Holstein lactating dairy cows (540 ± 25.3 kg body weight; 100 ± 8.5 days in milk, 33.0 ± 0.78 kg/d milk yield; mean ± SD) were selected as the donors of rumen fluid. Cows were maintained on a total mixed ration (calculated as % dry matter basis) of alfalfa hay (16.32%), whole corn silage (24.61%), 1 kg corn meal (3.95%), extruded corn (19.60%), soybean meal (14.38%), soybean hull (4.09%), extruded soybean (3.43%), whole cottonseed (8.25%), trace mineral, and vitamin premix (5.37%), according to the Chinese Feeding Standard of Dairy Cow (NY/T 34 2004).

2.3. In Vitro Experiment

In vitro fermentations in anaerobic glass bottles (volume capacity of 120 mL) incubated with rumen fluids were performed following the previous description of Zhang and Yang [11]. The treatments included the control (no additive treatment), 10 mmol/L of NE, 10 mmol/L of NEOH, and 10 mmol/L of NPOH. Corn meal and alfalfa hay (500 mg; 80:20, w/w) were used as the fermentation substrates.

Rumen fluids were collected before morning feeding from each rumen-cannulated donor cow into a pre-warmed thermos flask at 39 °C. After filtering through 4 layers of cheesecloth and mixed in equal proportion, 25 mL of rumen fluids were incubated into anaerobic glass bottles with 50 mL buffered medium (pH 6.8) [12]. The batch cultures were performed at 39 °C in both automated and manual systems. In the automated system, five bottles per treatment were connected to the gas inlets of an automated gas recording system (AGRS) and continuously incubated for 72 h to continuously record cumulative gas production (GP). In the manual system, five bottles per treatment were connected to pre-emptied air bags to collect fermentation gas samples and removed at 6, 12, 24, 48, and 72 h of incubation. The batch cultures were repeated and completed in three consecutive runs. One milliliter of gas sample was drawn out of the air bags using a syringe to measure the CH_4 concentration according to the gas chromatographic method.

2.4. Sampling

After 6, 12, 24, 48, and 72 h of incubation in the manual system, the contents of each bottle were filtered through a nylon bag (8 × 12 cm; 42 μm pore size) and dried at 105 °C to determine the in vitro dry matter disappearance (IVDMD). Then, the culture fluids (6 × 1.0 mL) were sampled into DNase-free polypropylene tubes and stored at −80 °C for later analysis of VFA, nitrocompounds, microbial populations, *mcr*A (methyl coenzyme-M reductase subunit A) gene expression, coenzyme F_{420} content, and coenzyme F_{430} content.

2.5. Measurement of VFA, Coenzyme, and Nitrocompound Contents

The culture fluids (1.0 mL) from each of the 5 aforementioned time stamps were mixed with 300 μL metaphosphoric acid solution (25%, w/v) for 30 min and centrifuged at 10,000× g for 15 min at 4 °C. Supernatants (0.5 mL) were injected into gas chromatography to determine the concentrations of acetate, propionate, isobutyrate, butyrate, isovalerate, and valerate [11].

Following the description of Reuter et al. [13], coenzyme F_{420} was determined and expressed as fluorescence intensity. Assays were performed at 37 °C anaerobically in the dark. Culture fluid samples (1.0 mL) were stirred continuously and boiled at 95 °C in water bath for 30 min. Fluid aliquots were then centrifuged at 10,000× g for 10 min, and a volume of 500 μL from supernatants was mixed with 1 mL of isopropanol. Subsequently, the mixture was precipitated for 2 h and centrifuged at 10,000× g again for 15 min. Finally, the fluorescence intensity of the supernatants was measured at 420 nm by the fluorescence spectrophotometer (Thermo Fisher Scientific Co., Ltd., shanghai, China). Coenzyme F_{430} content was examined via the ultraviolet/visible spectrum by determining the loss of absorbance [14]. Briefly, the culture fluids were quenched with equal volumes of methanol and centrifuged aerobically at 6153× g for 20 min in dim light. The precipitate was discarded and the supernatants were determined colorimetrically using a spectrophotometer at 430 nm (Laspec Technology Co., Ltd., shanghai, China). Coenzyme F_{430} content was expressed as the relative absorbance of coenzyme F_{430} at 430 nm.

The contents of NE, NEOH, and NPOH were determined colorimetrically by a spectrophotometer (Laspec Technology Co., Ltd., shanghai, China) [15]. The culture fluids (1 mL) were firstly centrifuged at $10,000 \times g$ for 15 min. Supernatants (50 μL) were then diluted with 2 mL of distilled water and mixed with 100 μL of NaOH (0.65 M) and 100 μL of diazotized p-nitroaniline. Finally, the absorbances of the culture fluids were measured at a wavelength of 405 nm.

2.6. Analysis of mcrA Gene Expression

Total RNA of culture fluid was extracted by RNeasy Mini kit (Tiangen Biotech, Beijing, China) using an RNase-Free DNase Set (Qiagen). The detailed procedures for analysis of *mcr*A gene expression were described by Guo et al. [16]. The specific primer sets for the *mcr*A gene (F: 5′-TTCG GTGG ATCD CARA GRGC-3′, R: 3′-GBAR GTCG WAWC CGTA GAAT CC-5′) and the 16S rRNA gene (F: 5′-CGGC AACG AGCG CAAC CC-3′, R: 3′-CCAT TGTA GCAC GTGT GT AG CC-5′) were applied as described by Denman et al. [17] and Denman and McSweeney [18]. The $2^{-\Delta\Delta Ct}$ method was used for the expression analysis of the *mcr*A gene, with the 16S rRNA set as the reference gene.

2.7. Microbial Population Analysis with Real-Time PCR

Total DNA of culture fluid (1 mL) was extracted with the FastDNA kit and FastPrep instrument (Tiangen®Biotech, Beijing, China) by a bead-beating method as described by Denman and McSweeney [18]. According to the real-time PCR method [18], enumeration of total bacteria, total methanogens, *Methanobacteriales*, *Methanococcales*, *Methanomicrobiales*, protozoa, fungi, *Ruminococcus flavefaciens*, *Ruminococcus albus,* and *Fibrobacter succinogenes* was measured using the Bio-Rad Multicolor Real-Time PCR Detection System (Bio-rad Company, CA, USA) with qReal Master Mix SYBR®Green (Tiangen®Biotech, Beijing, China). The primer sets for the detection and enumeration of the microbial populations were described by Denman and McSweeney [18], Zhou et al. [19], and Yu et al. [20]. The abundance of the microbial population was expressed as a proportion of total estimated rumen bacterial (16S rDNA) according to the following equation: relative quantification = $2^{-(CT\ target\ -\ CT\ total\ bacteria)}$, where CT represents the threshold cycle.

2.8. Data Calculation and Statistical Analysis

Cumulative gas production data from the AGRS were fitted according to an exponential model as described by France [21]. In addition, the average gas production rate (AGPR) (mL/h) and hydrogen recovery (2Hrec) were also calculated following the description by Zhang et al. [22].

The data for gas production, IVDMD, fermentation gas composition, coenzyme contents, and VFAs were subjected to analysis of variance with the MIXED model procedure of SAS (Statistical Analysis for Windows, SAS Institute Inc., Cary, NC, USA). The model was applied as $Y_{ijk} = \mu + R_i + N_j + T_k + (N \times T)_{jk} + e_{ijk}$, where Y_{ijk} is the dependent variable, μ represents the overall mean, R_i is the effect of the experimental run, N_j is the effect of the nitrocompound treatment, T_k is the effect of the incubation time, $N \times T$ are the interactions between the nitrocompounds and the incubation time, and e_{ijk} is the residual. The data for microbial abundance analysis were applied according to the model $Y_{ijk} = \mu + R_i + N_j + e_{ijk}$, where Y_{ijk} is the dependent variable, μ represents the overall mean, R_i is the effect of the experimental run, N_j is the effect of the nitrocompound treatment, and e_{ijk} is the residual. Least square means and standard error (SEM) were calculated, and treatment differences were estimated using a multiple comparisons test (Tukey/Kramer). Correlation analyses between variables were performed using the CORR procedure of SAS. Significance was declared at $p < 0.05$.

3. Results

3.1. IVDMD and Gas Production Kinetics

As the incubation time increased, both IVDMD (Figure 1a) and gas production (Figure 1b) continuously increased. Addition of NE, NEOH, and NPOH slowed down the fermentation process and caused IVDMD to decline (Figure 1a, $p < 0.01$).

Regarding the kinetics of gas production, NE, NEOH, and NPOH addition decreased asymptotic gas production (A) (Table 1, $p < 0.01$), while NE and NEOH addition increased the fractional gas production rate (c, h^{-1}). Neither NE nor NEOH addition altered AGPR, but NPOH decreased AGPR compared to the control.

Table 1. Effect of NE, NEOH, and NPOH addition (10 mmol/L) to culture fluids on gas production kinetics and fermentation gas composition during 72 h incubation.

Items [1]	Treatment [2]				SEM	p [3]		
	Control	NE	NEOH	NPOH		Treatment	Time	Trt × Time
Gas production kinetics								
A, mL/g DM	199 [a]	178 [b]	170 [c]	173 [bc]	1.96	<0.01	-	-
c, h^{-1}	0.09 [b]	0.11 [a]	0.12 [a]	0.10 [b]	0.02	<0.01	-	-
$T_{1/2}$, h	3.1	3.1	3.0	3.1	0.02	0.07	-	-
AGPR, mL/h	13.3 [a]	13.9 [a]	13.9 [a]	11.9 [b]	0.29	<0.01	-	-
Fermentation gas composition, %								
CH_4	15.1 [a]	0.5 [c]	0.5 [c]	8.8 [b]	0.39	<0.01	<0.01	<0.01
H_2	0.3 [d]	9.8 [b]	10.5 [a]	2.0 [c]	0.23	<0.01	<0.01	<0.01
CO_2	84.6 [b]	89.4 [a]	89.0 [a]	89.2 [a]	0.36	<0.01	<0.01	0.01

[a–d] Means with different superscripts within a row are significantly different ($p < 0.05$); [1] A: asymptotic gas production; c: fractional gas production rate; $T_{1/2}$: the time when half of A occurred; AGPR: average gas production rate. [2] NE: nitroethane; NEOH: 2-nitroalcohol; NPOH: 2-nitro-1-propanol. [3] Interaction effect between treatment and incubation time.

3.2. Fermentation Gas Composition

H_2 accumulation in the NE, NEOH, and NPOH groups was far greater than in the control group (Table 1, $p < 0.01$). The molar proportions of CO_2 were increased (5.7%, 5.2%, and 5.4%) by NE, NEOH, and NPOH compared to the control, whereas the molar proportions of CH_4 were notably decreased (96.7%, 96.7%, and 41.7%, Table 1). An interaction between the nitrocompound treatment and the incubation time was observed for gas composition ($p < 0.01$). The molar proportion of CH_4 constantly increased with increasing incubation time in the control and NPOH groups, but it was at a pretty low level in both the NE and NEOH groups ($p < 0.01$, Figure 2a). In contrast, molar H_2 production in the NE and NEOH groups continuously increased with increasing incubation time and was always far greater than the control and NPOH throughout the incubation ($p < 0.01$, Figure 2b). In addition, as the incubation time increased, the molar proportion of CO_2 gradually decreased in all groups, and it was greater in the nitrocompound-treated cultures than the control ($p < 0.01$, Figure 2c).

3.3. Coenzymes Related to CH_4 Production

Compared to the control, both coenzyme F_{420} fluorescence intensity and F_{430} ultraviolet (UV) absorbance declined in the NE, NEOH, and NPOH groups (Table 2, $p < 0.01$). In addition, NE, NEOH, and NPOH addition decreased *mcr*A gene expression by 83.1%, 79.7%, and 53.5%, respectively.

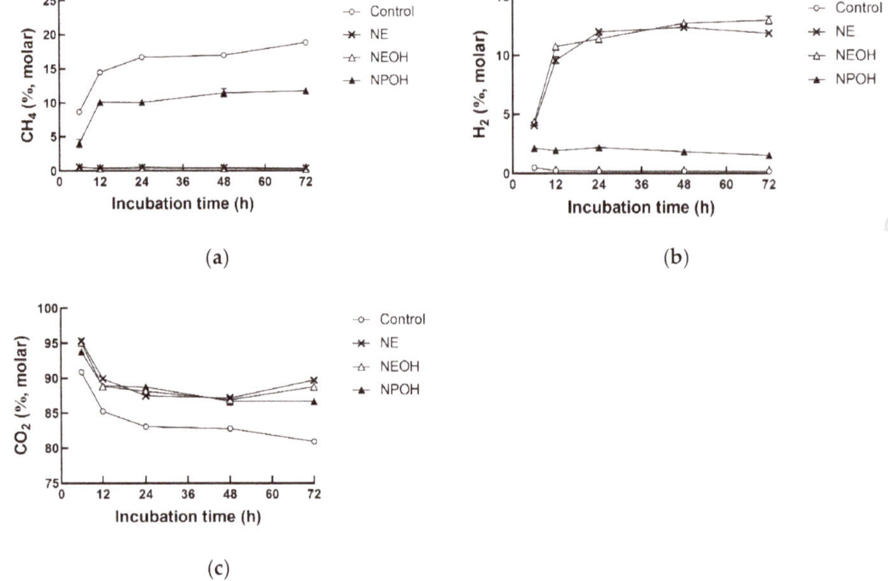

Figure 2. Effect of NE, NEOH, and NPOH addition to culture fluids on CH_4 (**a**), H_2 (**b**), and CO_2 (**c**) production of grain-rich feed incubated with rumen fluids obtained from lactating dairy cows. NE: nitroethane; NEOH: 2-nitroalcohol; NPOH: 2-nitro-1-propanol. Statistical analyses showed the effects of nitrocompounds on CH_4, H_2, and CO_2, with correlations significant at $p < 0.01$.

Table 2. Effect of NE, NEOH, and NPOH addition (10 mmol/L) on coenzyme content, *mcr*A gene expression, and volatile fatty acid (VFA) production in fermentation fluids across different incubation times (6, 12, 24, 48, and 72 h).

Items [1]	Treatment [2]				SEM	p [3]		
	Control	NE	NEOH	NPOH		Treatment	Time	Trt × Time
Coenzyme content								
F_{420}	16.7 [a]	11.9 [c]	11.4 [c]	15.2 [b]	0.23	<0.01	<0.01	<0.01
F_{430}	0.59 [a]	0.48 [b]	0.44 [c]	0.50 [b]	0.02	<0.01	<0.01	0.01
*mcr*A expression	1 [a]	0.11 [c]	0.12 [c]	0.42 [b]	0.02	<0.01	<0.01	<0.01
Total VFA, mmol/L	102.8 [a]	93.7 [c]	94.7 [bc]	97.4 [b]	1.64	<0.01	<0.01	<0.01
Acetate, mmol/L	60.9 [a]	52.1 [c]	53.1 [c]	54.7 [b]	1.31	<0.01	<0.01	0.02
Propionate, mmol/L	24.0	23.1	24.2	23.9	0.53	0.33	<0.01	0.02
Butyrate, mmol/L	12.8 [bc]	13.5 [a]	12.5 [c]	13.2 [ab]	0.30	<0.01	<0.01	<0.01
BCVFA, mmol/L	5.1	5.0	4.9	5.6	0.31	0.14	<0.01	0.14
2Hrec, %	67.6 [a]	47.2 [c]	48.5 [c]	59.3 [b]	0.87	<0.01	<0.01	<0.01

[a–d] Means with different superscripts within a row are significantly different ($p < 0.05$); [1] F_{420}: F_{420} fluorescence intensity; F_{430}: UV absorbance of coenzyme F_{430}; BCVFA: branch-chained volatile fatty acids; 2Hrec: hydrogen recovery. [2] NE: nitroethane, NEOH: 2-nitroalcohol; NPOH: 2-nitro-1-propanol. [3] Interaction effect between treatment and incubation time.

An interaction between the nitrocompound treatment and the incubation time was evident for *mcr*A gene expression and coenzyme F_{420} and F_{430} contents ($p < 0.01$). As the incubation time increased, *mcr*A gene expression in the NPOH group relative to the control peaked at 12 h during the 72 h incubation, and it was continuously far lower in the NE and NEOH groups than in the control and NPOH groups (Figure 3a). Coenzyme F_{420} fluorescence intensity gradually declined in both the NE and NEOH groups with increasing incubation time, and it was continuously lower in the nitrocompound-treated cultures than the control (Figure 3b). The F_{430} ultraviolet (UV) absorbance continuously decreased in the NE and NEOH groups with increasing incubation time (Figure 3c), but it fluctuated during 72 h incubation in the NPOH group. In addition, the coenzyme F_{430} content was constantly lower in the nitrocompound-treated cultures than the control.

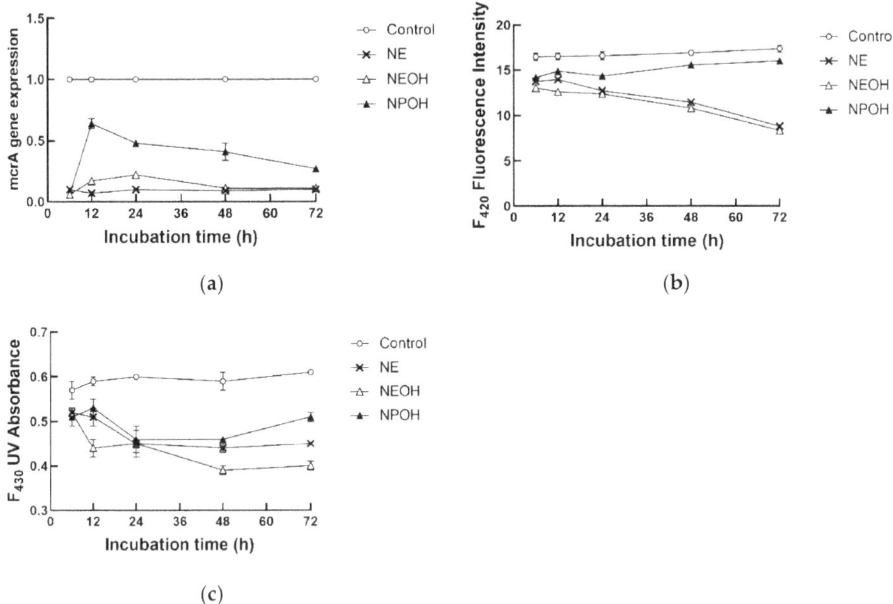

Figure 3. Effect of NE, NEOH, and NPOH addition to culture fluids on *mcr*A gene expression (**a**) and the contents of coenzyme F_{420} (**b**) and coenzyme F_{430} (**c**) of grain-rich feed incubated with rumen fluids obtained from lactating dairy cows. NE: nitroethane; NEOH: 2-nitroalcohol; NPOH: 2-nitro-1-propanol. Statistical analyses showed that the effects of nitrocompounds were significant at $p < 0.01$.

3.4. Fermentation Characteristics

Concentrations of total VFAs and acetate were lower in the NE, NEOH and NPOH groups than in the control (Table 2, $p < 0.01$), whereas the concentration of butyrate increased with NE addition (Table 2, $p < 0.01$). NE, NEOH, and NPOH addition had no significant influence on propionate and branched-chain VFAs (BCVFAs). Compared to the control, NE, NEOH, and NPOH addition decreased 2Hrec by 30.2%, 28.3%, and 12.3%, respectively (Table 2, $p < 0.01$). An interaction between the nitrocompound treatment and the incubation time was observed for total VFA, acetate, propionate, and butyrate. The concentrations of total VFAs, acetate, propionate, butyrate, and BCVFA were continuously increased in all groups with increasing incubation time (Figure 4). In addition, concentrations of acetate were continuously lower in nitrocompound-treated cultures than in the control, whereas the concentrations of propionate, butyrate, and BCVFAs fluctuated during 72 h incubation.

3.5. Microbial Populations

Real-time PCR results showed that NE, NEOH, and NPOH addition decreased the relative populations of total methanogens, *Methanobacteriales*, *Methanomicrobiales*, and *Fibrobacter succinogenes* (Table 3, $p < 0.05$). Compared to the control, the relative populations of total methanogens decreased by 49.2%, 36.9%, and 41.5% in the NE, NEOH, and NPOH groups, and the populations of *Methanobacteriales* decreased by 46.1%, 35.9%, and 17.9%, respectively. In addition, the relative populations of fungi tended to decrease according to nitrocompound treatment ($p = 0.09$). Compared with control, the populations of *R. albus* increased by 50.0% and 50.0% in the NEOH and NPOH groups, and the populations of

R. flavefaciens decreased by 84.5% and 70.7% in the NE and NEOH groups. However, the populations of protozoa were not affected ($p = 0.29$).

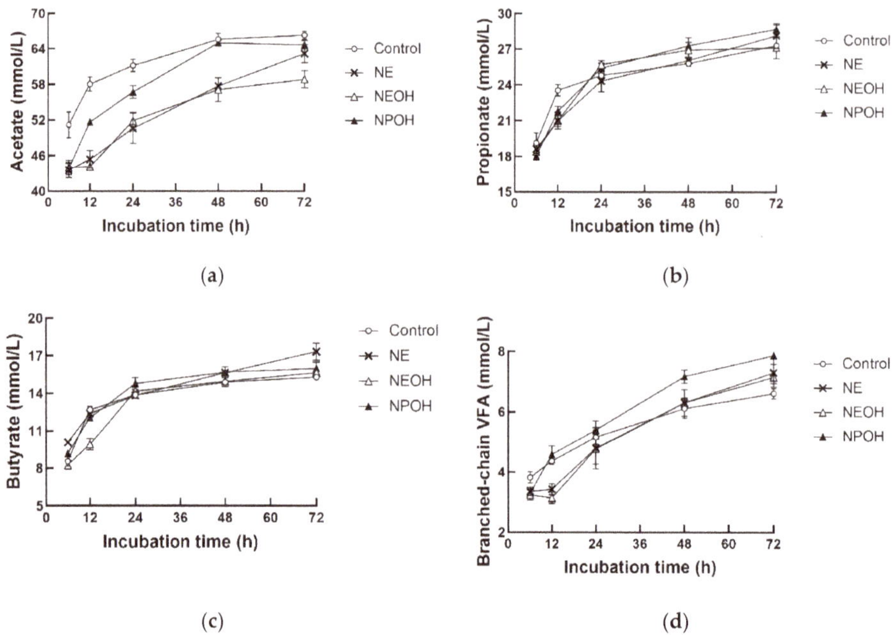

Figure 4. Effect of NE, NEOH, and NPOH addition to culture fluids on acetate (**a**), propionate (**b**), butyrate (**c**), and branched-chain VFA (**d**) of grain-rich feed incubated with rumen fluids obtained from lactating dairy cows. NE: nitroethane; NEOH: 2-nitroalcohol; NPOH: 2-nitro-1-propanol. Statistical analyses showed that the effects of nitrocompounds on acetate and butyrate were significant at $p < 0.01$, but the effects of nitrocompounds on propionate and branched-chain VFAs were not significant.

Table 3. Effects of NE, NEOH, and NPOH addition (10 mmol/L) to culture fluids on microorganism relative populations after 72 h incubation.

Items	Treatment					
	Control	NE	NEOH	NPOH	SEM	p
Total methanogen	0.65 [a]	0.33 [b]	0.41 [b]	0.38 [b]	0.041	0.04
Methanobacteriales	0.39 [a]	0.21 [c]	0.25 [bc]	0.32 [b]	0.013	0.02
Methanomicrobiales, $\times 10^{-3}$	14.2 [a]	4.6 [c]	9.5 [b]	5.1 [c]	0.985	0.01
Methanococcales, $\times 10^{-3}$	0.44 [bc]	0.56 [ab]	0.65 [a]	0.38 [c]	0.041	0.03
Fungus, $\times 10^{-3}$	1.82 [a]	0.37 [b]	1.25 [ab]	1.14 [ab]	0.240	0.09
Protozoa, $\times 10^{-3}$	0.28	0.29	0.43	0.22	0.068	0.29
Ruminococcus albus	0.12 [b]	0.10 [b]	0.18 [a]	0.18 [a]	0.012	<0.01
Ruminococcus flavefaciens	0.58 [a]	0.09 [b]	0.17 [b]	0.43 [a]	0.056	0.01
Fibrobacter succinogenes, $\times 10^{-3}$	3.33 [a]	1.39 [b]	0.97 [b]	0.56 [b]	0.232	<0.01

[a–c] Means with different superscripts within a row are significantly different ($p < 0.05$). NE: nitroethane; NEOH: 2-nitroalcohol; NPOH: 2-nitro-1-propanol.

3.6. Correlation between CH_4 Production and mcrA Gene Expression/Coenzyme Contents

CH_4 production was positively correlated with *mcr*A gene expression (Table 4; r = 0.88, $p < 0.05$), coenzyme F_{420} content (Table 4; r = 0.74, $p < 0.01$), and coenzyme F_{430} content (Table 4; r = 0.24, $p < 0.05$). In addition, there was a positive correlation between *mcr*A gene expression and the coenzyme

F_{420} content (r = 0.78, $p < 0.01$), and between *mcr*A gene expression and the coenzyme F_{430} content (r = 0.22, $p < 0.05$). A positive correlation was also observed between the coenzyme F_{420} content and the coenzyme F_{430} content (r = 0.26, $p < 0.05$).

Table 4. Correlations between CH_4 production, *mcr*A gene expression, coenzyme F_{420} content, and coenzyme F_{430} content, regardless of the type of nitrocompound added.

Items	*mcr*A	F_{420}	F_{430}
methane production	0.88 *	0.74 **	0.24 *
*mcr*A		0.78 **	0.22 *
F_{420}			0.26 *

Significance: * $p < 0.05$, ** $p < 0.01$.

3.7. Disappearance of Nitrocompounds

Concentrations of NE, NEOH, and NPOH decreased with increasing incubation time (Figure 5). The disappearance of nitrocompounds was fitted to a linear model, and the nitrocompound degradation rate was ranked as NEOH (−1.20) > NE (−0.80) > NPOH (−0.78).

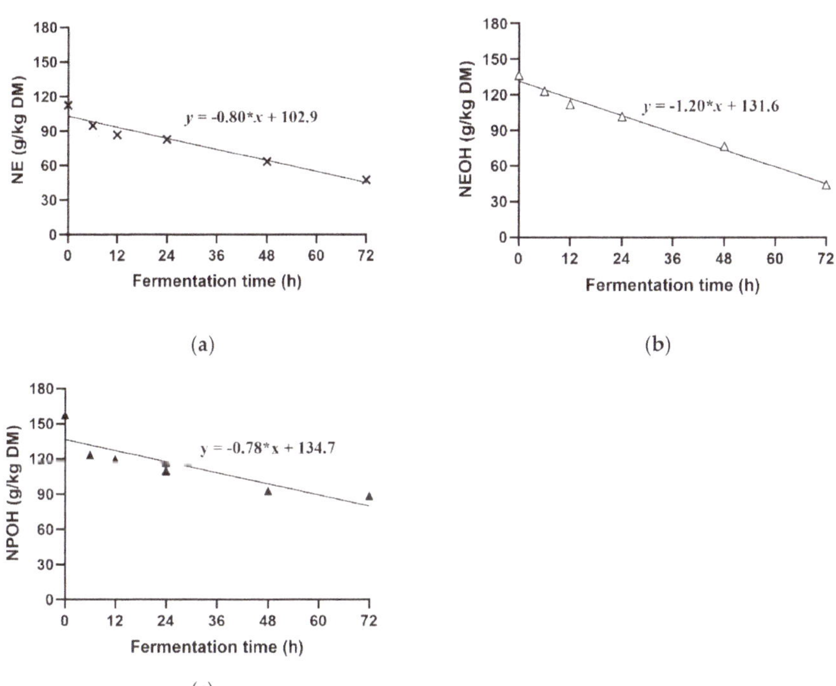

Figure 5. Variation of NE (**a**), NEOH (**b**), and NPOH (**c**) in culture fluids during 72 h incubation of grain-rich feed incubated in batch cultures of mixed rumen microorganisms. DM: dry matter; NE: nitroethane; NEOH: 2-nitroalcohol; NPOH: 2-nitro-1-propanol.

4. Discussion

4.1. Effects of Nitrocompounds on IVDMD and Gas Production

A majority of CH_4 inhibition strategies tend to compromise fermentative efficiency, resulting in the reduction of certain digestive processes [11,23]. In the present study, the addition of NE, NEOH, and NPOH caused notable reductions of IVDMD, gas production, and total VFAs, indicating that the activity of the microbes responsible for the degradation of substrates was inhibited by these nitrocompounds. The inclusive level of nitrocompounds was determined according to previous studies [11,24], however, a remarkable reduction in total VFA production occurred in the NE, NEOH, and NPOH groups, indicating that the fermentative bacterial population may be sensitive to a 10 mM dose level of nitrocompounds.

4.2. Effects of Nitrocompounds on CH_4 Production

The antimethanogenic activity of NE, NEOH, and NPOH was previously observed. For instance, more than 90% of CH_4 production was inhibited by NE, NEOH, and NPOH at a concentration of 9~24 mM after 24 h incubation [5,24]. The addition of NE, NEOH, and NPOH in the present study caused reductions in CH_4 production of 96.7%, 97.2%, and 39.5%, respectively, which agreed with the results of earlier studies [24,25]. In addition, both NE and NEOH were more effective in reducing ruminal CH_4 production than NPOH.

The nitro group is strongly electron-withdrawing. Because of this property, nitrocompounds may be able to serve as electron acceptors within ruminal microbes to reduce CH_4 production. Early studies done by Anderson et al. [26] revealed that nitrocompounds could serve as electron acceptors within ruminal microbes to inhibit CH_4 production. However, nitrocompounds also directly inhibit ruminal methanogenesis [27]. The first author in a review summarized the anti-methanogenic roles of nitrocompounds and their potential inhibitory action modes in terms of VFA production, hydrogen accumulation, formate oxidation and ferredoxin-linked hydrogenase activity [28]. Notable accumulation of H_2 in NE-, NEOH-, and NPOH-treated cultures and CH_4 production inhibition were observed in the present study. Similarly, Anderson et al. [25] also found that accumulations of H_2 were higher in nitrocompound-supplemented cultures than in controls after 24 h of incubation. In the unperturbed rumen, H_2 is usually present at approximately 1 µmol/L (0.1 kPa); however, it often increases to concentrations that inhibit hydrogenase activity (1 kPa), while ruminal CH_4 production is inhibited due to declined H_2 consumption by methanogens [23]. Considering that hydrogenases are reversible enzymes that can catalyze either the production or oxidation of H_2, the authors in the present study speculated that nitrocompounds may have inhibited H_2-oxidation hydrogenase activity as well, which was in accordance with Angermaier and Simon [29], who reported that NEOH inhibited ferredoxin-linked hydrogenase uptake activity. Likewise, Anderson et al. [30] also noted that NE, NEOH, NPOH, and NPA each inhibited oxidation of H_2 or formate when these reductants were added in excess (60 mM) to mixed cultures of rumen microbial populations, thereby implicating a possible mechanism of activity against ruminal methanogenesis.

Consistent with the different levels of CH_4 inhibition efficiency, NE and NEOH were shown to be nearly equally effective in promoting H_2 accumulation in vitro, with both of them promoting H_2 accumulation more effectively than NPOH. The molar proportions of H_2 in total fermentation gas production were 9.8%, 10.5%, and 2.0% in the NE, NEOH, and NPOH groups, respectively. However, the decreasing extent of 2Hrec in the present study was far less than that of CH_4 production. The fate of the remaining H_2 is not known with certainty, however, the consumption of reducing equivalents may occur during anabolic processes, including cell growth, intracellular polyhydroxyalkonoate, or extracellular polysaccharide production [25].

4.3. Effects of Nitrocompounds on VFA Production

In order to compensate for the disruption of electron flow to ruminal methanogenesis, some CH_4 inhibitors, such as sodium sulfite, organic halides, and monensin, cause notable increases in propionate production during fermentation, which is frequently accompanied by decreased acetate [23]. However, NE, NEOH, and NPOH addition in the present study had little effect on propionate produced by ruminal populations. Therefore, the results indicate that the reducing equivalents produced during ruminal fermentation were not necessarily directed toward increased production of propionate. In addition, the concentrations of acetate in the present study were decreased in the nitrocompound-treated cultures, along with the reduction of CH_4 production. It is well recognized that acetate accompanies H_2 production, and the latter can be used for CH_4 formation by methanogenic archaea, with CH_4 being positively associated with the acetate to propionate ratio [31].

4.4. Effects of Nitrocompounds on Methane Production and Related Changes in Microbial Populations and Coenzyme Contents

Methanogens are a unique group of ruminal microbes that generate CH_4 as a stoichiometric end-product of their metabolism. Methanogen populations are generally closely associated with CH_4 production [32]. In the present study, NE, NEOH, and NPOH addition decreased total methanogens by 49.2%, 36.9%, and 41.5%, respectively, and the authors speculated that NE, NEOH, and NPOH likely exerted a direct inhibition of ruminal methanogenesis via direct suppression of methanogens. *Methanobacteriales* are the predominant populations and constitute the major portion of methanogen community in ruminants, being the second most prevalent archaea in the rumen ecology [33]. Presently, both the relative abundances of *Methanobacteriaceae* and *Methanomicrobiales* were reduced with NE, NEOH, and NPOH addition, but different sensitivity responses to these nitrocompounds were observed. *Methanobacteriaceae* reduction varied in the order of NE (−46.2%) > NEOH (−35.9%) > NPOH (−17.9%), whereas *Methanomicrobiales* reduction varied in the order of NE (−67.6%) > NPOH (−64.1%) > NEOH (−33.1%).

Rumen protozoa constitute only a small portion of ruminal microorganisms, however, they play important roles in feed degradation and making energy and protein available to the hosts. The RT-PCR results showed no apparent decrease in the protozoal population in the NE-, NEOH- and NPOH-treated cultures, indicating that there were no adverse effects of the nitrocompounds on the rumen protozoa populations.

Few studies have determined microbial responses to nitrocompounds. Anderson et al. [26] noted that 3-NPOH and NPA modestly inhibited total culturable anaerobes from bovine rumens, but inhibited microbes were not characterized. Rumens harbor different types of bacteria, which are most actively involved in plant fiber degradation. It is well recognized that the depression of cellulolysis can decrease the rate and extent of neutral detergent fiber digestion [34], and the inhibition to cellulolytic microorganisms (*R. flavefaciens*, *F. succinogenes*) by nitrocompounds may explain why decreased IVDMD was observed in the cultures supplemented with NE, NEOH, or NPOH. Fibrolytic bacteria families, including *R. flavefaciens*, *R. albus*, and *F. succinogenes*, have vbeen shown to release many fibrolytic enzymes and promote H_2, acetate and formic acid production for methanogen utilization [35]. Real-time PCR results showed that the relative populations of the methanogens *R. Flavefaciens* and *F. succinogenes* decreased with nitrocompound addition, suggesting that the mutual-aid interaction between methanogens and fibrolytic bacteria might be one reason why *R. Flavefaciens* and *F. succinogenes* decreased along with methanogen inhibition by NE, NEOH, and NPOH. The rumen protozoa produce fermentation end-products similar to those made by the bacteria, particularly acetate, butyrate, and H_2. They utilize large amounts of starch at one time and can store it in their bodies. The corn-rich substrate applied in the present study may have been adequate to maintain the growth of protozoa, which may explain why the nitrocompounds had no negative effect on the abundance of protozoa. In the present study, the abundance of *R. albus* increased with NEOH and NPOH addition, while anaerobic rumen fungi decreased with nitrocompound supplementation. This phenomenon could be attributed to the antagonistic association between ruminal fungi and cellulolytic bacteria [36]. In summary, although nitrocompounds could change the relative abundance of some microbial populations, the differences in diversity and metabolic activity in response to NE, NEOH, and NPOH need further investigation in order to determine the maximal inhibitory effect on CH_4 production with minimal adverse effects on rumen fermentation.

The hydrogenotrophic, methylotrophic, and acetoclastic pathways are the three major pathways for ruminal CH_4 production. The biochemical reactions and enzyme profiles involved in methanogenesis are well identified and described [37]. Methyl-coenzyme M reductase (*mcr*) is a key enzyme responsible for catalyzing the CH_4-producing step in the process of methanogenesis. As a gene encoding the alpha-subunit of *mcr*, *mcr*A was evolutionarily highly conserved, probably due to functional constraints on the catalytic activity of *mcr* [35]. Recently, the determination of *mcr*A gene expression was accepted for *mcr* activity measurement [16]. A positive correlation between decreased *mcr*A gene expression and decreased molar CH_4 proportions was found in the present study, suggesting that NE, NEOH, and NPOH inhibited CH_4 production by decreasing *mcr* activity. In addition, the activity of *mcr* is dependent mainly on the unique nickel-containing tetrapyrrole known as coenzyme F_{430} [38]. As reported by Gunsalus and Wolfe [39], coenzyme F_{430} is a yellow nonfluorescent compound released from *mcr*, with an absorption maximum at 430 nm on its UV-Vis absorption spectrum. In the present study, NE, NEOH, and NPOH addition also reduced the content of coenzyme F_{430}. Coenzyme F_{420} is of importance for methanogenesis and can act as an indicator for methanogenic activity [8,40]. The coenzyme F_{420} content in this study also decreased in the nitrocompound-treated cultures and was accompanied by CH_4 reduction, with the anti-methanogenic activity ranked as NEOH > NE > NPOH. A possible reason for these results might be due to the toxic action of nitrocompounds to methanogens [24,27]; this needs further clarification.

4.5. Degradation of NE, NEOH, and NPOH

Decreased concentrations of NE, NEOH, and NPOH were observed with increasing incubation time in all of the nitrocompound-treated cultures, thus confirming the presence of competent nitrocompound-degrading microbes within the incubations. A previous study by Anderson et al. [26] revealed that most of the rumen microorganisms tolerated nonlethal concentrations of naturally occurring nitrocompounds. In addition, the degradation rate of nitrocompounds can be enhanced via exposure to nitropropionic acid-containing forages. This phenomenon could be ascribed the improvement of nitro-degrading activity or enrichment in numbers of nitro-degrading microorganisms. Until now, *Denitrobacterium detoxificans* was recognized as a unique nitrocompound-reducing microbe which oxidizes reducing substrates, including H_2, formate, and lactate, to reduce nitrocompounds to their respective amines and minor nitrites [41]. Consequently, *D. detoxificans* has the potential to outcompete ruminal methanogens for available reductants [15]. In the present study, the nitrocompound disappearance rate was ranked as NEOH (−1.20) > NE (−0.80) > NPOH (−0.78), suggesting that the rumen microbes presented divergent metabolic capabilities regarding nitrocompound degradation, thus partially explaining why NE, NEOH, and NPOH showed different competition for methanogenesis-produced reductants.

5. Conclusions

Along with a dramatic increase in H_2 accumulation, both NE and NEOH were shown to be more effectivene in inhibiting methanogenesis than NPOH. Although nitrocompound addition decreased acetate and total VFA production, it had no negative effect on propionate. In addition, NE, NEOH, and NPOH addition decreased the population abundance of total methanogens, *Methanobacteriales*, and *Methanomicrobiales*, also causing decreases in *mcr*A gene expression and coenzyme F_{420} and F_{430} contents. The results provided evidence that NE, NEOH, and NPOH could reduce methanogen populations and dramatically decrease *mcr*A gene expression and coenzyme F_{420} and F_{430} contents with different magnitudes to reduce overall ruminal CH_4 production.

Author Contributions: Conceptualization, Z.Z. and H.Y.; Data curation, H.Y.; formal analysis, Z.Z., Y.W., X.S., and H.Y.; investigation, Z.Z., Y.W., and X.S.; writing—original draft preparation, Z.Z.; writing—review and editing, Z.Z., Z.C., S.L. and H.Y.; funding acquisition, H.Y.; project administration, H.Y. All authors have read and agreed to the published version of the manuscript.

Funding: The authors acknowledge funding support from National Natural Science Foundation of China (Grant Number: 31572432) and National Key Research and Development Project of China (Grant Number: 2018YFD0502104-3).

Conflicts of Interest: We declare no conflicts of interest.

References

1. Anderson, R.C.; Majak, W.; Rassmussen, M.A.; Callaway, T.R.; Beier, R.C.; Nisbet, D.J.; Allison, M.J. Toxicity and metabolism of the conjugates of 3-nitropropanol and 3-nitropropionic acid in forages poisonous to livestock. *J. Agric. Food Chem.* **2005**, *53*, 2344–2350. [CrossRef] [PubMed]
2. Latham, E.A.; Anderson, R.C.; Pinchak, W.E.; Nisbet, D.J. Insights on alterations to the rumen ecosystem by nitrate and nitrocompounds. *Front. Microbiol.* **2016**, *7*, 228. [CrossRef] [PubMed]
3. Anderson, R.C.; Carstens, G.E.; Miller, R.K.; Callaway, T.R.; Schultz, C.L.; Edrington, T.S.; Harvey, R.B.; Nisbet, D.J. Effect of oral nitroethane and 2-nitropropanol administration on methane-producing activity and volatile fatty acid production in the ovine rumen. *Bioresour. Technol.* **2006**, *97*, 2421–2426. [CrossRef] [PubMed]
4. Gutierrez-Bañuelos, H.; Anderson, R.C.; Carstens, G.E.; Slay, L.J.; Ramlachan, N.; Horrocks, S.M.; Callaway, T.R.; Edrington, T.S.; Nisbet, D.J. Zoonotic bacterial populations, gut fermentation characteristics and methane production in feedlot steers during oral nitroethane treatment and after the feeding of an experimental chlorate product. *Anaerobe.* **2007**, *13*, 21–31. [CrossRef]
5. Gutierrez-Banuelos, H.; Anderson, R.C.; Carstens, G.E.; Tedeschi, L.O.; Pinchak, W.E.; Cabrera-Diaz, E.; Krueger, N.A.; Callaway, T.R.; Nisbet, D.J. Effects of nitroethane and monensin on ruminal fluid fermentation characteristics and nitrocompound-metabolizing bacterial populations. *J. Agric. Food Chem.* **2008**, *56*, 4650–4658. [CrossRef]
6. Brown, E.G.; Anderson, R.C.; Carstens, G.E.; Gutierrez-Bañuelos, H.; McReynolds, J.L.; Slay, L.J.; Callaway, T.R.; Nisbet, D.J. Effects of oral nitroethane administration on enteric methane emissions and ruminal fermentation in cattle. *Anim. Feed Sci. Technol.* **2011**, *166–167*, 275–281. [CrossRef]
7. Hristov, A.N.; Oh, J.; Giallongo, F.; Frederick, T.W.; Harper, M.T.; Weeks, H.L.; Brancob, A.F.; Moatec, P.J.; Deightonc, M.H.; Williamsc, R.O.S.; et al. An inhibitor persistently decreased enteric methane emission from dairy cows with no negative effect on milk production. *Proc. Natl. Acad. Sci. USA* **2015**, *112*, 10663–10668. [CrossRef]
8. Hendrickson, E.L.; Leigh, J.A. Roles of coenzyme F420-reducing hydrogenases and hydrogen- and F420-dependent methylenetetrahydromethanopterin dehydrogenases in reduction of F420 and production of hydrogen during methanogenesis. *J. Bacteriol.* **2008**, *190*, 4818. [CrossRef]
9. Wongnate, T.; Sliwa, D.; Ginovska, B.; Smith, D.; Wolf, M.W.; Lehnert, N.; Raugei, S.; Ragsdale, S.W. The radical mechanism of biological methane synthesis by methyl-coenzyme M reductase. *Science* **2016**, *352*, 953–958. [CrossRef]
10. Zheng, K.; Ngo, P.D.; Owens, V.L.; Yang, X.P.; Mansoorabadi, S.O. The biosynthetic pathway of coenzyme F430 in methanogenic and methanotrophic archaea. *Science* **2016**, *351*, 339–342. [CrossRef]
11. Zhang, D.F.; Yang, H.J. Combination effects of nitrocompounds, pyromellitic diimide, and 2-bromoethanesulfonate on *in vitro* ruminal methane production and fermentation of a grain-rich feed. *J. Agric. Food Chem.* **2012**, *60*, 364–371. [CrossRef] [PubMed]
12. Menke, K.H.; Steingass, H. Estimation of the energetic feed value obtained by chemical analysis and in vitro gas production using rumen fluid. *Anim. Res. Dev.* **1988**, *28*, 7–55.
13. Reuter, B.W.; Egeler, T.; Schneckenburger, H.; Schoberth, S.M. *In vivo* measurement of F420 fluorescence in cultures of *Methanobacterium thermoautotrophicum*. *J. Bacteriol.* **1986**, *4*, 325–332. [CrossRef]
14. Ellefson, W.L.; Whitman, W.B.; Wolfe, R.S. Nickel-containing factor F430: Chromophore of the methylreductase of *Methanobacterium*. *Proc. Natl. Acad. Sci. USA* **1982**, *79*, 3707–3710. [CrossRef] [PubMed]
15. Ochoa-Garcia, P.A.; Arevalos-Sanchez, M.M.; Ruiz-Barrera, O.; Anderson, R.C.; Maynez-Perez, A.O.; Rodriguez-Almeida, F.A.; Chavez-Martinez, A.; Gutierrez-Banuelos, H.; Corral-Luna, A. *In vitro* reduction of methane production by 3-nitro-1-propionic acid is dose-dependent. *J. Anim. Sci.* **2019**, *97*, 1317–1324. [CrossRef]
16. Guo, Y.Q.; Liu, J.X.; Lu, Y.; Zhu, W.Y.; Denman, S.E.; McSweeney, C.S. Effect of tea saponin on methanogenesis, microbial community structure and expression of *mcr*A gene, in cultures of rumen micro-organisms. *Lett. Appl. Microbiol.* **2008**, *47*, 421–426. [CrossRef]

17. Denman, S.E.; Tomkins, N.W.; McSweeney, C.S. Quantitation and diversity analysis of ruminal methanogenic populations in response to the antimethanogenic compound bromochloromethane. *FEMS Microbiol. Ecol.* **2007**, *62*, 313–322. [CrossRef]

18. Denman, S.E.; McSweeney, C.S. Development of a real-time PCR assay for monitoring anaerobic fungal and cellulolytic bacterial populations within the rumen. *FEMS Microbiol. Ecol.* **2006**, *58*, 572–582. [CrossRef]

19. Zhou, M.; Hernandez-Sanabria, E.; Guan, L.L. Assessment of the microbial ecology of ruminal methanogens in cattle with different feed efficiencies. *Appl. Environ. Microbiol.* **2009**, *75*, 6524–6533. [CrossRef]

20. Yu, Y.; Lee, C.; Kim, J.; Hwang, S. Group-specific primer and probe sets to detect methanogenic communities using quantitative real-time polymerase chain reaction. *Biotechnol. Bioeng.* **2005**, *89*, 670–679. [CrossRef]

21. France, J.; Dijkstra, J.; Dhanoa, M.S.; Lopez, S.; Bannink, A. Estimating the extent of degradation of ruminant feeds from a description of their gas production profiles observed *in vitro*: Derivation of models and other mathematical considerations. *Br. J. Nutr.* **2000**, *83*, 143–150. [CrossRef] [PubMed]

22. Zhang, Z.W.; Wang, Y.L.; Wang, W.K.; Cao, Z.J.; Li, S.L.; Yang, H.J. The inhibitory action mode of nitrocompounds on in vitro rumen methanogenesis: A comparison of nitroethane, 2-nitroethanl and 2-nitro-1-propanol. *J. Agric. Sci.* **2019**, *9*, 1–9. [CrossRef]

23. Van Nevel, C.J.; Demeyer, D.I. Control of rumen methanogenesis. *Environ. Monit. Assess.* **1996**, *42*, 73–97. [CrossRef] [PubMed]

24. Anderson, R.C.; Callaway, T.R.; Kessel, J.A.S.V.; Soo, J.Y.; Edrington, T.S.; Nisbet, D.J. Effect of select nitrocompounds on ruminal fermentation; an initial look at their potential to reduce economic and environmental costs associated with ruminal methanogenesis. *Bioresour. Technol.* **2003**, *90*, 59–63. [CrossRef]

25. Anderson, R.C.; Huwe, J.K.; Smith, D.J.; Stanton, T.B.; Krueger, N.A.; Callaway, T.R.; Edrington, T.S.; Harvey, R.B.; Nisbet, D.J. Effect of nitroethane, dimethyl-2-nitroglutarate and 2-nitro-methyl-propionate on ruminal methane production and hydrogen balance *in vitro*. *Bioresour. Technol.* **2010**, *101*, 5345–5349. [CrossRef]

26. Anderson, R.C.; Rasmussen, M.A.; Allison, M.J. Metabolism of the plant toxins nitropropionic acid and nitropropanol by ruminal microorganisms. *Appl. Environ. Microbiol.* **1993**, *59*, 3056–3061. [CrossRef]

27. Anderson, R.C.; Rasmussen, M.A. Use of a novel nitrotoxin-metabolizing bacterium to reduce ruminal methane production. *Bioresour. Technol.* **1998**, *64*, 89–95. [CrossRef]

28. Zhang, Z.W.; Cao, Z.J.; Wang, Y.L.; Wang, Y.J.; Yang, H.J.; Li, S.L. Nitrocompounds as potential methanogenic inhibitors in ruminant animals: A review. *Anim. Feed Sci. Technol.* **2018**, *236*, 107–114. [CrossRef]

29. Angermaier, L.; Simon, H. On the reduction of aliphatic and aromatic nitro compounds by Clostridia, the role of ferredoxin and its stabilization. *Hoppe-Seyler's Z. Physiol. Chem.* **1983**, *364*, 961–975. [CrossRef]

30. Anderson, R.C.; Krueger, N.A.; Stanton, T.B.; Callaway, T.R.; Edrington, T.S.; Harvey, R.B.; Jung, Y.S.; Nisbet, D.J. Effects of select nitrocompounds on *in vitro* ruminal fermentation during conditions of limiting or excess added reductant. *Bioresour. Technol.* **2008**, *99*, 8655–8661. [CrossRef]

31. Soltan, Y.A.; Hashem, N.M.; Morsy, A.S.; El-Azrak, K.M.; Sallam, S.M. Comparative effects of Moringa oleifera root bark and monensin supplementations on ruminal fermentation, nutrient digestibility and growth performance of growing lambs. *Anim. Feed Sci. Technol.* **2018**, *235*, 189–201. [CrossRef]

32. Machmüller, A. Medium-chain fatty acids and their potential to reduce methanogenesis in domestic ruminants. *Agric. Ecosyst. Environ.* **2006**, *112*, 107–114. [CrossRef]

33. Janssen, P.H.; Kirs, M. Structure of the archaeal community of the rumen. *Appl. Environ. Microbiol.* **2008**, *74*, 3619–3625. [CrossRef] [PubMed]

34. Wina, E.; Muetzel, S.; Hoffmann, E.; Makkar, H.P.S.; Becker, K. Saponins containing methanol extract of sapindus rarak affect microbial fermentation, microbial activity and microbial community structure in vitro. *Anim. Feed Sci. Technol.* **2005**, *121*, 159–174. [CrossRef]

35. Vogels, G.D.; Hoppe, W.F.; Stumm, C.K. Association of methanogenic bacteria with rumen ciliates. *Appl. Environ. Microbiol.* **1980**, *40*, 608–612. [CrossRef]

36. Dehority, B.A.; Tirabasso, P.A. Antibiosis between ruminal bacteria and ruminal fungi. *Appl. Environ. Microbiol.* **2000**, *66*, 2921–2927. [CrossRef]

37. Lessner, D.J. Methanogenesis Biochemistry. In *eLS*; John Wiley & Sons, Ltd.: Hoboken, NJ, USA, 2009.

38. Hallam, S.J.; Peter, R.G.; Christina, M.P.; Paul, M.R.; Edward, F.D. Identification of methyl coenzyme M reductase A (mcrA) genes associated with methane-oxidizing archaea. *Appl. Environ. Microbiol.* **2003**, *69*, 5483–5491. [CrossRef]

39. Gunsalus, R.P.; Wolfe, R.S. Chromophoric factors F342 and F430 of Methanobacterium thermoautotrophicum. *FEMS Microbiol. Ecol.* **1978**, *3*, 191–193. [CrossRef]

40. Schönheit, P.; Keweloh, H.; Thauer, R.K. Factor F420 degradation in *Methanobacterium thermoautotrophicum* during exposure to oxygen. *FEMS Microbiol. Ecol.* **2010**, *12*, 347–349. [CrossRef]

41. Anderson, R.C.; Rasmussen, M.A.; Jensen, N.S.; Allison, M.J. Denitrobacterium detoxificans gen. nov.; sp. nov.; a ruminal bacterium that respires on nitrocompounds. *Int. J. Syst. Evol. Microbiol.* **2000**, *50*, 633–638. [CrossRef]

Article

The Role of Condensed Tannins in the In Vitro Rumen Fermentation Kinetics in Ruminant Species: Feeding Type Involved?

Ives C. S. Bueno [1], Roberta A. Brandi [1], Gisele M. Fagundes [2,*], Gabriela Benetel [1] and James Pierre Muir [3]

1   Department of Animal Science, University of São Paulo, Pirassununga, São Paulo 1365-900, Brazil; ivesbueno@usp.br (I.C.S.B.); robertabrandi@usp.br (R.A.B.); gabriela.benetel@usp.br (G.B.)
2   Department of Animal Science, Federal University of Roraima—UFRR, BR 174, km 12, Boa Vista, Roraima 69300-000, Brazil
3   Texas A&M AgriLife Research, Texas A&M University, Stephenville, TX 76401, USA; Jim.Muir@ag.tamu.edu
*   Correspondence: gisele.fagundes@ufrr.br; Tel.: +55-95-3627-2898

Received: 20 February 2020; Accepted: 26 March 2020; Published: 7 April 2020

Simple Summary: Inoculum from different feeding types of the ruminant species host has unequal tolerance and effects to condensed tannin (CT) due to their respective feeding strategies behavior producing different ruminal microbiota profiles. This paper describes that in long term incubation, CT plant extract addition affects in vitro fermentation kinetics more severely in grazing ruminant than browsing ruminants.

Abstract: Animal feeding behavior and diet composition determine rumen fermentation responses and its microbial characteristics. This study aimed to evaluate the rumen fermentation kinetics of domestic ruminants feeding diets with or without condensed tannins (CT). Holstein dairy cows, Nelore beef cattle, Mediterranean water buffalo, Santa Inês sheep and Saanen goats were used as inoculum donors (three animals of each species). The substrates were maize silage (*Zea mays*), fresh elephant grass (*Pennisetum purpureum*), Tifton-85 hay (*Cynodon* spp.) and fresh alfalfa (*Medicago sativa*). Acacia (*Acacia molissima*) extract was used as the external CT source. The in vitro semi-automated gas production technique was used to assess the fermentation kinetics. The experimental design was completely randomized with five inoculum sources (animal species), four substrates (feeds) and two treatments (with or without extract). The inclusion of CT caused more severe effects in grazing ruminants than selector ruminants.

Keywords: condensed tannins; fermentability; gas production; grazing ecology; ruminant; microbial responses

---

1. Introduction

Tropical shrubs and trees are important feed for livestock because they are sources of protein [1], minerals and vitamins as well as playing important roles in ruminant feeding systems. However, many of these plant species have secondary compounds capable of changing the utilization of nutrients by mammalian herbivores.

Because plants developed defense mechanisms against herbivores and pathogens, animals have developed mechanisms to nullify or restrict the toxic and negative effects of ingested plant secondary compounds such as condensed tannins (CT) [2,3]. Ruminant herbivores and plant CT coexist and adapt natural evolutionary processes. Some ruminant feeders, especially goats, developed physiological adaptations, and even dependence on CT-rich legumes, selectively including such plants in their selector habits [4]. In the case of some browsers such as goats, this adaptation takes the form of salivary glands that produce large amounts of mucus-containing enzymes that can bind to CT to increase palatability and leaving plant proteins more available for digestion [5].

The evolution of different feeding strategies among domestic ruminant species implies differing microbial interactions with CT and, consequently, the diversity of rumen microorganisms and digestive capacity. Thus, we hypothesize that the role of CT on ruminal fermentability may vary depending on the species of the ruminant and, in the case of in vitro fermentation, the species of the rumen fluid donor. Comparative studies with different rumen-fluid donor species with previous pre-adapted to tannin feeding have been published [6–9]. However, the ability of the rumen microbes of different non-CT-adapted ruminant species to adapt to CT has not been fully investigated. Therefore, the objective of our study was to compare the effects of plant extracts containing CT on ruminal fermentation kinetics of taurine dairy cattle (*Bos taurus taurus*), zebu beef cattle (*Bos taurus indicus*), water buffaloes (*Bubalus bubalis*), sheep (*Ovis aries*) and goats (*Capra hircus*) without previous CT feeding exposure.

2. Material and Methods

2.1. Animals

Rumen fluid donors included three Holstein dairy cows (*B. taurus taurus*), three Nelore beef cows (*B. taurus indicus*), three Mediterranean water buffalo cows, three Santa Inês ewes and three Saanen goats. Diets were formulated to meet the nutritional requirements of each animal species/breed and consisted of 60% to 80% forage and 20% to 40% concentrate with ground maize and soybean meal. All animals were allowed free access to water and mineral mixture. All methods and animal care were performed in accordance with the relevant guidelines and regulations of the Ethic Committee on Animal Use of the School of Animal Science and Food Engineering, (São Paulo University).

2.2. Substrates

Four forages were evaluated as substrates: maize silage (*Zea mays*), fresh elephant grass (*Pennisetum purpureum*), Tifton-85 hay (*Cynodon* spp.) and fresh alfalfa (*Medicago sativa*). Forage samples were dried at 55 °C and ground through a 1-mm sieve. All were analyzed (Table 1) for dry matter (DM), mineral matter (MM), crude protein (CP) and acid-detergent fiber (ADF) with residual ash according to AOAC [10]. Neutral-detergent fiber (NDF) was estimated according to the methodology described by Mertens [11]. Total phenol (TP) concentrations were determined by the Folin-Ciocalteau reagent method [12] and total tannins (TT) were estimated as the difference in TP concentration before and after the treatment with insoluble polyvinylpolypirrolidone [13], using tannic acid as standard. Condensed tannin concentrations were determined by the butanol-HCl method [12], using leucocyanidin as a standard.

Table 1. Chemical composition of substrates used during the assay of in vitro fermentation kinetics.

Composition	Substrates [1]				
	ALF	**ELE**	**TIF**	**SIL**	**ACA**
organic matter [2]	916.82	897.75	936.45	964.64	978.86
ether extract [2]	84.04	46.72	57.92	62.98	n.d. [5]
crude protein [2]	278.97	60.28	158.02	82.02	n.d. [5]
neutral-detergent fiber [2]	735.62	770.03	795.29	563.28	n.d. [5]
acid-detergent fiber [2]	510.25	519.52	428.92	332.30	n.d. [5]
acid-detergent lignin [2]	126.69	121.63	133.08	71.35	n.d. [5]
total phenols [3]	13.60	5.47	5.32	10.18	558.63
total tannins [3]	8.14	3.05	2.82	6.58	519.58
condensed tannins [4]	0.25	0.10	0.10	0.15	235.87

[1] ALF: fresh alfalfa; ELE: fresh elephant grass; TIF: Tifton-85 hay; SIL: maize silage; ACA: Acacia tannin extract. [2] expressed as g/kg DM. [3] expressed as eq-g tannic acid/kg DM. [4] expressed as eq-g leucocyanidin/kg DM. [5] n.d.: not determined.

Acacia (*Acacia molissima*) extract (Seta® Estância Velha, Brazil) was added to the diets to raise the CT concentration to 50 eq-g of leucocyanidin per kg of feed DM. This CT concentration has been appointed as the minimum to cause harmful effects to ruminants [14]. The chemical characterization of substrates and *Acacia* extract were performed at the Laboratory of Animal Nutrition, Center for Nuclear Energy in Agriculture, University of São Paulo.

2.3. In Vitro Gas Production Assay

Ruminal contents were collected through permanent ruminal cannulas from each animal. Equal volumes of liquid and solid phases sample were homogenized in a blender for 10 s. The resulting material was filtered through three layers of cotton (cheese cloth) tissue [15]. Filtered fractions were kept in a water bath at 39 °C and CO_2 saturation until introduced into the in vitro system.

The in vitro gas production described by Theodorou et al. [16] and Maurício et al. [17] was used to compare disappearance rates. Each 160-mL fermentation flask received 0.5 g of the substrate with or without 0.3 g of extract CT. The inoculation was performed by injecting 10 mL of inoculum in 90 mL of buffered mineral solution [16] into each fermentation flask. The flasks were sealed with rubber stoppers, then shaken and incubated in an oven with forced-air circulation at 39 °C, thus allowing gas accumulation within each flask.

The pressure of the generated gases was measured at 4, 8, 12, 16, 24, 30, 46, 59, 72 and 96 h after inoculation, using a "transducer" (PressData 800®). After each measuring, the accumulated gas was released from all bottles. These values were used to calculate the volume of gas produced. At the end of this bioassay (96 h), 2 mL of liquid phase were sampled with a syringe and frozen until analysis of short-chain fatty acids (SCFA). After 96 h, residual material was filtered through sintered crucibles to determine in vitro dry matter degradability (IVDMD) and in vitro organic matter degradability (IVOMD). The partitioning factor (PF), calculated by relating DM degradation and OM degradation to total gas production, was used to compare microbial efficiency [18].

Gas production data were used to determine fermentation kinetics based on the model of Ørskov and McDonald [19] modified by McDonald [20] as:

$$p = 0 \text{ to } t < t_0 \tag{1}$$

$$p = a + b \left(1 - exp^{-ct}\right) \text{ to } t \geq t_0 \tag{2}$$

where p is gas production (ml) in time t; a and b are constants of the model; c is the gas production rate (h^{-1}); $a + b$ is the potential gas production (mL); t_0 is the lag time (h).

2.4. Short-Chain Fatty Acids Determination

Short-chain fatty acid analysis was measured by gas chromatography (GC-2014, Shimadzu®, Tokyo, Japan), split-injector, flame ionization detector and capillary column (Stabilwax®, Restek, Bellefonte, PA, USA) at 145 °C (isothermal) according to Erwin et al. (1961) [21], with adaptations by Getachew et al. [22]. Acetic, propionic, isobutyric, butyric, isovaleric and valeric acid (99.5% purity, Chem service, USA) were used as a quantitative external standard. The operational conditions were: injector and detector temperatures were 250 °C; helium was the carrier gas at 8.01 mL/min; hydrogen flow to the flame jet at 60 kPa and synthetic air at 40 kPa. The samples were thawed at room temperature and centrifuged at 14,500× g for 10 min. The supernatant (800 µL) was transferred to a dry and clean flask with 200 µL formic acid (98–100%) and 100 µL of the internal standard (100 mM 2-ethyl butyric acid, Chem service, USA).

2.5. Experimental Design and Statistical Analysis

The experiment tested three factors and their interactions. These included four forage substrates, five sources of inoculum (animal species) and two levels of CT extract. Individuals ($n = 3$) of each species constituted the experimental units for three replications.

The statistical design was completely randomized, according to the following model:

$$Y_{ijk} = \mu + F_i + S_j + T_k + (F_i \times S_j) + (S_j \times T_k) + (F_i \times T_k) + e_{ijk} \tag{3}$$

where Y_{ijk} is the dependent variable; μ the overall mean; F the effect of feeds (substrates) (i = 1 to 4); S the effect of animal species (j = 1 to 5); T the effect of CT (k = 1 to 2); F × S the interaction of feeds and animal species; S × T the interaction of animal species and CT; F×T the interaction of feeds and CT; e_{ijk} the residual error of the model. Results were compared by a Tukey test, using the software SAS for Windows® [23]. Results were considered different at $p \leq 0.05$.

3. Results

3.1. Tannin Extract Effects

There was an effect ($p \leq 0.001$) of CT inclusion on IVDMD and IVOMD regardless of the rumen fluid source. The inclusion of CT inhibited fermentation as indicated by the model parameters (Table 2), visualized in Figure 1. The partitioning factor is a measure created to bring together two variables: degradability and gas production. It was greater ($p \leq 0.001$) when CTs were included in the diet (Table 2).

Table 2. Effect of condensed tannins (CT) on degradability, fermentability and microbial efficiency during in vitro rumen fermentation in four ruminant species.

Variables	no CT	CT	SEM [1]	*p*-Value [2]
in vitro DM degradability	0.646 [a]	0.466 [b]	0.0090	***
in vitro OM degradability	0.644 [a]	0.458 [b]	0.0095	***
partitioning factor (mg DMD/mL)	3.40 [b]	7.68 [a]	0.234	***
partitioning factor (mg OMD/mL)	3.14 [b]	6.94 [a]	0.206	***
Model parameters [3]				
a	−27.39 [b]	−9.06 [a]	0.816	***
b	281.89 [a]	133.47 [b]	7.023	***
c	0.0281	0.0258	0.00130	ns
a + b	254.50 [a]	124.40 [b]	7.097	***
t_0	4.39 [a]	3.06 [b]	0.149	***

[1] SEM: standard error of means. [2] ns: not significant ($p > 0.05$); *: $p \leq 0.05$; **: $p \leq 0.01$; ***: $p \leq 0.001$. [3] Model of Ørskov and McDonald (1979), modified by McDonald (1981): $p = a + b\,(1 - exp^{-ct})$, where p = gas production (mL), in time t; a and b = constants of model; c = production gas rate (h^{-1}); $a + b$ = potential gas production (mL); t_0 = *lag time* (h) [a,b] means followed by distinct superscripts, within rows, are different (Tukey test at $p \leq 0.05$).

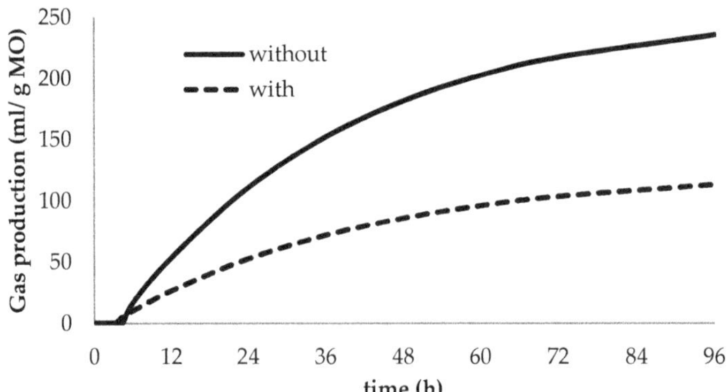

Figure 1. Effects of condensed tannin in diet (with and without) on gas production profiles during in vitro rumen fermentation (pooled over ruminant species).

Short-chain fatty acids production was also affected by the inclusion of CT extracts (Table 3). The addition of CT promoted a greater ($p \leq 0.001$) proportion of propionic acid and less ($p \leq 0.001$) acetic acid.

Table 3. **Effect** of tannins on short-chain fatty acids (SCFA) production, evaluated during in vitro organic matter rumen fermentation with and without condensed tannins (CT).

Variables	no CT	CT	SEM [1]	*p*-Value [2]
SCFA production (mmol/g OMD)				
acetic acid	6.57 [a]	2.83 [b]	0.26	***
propionic acid	2.46 [a]	1.30 [b]	0.09	***
iso-butyric acid	0.11 [a]	0.02 [b]	0.01	***
butyric acid	1.07 [a]	0.42 [b]	0.04	***
iso-valeric acid	0.15 [a]	0.05 [b]	0.01	***
valeric acid	0.19 [a]	0.10 [b]	0.01	***
total SCFA	10.51 [a]	4.69 [b]	0.39	***

[1] SEM: standard error of means. [2] ns: not significant ($p > 0.05$); *: $p \leq 0.05$; **: $p \leq 0.01$; ***: $p \leq 0.001$. [a,b] means followed by distinct superscripts, within rows, are different (Tukey test at $p \leq 0.05$).

3.2. Animal Species Effects

At the end of the 96-h in vitro fermentation, no differences ($p > 0.05$) in degradability and partitioning factor were observed among animals (Table 4). All species degraded similar proportions of feeds (IVDMD and IVOMD from 0.54 to 0.56) with similar microbial efficiency (PF). However, differences ($p \leq 0.01$) among model parameters were observed (Figure 2). The production of acetic, propionic and butyric acids by each unit of the degraded substrate was greater ($p \leq 0.001$) in large than in small ruminants and this was reflected in the total amount of SCFA produced (Table 5). Acetate and butyrate to total acid production were not different.

Table 4. Effect of animal species on degradability, fermentability and microbial efficiency assayed in four forages during in vitro rumen fermentation.

Variable [1]	Animal Species [2]					SEM [3]	*p* Value [4]
	Goats	Sheep	Buffalo	Taurine Cattle	Zebu Cattle		
IVDMD	0.556	0.565	0.543	0.564	0.554	0.014	Ns
IVOMD	0.551	0.560	0.538	0.559	0.547	0.015	Ns
PF (mg DMD/mL)	5.15	4.94	6.19	5.70	5.69	0.37	Ns
PF (mg OMD/mL)	4.65	4.48	5.70	5.31	5.07	0.33	Ns
Model parameters [5]							
A	−15.15 [a]	−22.01 [c]	−15.92 [ab]	−20.92 [bc]	−17.13 [abc]	1.29	***
B	185.82 [b]	197.70 [ab]	238.02 [a]	194.83 [ab]	222.00 [ab]	11.10	**
C	0.0348 [a]	0.0314 [a]	0.0210 [b]	0.0267 [ab]	0.0209 [b]	0.0021	***
a + b	170.67 [b]	175.69 [b]	222.10 [a]	173.91 [b]	204.88 [ab]	11.22	**
t_0	2.67 [b]	3.42 [ab]	4.24 [a]	4.27 [a]	4.03 [a]	0.24	***

[1] IVDMD: in vitro dry matter degradability; IVOMD: in vitro organic matter degradability; GP: gas production; PF: partitioning factor. [2] breeds—goats: Saanen; sheep: Santa Inês; buffalo: Mediterranea; taurine cattle: Holstein; zebu cattle: Nelore. [3] SEM: standard error of means. [4] ns: not significant ($p > 0.05$); *: $p \leq 0.05$; **: $p \leq 0.01$; ***: $p \leq 0.001$. [5] Model of Ørskov and McDonald (1979), modified by McDonald (1981): $p = a + b\,(1 - exp^{-ct})$, where p = gas production (mL), in time t; a and b = constants of model; c = production gas rate (h^{-1}); $a + b$ = potential gas production (mL); t_0 = *lag time* (h). [a,b,c] means followed by distinct superscripts, within rows, are different (Tukey test at 5%).

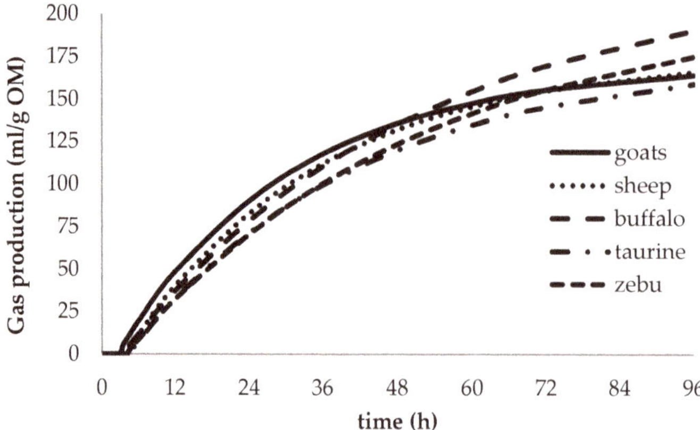

Figure 2. Effects of animal species on gas production profiles during in vitro rumen fermentation (pooled over condensed tannin treatments in four forages).

Table 5. Effect of animal species on short-chain fatty acids (SCFA) production evaluated during in vitro rumen fermentation.

Variables	Animal Species [1]					SEM [2]	*p*-Value [3]
	Goats	Sheep	Buffalo	Taurine Cattle	Zebu Cattle		
SCFA production (mmol/g OMD)							
acetic acid	3.37 c	3.92 bc	5.44 ab	5.70 a	5.06 ab	0.41	***
propionic acid	1.30 b	1.45 b	2.25 a	2.28 a	2.11 a	0.14	***
iso-butyric acid	0.09 a	0.06 abc	0.07 ab	0.02 c	0.04 bc	0.01	***
butyric acid	0.53 c	0.61 bc	0.90 a	0.88 a	0.82 ab	0.07	***
iso-valeric acid	0.10	0.10	0.10	0.12	0.10	0.01	ns
valeric acid	0.12 bc	0.10 c	0.15 abc	0.19 a	0.17 ab	0.02	***
total SCFA	5.41 c	6.21 bc	8.89 a	9.17 a	8.30 ab	0.61	***

[1] breeds—goats: Saanen; sheep: Santa Inês; buffalo: Mediterranean; taurine cattle: Holstein; zebu cattle: Nelore. [2] SEM: standard error of means. [3] ns: not significant ($p > 0.05$); *: $p \leq 0.05$; **: $p \leq 0.01$; ***: $p \leq 0.001$. a,b,c means followed by distinct superscripts, within rows, are different (Tukey test at $p \leq 0.05$).

3.3. Interactions

CT extract was effective ($p \leq 0.001$) in reducing IVDMD and IVOMD in all four ruminant species evaluated (Figure 3). However, there were no differences in degradability among animal species between the paired treatment groups receiving CT and those which did not ($p > 0.05$). Among the treatment combinations including CT, buffalo and cattle showed greater microbial efficiency ($p \leq 0.01$) than sheep or goats (Figure 3). The same was not observed for the partitioning factor by species not receiving CT because all species showed similar values. Condensed tannin addition to the diet increased the partitioning factor in large and small ruminant species.

The non-CT diet showed greater ($p \leq 0.001$) levels of SCFA in cattle and buffalo than in sheep and goats. However, for animals receiving CT, there were no differences ($p > 0.05$) in SCFA between large and small ruminants (Figure 4).

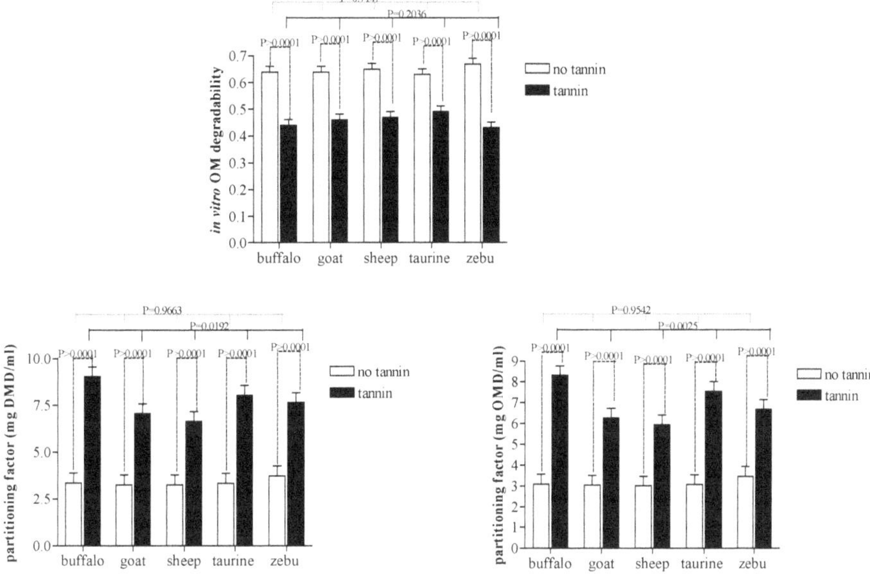

Figure 3. Effects of condensed tannin and animal species interactions on the organic matter degradability and partitioning factor during in vitro rumen fermentation.

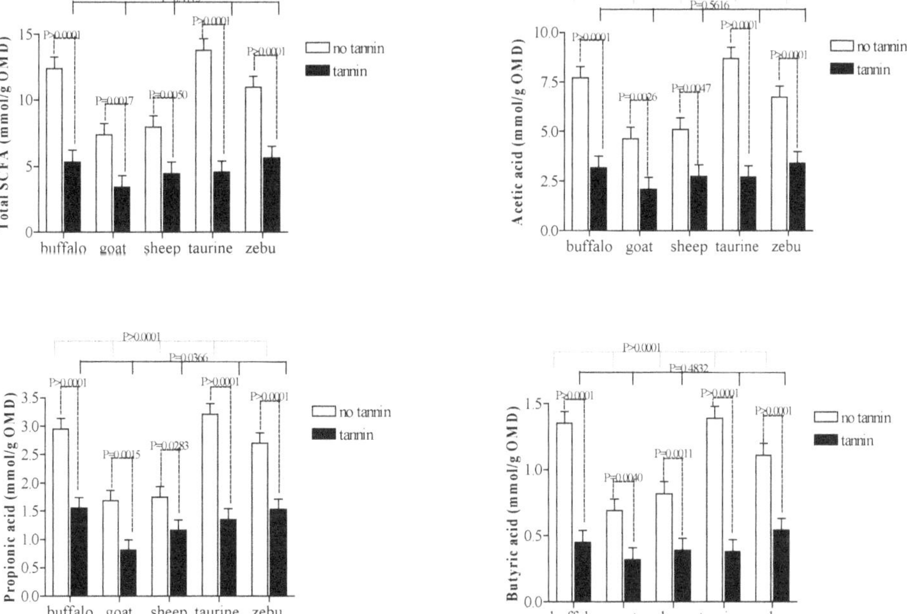

Figure 4. Effects of condensed tannin and animal species interaction on short-chain fatty acids (SCFA) production during in vitro rumen fermentation.

4. Discussion

4.1. Condensedtannin Extract Effects

Flavonoids from *Acacia* spp. have potential biological effects on rumen fermentation [24]; thus, gas production reduction from diets containing CT extract was expected due to its effect on rumen microbiota. The regression equation for *Acacia* CT extract had a slow initial slope (12 h) followed by a slight increase, which allowed for a constant long-term assay up to 96 h. The slow but steady gas production observed throughout the total incubation period reflects the lower ruminal digestion of organic matter in diets containing CT as reported elsewhere in the literature [25,26]. The inhibitory effects of CT on rumen gas production can be explained by a direct effect of CT-protein complexes on the rumen fibrolytic microbe and consequently lower cell wall degradation. The greater SCFA production and organic matter degradability for tannin-free diets also highlights the interference of CT on cell wall degradability. Previous in vitro studies support our findings that CT may reduce or slow organic matter degradability in the rumen, as reflected in less total SCFA production [27–29].

The addition of CT resulted in greater partitioning factors, expressed as mg OM degraded per mL of produced gases. The greater partitioning factor observed in this trial could be due to the initial washing losses by a substrate that might provide greater values of organic matter degradability in poorly substrate degradation samples with CT and therefore with less gas production. However, this assumption needs to be confirmed by future in vitro trials.

4.2. Animal Species Effects

Despite their different feeding behavior and digestive capacity, there were similar nutrient degradability and microbial efficiencies among ruminant species studied, which can be explained by the prolonged incubation period of 96 h. The incubation duration used in our study may have allowed the adaptation of the rumen bacterial community to diets, eventually resulting in equivalent substrate fermentation. Contrary to our observations, Calabrò et al. [30] found differences in rumen fermentation between buffaloes and cattle; however, we believe that differences in microbial ecology and their donors probably are strongly reflected in short term incubation [31]. This emphasizes the importance of appropriate incubation duration in studies comparing substrates exposed to inoculants from donors fed different diets.

Due to better fiber digestion capacity, buffaloes, taurine and zebu cattle produced greater amounts of SCFA (per mmol MOD) than small ruminants, regardless of donor diet. Within the ruminant species, goats showed lower production of the major SCFA, including acetic, propionic and butyric acids as well as total SCFA. These results confirm the lower ruminal and omasal capacity and its less efficient degradability of poor quality forages by small-bodied browsers [2,3].

4.3. Interactions

The evolution of different feeding behaviors in ruminant species has required anatomic, physiologic and microbial gastrointestinal adaptations to their respective dietary niches [2,32]. In accordance with this hypothesis, Clauss et al. [33] reported clear effects of the ecology of wild ruminants on the dynamics of their protozoal fauna. In this context, the interaction between tanniniferous plants and browsers likely resulted in greater tolerance to flavonoid compounds than in grazers.

The major objective of our study was to assess possible CT-tolerant microbes present in rumen fluid from browsers or grazers. As a result, our 96-h incubation produced different results than those reported for 24 h by Bueno et al. [31], with CT addition resulting in similar total degradation of substrates in vitro during rumen fermentation among grazing and browsing ruminants. In other words, CT was effective in reducing cell wall degradability rates and SCFA production in all species evaluated. This finding suggests that an incubation time of 96 h was insufficient to alter rumen microbial dynamics associated with changing diets, resulting in an incomplete adaptation of ruminal microbial community to the new diet.

Comparisons across CT treatment groups revealed greater propionic acid production and microbial efficiency in buffalo compared to cattle/taurine. Both results suggest that microbial populations in buffalo appear to be less sensitive to dietary CT. Although this contradicts most published research on the topic, it agrees with results reported by Salem [34] that buffalos were more tolerant of CT than other large grazers.

Despite the few differences observed as a result of our longer assay duration, results reported from in vivo animal trials must consider factors other than simply diet. These include anatomical adaptions of the mouth, teeth, salivary glands, body mass and digestive system altering selectivity, forage intake, passage rate and rumination time among the different feeding categories. As a result, degradability trials should account for differentiated rumen fermentation responses in browsers and grazers and selective versus bulk feeders when assaying CT-rich substrates.

5. Conclusions

The initial differences in the microbial communities resulting from feeding of donor species provide different responses between large and small ruminants in vitro tannin-rich diet fermentation. Inoculum from sheep and goats is less affected by the addition of CT than buffalo, zebu and taurine.

Author Contributions: Designed the study I.C.S.B.; performed the experiment and collected the samples, R.A.B. and I.C.S.B.; performed the laboratory analyses, G.M.F., G.B., and J.P.M.; analyzed and interpreted the data, G.M.F., R.A.B. and I.C.S.B.; wrote the manuscript: G.M.F. All authors have read and agreed to the published version of the manuscript.

Funding: Please add: This research was funded by Fundação de Amparo à Pesquisa do Estado de São Paulo (FAPESP), project number: 10/19654-2.

Acknowledgments: The authors acknowledge the financial support of Fundação de Amparo à Pesquisa do Estado de São Paulo (FAPESP), Cristina Teresa Alves (Embrapa, São Carlos, Brazil), Francisco Palma Rennó (FMVZ-Pirassununga, Brazil) and Paulo Henrique Mazza Rodrigues (FMVZ, Pirassununga, Brazil), who kindly collaborated with this study, and Seta® (Estância Velha, Brazil) for the CT extract.

Conflicts of Interest: The authors (I.C.S. Bueno, R. A. Brandi, G.M. Fagundes, G. Benetel, and J.P. Muir) have no financial or personal relationship with other people or organizations that could inappropriately influence or bias the research.

References

1. Fagundes, G.M.; Modesto, E.C.; Fonseca, C.E.M.; Lima, H.R.P.; Muir, J.P. Intake, digestibility and milk yield in goats fed *Flemingia macrophylla* with or without polyethylene glycol. *Small Rumin. Res.* **2014**, *116*, 88–93. [CrossRef]

2. Van Soest, P.J. *Nutritional Ecology of the Ruminant*, 2nd ed.; Cornell University Press: Ithaca, NY, USA, 1994; 476p.

3. Lamy, E.; Rawel, H.; Schweigert, F.J.; Capela e Silva, F.; Ferreira, A.; Costa, A.R.; Antunes, C.; Almeida, A.M.; Coelho, A.V.; Sales-Baptista, E. The effect of tannins on Mediterranean ruminant ingestive behavior: The role of the oral cavity. *Molecules* **2011**, *16*, 2766–2784. [CrossRef] [PubMed]

4. Muir, J.P. The multi-faceted role of condensed tannins in the goat ecosystem. *Small Rumin. Res.* **2011**, *98*, 115–120. [CrossRef]

5. Hofmann, R.R. Evolutionary steps of ecophysiological adaptation and diversification of ruminants: A comparative view of their digestive system. *Oecologia* **1989**, *78*, 443–457. [CrossRef] [PubMed]

6. Osawa, R.O.; Sly, L.I. Occurence of tannin-protein complex degrading Streptococcus sp. in feces of various animals. *Syst. Appl. Microbiol.* **1992**, *15*, 144–147. [CrossRef]

7. Odenyo, A.A.; McSweeney, C.S.; Palmer, B.; Negassa, D.; Osuji, P.O. In vitro screening of rumen fluid samples from indigenous African ruminants provides evidence for rumen fluid with superior capacities to digest tannin-rich fodders. *Aust. J. Agric. Res.* **1999**, *50*, 1147–1157. [CrossRef]

8. Jones, R.J.; Meyer, J.H.F.; Bechaz, F.M.; Stolzt, M.A.; Palmer, B.; van der Merwe, G. Comparison of rumen fluid from South African game species and from sheep to digest tanniniferous browse. *Aust. J. Agric. Res.* **2001**, *52*, 453–460. [CrossRef]

9. Odenyo, A.A.; Bishop, R.; Asefa, G.; Jamnadass, R.; Odongo, D.; Osuji, P.O. Characterization of tannin-tolerant bacterial isolates from East African ruminants. *Anaerobe* **2001**, *7*, 5–156. [CrossRef]

10. Association of Official Analytical Chemists (AOAC). *Official Methods of Analysis*, 16th ed.; AOAC: Arlington, VA, USA, 1995.

11. Mertens, D.R. Gravimeric determination of amylase-treated neutral detergent fibre in feeds with refluxing beakers or crucibles: Collaborative study. *J. Assoc. Off. Anal. Chem.* **2002**, *85*, 1217–1240.

12. Makkar, H.P. *Quantification of Tannins in Tree and Shrub Foliage: A Laboratory Manual*; Springer: Berlin/Heidelberg, Germany, 2003.

13. Makkar, H.P.S.; Blümmel, M.; Borowy, N.K.; Becker, K. Gravimetric determination of tannins and their correlations with chemical and protein precipitation methods. *J. Sci. Food Technol.* **1993**, *61*, 161–165. [CrossRef]

14. Minho, A.P.; Bueno, I.C.S.; Louvandini, H.; Jackson, F.; Gennari, S.M.; Abdalla, A.L. Effect of *Acacia molissima* tannin extract on the control of gastrointestinal parasites in sheep. *Anim. Feed Sci. Technol.* **2008**, *147*, 172–181. [CrossRef]

15. Bueno, I.C.S.; Cabral Filho, S.L.S.; Gobbo, S.P.; Louvandini, H.; Vitti, D.M.S.S.; Aballa, A.L. Influence of inoculum source in a gas production method. *Anim. Feed Sci. Technol.* **2005**, *123–124*, 96–105. [CrossRef]

16. Theodorou, M.K.; Williams, B.A.; Dhanoa, M.S.; McAllan, A.B.; France, J. A simple gas production method using a pressure transducer to determine the fermentation kinetics of ruminant feeds. *Anim. Feed Sci. Technol.* **1994**, *48*, 185–197. [CrossRef]

17. Mauricio, R.M.; Mould, F.L.; Dhanoa, M.S.; Owen, E.; Channa, K.S.; Theodorou, M.K. A semi-automated in vitro gas production technnique for ruminant feedstuff evaluation. *Anim. Feed Sci. Technol.* **1999**, *79*, 321–330. [CrossRef]

18. Blümmel, M.; Makkar, H.P.S.; Becker, K. In vitro gas production: A technique revisited. *J. Anim. Physiol. Anim. Nutr.* **1997**, *77*, 24–34. [CrossRef]

19. Ørskov, E.R.; Mcdonald, I. The estimation of protein degradability in the rumen from incubation measurements weighted according to rate of passage. *J. Am. Sci.* **1979**, *92*, 449–453. [CrossRef]

20. Mcdonald, I. A revised model for estimation of protein degradability in the rumen. *J. Agric. Sci.* **1981**, *96*, 251–252. [CrossRef]

21. Erwin, E.S.; Marco, G.J.; Emery, E.M. Volatile fatty acid analyses of blood and rumen fluid by gas chromatography. *J. Dairy Sci.* **1961**, *44*, 1768–1771. [CrossRef]

22. Getachew, G.; Makkar, H.P.S.; Becker, K. Tropical browses: Contents of phenolic compounds, in vitro gas production and stoichiometric relationship between short chain fatty acid and in vitro gas production. *J. Agric. Sci.* **2002**, *139*, 341–352. [CrossRef]

23. SAS INSTITUTE. *The SAS System for Windows*; Release 8.01; Sas Institute: Cary, NC, USA, 2000.

24. Bueno, I.C.S.; Vitti, D.M.; Louvandini, H.; Abdalla, A.L. A new approach for in vitro bioassay to measure tannin biological effects based on a gas production technique. *Anim. Feed Sci. Technol.* **2008**, *141*, 153–170. [CrossRef]

25. Patra, A.K.; Saxena, J. Exploitation of dietary tannins to improve rumen metabolism and ruminant nutrition. *J. Sci. Food Agric.* **2011**, *9*, 24–37. [CrossRef]

26. Tedeschi, L.O.; Ramírez-Restrepo, C.A.; Muir, J.P. Developing a conceptual model of possible benefits of condensed tannins for ruminant production. *Animal* **2014**, *8*, 1095–1105. [CrossRef] [PubMed]

27. Hervás, G.; Frutos, P.; Serrano, E.; Mantecón, A.R.; Giráldez, F.J. Effect of tannic acid on rumen degradation and intestinal digestion of treated soya bean meals in sheep. *J. Agric. Sci.* **2000**, *135*, 305–310.

28. Tiemann, T.T.; Lascano, C.E.; Wettstein, H.R.; Mayer, A.C.; Kreuzer, M.; Hess, H.D. Effect of the tropical tannin-rich shrub legumes *Calliandra calothyrsus* and *Flemingia macrophylla* on methane emission and nitrogen and energy balance in growing lambs. *Animal* **2008**, *2*, 790–799. [CrossRef] [PubMed]

29. Hariadi, B.T.; Santoso, B. Evaluation of tropical plants containing tannin on in vitro methanogenesis and fermentation parameters using rumen fluid. *J. Sci. Food Agric.* **2010**, *90*, 456–461. [PubMed]

30. Calabrò, S.; Moniello, G.; Piccolo, V.; Bovera, F.; Infascelli, F.; Tudisco, R.; Cutrignelli, M.I. Rumen fermentation and degradability in buffalo and cattle using the in vitro gas production technique. *J. Anim. Physiol. Anim. Nutr.* **2008**, *92*, 356–362.

31. Bueno, I.C.; Brandi, R.A.; Franzolin, R.; Benetel, G.; Fagundes, G.M.; Abdalla, A.L.; Louvandini, H.; Muir, J.P. In vitro methane production and tolerance to condensed tannins in five ruminant species. *Anim. Feed Sci. Technol.* **2015**, *205*, 1–9. [CrossRef]

32. Gordon, I.J. Browsing and grazing ruminants: Are they different beasts? *Forest. Ecol. Manag.* **2003**, *181*, 13–21. [CrossRef]

33. Clauss, M.; Müller, K.; Fickel, J.; Streich, W.J.; Hatt, J.-M.; Südekum, K.-H. Macroecology of the host determines microecology of endobionts: Protozoal faunas vary with wild ruminant feeding type and body mass. *J. Zool.* **2011**, *283*, 169–185. [CrossRef]

34. Salem, A.F.Z. Impact of season of harvest on in vitro gas production and dry matter degradability of *Acacia saligna* leaves with inoculum from three ruminant species. *Anim. Feed Sci. Technol.* **2005**, *123*, 67–79. [CrossRef]

Article

Fermentation Pattern of Several Carbohydrate Sources Incubated in An in Vitro Semicontinuous System with Inocula From Ruminants Given Either Forage or Concentrate-Based Diets

Zahia Amanzougarene, Susana Yuste and Manuel Fondevila *

Departamento de Producción Animal y Ciencia de los Alimentos, Instituto Agroalimentario de Aragón (IA2), Universidad de Zaragoza-CITA, M. Servet 177, 50013 Zaragoza, Spain; zahiaagro@yahoo.fr (Z.A.); susana_monre@hotmail.com (S.Y.)
* Correspondence: mfonde@unizar.es; Tel.: +34-876554171

Received: 27 December 2019; Accepted: 3 February 2020; Published: 6 February 2020

Simple Summary: A sudden change from a milk/forage diet to a high concentrate diet in young ruminants increases the rate and extent of rumen microbial fermentation, leading to digestive problems, such as acidosis. The magnitude of this effect depends on the nature of the ingredients. Six carbohydrate sources were tested: three cereal grains (barley, maize and brown sorghum), as high starch sources of different availability, and three byproducts (sugarbeet pulp, citrus pulp and wheat bran), as sources of either insoluble or soluble fibre. An in vitro semicontinuous incubation system was used to compare the fermentation pattern of substrates incubated with inocula-simulating concentrate or forage diets, under the pH and liquid outflow rate conditions of intensive feeding systems. The magnitude of microbial fermentation was higher with the concentrate than the forage inoculum, and the drop in pH in the first part of incubation was more profound. Among the substrates, citrus pulp had a greater acidification potential and was fermented at a higher extent, followed by wheat bran and barley. In conclusion, the acidification capacity of substrates plays an important role in the environmental conditions, depending on the type of diet given to the ruminant. This in vitro system allows us to compare the substrates under conditions simulating high-concentrate feeding.

Abstract: The fermentation pattern of several carbohydrate sources and their interaction with the nature of microbial inoculum was studied. Barley (B), maize (M), sorghum, (S), sugarbeet pulp (BP), citrus pulp (CP) and wheat bran (WB) were tested in an in vitro semicontinuous system maintaining poorly buffered conditions from 0 to 6 h, and being gradually buffered to 6.5 from 8 to 24 h to simulate the rumen pH pattern. Rumen fluid inoculum was obtained from lambs fed with either concentrate and barley straw (CI) or alfalfa hay (FI). The extent of fermentation was higher with CI than FI throughout the incubation ($p < 0.05$). Among the substrates, S, BP and M maintained the highest pH ($p < 0.05$), whereas CP recorded the lowest pH with both inocula. Similarly, CP recorded the highest gas volume throughout the incubation, followed by WB and B, and S recorded the lowest volume ($p < 0.05$). On average, the total volatile fatty acid (VFA), as well as lactic acid concentration, was higher with CP than in the other substrates ($p < 0.05$). The microbial structure was more affected by the animal donor of inoculum than by the substrate. The in vitro semicontinuous system allows for the study of the rumen environment acidification and substrate microbial fermentation under intensive feeding conditions.

Keywords: cereals; fibrous byproducts; gas volume; pH; volatile fatty acids; in vitro fermentation

---

1. Introduction

Reaching a high productive performance in the fattening of young ruminants requires high-energy diets that promote a high rate and extent of rumen microbial fermentation, with acidosis as a frequent consequence [1]. In practice, ruminants reared at pasture are often abruptly introduced to intensive feeding systems without being previously adapted to high-concentrate diets, promoting variable responses in the rate and extent of fermentation [2]. Cereals are commonly considered as ingredients of concentrate diets for ruminants. Their energy value depends on starch availability, which differs according to their chemical structure, protein matrix or, in some cases, the presence of phenolic compounds [3,4]. Fibrous byproducts with either insoluble (cellulose, hemicelluloses) or soluble (mostly pectin) polysaccharides and variable proportions of either starch or sugars [5,6] are also included among the carbohydrate sources currently used. Fitting substrate characteristics to the fermentative ability of rumen microbiota while environmental conditions are maintained at an optimal range is a key factor for maximising the efficiency of energy utilisation, and the risk of physiological impairment is also reduced. The characteristics of the specific rumen microbial community promoted by a certain diet also affect substrate utilisation [7], as the activity of the bacterial species able to fermenting starch or fibrous polysaccharides depends on environmental characteristics [8,9].

The comparison of these energy sources and their effects in the rumen under in vivo conditions is laborious and expensive, and is often biased by the feeding pattern and hardly controlled fermentation conditions [10]. On the other hand, in vitro studies are cheaper and faster and allow for a good insight into rumen fermentation processes [11]. However, most of these in vitro methods are designed to mimic the environment promoted by high forage diets, including the use of inoculum from forage-fed animals [12], and it is not easy to adapt the main physiological conditions, such as pH and rate of passage to conditions promoted by high-concentrate diets [13]. Amanzougarene and Fondevila [14] succeeded in maintaining a low incubation pH in an in vitro closed-batch system by reducing the bicarbonate concentration in the incubation solution, allowing to compare the fermentation of different carbohydrate sources under conditions simulating high-concentrate feeding [15,16]. However, this is not the real physiological situation in vivo, as pH changes across a wide range throughout the day [17] and, besides, rumen outflow rate cannot be assessed in this system. In this regard, the semicontinuous incubation system [18] modified by Prates et al. [19], applying the procedure proposed by Amanzougarene and Fondevila [14] for controlling incubation pH, appears to be a useful tool to mimic the rumen pH pattern and liquid outflow rates under in vitro conditions.

Therefore, in a semicontinuous in vitro incubation system, we compared the acidification potential and the rumen microbial fermentation pattern of several carbohydrate sources of variable composition when a different rumen environment is promoted by either high-forage or high-concentrate diets, aiming to minimise, where possible, the risk of acidosis during feeding transition from a fibrous to a high-concentrate diet.

2. Materials and Methods

2.1. Substrates and Inocula

Six carbohydrate sources were chosen as substrates: three cereal grains (barley var. Gustav (B), maize var. Dekalb 6667Y (M), and a brown sorghum of unknown variety (S)) and three by-product feeds (sugarbeet pulp (BP), citrus pulp (CP) and wheat bran (WB)). All substrates were ground in a hammer mill (Retsch Gmbh/SK1/417449, Haan, Germany) through a 1 mm sieve. The chemical compositions of the substrates are given in Table 1.

Table 1. Chemical composition (g/kg DM) of feeds used as incubation substrates.

Component	B	M	S	BP	CP	WB
OM	978	986	979	953	940	944
CP	105	75	113	107	59	161
EE	24	34	11	5	14	31
Starch	672	706	647	-	-	245
aNDFom	173	91	97	437	207	499
ADF	56	25	60	272	192	145
ADL	18	2	5	75	21	37
NDSF	4	77	110	457	423	155
Sugars	1.6	13	1.3	9	243	31
TP	-	-	2.6	-	-	-
TT	-	-	1.3	-	-	-

Barley (B); maize (M); sorghum (S); sugar beet pulp (BP); citrus pulp (CP); wheat bran (WB). Dry matter (DM); organic matter (OM); crude protein (CP); ether extract (EE); neutral detergent fibre (aNDFom); acid detergent fibre (ADF); acid detergent lignin (ADL); neutral detergent soluble fibre (NDSF). Total phenolics (TP); total tannins (TT).

Rumen fluid was obtained from six lambs housed in the facilities of the Servicio de Apoyo a la Experimentación Animal of the Universidad de Zaragoza. The animal care and procedures for extraction of rumen inoculum were approved by the Ethics Committee for Animal Experimentation. The care and management of animals agreed with the Spanish Policy for Animal Protection RD 53/2013, which complies with EU Directive 2010/63 on the protection of animals used for experimental and other scientific purposes. The lambs were weaned at 49 ± 8 days (average weight 13.6 ± 0.78 kg) and, thereafter, three lambs (1, 2 and 3) were fed ad libitum with a concentrate mixture (composed of barley, maize, wheat, and soybean meal) and barley straw (88:12 concentrate to straw ratio) for 35 days, and then slaughtered (average weight 20.6 ± 1.85 kg) to obtain concentrated inoculum (CI). The other three lambs (4, 5 and 6) were fed ad libitum with alfalfa hay and slaughtered after 45 days (average weight 16.5 ± 0.33 kg) to obtain forage inoculum (FI). The rumen contents of each animal were individually filtered through a cheesecloth and dispensed in 16 mL aliquots into 110×16 mm tubes, which were immediately frozen in liquid nitrogen and preserved at -80 °C until use [19]. Immediately before incubation, the rumen inoculum was thawed in a water bath at 39 °C (about 2 min).

2.2. Experimental Conditions

The in vitro semicontinuous system of Fondevila and Pérez-Espés [18], modified by Prates et al. [19], was used. The substrate samples (800 mg) were dispensed into 4×4 cm nylon bags (45 μm pore size) that were sealed and introduced into duplicated bottles (116 mL total volume). The bottles were filled under CO_2 flux with 80 mL of incubation solution, including 16 mL (0.20 of total volume) thawed rumen inoculum, without resazurin and microminerals [20], and were incubated in a water bath at 39 °C for 24 h in three incubation series, each corresponding to a different donor animal, for each type of inoculum. The buffer solution was modified to include 0.006 M bicarbonate ion in order to get a poorly buffered medium [14].

The pressure produced on each bottle was measured every 2 (from 0 to 12 h incubation) or 4 h (from 12 to 24 h) with a HD8804 manometer provided with a TP804 pressure gauge (DELTA OHM, Caselle di Selvazzano, Italy). The readings, corrected for the atmospheric pressure, were converted to volume (ml) using a pre-established linear regression recorded in this type of bottle ($n = 48$, $R^2 = 0.993$), and were expressed per unit of incubated organic matter (OM). Along the incubation, an aliquot volume of the medium was extracted immediately after each gas measurement and replaced anaerobically by the same volume of incubation solution (without microbial inoculum) to simulate an approximate liquid turnover rate of 0.08/h. In order to simulate daily rumen pH fluctuations, from 0 to 6 h, the incubation solution was poorly buffered, as explained above, to allow the incubation pH to drop as fermentation proceeded, whereas, from 8 h onwards, the replacing incubation solution was made up with 0.058 M bicarbonate ion to allow the pH to increase to around 6.5.

The incubation pH was recorded on every extraction. In addition, the medium was sampled at 6 and 10 h for determination of volatile fatty acids concentration (VFA; 2 mL on a 0.5 mL solution of 0.5 M phosphoric acid with 1 mg 4-methyl-valeric acid as the internal standard) and at 6 h for the determination of the lactic acid concentration (2 mL). The samples were stored at -20 °C until analysis. Moreover, another sample (6 mL) was also taken at 8 h and immediately frozen (−80 °C) for the determination of microbial biodiversity by terminal restriction fragment length polymorphism (tRFLP). At the end of incubation, the substrate bags were removed from the bottles, rinsed and dried at 60 °C for 48 h for the determination of dry matter disappearance (DMd).

2.3. Chemical and Microbiological Analyses

The dry matter (DM) and OM content in the substrates and the incubation residues were analysed following the AOAC [21] procedures (methods ref. 934.01 and 942.05). The substrates were also analysed for crude protein (CP) and ether extract (EE) (ref. 976.05 and 2003.05) [21], and their concentration of neutral detergent fibre (aNDFom) was analysed as described by Mertens [22] in an Ankom 200 Fibre Analyser (Ankom Technology, New York, NY, USA), using α–amylase and sodium sulphite, with results being expressed exclusive of residual ashes. The acid detergent fibre (ADF) (ref. 973.18) and acid detergent lignin (ADL) were determined as described by AOAC [21] and Robertson and Van Soest [23], respectively. Neutral detergent soluble fibre (NDSF) was estimated following Hall et al. [24], discounting the aNDFom and the ethanol insoluble EE, CP and starch fractions from the insoluble OM. The total starch content in B, M, S and WB substrates was determined enzymatically from samples ground to 0.5 mm using a commercial kit (Total Starch Assay Kit K-TSTA 07/11, Megazyme, Bray, Ireland). The total phenolic (TP) content in S was analysed following the colourimetric method of Makkar et al. [25] using the Folin–Ciocalteu reagent and with tannic acid (MERCK Chemicals, Madrid, Spain) as the reference standard. The total tannins (TT) were estimated as the difference between TP before and after treatment with polyvinyl polypyrrolidone.

The frozen samples of medium incubation were thawed and centrifuged at 13,000 g for 15 minutes at 4 °C in order to analyse their lactic acid and VFA. The VFA were determined by gas chromatography on an Agilent 6890 apparatus equipped with a flame detector and a capillary column (HP-FFAP Polyethylene glycol TPA, 30 m × 530 μm id). The lactic acid concentration was determined by the colourimetric method proposed by Barker and Summerson [26]. For the microbial diversity analysis, frozen microbial samples were freeze-dried, thoroughly mixed and disrupted (Mini-Bead Beater, Biospec Products, Bartlesville, OK, USA). The DNA was extracted using the Qiagen QIAmp DNA Stool Mini Kit (Qiagen Ltd., West Sussex, UK) following the manufacturer's recommendations, except that the samples were initially heated at 95 °C for 5 min to maximise the lysis of the bacterial cells. The concentration of extracted DNA was tested in a Nanodrop ND-1000 (Nano-Drop Technologies, Inc., Wilmington, DE, USA). PCR was performed using a 16S rRNA bacteria-specific primer (cyanine-labelled forward 27F, 5′-AGA GTT TGA TCC TGG CTCAG-3′ and unlabelled reverse 1389R, 5′-AGG GGG GGT GTG TAG AAG-3′; [27]) using a DNAEngine® Gradient Cycler (Bio-Rad, Spain). The PCR product was purified using a Purelink PCR purification kit (ref. K3100-01; Invitrogen) and diluted to 10 μL. The DNA concentration of each amplified and purified sample was obtained by spectrophotometry (Nanodrop® ND-1000 spectrophotometer) to enable a standardised quantity of 50 ng DNA to be used per restriction enzyme digest reaction. The digestion of samples was carried out using HhaI, HaeIII and MspI (Promega, Spain), following the manufacturer recommendations, except for HhaI, where the recommended addition of bovine serum albumin was omitted. The restriction digests were purified by ethanol precipitation [28] in 35 μL sample loading solution buffer, including a 600 bp size standard (Beckman Coulter Inc., Fullerton, CA, USA) before being applied to a 3500 × L Genetic Analyzer (Applied Biosystems). Once the size and height of every peak was obtained, 1% of the second highest peak was used as the criteria for the lower threshold for peaks, to detect and eliminate smaller, broader peaks that would not be indicative of single true OTUs.

2.4. Calculations And Statistical Analyses

The tRFLP results were analysed from a matrix generated for each data list obtained, and results were presented in the form of relative abundance. The three matrices resulting from each series and enzyme were concatenated and analysed with R statistical software (https://cran.r-project.org/bin/windows/base/, version 3.5.0, R Foundation for Statistical Computing, Vienna, Austria). FactoMineR, Factoextra, MixOmics, Vegan, MASS, and Ggplot2 packages were used to carry out the analysis of hierarchical classification on the principal components for obtaining the cluster dendrogram.

The results were analysed statistically by ANOVA using the Statistix 10 package [29]. On each sampling time, the effect of the incubation series (equivalent to the donor animal; interaction inoculum x incubation series, random effect), the type of inoculum, the type of substrate, and the interaction of both factors on pH, gas production, total VFA and lactic acid concentration, and VFA profile were studied as factors. The treatment differences among the means with $p < 0.05$ and $0.05 < p < 0.10$ were accepted as representing statistically significant differences and a trend to the differences, respectively. When significant, the differences were contrasted by the Tukey *t*-test. Simple and multiple linear regressions were established to study the relationships among the different parameters studied.

3. Results

3.1. Pattern of Incubation pH

The mean inoculum pH at the start of the incubation series was 6.45 ± 0.15 and 6.87 ± 0.02 for CI and FI, respectively ($n = 3$). The average minimum pH was recorded at 6 h incubation (5.96) for CI, and at 8 h (6.22) for FI. Thereafter, the pH increased to reach its maximum (6.64 for both inocula) at 24 and 20 h for CI and FI. The pH differences in the incubation medium among inocula ($p < 0.05$) were ± 0.3 units from 2 to 6 h, decreasing gradually to ± 0.1 at 12 h. A significant interaction inoculum × substrate ($p < 0.05$) observed on pH at 4, 8, 10, 12, 16 and 20 h and a tendency ($p = 0.052$) at 2 h incubation indicates the different behaviour of the substrates depending on the inoculum. Therefore, a comparison of the pH pattern among the incubated substrates is presented in Figure 1 separately for each inoculum. With CI (Figure 1a), the lowest incubation pH from 2 to 12 h was recorded with CP ($p < 0.05$), reaching its minimum at 6 h (5.60), although recovered thereafter to 6.63 at 24 h incubation. In ascending order, WB and B reached their minimum pH at 8 h (5.89 and 5.97, respectively), whereas BP, M, and S maintained a higher medium pH from 4 to 8 h ($p < 0.05$). The differences among M, S, BP, and B disappeared from 10 to 16 h ($p > 0.05$), and no differences were detected among the substrates ($p > 0.05$) at the end of incubation. When the substrates were incubated using FI (Figure 1b), CP recorded the lowest pH from 4 to 10 h incubation ($P < 0.05$), and its minimum value was 5.90, whereas S, M and BP maintained the highest medium pH during this period (6.30 to 6.46), and B and WB were grouped at intermediate values ($p < 0.05$). At 16 and 20 h incubation only, B recorded a lower value (6.44 and 6.57; $p < 0.05$) and, again, no differences were detected among the substrates at the end of incubation ($p > 0.05$).

3.2. Pattern of in Vitro Gas Production

The volume of gas produced with the CI inoculum was higher than that obtained with FI at all incubation times ($p < 0.05$). Because of the interaction inoculum x substrate at 4 h and from 8 to 24 h ($p < 0.05$), for an easier understanding, the gas production is presented separately for CI and FI (Figure 2). The major difference among substrate fermentative behaviour between the inocula is manifested in the magnitude of differences among them. Thus, with CI (Figure 2a), CP recorded the highest gas volume from 4 h onwards, at 12 h being on average 0.42 times higher than the other substrates, while also recorded differences at 2 h with BP and S ($p < 0.05$). The gas volume with WB was higher than BP and S from 4 h onwards, and higher than M from 6 h and B from 8 to 20 h ($p < 0.05$). Differences were also recorded between B and S from 8 to 16 h and at 24 h ($p < 0.05$). A similar pattern was observed with FI (Figure 2b), but the magnitude of differences was lower. Thus, CP was higher than B, M, BP and S from 6 to 24 h ($p < 0.05$), with differences at 12 h reaching 0.59 of their average, but did not differ from

WB, which was higher than BP and S in that period and also higher than M from 6 to 10 h ($p < 0.05$). Differences between B and M respect to S were also detected from 16 and 20 h onwards, respectively ($p < 0.05$).

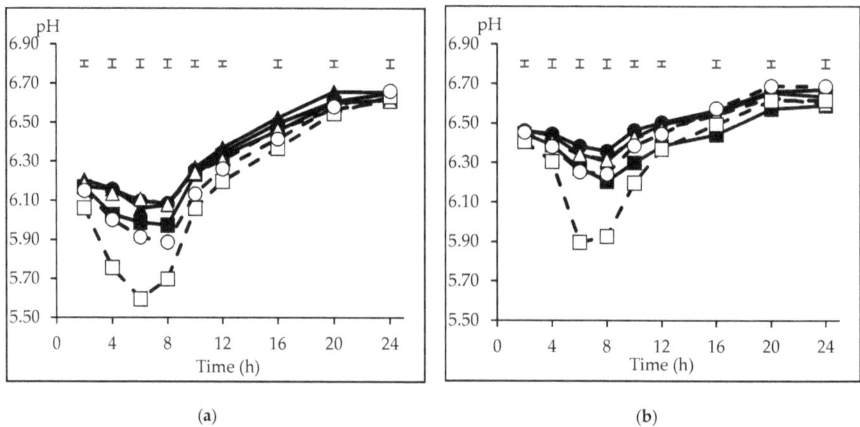

Figure 1. Pattern of medium pH of carbohydrate substrates (barley (B) □, maize (M) ▲, sorghum (S) ●; solid lines, citrus pulp (CP) ■, sugar beet pulp (BP) △, wheat bran (WB) ○; dashed lines) incubated with inoculum from concentrate (CI, Figure 1a) or forage (FI, Figure 1b) diets. The initial pH was 6.45 (Figure 1a) and 6.87 (Figure 1b). The upper bars show the standard error of the means ($n = 3$).

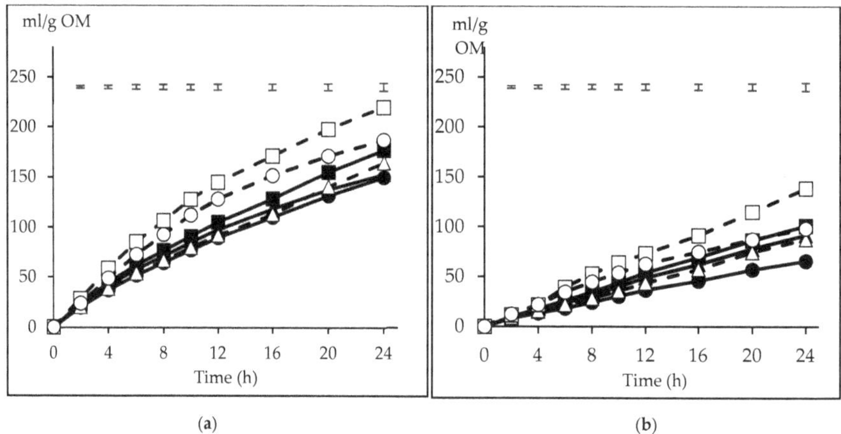

Figure 2. Pattern of gas production from the carbohydrate substrates (barley (B) ■, maize (M) ▲, sorghum (S) ●; solid lines, citrus pulp (CP) □, sugar beet pulp (BP) △, wheat bran (WB) ○; dashed lines) incubated with inoculum from concentrate (Figure 2a) or forage (Figure 2b) diets. The upper bars show the standard error of the means ($n = 3$).

3.3. Dry matter Disappearance (DMd)

Inoculum differences in DMd after 24 h of incubation were not detected, although CI was numerically higher than FI (proportions of 0.382 *vs.* 0.339 from the substrate weight; $p > 0.05$). The substrates ranked according to the proportion of DMd as follows: CP, 0.502 > B, 0.449 > WB, 0.360, M, 0.343 > BP, 0.265, S, 0.243 ($p < 0.001$; SEM = 0.0120). The interaction inoculum x substrate was not significant ($p = 0.21$), indicating that the substrates behaved similarly with both inocula.

3.4. Volatile Fatty Acids and Lactic Acid Production

Tables 2 and 3 show that CI promoted a higher ($p < 0.05$) concentration of total VFA than FI at both 6 (23.2 *vs.* 9.8 mM) and 10 h (22.2 *vs.* 9.3 mM). The molar proportions of acetate, propionate, and butyrate did not manifest the differences between inocula ($p > 0.05$), whereas with CI, valerate was higher and branched-chain volatile fatty acids (BCVFA, sum of isobutyrate and isovalerate) lower than with FI at both incubation times ($p < 0.05$).

Table 2. Average of total volatile fatty acids concentration (VFA, mM) and molar VFA proportions (mmol/mmol), together with lactate concentration (mM) recorded at 6 h of the different carbohydrate sources incubated as substrates with concentrate (CI) or forage (FI) inoculum.

Substrates	VFA	Acetate	Propionate	Butyrate	Valerate	BCVFA	Lactic acid
			with CI				
B	22.34 [ab]	0.570	0.240	0.153	0.022	0.014	3.83 [b]
M	20.18 [bc]	0.578	0.235	0.149	0.023	0.016	2.35 [c]
S	21.72 [abc]	0.588	0.245	0.131	0.022	0.015	1.93 [c]
BP	21.66 [abc]	0.593	0.235	0.135	0.021	0.016	0.90 [c]
CP	26.55 [a]	0.595	0.238	0.135	0.020	0.012	8.70 [a]
WB	26.94 [a]	0.590	0.245	0.132	0.020	0.013	2.95 [c]
			With FI				
B	10.31 [xyz]	0.633	0.226	0.102	0.009	0.030 [y]	3.05 [y]
M	8.83 [yz]	0.632	0.229	0.095	0.010	0.034 [xy]	2.69 [y]
S	8.31 [z]	0.642	0.216	0.094	0.009	0.039 [x]	0.84 [y]
BP	9.52 [yx]	0.665	0.207	0.088	0.008	0.032 [y]	1.64 [y]
CP	12.12 [yz]	0.686	0.209	0.075	0.008	0.023 [z]	8.67 [x]
WB	9.83 [yz]	0.653	0.225	0.083	0.009	0.030 [y]	2.97 [y]
SEM	1.065	0.0156	0.0076	0.0085	0.0010	0.0009	0.588
			p-Value				
Inoculum	0.002	0.077	NS	NS	0.005	<0.001	NS
Substrate	<0.001	NS	NS	NS	NS	<0.001	<0.001
Inoc. × Subs.	NS	NS	NS	NS	NS	<0.001	NS

Means within a column with different superscripts for CI ([a,b,c]) or FI ([x,y,z]) differ ($p < 0.05$). Standard error of the means (SEM). Branched-chain volatile fatty acids (BCVFA) (sum of isobutyrate + isovalerate). NS: $p > 0.10$.

Table 3. Average of total volatile fatty acids concentration (VFA, mM) and molar VFA proportions (mmoL/mmoL), recorded at 10 h of the different carbohydrate sources incubated as substrates with concentrate (CI) or forage (FI) inoculum.

Substrates	VFA	Acetate	Propionate	Butyrate	Valerate	BCVFA
			With CI			
B	21.16 [b]	0.561	0.225	0.172	0.028	0.014
M	19.42 [b]	0.548	0.227	0.178	0.031	0.017
S	19.87 [b]	0.557	0.249	0.152	0.026	0.017
BP	20.25 [b]	0.604	0.229	0.130	0.022	0.016
CP	27.32 [a]	0.553	0.240	0.164	0.031	0.012
WB	25.10 [ab]	0.537	0.261	0.158	0.029	0.015
			With FI			
B	10.46	0.597	0.256 [xy]	0.117	0.009	0.022
M	8.51	0.620	0.236 [xy]	0.109	0.009	0.026
S	8.48	0.620	0.222 [y]	0.120	0.009	0.029
BP	8.34	0.611	0.227 [y]	0.126	0.009	0.027
CP	10.47	0.579	0.267 [x]	0.124	0.010	0.019
WB	9.76	0.575	0.274 [x]	0.117	0.010	0.025
SEM	1.137	0.0233	0.0078	0.0169	0.0019	0.0024
			p-Value			
Inoculum	0.011	NS	NS	NS	0.008	0.008
Substrate	0.001	NS	<0.001	NS	NS	0.046
Inoc. × Subs.	0.058	NS	0.017	NS	NS	NS

Means within a column with different superscripts for CI ([a,b,c]) or FI ([x,y,z]) differ ($p < 0.05$). SEM: standard error of the means. Branched-chain volatile fatty acids (BCVFA) (sum of isobutyrate + isovalerate).

Among the substrates, at 6 h (Table 2) CP recorded a higher total VFA concentration than BP, M and S (average values of 19.3, 15.6, 14.5 and 15.0 mM, respectively), whereas WB (15.9 mM) was also higher than M and S ($p < 0.05$). Differences in the molar VFA profile were only recorded for BCVFA, with the highest proportions in S and M and the lowest with CP ($p < 0.05$); however, the interaction inoculum x substrate ($p < 0.001$) indicates that differences in BCVFA proportion were only observed with FI. Regarding the concentration of lactic acid at 6 h among the substrates, CP recorded the highest concentration and BP and S the lowest (8.7 *vs.* 1.3 and 1.4 mM; $p < 0.05$). Similar trends were observed at 10 h in total VFA concentration (Table 3), with CP rendering a higher concentration than BP, M and S, but tending to be significant only with CI interaction inoculum x substrate, ($p = 0.058$). The interaction inoculum x substrate in the proportion of propionate ($p = 0.017$) indicates that values recorded with WB and CP were higher than those with BP and S, but differences were only manifested with FI, whereas no differences ($p > 0.05$) among the substrates were recorded on acetate, butyrate and valerate proportions. The highest proportion of BCVFA was promoted by S and the lowest by CP ($p < 0.05$).

3.5. Bacterial Biodiversity

Bacterial biodiversity after 8 h of incubation was markedly affected by the source of rumen inoculum. Thus, the substrates incubated with rumen inoculum from lambs fed the high-concentrate diet clustered together, except for WB in the first incubation run, as well as substrates that were incubated with FI (Figure 3). Bacterial biodiversity was also markedly affected by the incubation series—that is, the donor animal—for both inocula.

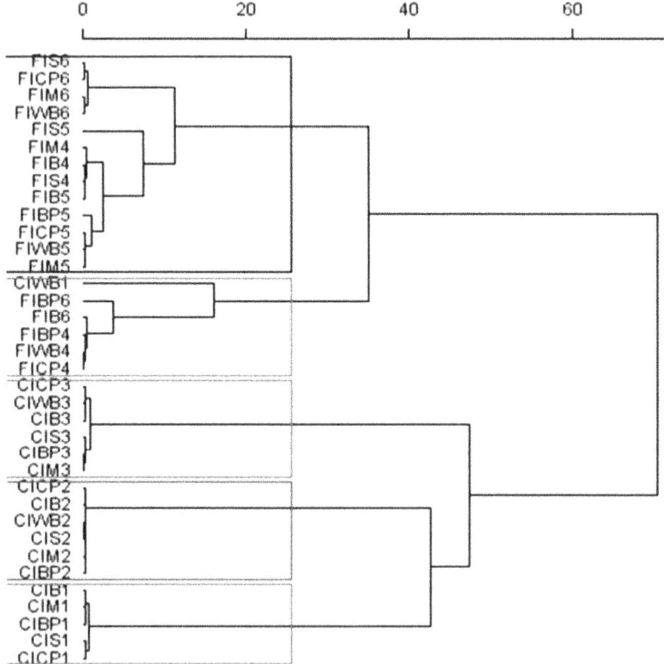

Figure 3. Dendrogram of bacteria diversity from terminal restriction fragment length polymorphism (tRFLP) data generated by enzyme digestion (HhaI, MspI, and HaeIII) for the carbohydrate substrates (B, M, S, BP, CP, and WB) incubated for 8 h with inoculum from concentrate (CI) or forage (FI) diets. The scale bar shows the Euclidean distances, "Ward method".

4. Discussion

Conventional in vitro closed batch systems are adapted for the study of microbial fermentation under conditions mimicking high forage diets, which is not applicable to evaluate diets given in intensive ruminant-fattening systems. An in vitro semicontinuous incubation system [18,19], adapted to control of the pH by modifying the bicarbonate ion concentration [14] allows us to approach the ruminal fermentation pattern of the different carbohydrate sources to the rumen physiological conditions that occur in intensive feeding systems, either during a transition process to high-concentrate diets (i.e., when rumen conditions are still modulated by a forage diet) or when animals are adapted to such feeding conditions (as promoted by a concentrate diet). The pH pattern obtained along the in vitro incubation with CI and FI, reaching a minimum value at 6–8 h after substrate availability and then progressively increasing to final pH values of around 6.4–6.5, fitted well with the circadian evolution of rumen pH observed with practical forage or concentrate feeding of ruminants [30]. Thus, we can assume it allows the different substrates to express their acidification potential at the time that their fermentation is compared under more realistic conditions.

4.1. Effect of the Inoculum Source on the in Vitro Fermentation Pattern

The source of rumen fluid has an important role in the pattern of in vitro fermentation [15,31,32], with an inoculum promoted by a concentrate diet having a higher fermentative potential than another from a forage diet. In our experiment, the lower buffering of the incubation medium during the first 6–8 hours allowed for a clear expression of the acidification potential of the incubated substrates, which was expressed at a higher extent with CI than FI (average pH along the 24 h incubation period ranging from 6.45 to 5.96 *vs.* 6.87 to 6.22) as pH dropped to values close to those considered as a threshold for microbial activity [33], whereas a higher pH was maintained with FI. Despite this, substrates incubated with CI rendered almost two-fold gas volume more than with FI, irrespective of the chemical nature (starch- or fibre-rich) of those substrates. Despite the more pronounced drop of pH with CI, the incubation environment promoted by a concentrate diet given to the donor animals was more favourable for fermentation of non-fibrous carbohydrates than that induced by a fibrous diet [20,34], probably because of the lack of adaptation of microbiota to ferment starch and sugar substrates with a forage inoculum [15,35] and the inherent buffering capacity of forage legumes such as alfalfa. However, assuming that a part of the gas produced comes from the activity of bicarbonate ion in the buffering of fermentation acids produced, such differences in gas production could be partly associated with the lower pH promoted by CI inoculum, although the contribution of this indirect gas is hard to quantify [14]. In the case of the byproducts, characterised by their richness in rapidly fermentable fibre, microbiota might easily counterbalance the lack of adaptation for their degradation [36,37]. In contrast, the low pH occurring during the initial part of incubation may affect, at a higher extent, the activity of the bacterial species adapted to fibre degradation, causing a lower magnitude of fermentation of structural polysaccharides like cellulose and hemicelluloses [12,38,39].

Contrary to what might be expected, the results of the gas production were not supported by those of DMd. This parameter was especially low compared to the extent of the rumen degradation of starch-rich sources (around 0.70–0.80 [40]) or fibrous sources (ranging from 0.40 to 0.70 [41]). This is difficult to explain, but we have also observed this low response in previous in vitro experiments [42], partly associated with a low pH [13]. Calsamiglia et al. [43] justified similar results by the differences between rumen and in vitro microbial ecosystems, partly because the dilution of inoculum in the latter reduces the extent of the degradation. In contrast to DMd, the concentration of total VFA followed a similar trend than that of gas production, being higher for CI at both sampling times, as has been observed by others [15,32,43]. Calsamiglia et al. [43] did not observe any inoculum effect on the acetate and butyrate proportions, and propionate proportion was higher with the concentrate inoculum, as observed in our study at 10 h incubation. However, differences in the proportion of BCVFA, which resulted from fermentation of protein and branched-chain amino acids [44] were higher with FI between 6 and 10 h, probably because of the fermentation of protein from the alfalfa hay fed to the donor lambs.

The effect of the inoculum source was also observed in microbial diversity, suppporting the recent findings reported by Tapio et al. [45] and Nagata et al. [46] who showed the difference in the rumen microbial population when bulls were fed with forage or concentrate diets.

4.2. Effect of Different Substrates on the In Vitro Fermentation Pattern

Despite the marked differences in the magnitude of fermentation between CI and FI, the fermentation pattern among the substrates was almost the same between both inocula. The results of the measured parameters showed a strong correlation between gas production and the other parameters (pH, VFA and lactic acid concentrations) at 6 h ($n = 36$; adjusted $R^2 = 0.90$; $p < 0.001$). Similarly, at 10 h incubation, the volume of gas produced was strongly correlated with incubation pH and VFA ($n = 36$; adjusted $R^2 = 0.84$; $p < 0.001$). These results confirm that the gas production and, equally, the concentration of total VFA and lactic acid, are the main factors indicating the acidification potential of the incubated substrates [31,47,48]. Citrus pulp had a higher acidification capacity than the other substrates, which is associated with a higher magnitude of fermentation that is manifested in high gas production, as well as VFA and lactic acid concentration. Despite the high concentration of lactic acid with CP at 6 h (Table 2), it did not achieve the range considered as a risk of acidosis in vivo [30] and, in fact, did not promote the values of incubation pH below 5.5 that are considered as a threshold for the onset of subacute acidosis [49]. These results were in agreement with those found by Amanzougarene et al. [42] in a batch culture with a minimum buffer concentration, and could be associated with its richness in soluble sugars [50,51], estimated as 0.24 g/kg DM (Table 1), which are fermented at a very fast rate. Although CP also has a high proportion of soluble fibre (0.42, Table 1), this response cannot be directly associated with the fast fermentation of pectin [37,52], since BP includes a similar NDSF proportion (0.46) and it was fermented at a slower rate and magnitude. In fact, Strobel and Russel [53] reported that at a pH of 6.00, the extent of pectin fermentation was reduced with respect to a higher pH. The lower fermentation rate of BP and, thus, its lower acidification potential can also be related to its high NDF content, which does not ensure its maximum fermentation in the 24 h incubation period [54]. Considering the aforementioned characteristics of BP composition, mainly its high NDF and NDSF proportions as well as its low sugar content, its lower concentration of lactic acid produced with respect to the other incubated substrates could be expected. Others [52,53] have also stated that the yield of lactic acid production from pectins fermentation is very low.

The extent of fermentation of WB and B was lower than that of CP, but higher than those of the remaining substrates, probably linked to the high proportion of rapidly fermentable starch in these substrates, compared with those of M and S. Nocek and Tamminga, [55] indicated that 0.80 to 0.90 of barley or wheat starch is digested in the rumen, compared to only 0.55 to 0.70 of that of corn and sorghum. In addition, WB have a considerable amount of NDSF and highly fermentable NDF (Table 1). The structure of the starch endosperm of maize and sorghum, together with their different proportions of amylose [4], as well as the protein matrix in the endosperm in these cereal species [56] and the presence of phenolic compounds in the brown sorghum [16] explain why the fermentation of starch of barley and wheat bran by ruminal bacteria was higher [4,57,58]. Consequently, the differences in the starch characteristics and fermentation rate promote the response in a medium pH [15]. When incubating several grains in a well buffered medium, Lanzas et al. [59] observed a higher fractional rate of 48 h gas production with barley than maize and sorghum varieties (on average, 0.24, 0.15 and 0.06/h). Opatpatanakit et al. [60], modified the incubation pH similarly to the present work, and also observed the highest gas production with barley, intermediate with maize and lowest with sorghum (on average, 222, 138 and 104 mL/g DM, respectively), under pH values at 7 h incubation ranging from 5.7 to 6.1 for barley, 6.5 to 6.9 for maize and 6.5 to 6.8 for sorghum.

From our findings, it can be indicated that citrus pulp and, to a lower extent, wheat bran had an acidic capacity of an even higher magnitude than cereal sources, including barley. Despite the differences on the magnitude and extent of fermentation between the different incubated substrates, the results of microbial diversity with both inocula showed the major effect of the donor animal on this parameter, partly because of aspects related to in vitro methodology, such as the short period of incubation. These results were in accordance with those reported by Taxis et al. [61] and Söllinger et al. [62], explaining the differences in microbial diversity from one animal to another. However, within each series (donor animal), our results did not demonstrate differences between the substrates.

5. Conclusions

Under the fermentation conditions of high-concentrate feeding, some sources of highly fermentable fibre, such as citrus pulp and, to a lower extent, wheat bran may create a more acidic environment than cereals. Among these, barley promotes a lower pH than maize or sorghum, as this grain is associated with a higher rate and extent of fermentation. The rumen environment promoted by high forage/fibre diets is not adapted for non-fibrous carbohydrates, and fermentation of soluble fibre is not differentially enhanced, producing a lower extent of substrate fermentation than concentrate diets. Therefore, the choosing of ingredients is important when ruminants are changed from a forage to a high-concentrate diet, although this cannot be inferred from this study. In any case, in this experiment, acidification levels did not reach those that may change the fermentation pattern. Care must be taken in substrate comparison in terms of gas production, since the buffering of the medium under low pH conditions may overestimate the fermentation differences by increasing the indirect gas production. The in vitro semicontinuous system, adapted to a variable medium pH, has proven to be useful for the study of rumen microbial fermentation under intensive feeding conditions.

Author Contributions: Conceptualization, M.F.; methodology, Z.A. and M.F.; software, Z.A. and M.F.; validation, Z.A., S.Y. and M.F.; formal analysis, Z.A. and M.F.; investigation, Z.A., M.F. and S.Y.; resources, M.F.; data curation, M.F.; writing—original draft preparation, Z.A.; writing—review and editing, M.F.; visualization, Z.A.; supervision, M.F.; project administration, M.F.; funding acquisition, M.F. All authors have read and approved the final manuscript.

Funding: Work financed by the Spanish Government (MINECO project AGL 2013-46820), with the participation of the Department of Industry and Innovation of the Government of Aragón and the European Social Fund. S. Yuste was supported by a FPU fellowship from the Ministerio de Educación, Cultura y Deporte (Spanish Government).

Acknowledgments: Sincere appreciation to the lab team for their support in the laboratory work.

Conflicts of Interest: There were no conflicts of interest.

References

1. Bevans, D.W.; Beauchemin, K.A.; Schwartzkopf-Genswein, K.S.; McKinnon, J.J.; McAllister, T.A. Effect of rapid or gradual grain adaptation on subacute acidosis and feed intake by feedlot cattle. *J. Anim. Sci.* **2005**, *83*, 1116–1132. [CrossRef] [PubMed]
2. Fernando, S.C.; Purvis, H.T.; Najar, F.Z.; Sukharnikov, L.O.; Krehbiel, C.R.; Nagaraja, T.G.; Roe, B.A.; de Silva, U. Rumen microbial population dynamics during adaptation to a high-grain diet. *Appl. Environ. Microbiol.* **2010**, *76*, 7482–7490. [CrossRef] [PubMed]
3. O'Brien, L. Genotype and environment effects on feed grain quality. *Aust. J. Agric. Res.* **1999**, *50*, 703–719. [CrossRef]
4. Offner, A.; Bach, A.; Sauvant, D. Quantitative review of *in situ* starch degradation in the rumen. *Anim. Feed Sci. Technol.* **2003**, *106*, 81–93. [CrossRef]
5. De Peeters, E.J.; Fadel, J.G.; Arosemena, A. Digestion kinetics of neutral detergent fiber and chemical composition within some selected by-product feedstuffs. *Anim. Feed Sci. Technol.* **1997**, *67*, 127–140. [CrossRef]
6. Maes, C.; Delcour, J.A. Alkaline hydrogen peroxide extraction of wheat bran non-starch polysaccharides. *J. Cereal Sci.* **2001**, *34*, 29–35. [CrossRef]
7. Mould, F.L.; Kliem, K.E.; Morgan, R.; Mauricio, R.M. *In vitro* microbial inoculum: A review of its function and properties. *Anim. Feed Sci. Technol.* **2005**, *123–124*, 31–50. [CrossRef]
8. McAllister, T.A.; Cheng, K.J.; Rode, L.M.; Forsberg, C.W. Digestion of barley, maize, and wheat by selected species of ruminal bacteria. *Appl. Environ. Microbiol.* **1990**, *56*, 3146–3153. [CrossRef]

9.   Klieve, A.V.; Hennessy, D.; Ouwerkerk, D.; Forster, R.J.; Mackie, R.I.; Attwood, G.T. Establishing populations of Megasphaera elsdenii YE34 and Butyrivibrio fibrisolvens YE44 in the rumen of cattle fed high grain diets. *J. Appl. Microbiol.* **2003**, *95*, 621–630. [CrossRef]

10. Dijkstra, J.; Kebreab, E.; Bannink, A.; France, J.; López, S. Application of the gas production technique to feed evaluation systems for ruminants. *Anim. Feed Sci. Technol.* **2005**, *123–124*, 561–578. [CrossRef]

11. Raab, L.; Cafantaris, B.; Jilg, T.; Menke, K.H. Rumen protein degradation and biosynthesis. A new method for determination of protein degradation in rumen fluid *in vitro*. *Br. J. Nutr.* **1983**, *50*, 569–582. [CrossRef]

12. Sari, M.; Ferret, A.; Calsamiglia, S. Effect of pH on *in vitro* microbial fermentation and nutrient flow in diets containing barley straw or non-forage fiber sources. *Anim. Feed Sci. Technol.* **2015**, *200*, 17–24. [CrossRef]

13. Bertipaglia, L.M.A.; Fondevila, M.; van Laar, H.; Castrillo, C. Effect of pelleting and pellet size of a concentrate for intensively reared beef cattle on *in vitro* fermentation by two different approaches. *Anim. Feed Sci. Technol.* **2010**, *159*, 88–95. [CrossRef]

14. Amanzougarene, Z.; Fondevila, M. Fitting of pH conditions for the study of concentrate feeds fermentation by the *in vitro* gas production technique. *Anim. Prod. Sci.* **2018**, *58*, 1751–1757. [CrossRef]

15. Amanzougarene, Z.; Yuste, S.; Castrillo, C.; Fondevila, M. *In vitro* acidification potential and fermentation pattern of different cereal grains incubated with inoculum from animals given forage or concentrate-based diets. *Anim. Prod. Sci.* **2018**, *58*, 2300–2307. [CrossRef]

16. Amanzougarene, Z.; Yuste, S.; de Vega, A.; Fondevila, M. Differences in nutritional characteristics of three varieties of sorghum grain determine their *in vitro* rumen fermentation. *Span. J. Agric. Res.* **2018**, *16*. [CrossRef]

17. Krause, K.M.; Combs, D.K. Effects of forage particle size, forage source and grain fermentability on performance and ruminal pH in mid lactation cows. *J. Dairy Sci.* **2003**, *86*, 1382–1397. [CrossRef]

18. Fondevila, M.; Pérez-Espés, B. A new in vitro system to study the effect of liquid phase turnover and pH on microbial fermentation of concentrate diets for ruminants. *Anim. Feed Sci. Technol.* **2008**, *144*, 196–211. [CrossRef]

19. Prates, A.; de Oliveira, J.A.; Abecia, L.; Fondevila, M. Effects of preservation procedures of rumen inoculum on *in vitro* microbial diversity and fermentation. *Anim. Feed Sci. Technol.* **2010**, *155*, 186–193. [CrossRef]

20. Mould, F.L.; Morgan, R.; Kliem, K.E.; Krystallidou, E. A review and simplification of the *in vitro* incubation medium. *Anim. Feed Sci. Technol.* **2005**, *123–124*, 155–172. [CrossRef]

21. AOAC. *Official Methods of Analysis*, 18th ed.; Horwitz, W., Latimer, G.W., Eds.; Association of Official Analytical Chemists: Gaithersburg, MD, USA, 2005.

22. Mertens, D.R. Gravimetric determination of amylase-treated neutral detergent fiber in feeds with refluxing in beakers or crucibles: Collaborative study. *J. AOAC Int.* **2002**, *85*, 1217–1240. [PubMed]

23. Robertson, J.B.; Van Soest, P.J. The detergent system of analysis and its application to human foods. In *The Analysis of Dietary Fiber in Foods*; James, W.P.T., Theander, O., Eds.; Marcel Dekker: New York, NY, USA, 1981; pp. 123–158.

24. Hall, M.B.; Lewis, B.A.; Van Soest, P.J.; Chase, L.E. A simple method for estimation of neutral detergent-soluble fibre. *J. Sci.Food Agric.* **1997**, *74*, 441–449. [CrossRef]

25. Makkar, H.P.S.; Blummel, M.; Borowy, N.K.; Becker, K. Gravimetric determination of tannins and their correlations with chemical and protein precipitation methods. *J. Sci. Food Agric.* **1993**, *61*, 161–165. [CrossRef]

26. Barker, S.B.; Summerson, W.H. The colorimetric determination of lactic acid in biological material. *J. Biol. Chem.* **1941**, *138*, 535–554.

27. Hongoh, Y.; Yuzawa, H.; Ohkuma, M.; Kudo, T. Evaluation of primers and PCR conditions for the analysis of 16S rRNA genes from a natural environment. *FEMS Microbiol. Lett.* **2003**, *221*, 299–304. [CrossRef]

28. De la Fuente, G.; Belanche, A.; Girwood, S.E.; Pinloche, E.; Wilkinson, T.; Newbold, C.J. Pros and cons of ion-torrent next generation sequencing versus terminal restriction fragment length polymorphism T-RFLP for studying the rumen bacterial community. *PLoS ONE* **2014**, *9*, e101435. [CrossRef]

29. Analytical Software. *Statistix 10 for Windows*; Analytical Software: Tallahasee, FL, USA, 2010.

30. Nagaraja, T.G.; Titgemeyer, E.C. Ruminal acidosis in beef cattle: The current microbiological and nutritional outlook. *J. Dairy Sci.* **2007**, *90*, E17–E38. [CrossRef]

31. Broudiscou, L.P.; Offner, A.; Sauvant, D. Effects of inoculum source, pH, redox potential and headspace di-hidrogen on rumen *in vitro* fermentation yields. *Animal* **2014**, *8*, 931–937. [CrossRef]

32. Kim, S.H.; Mamuad, L.L.; Kim, E.J.; Sung, H.G.; Bae, G.S.; Cho, K.K.; Lee, C.; Lee, S.S. Effect of different concentrate diet levels on rumen fluid inoculum used for determination of in vitro rumen fermentation, methane concentration, and methanogen abundance and diversity. *Ital. J. Anim. Sci.* **2018**, *17*, 359–367. [CrossRef]

33. Hiltner, P.; Dehority, B.A. Effect of soluble carbohydrates on digestion of cellulose by pure cultures of rumen bacteria. *Appl. Environ. Microbiol.* **1983**, *46*, 642–648. [CrossRef]

34. Menke, K.H.; Steingass, H. Estimation of the energy feed value obtained from chemical analysis and in vitro gas production using rumen fluid. *Anim. Res. Dev.* **1988**, *28*, 7–55.

35. Nagadi, S.; Herrero, M.; Jessop, N.S. The influence of diet of the donor animal on the initial bacterial concentration of ruminal fluid and in vitro gas production degradability parameters. *Anim. Feed Sci. Technol.* **2000**, *87*, 231–239. [CrossRef]

36. Hatfield, R.D.; Weimer, P.J. Degradation characteristics of isolated and in situ cell wall lucerne pectic polysaccharides by mixed ruminal microbes. *J. Sci. Food Agric.* **1995**, *69*, 185–196. [CrossRef]

37. Barrios-Urdaneta, A.; Fondevila, M.; Castrillo, C. Effect of supplementation with different proportions of barley grain or citrus pulp on the digestive utilization of ammonia-treated straw by sheep. *Anim. Sci.* **2003**, *76*, 309–317. [CrossRef]

38. Mould, F.L.; Ørskov, E.R. Manipulation of rumen fluid pH and its influence on cellulolysis *in sacco*, dry matter degradation and the ruminal microflora of sheep offered either hay or concentrate. *Anim. Feed Sci. Technol.* **1983**, *10*, 1–14. [CrossRef]

39. Grant, R.H.; Mertens, D.R. Influence of buffer pH and raw corn starch addition on in vitro fiber digestion kinetics. *J. Dairy Sci.* **1992**, *75*, 2762–2768. [CrossRef]

40. Cerneau, P.; Michalet-Doreau, B. In situ starch degradation of different feeds in the rumen. *Reprod. Nutr. Dev.* **1991**, *31*, 65–72. [CrossRef]

41. Demarquilly, C.; Andrieu, J. Les fourrages. In *Alimentation des Bovins Ovins et Caprins*; INRA: Paris, France, 1988; Volume 3, pp. 57–74.

42. Amanzougarene, Z.; Yuste, S.; de Vega, A.; Fondevila, M. *In vitro* fermentation pattern and acidification, potential of different sources of carbohydrates by ruminants giving high concentrate diets. *Span. J. Agric. Res.* **2017**, *15*, e0602. [CrossRef]

43. Calsamiglia, S.; Cardozo, P.W.; Ferret, A.; Bach, A. Changes in rumen microbial fermentation are due to a combined effect of type of diet and pH. *J. Anim. Sci.* **2008**, *86*, 702–711. [CrossRef]

44. Saro, C.; Ranilla, M.J.; Tejido, M.L.; Carro, M.D. Influence of forage type in the diet of sheep on rumen microbiota and fermentation characteristics. *Livest. Sci.* **2014**, *160*, 52–59. [CrossRef]

45. Tapio, I.; Fischer, D.; Blasco, L.; Tapio, M.; Wallace, R.J.; Bayat, A.R.; Ventto, L.; Kahala, M.; Negussie, E.; Shingfield, K.J.; et al. Taxon abundance, diversity, co-occurrence and network analysis of the ruminal microbiota in response to dietary changes in dairy cows. *PLoS ONE* **2017**, *12*, e0180260. [CrossRef] [PubMed]

46. Nagata, R.; Kim, Y.H.; Ohkubo, A.; Kushibiki, S.; Ichijo, T.; Sato, S. Effects of repeated subacute ruminal acidosis challenges on the adaptation of the rumen bacterial community in Holstein bulls. *J. Dairy Sci.* **2018**, *101*, 4424–4436. [CrossRef] [PubMed]

47. Russell, J.B.; Hino, T. Regulation of lactate production in streptococcus bovis—A spiraling effect that contributes to rumen acidosis. *J. Dairy Sci.* **1985**, *68*, 1712–1721. [CrossRef]

48. Sauvant, D.; Giger-Reverdin, S.; Meschy, F. Le contrôle de l'acidose ruminale latente. *INRA Prod. Anim.* **2006**, *19*, 69–78.

49. Krause, K.M.; Oetzel, G.R. Understanding and preventing subacute ruminal acidosis in dairy herds: A review. *Anim. Feed Sci. Technol.* **2006**, *126*, 215–236. [CrossRef]

50. Hall, M.B.; Pell, A.N.; Chase, L.E. Characteristics of neutral detergent-soluble fiber fermentation by mixed ruminal microbes. *Anim. Feed Sci. Technol.* **1998**, *70*, 23–39. [CrossRef]

51. Ariza, P.; Bach, A.; Stern, M.D.; Hall, M.B. Effects of carbohydrates from citrus pulp and hominy feed on microbial fermentation in continuous culture. *J. Anim. Sci.* **2001**, *79*, 2713–2718. [CrossRef]

52. Bampidis, V.A.; Robinson, P.H. Citrus by-products as ruminant feeds: A review. *Anim. Feed Sci. Technol.* **2006**, *128*, 175–217. [CrossRef]

53. Strobel, H.J.; Russel, J.B. Effects of pH and energy spilling on bacterial protein synthesis by carbohydrate-limited cultures of mixed rumen bacteria. *J. Dairy Sci.* **1986**, *69*, 2941–2947. [CrossRef]

54. Sauvant, D.; Dorleans, M.; Delacour, C.; Bertrand, D.; Giger, S. La modélisation des cinétiques de dégradation des constituants pariétaux et cellulaires des aliments dans le rumen. *Reprod. Nutr. Dev.* **1986**, *26*, 303–304. [CrossRef]

55. Nocek, J.E.; Tamminga, S. Site of digestion of starch in the gastrointestinal tract of dairy cows and its effect on milk yield and composition. *J. Dairy Sci.* **1991**, *74*, 3598–3629. [CrossRef]

56. McAllister, T.A.; Philippe, R.C.; Rode, L.M.; Cheng, K.J. Effect of the protein matrix on the digestion of cereal grains by ruminal microorganisms. *J. Anim. Sci.* **1993**, *71*, 205–212. [CrossRef] [PubMed]

57. Overton, T.R.; Cameron, M.R.; Elliott, J.P.; Clark, J.H.; Nelson, D.R. Ruminal fermentation and passage of nutrients to the duodenum of lactating cows fed mixtures of corn and barley. *J. Dairy Sci.* **1995**, *78*, 1981–1998. [CrossRef]

58. Firkins, J.L.; Eastridge, M.L.; St-Pierre, N.R.; Noftsger, S.M. Effects of grain variability and processing on starch utilization by lactating dairy cattle. *J. Anim. Sci.* **2001**, *79*, E218–E238. [CrossRef]

59. Lanzas, C.; Fox, D.G.; Pell, A.N. Digestion kinetics of dried cereal grains. *Anim. Feed Sci. Technol.* **2007**, *136*, 265–280. [CrossRef]

60. Opatpatanakit, Y.; Kellaway, R.C.; Lean, I.J.; Annison, G.; Kirby, A. Microbial fermentation of cereal grains in vitro. *Aust. J. Agric. Res.* **1994**, *45*, 1247–1263. [CrossRef]

61. Taxis, T.M.; Wolff, S.; Gregg, S.J.; Minton, N.O.; Zhang, C.; Dai, J.; Schnabel, R.D.; Taylor, J.F.; Kerley, M.S.; Pires, J.C.; et al. The players may change but the game remains: Network analyses of ruminal microbiomes suggest taxonomic differences mask functional similarity. *Nucleic Acids Res.* **2015**, *43*, 9600–9612. [CrossRef]

62. Söllinger, A.; Tveit, A.T.; Poulsen, M.; Noel, S.J.; Bengtsson, M.; Bernhardt, J.; Frydendahl Hellwing, A.L.; Lund, P.; Riedel, K.; Schleper, C.; et al. Holistic assessment of rumen microbiome dynamics through quantitative metatranscriptomics reveals multifunctional redundancy during key steps of anaerobic feed degradation. *mSystems* **2018**, *3*. [CrossRef]

Article

Exploring Additive, Synergistic or Antagonistic Effects of Natural Plant Extracts on In Vitro Beef Feedlot-Type Rumen Microbial Fermentation Conditions

Ignacio Fandiño [1,†], Gonzalo Fernandez-Turren [2], Alfred Ferret [1], Diego Moya [1], Lorena Castillejos [1] and Sergio Calsamiglia [1,*]

[1] Departamento de Ciència Animal i dels Aliments, Universitat Autònoma de Barcelona, Servicio de Nutrición y Bienestar Animal (SNiBA), 08193 Bellaterra, Spain; alfred.ferret@uab.cat (A.F.); diego.moya@usask.ca (D.M.); Lorena.castillejos@uab.cat (L.C.)

[2] Departamento de Producción Animal, Instituto de Producción Animal, Facultad de Veterinaria, Universidad de la República, Ruta 1 km 42, CP 80100 San José, Uruguay; gonzalofernandezt@gmail.com

* Correspondence: sergio.calsamiglia@uab.cat; Tel.:+34-9358-1495; Fax: +34-935-811-494

† Deceased.

Received: 22 December 2019; Accepted: 14 January 2020; Published: 20 January 2020

Simple Summary: Essential oils (EO) can be used as natural alternatives to in-feed antibiotics. Most EO products in the market are based on a combination of EO or their active molecules but prove of additivity or synergy is lacking. The effect of six EO (tea tree oil—TeTr, oregano oil—Ore, clove bud oil—Clo, thyme oil—Thy, rosemary oil—Ros and sage oil—Sag) and different mixes on in vitro microbial fermentation profile of a feedlot beef cattle type fermentation were evaluated for their additive, synergistic or antagonistic effects. Mixing TeTr with Thy, Ore or Thy + Ore modified rumen microbial fermentation profile, but the size of the effect was similar to that obtained with TeTr alone, suggesting that the effects were not additive. When Thy, Ore or Thy + Ore were mixed with Clo, most effects on rumen fermentation profile disappeared even when TeTr was part of the mix, suggesting an antagonistic interaction of Clo with Thy and Ore. Results do not support the hypothesis of additivity among the EO tested, and antagonistic effects may occur among some of them, at least in a low pH, beef-type fermentation conditions.

Abstract: Six Essential oils (EO) (tea tree oil—TeTr, oregano oil—Ore, clove bud oil—Clo, thyme oil—Thy, rosemary oil—Ros, and sage oil—Sag) in Experiment 1; and different combinations of selected oils in Experiment 2, were evaluate at four doses in an in vitro microbial fermentation system using ruminal fluid from beef cattle fed a 10:90 straw: Concentrate diet. In Experiment 1, TeTr, Ore, Clo and Thy improved rumen fermentation profile in a direction consistent with better feed utilization. In Experiment 2, TeTr mixed with Thy, Ore, Thy + Ore or Clo at 200 and 400 mg/L increased the molar proportion of propionate and decreased that of acetate, and the acetate to propionate ratio. However, the size of the effect was similar to that obtained with TeTr alone, suggesting that effects were not additive. When Thy, Ore or Thy + Ore where mixed with Clo, most effects on rumen fermentation profile disappeared, suggesting an antagonistic interaction of Clo with Thy and Ore. Results do not support the hypothesis of additivity among the EO tested, and antagonistic effects of Clo mixed with Thy or Ore were demonstrated at least in a low pH, beef-type fermentation conditions.

Keywords: essential oils; rumen microbial fermentation; synergies

---

1. Introduction

Essential oils (EO) are aromatic oily liquids obtained from plants that can be used as natural alternatives to in-feed antibiotics [1,2]. Many different plant EO and their active components have been tested for their effects on ruminal microbial fermentation profile [3–5]. Because of the diverse mechanisms of action and effects, Calsamiglia et al. [1] suggested that the wise selection and combination of different EO may enhance ruminal fermentation. When combinations of EO are used, the effect may be additive, synergistic or antagonistic. Most EO products for ruminants in the market are based on combination of different EO or their active molecules (i.e., CRINA-Ruminants®, DSM, Switzerland; AGOLIN®, Agolin Sa, Switzerland, XTRACT®, Pancosma, Switzerland), but proof of additivity or synergy is lacking. Some studies attributed additive and synergistic effects to phenolic and alcohol compounds. In general, compounds with similar structures exhibit additive rather than synergistic effects. Antagonistic effects have been attributed to the interaction between non-oxygenated and oxygenated monoterpene hydrocarbons [6]. All these interactions have been demonstrated in cosmetic and food industry [6–8]. However, it is surprising that no studies have been specifically designed to prove additivity or synergies of different EO on rumen microbial fermentation. The mechanism of action of most EO is mediated through their interaction with the cell membrane, and this interaction is dependent on the fermentation conditions like the source of rumen fluid, the substrate of fermentation and pH [1,5]. The additive, synergistic or antagonistic effects in EO mixes may also be dependent on these fermentation conditions. While most research in vitro has been conducted using rumen fluid from dairy cattle, forage as a main fermentation substrate and pH above 6.5, research in high concentrate feedlot-type beef fermentation conditions is more limited.

We hypothesize that the combination of different EO, in particular simple phenolic-type molecules of the family of monoterpenoids and phenylpropanes, will result in additive or synergistic effects. The objective of the present study was to evaluate the effects of different doses of six EO and the additive, synergistic or antagonistic effects of their combinations under feedlot beet-type ruminal microbial fermentation conditions in vitro.

2. Materials and Methods

2.1. Experimental Protocol and Treatments

The research protocol was approved by Campus Laboratory Animal Care Committee of the Universitat Autónoma of Barcelona (Spain) (reference CEAAH 00604). The effects of six EO and their combinations were evaluated in an in vitro batch fermentation system. Rumen fluid was obtained from four beef heifers fed a 10:90 barley straw:concentrate diet (161 g/kgcrude protein, 321 g/kg neutral detergent fiber, 168 g/kg acid detergent fiber on a dry matter basis) designed to meet or exceed nutrient recommendations [9] of growing beef cattle. The concentrate consisted of (DM basis) ground barley grain (444 g/kg), corn gluten feed (155 g/kg), ground corn grain (133 g/kg), soybean meal (72 g/kg), sunflower meal (72 g/kg), ground wheat grain (69 g/kg), palm kernel (52 g/kg), vitamin A (4600 IU/kg), vitamin D (450 IU/kg), vitamin E (7.5 mg/kg), zinc (10.5 mg/kg), iron (6 mg/kg), manganese (1.2 mg/kg), copper (0.75 mg/kg), cobalt (0.15 mg/kg), iodine (0.11 mg/kg) and selenium (0.08 mg/kg). Rumen fluid was strained through two layers of cheesecloth and mixed in a 1:1 proportion with a phosphate-bicarbonate buffer [10]. The pH of the buffered fluid was adjusted to 6.20 with hydrochloric acid. Incubations were conducted in a 75 ml tubes containing 50 ml of the diluted rumen fluid and 0.5 g of the same diet fed to the donor animals. The diet was ground through a 2 mm screen, and each tube was gassed with CO_2 before sealing with rubber corks with a gas release valve. Incubations were conducted in a water bath at 39 °C for 24 h in two consecutive periods with triplicates within period.

In Experiment 1, four doses of different EO (10, 50, 200 and 400 mg/L of the culture fluid) were evaluated. Most EO have shown to be effective at doses around 200–400 mg/L [1,3,5]. Doses were selected based on this evidence and at lower doses to show potential additivity and synergies of effects. Treatments were a negative control without EO (CTR), a positive control with Monensin at

10 mg/L (M5273 Sigma-Aldrich Chemical, St Louis, MO, USA) and six different phenolic-derived monoterpene or phenylpropane EO, of which two were oxygenated and four non-oxygenated: 1) Tea tree oil (*Melaleuca alternifolia*; TeTr; containing 42% terpinen-4-ol as active component), Oregano oil (*Origanum vulgare*; Ore; containing 53% carvacrol as active components), Clove bud oil (*Syzygium aromaticum*; Clo; containing 82% eugenol as main active component), Thyme oil (*Thymus vulgaris*; Thy; containing 47% thymol as main active component), Rosemary oil (*Rosmarinus officinalis*; Ros; containing 61% 1,8-cineole as main active component) and Sage oil (*Salvia officinalis*; Sag; containing 65% thujone as main active component). In Experiment 2, four EO selected from Experiment 1 (TeTr, Thy, Ore and Clo) were used in 13 different combinations at four concentrations (10, 50, 200 and 400 mg/L of the culture fluid; Table 1) to test additive, synergistic or antagonist effects.

Table 1. Treatments and combination of essential oils evaluated in *in vitro* fermentation.

Treatment Number	Combination of Essential Oils (%)			
	Tea-tree	Clove Bud	Thyme	Oregano
1	75	25	-	-
2	75	-	25	-
3	75	-	-	25
4	50	50	-	-
5	50	-	25	25
6	50	25	25	-
7	50	25	-	25
8	25	50	25	-
9	25	50	-	25
10	25	75	-	-
11	-	75	25	-
12	-	75	-	25
13	-	50	25	25

2.2. Sample Collection and Chemical Analyses

After 24 h, the fermented fluid pH was measured immediately (Model 507; Crison Instruments, Alella, Barcelona, Spain) and liquid samples were withdrawn from each tube to analyze ammonia-N and volatile fatty acid (VFA) concentrations. Samples for ammonia-N analysis (5 mL of fermentation fluid preserved with 1 mL of a 50 g/L orthophosphoric acid solution) were centrifuged at 15,000× *g* for 15 min at 4 °C, and the supernatant was analyzed for ammonia-N by colorimetry [11]. Samples for VFA analysis were prepared as described by Jouany [12]. Samples of fermented fluid (5 mL) were preserved with 1 mL of a solution made of 2 g/L of mercuric chloride, 35 g/L orthophosphoric acid and 2 g/L of 4-methylvaleric acid as an internal standard, and frozen. After thawing, samples were centrifuged at 3000× *g* for 30 min, and the supernatant fraction was analyzed by gas chromatography (Model 6890; Hewlett Packard, Palo Alto, CA, USA) using a polyethylene glycol nitroterephthalic acid-treated capillary column (BP21; SGE, Europe Ltd., Buckinghamshire, UK) at 275 °C in the injector and a gas flow rate of 29.9 mL/min.

2.3. Statistical Analyses

Results of the batch fermentation experiment were analyzed as a randomized complete block design using the PROC MIXED procedure (SAS, version 9.4, SAS Institute Inc., Cary, NC, USA), where the essential oils or mixes were the fixed effect and the period was considered a random effect. Because the objective was to identify the lowest dose of EO at which effects were observed, the Dunnett test was used to identify significant differences ($p < 0.05$). When required, the synergistic, additive or antagonistic effects of mixes were evaluated comparing the relative response of the mixes vs. the corresponding effects of the main essential oil using a paired t-test at a significant level of $p < 0.05$.

3. Results

Because of the large number of treatments tested, two runs in Experiment 1, and six runs in Experiment 2 were conducted. To account for possible run-to-run variation, a negative (no additive) control was used, and results are reported as percent change compared with this negative control (Tables 2 and 3).

3.1. Experiment 1 (Individual Essential Oils)

Results from Experiment 1 are shown in Table 2. Monensin was used as a positive control and always increased ($p < 0.05$) the proportion of propionate and decreased ($p < 0.05$) the molar proportion of acetate and the acetate to propionate (A:P) ratio. In most cases, monensin also reduced the molar proportion of butyrate and brach-chain VFA (BCVFA), and the concentration of ammonia-N when compared with control.

None of the EO treatments affected rumen fermentation profile at 10 mg/L compared with control. Tea tree oil at 50 and 200 mg/L decreased ($p < 0.05$) the molar proportion of acetate (−9 to −15%) and the A:P ratio (−22 to −36%), and increased ($p < 0.05$) the molar proportion of propionate (+18 to +34%) and butyrate (+14 to +21%). At 400 mg/L TeTr only increased ($p < 0.05$) the molar proportion of butyrate (+14%) compared with control. Oregano oil at 50 and 200 mg/L decreased ($p < 0.05$) the molar proportion of acetate (−9 to −11%) and the A:P ratio (−26 to −9%) and increased ($p < 0.05$) the molar proportion of propionate (+24 to +25%) and the molar proportion of butyrate (+12 to +16%), but at 400 mg/L increased ($p < 0.05$) the molar proportion of butyrate (+17%) and the pH (+2%) compared with control. Clove bud oil at 50 mg/L had no effects on rumen microbial fermentation, but at 200 and 400 mg/L increased ($p < 0.05$) the molar proportion of butyrate (+10 to +18%), and at 400 mg/L, decreased ($p < 0.05$) the molar proportion of propionate (−13%), and increased ($p < 0.05$) the A:P ratio (+15%) and the pH (+2%) compared with control. Thyme oil at 50 and 200 decreased ($p < 0.05$) the molar proportion of acetate (−6%) and the A:P ratio (−25 to −28%), and increased ($p < 0.05$) the molar proportion of propionate (+24 to +30%). At 200 and 400 mg/L Thy decreased ($p < 0.05$) the molar proportion of butyrate (−15 to −21%), but only at 400 mg/L decreased ($p < 0.05$) the total VFA concentration (−4%), the BCVFA concentration (−16%) and the concentration of ammonia-N (−17%), and increased ($p < 0.05$) the molar proportion of acetate (+5%). Rosemary oil at 50 mg/L had no effect on rumen microbial fermentation, but at 200 and 400 mg/L decreased ($p < 0.05$) the molar proportion of butyrate (−14 to −19%) and at 400 mg/L increased ($p < 0.05$) the molar proportion of acetate (−4%) and decreased the total VFA concentration (−5%) compared with control. Sage oil at 50 mg/L had no effect on rumen microbial fermentation, but at 200 mg/L decreased ($p < 0.05$) the molar proportion of butyrate (−17%) and at 400 mg/L, increased ($p < 0.05$) the molar proportion of propionate (+12%), reduced ($p < 0.05$) the A:P ratio (−13%) and the concentration of ammonia-N (−16%), and tended to reduce ($p < 0.10$) the total VFA concentration (−4%) compared with control.

Table 2. Effect of Tea Tree (TeTr), Oregano (Ore), Clove bud (Clo), Thyme (Thy), Rosmarinus officinalis (Ros), Sage oil (Sag) and Monensin (MON) on pH, ammonia-N and total and individual VFA concentrations in vitro fermentation of 10:90 forage to concentrate diet ($n = 6$).

Treatment	pH	NH_3, mg/dL	Total VFA, mM	Acetate, %	Propionate, %	Butyrate, %	Valerate, %	BCVFA [2], mM	A:P [3]
CTR	5.33	13.2	133.2	61.8	18.8	13.4	3.46	3.47	3.28
TeTr10 [a]	5.32	13.2	132.7	61.3	18.1	14.4	3.53	3.41	3.38
TeTr50 [b]	5.33	13.2	129.0	56.5 *	22.2 *	15.2 *	3.56	3.26	2.55 *
TeTr200 [c]	5.33	13.6	127.0	52.4 *	25.2 *	16.2 *	3.53	3.54	2.11 *
TeTr400 [d]	5.38	13.2	127.8	60.2	18.3	15.2 *	3.81	3.15	3.28
Ore10 [a]	5.33	13.7	135.9	59.9	19.9	14.4	3.36	3.32	3.07
Ore50 [b]	5.33	13.5	132.6	56.0 *	23.3 *	15.0 *	3.25	3.29	2.43 *
Ore200 [c]	5.33	13.6	128.7	54.8 *	23.5 *	15.5 *	3.53	3.32	2.33 *
Ore400 [d]	5.44 *	13.6	131.1	60.8	17.6	15.6 *	3.49	3.19	3.44
Clo10 [a]	5.31	13.3	135.9	60.6	19.0	14.6	3.23	3.33	3.19
Clo50 [b]	5.33	13.5	133.0	60.9	18.8	14.6	3.12	3.40	3.24
Clo200 [c]	5.33	13.5	133.1	60.3	19.2	14.7 *	3.22	3.41	3.14
Clo400 [d]	5.44 *	13.9	127.2	61.8	16.4 *	15.8 *	3.46	3.22	3.77 *
MON	5.44 *	11.1 *	130.0	53.9 *	28.2 *	12.5	3.39	2.64 *	1.99 *
SEM [1]	0.02	0.80	2.63	0.95	0.83	0.28	0.14	0.20	0.10
CTR	5.26	17.4	131.5	61.8	20.0	14.3	1.60	2.90	3.14
Thy10 [a]	5.13	17.3	130.1	63.1	20.4	13.0	1.46	2.48	3.14
Thy50 [b]	5.18	17.9	127.0	58.4 *	24.8 *	13.3	1.49	2.59	2.37 *
Thy200 [c]	5.25	16.8	127.0	58.3 *	26.0 *	12.2 *	1.37	2.59	2.26 *
Thy400 [d]	5.37	14.4 *	125.7 *	64.9 *	20.0	11.2 *	1.59	2.44 *	3.30
Ros10 [a]	5.16	16.9	128.2	63.1	20.6	12.9	1.44	2.47	3.11
Ros50 [b]	5.15	17.1	128.0	63.1	20.4	12.9	1.66	2.59	3.14
Ros200 [c]	5.31	17.2	127.1	63.7	20.6	12.3 *	1.39	2.56	3.17
Ros400 [d]	5.31	17.4	125.3 *	64.3 *	20.2	11.6 *	1.49	2.85	3.30
Sag10 [a]	5.22	17.4	128.2	63.7	20.0	12.4	1.37	3.28	3.11
Sag50 [b]	5.19	17.3	129.3	63.7	20.4	12.3	1.37	3.31	3.14
Sag200 [c]	5.13	17.4	127.5	63.7	20.8	11.8 *	1.33	3.34	3.08
Sag400 [d]	5.16	14.5 *	125.9 *	60.7	22.4 *	12.9	1.43	3.16	2.72 *
MON	5.24	11.3 *	134.1	55.9 *	25.8 *	14.6	1.73	2.84	2.20 *
SEM [1]	0.05	0.71	2.00	0.71	0.44	0.70	0.14	0.16	0.08

[1] SEM—standard error of the mean. [2] BCVFA—Branch-chained VFA. [3] A:P—Acetate to propionate ratio. * Means within a column differ from control ($p < 0.05$). [a] 10 mg/L. [b] 50 mg/L. [c] 200 mg/L. [d] 400 mg/L.

Table 3. Effect of different doses of Treatment 1 (T1: 75% Tea tree, 25% Thyme), Treatment 2 (T2: 75% Tea tree, 25% Clove bud), Tretament 3 (T3: 75% Tea tree, 25% Oregano), Treatment 4 (T4; 50% Tea tree, 50% Clove bud), Treatment 5 (T5: 50% Tea tree, 25% Thyme, 25% Oregano), Treatment 6 (T6: 50% Tea tree, 25% Clove bud, 25% Thyme) and Monensin (MON) on pH, ammonia-N and total and individual VFA concentrations in vitro fermentation of 10:90 forage to concentrate diet ($n = 6$).

Treatment	pH	NH$_3$, mg/dL	VFA, mM	Acetate, %	Propionate, %	Butyrate, %	Valerate, %	BCVFA [2], mM	A:P [3]
CTR	5.33	27.5	130.8	62.0	21.0	14.0	1.08	2.47	2.94
T1 (10) [a]	5.32	25.2	129.1	61.5	21.3	14.3	1.05	2.54	2.91
T1 (50) [b]	5.30	25.2	128.2	61.2	21.3	14.7	1.10	2.41	2.87
T1 (200) [c]	5.30	26.5	128.6	58.6 *	23.6 *	14.3	1.20	2.36	2.48 *
T1 (400) [d]	5.28+	25.2	126.5	58.3 *	24.4 *	14.4	1.14	2.40	2.39 *
MON	5.29+	25.2 *	123.5 +	58.9 *	25.9 *	12.2 +	1.29	2.10 *	2.28 *
SEM [1]	0.01	1.34	2.70	0.68	0.53	0.51	0.09	0.05	0.09
CTR	5.38	27.0	116.9	60.1	20.8	15.4	1.24	2.64	2.89
T2 (10) [a]	5.39	24.9	116.7	60.1	21.5	14.6	1.30	2.69	2.82
T2 (50) [b]	5.36	26.3	116.3	59.6	21.5	15.6	1.19	2.58	2.80
T2 (200) [c]	5.38	26.1	114.6	59.6	21.7	15.0	1.38	2.60	2.75 *
T2 (400) [d]	5.37	27.0	115.0	56.9 *	24.2 *	15.4	1.36	2.59	2.36 *
T3 (10) [a]	5.36	27.3	115.7	60.0	20.6	15.6	1.20	2.64	2.88
T3 (50) [b]	5.38	26.6	116.2	59.7	21.5	15.3	1.33	2.62	2.78
T3 (200) [c]	5.36	26.0	114.9	59.3	21.5	14.8	1.16	2.59	2.81
T3 (400) [d]	5.37	26.7	116.5	57.3 *	24.0 *	15.1	1.48	2.58	2.40 *
MON	5.36	22.7	114.2	57.5 *	26.5 *	2.03 *	0.98	2.29 *	2.17 *
SEM [1]	0.01	1.34	2.70	0.68	0.53	0.51	0.09	0.05	0.09
CTR	5.40	24.9	126.0	63.8	18.1	14.6	1.09	2.95	3.53
T4 (10) [a]	5.39	25.6	127.2	62.7	19.0	14.7	1.10	3.01	3.30
T4 (50) [b]	5.34	25.9	127.2	62.3	19.6	14.5	1.26	3.04	3.19
T4 (200) [c]	5.38	24.9	127.2	63.2	18.7	14.6	1.18	2.90	3.41
T4 (400) [d]	5.31	23.0	125.3	59.8 *	22.5 *	14.1	1.14	3.10	2.66 *
T5 (10) [a]	5.37	24.7	124.9	62.5	19.6	14.6	1.11	2.74	3.20
T5 (50) [b]	5.33	25.1	127.2	63.0	18.7	14.9	1.14	2.84	3.37
T5 (200) [c]	5.38	24.2	125.6	60.2 *	21.9 *	14.1	1.28	2.98	2.75 *
T5 (400) [d]	5.35	21.4	127.2	58.8 *	22.7 *	15.2	1.20	2.93	2.60 *
T6 (10) [a]	5.41	27.6	129.2	61.6	20.6	14.9	1.22	2.28	3.02
T6 (50) [b]	5.37	27.7	125.9	61.1	20.8	14.1	1.20	2.37	3.05
T6 (200) [c]	5.39	28.1	128.6	62.1	21.2	14.7	1.21	2.34	2.92
T6 (400) [d]	5.38	27.1	127.2	59.3 *	21.0	14.6	1.28	2.39	2.58
MON	5.37	24.1 *	123.5 +	59.6 *	25.1 *	12.3 +	1.31	2.03 *	2.38 *
SEM [1]	0.01	1.12	2.70	0.90	0.69	0.48	0.10	0.08	0.13

[1] SEM—standard error of the mean. [2] BCVFA—Branch-chained VFA. [3] A:P—Acetate to propionate ratio. * Means within a column differ from control ($p < 0.05$). [a] 10 mg/L. [b] 50 mg/L. [c] 200 mg/L. [d] 400 mg/L.

3.2. Experiment 2 (Combination of Essential Oils)

Treatments 7 to 13 had no effect on rumen fermentation profile and are not reported. The effects of treatments 1 to 6 are shown in Table 3. Monensin reduced ($p < 0.05$) the molar proportion of acetate, the molar proportion of butyrate, the BCVFA concentrations, the A:P ratio and the concentration of ammonia-N, and increased ($p < 0.05$) the molar proportion of propionate compared with control. At 10 and 50 mg/L, none of the treatments had effects on rumen fermentation profile compared with control.

Treatment 1 at 200 and 400 mg/L decreased ($p < 0.05$) the molar proportion of acetate (-5 to -6%) and the A:P ratio (-16 to -19%), and increased ($p < 0.05$) the molar proportion of propionate ($+12$ to $+16\%$). Treatment 1 at 400 mg/L tended to reduce ($p < 0.10$) the pH compared with control Treatment 2 only at 400 mg/L decreased ($p < 0.05$) the molar proportion of acetate (-5%) and the A:P ratio (-18%), and increased ($p < 0.05$) the propionate proportion ($+16\%$) when compared with control. Treatment 3 only at 400 mg/L decreased ($p < 0.05$) the molar proportion of acetate (-4%) and the A:P ratio (-17%), and increased ($p < 0.05$) the molar proportion of propionate ($+15\%$) compared with control. Treatment 4 had no effect at 200 mg/L but at 400 mg/L decreased ($p < 0.05$) the molar proportion of acetate (-6%) and the A:P ratio (-25%), and increased ($p < 0.05$) the molar proportion of propionate ($+24\%$) compared with control. Treatment 5 at 200 and 400 mg/L decreased ($p < 0.05$) the molar proportion of acetate (-6 to -8%) and the A:P ratio (-22 to $+26\%$), and increased ($p < 0.05$) the molar proportion of propionate ($+21$ to $+25\%$) when compared with control. Treatment 6 had no effect at 200 mg/L, and only the molar proportion of acetate decreased (-5%; $p < 0.05$) at 400 mg/L compared with the control. When mixes modified rumen microbial fermentation (T1 to T6), the relative effects of the mixes of EO were compared with the closest doses of TeTr oil in Experiment 1, and the size of the effect increased, suggesting no additive or synergistic effects.

4. Discussion

In the last two decades, there has been a wealth of data produced by research on the effect of EO on rumen microbial fermentation, most of it in in vitro conditions [1,2]. However, most of it has focused on dairy cattle, where fermentation was conducted in high forage diets at pH 6.5–7.0. There are few reports on the effects of EO on ruminal microbial fermentation in high concentrate diets fed to feedlot beef cattle [5,13,14]. The antimicrobial activity of EO is affected by pH, and ruminal pH in cattle fed high-concentrate diets is different from that of dairy cattle diets and often below 6.0 [5]. These effects may also affect potential interactions among different EO in mixes. The selection criteria to identify EO with positive effects on rumen microbial fermentation in these conditions were an increase or no change in total VFA, an increase in the proportion of propionate, and/or a decrease in the proportion of acetate, the A:P ratio and/or the ammonia-N concentration. These effects would improve the efficiency of energy and N utilization [1]. On the other hand, Calsamiglia et al. [1] suggested that, because of different mechanisms of action, the combination of EO may result in additive or synergistic effects. In fact, most EO products in the market are based on combination of different EO or their active molecules (i.e., CRINA-Ruminants®, DSM, Switzerland; AGOLIN®, Agolin Sa, Switzerland, XTRACT®, Pancosma, Switzerland), but scientific evidence of their additive or synergistic effects is lacking.

Monensin has been recognized as an effective in-feed antibiotic that results in an increase in propionate at the expenses of acetate, and reduces amino acid deamination [15,16] and was used as a positive control. The effects of monensin in Experiments 1 and 2 show these effects and confirm the reliability of the method to measure these changes in the selected EO. At similar concentrations as monensin, EO in Experiment 1 and EO mixes in Experiment 2 had no effects on rumen fermentation, even when considering the concentration of active components. This observation indicates that the activity of EO is weaker compared with monensin, and higher doses are required to provoke a change in rumen microbial fermentation.

4.1. Experiment 1 (Individual Essential Oils)

Ammonia-N was only reduced in Thy and Sag at the highest dose, and the magnitude of the reduction was much lower than that observed for monensin. The effect of Thy on ammonia-N concentration agrees with previous reports where Thy was tested in similar conditions and doses [17]. Because Sag had negative effects on VFA and Thy did not affect VFA at the high dose, the effects of EO on ammonia-N could not be considered as a criterion to select EO for Experiment 2.

Thyme and Ore contain high concentrations of thymol and carvacrol as major active components, respectively. Thymol and carvacrol are phenolic monoterpenes with a wide-spectrum antimicrobial activity due to its capacity to act as membrane permeabilizer [18–20]. Juven et al. [21] observed that their antibacterial activity was greater at pH 5.5 compared with the activity at pH 6.5. In the beef-type diet conditions of this experiment, Ore and Thy at 50 and 200 mg/L were affected in the same way as total and individual VFA, reflecting a similar mechanism of action, and are in agreement with previous results in similar conditions [5,17]. The lack or negative effect of these EO up to 500 mg/L also agrees with previous reports [22,23], but higher doses inhibit rumen microbial fermentation. These results indicate that Ore and Thy should be included at moderate doses to modify rumen fermentation profile without inhibiting microbial activity. Because phenolic compounds with similar structure may have additive effects [6], these two EO were selected for further exploration of possible interactions in Experiment 2.

Terpinen-4-ol is the main active component in TeTr and has a wide-spectrum antimicrobial activity [24]. The TeTr at 50 and 200 mg/L decreased the molar proportion of acetate and the A:P ratio and increased the molar proportion of propionate and the molar proportion of butyrate, a fermentation profile consistent with better energy utilization [1]. In fact, at 200 mg/L, the effect on the A:P ratio was the largest among all EO tested in Experiment 1. Results agree with Castillejos et al. [22] using similar fermentation conditions and doses. Although some reports have shown negative effects of TeTr on rumen microbial fermentation [25–27], with a significant reduction in total VFA and propionate concentrations, these trials were carried out with ruminal liquid and conditions for dairy cows and at pH > 6.0. Several authors have already recognized the role of media pH and type of diet on the antimicrobial activity of EO [5,21]. Therefore, results obtained in the present report with high concentrate beef-type fermentation conditions justify the selection of TeTr for further study in Experiment 2.

Clove bud oil at 200 mg/L increased the molar proportion of butyrate, and at 400 mg/l decreased the molar proportion of propionate and increased the molar proportion of butyrate, the A:P ratio and the pH compared with the control. Castillejos et al. [22] also reported that Clo at 500 mg/L reduced the molar proportion of propionate and increased the molar proportion of butyrate and the A:P ratio. This fermentation profile is not adequate to improve energy utilization in beef cattle diets. In the present trial, Clo at moderate dose (10 and 50 mg/L) increased the ammonia-N concentrations in 1% and 2% vs. control, but differences were not significant. In contrast, at lower doses (30 mg/L), Cardozo et al. [5] reported that eugenol increased total VFA by 20% and increased the concentration of ammonia-N, which could be desirable if ammonia-N concentration limits microbial protein synthesis in feedlot cattle fed high percentages of concentrate [28].

Rosemary oil only had effects at the highest dose but reduced total VFA concentrations and the proportion of butyrate, and increased the acetate proportion. This fermentation profile is not desirable in beef cattle conditions, where an increase in propionate is desired [1]. Sage oil at the highest dose decreased the A:P ratio, but at the same time total VFA was also reduced, suggesting that microbial fermentation was inhibited. It is interesting to observe that the active components of Ros and Sag are 1,8-cineole and thujone, respectively, and both are oxygenated phenols. Oxygenated phenols have been reported to have negative or even antagonistic effects when mixed with other phenolic compounds [6], and therefore, were discarded for Experiment 2.

4.2. Experiment 2 (Combination of Essential Oils)

In Experiment 1, TeTr, Ore and Thy at 50 and 200 mg/L had the most appropriate fermentation profile to improve energy utilization in beef cattle diets. The main active components of these EO are phenolic monoterpens. Bassole and Juliani [6] indicated that phenolic compounds tend to have additive effects, while synergistic or antagonistic effects would occur with other chemical compounds and vary depending on the microbial ecosystem. Clove bud oil was selected because previous research suggested that at pH 5.5 and at moderate doses stimulated deamination activity and inhibited peptide lyses [3,5]. Because of the different mechanisms of action involved, synergies may be expected. Furthermore, the main active component, eugenol, is a phenylpropanoid, which may have synergies or antagonistic effects with phenolic monoterpenes [6]. There are no previous reports designed to explore interactions between the selected EO on rumen fermentation profile in beef feedlot-type fermentation conditions.

In Experiment 1, TeTr was the EO with the strongest response at doses of 50 to 200 mg/L. For this reason, most of the mixes proposed contained TeTr at different effective concentrations. Treatments at 200 or 400 mg/L and with 50 or 75% of TeTr (corresponding to an effective dose of TeTr of 100 to 300 mg/L) reduced the proportion of acetate and the A:P ratio, and increased the proportion of propionate without affecting the total concentration of VFA (T1 to T5). The effective doses of TeTr in the mixes were within the range of those tested in Experiment 1, and the magnitude of the response was never higher that the effect observed when TeTr oils was supplemented alone in Experiment 1. Therefore, there is no evidence that the combination of different EO had an additive or synergistic effect neither with monoterpenes, nor with phenolpropanoids. It is also interesting to observe that whenever Ore, Thy or both were mixed with Clo (see T7 to T13, Table 1), the fermentation profile was not affected. In treatments T7 to T13 (Table 1), 100 mg of Thy, Ore or the combination of both resulted in no effect if mixed with 100, 200 or 300 mg of Clo. This is surprising because both Thy and Ore reduced acetate, increase propionate and reduced the A:P ratio when incubated alone in Experiment 1, but this effect was inhibited by the presence of Clo. Treatment T6 contained TeTr, and even in cases where the effective concentration of TeTr was 100 and 200 mg/L that resulted in improved fermentation profile in Experiment 1, there was only a small reduction in the acetate proportion (−5%), no effect on propionate and a non-significant reduction of the A:P ratio. There seem to be an antagonistic effect by which the combination of either Thy, Ore or both with Clo resulted in no effect on rumen fermentation profile.

None of the mixes in Experiment 2 affected ammonia-N concentration or BCVFA, suggesting that deamination was not affected. This is not surprising, because only Thy and Sag at 400 mg in Experiment 1 had effects, and in none of the treatments in Experiment 2 the effective dose of these EO was reached. Clove bud oil was selected because Busquet et al. [27] indicated that it reduced peptide degradation. We hypothesized that the combination of the inhibition of peptide degradation by eugenol in Clo may be synergistic with inhibition of deaminnation shown in Thy. However, the combination of different EO appear no to have this additive or synergistic effect in protein degradation. These results question the common practice of mixing different EO based on potential additive or synergistic effects.

5. Conclusions

Tea tree oil, oregano oil and thyme oil can be effective as rumen fermentation modifiers under feedlot beef-type fermentation conditions. The combinations of TeTr oil with Ore oil, Thy oil or Ore + Thy did not have additive effects over those observed with TeTr oil alone. When Thy, Ore or Thy + Ore where mixed with Clo, all effects on rumen fermentation profile disappeared even when TeTr was part of the mix, suggesting an antagonistic interaction of Clo with Thy and Ore. Results do not support the hypothesis of additivity among the EO tested, and antagonistic effects of Clo mixed with Thy or Ore were demonstrated at least in a low pH, beef-type fermentation conditions.

Author Contributions: Conceptualization, I.F., A.F., D.M., L.C. and S.C.; methodology, I.F., D.M. and S.C.; formal analysis, I.F. and D.M.; resources, S.C. and L.C.; data curation, I.F., G.F.-T. and S.C.; writing—Original draft preparation, I.F and G.F.-T.; writing—Review and editing, A.F., D.M., L.C. and S.C.; supervision, S.C.; project administration, S.C. All authors have read and agreed to the published version of the manuscript.

Funding: This research received no external funding.

Conflicts of Interest: The authors declare no conflict of interest.

References

1. Calsamiglia, S.; Busquet, M.; Cardozo, P.W.; Castillejos, L.; Ferret, A. Invited Review: Essential oils as modifiers of rumen microbial fermentation. *J. Anim. Sci.* **2007**, *90*, 2580–2595. [CrossRef]
2. Benchaar, C.; McAllister, T.A.; Chouinard, P.Y. Digestion, ruminal fermentation, ciliate protozoal populations, and milk production from dairy cows fed cinnamaldehyde, quebracho condensed tannin, or *Yucca schidigera* saponin extracts. *J. Dairy Sci.* **2008**, *91*, 4765–4777. [CrossRef]
3. Busquet, M.; Calsamiglia, S.; Ferret, A.; Kamel, C. Screening for effects of plant extracts and active compounds of plants on dairy cattle rumen microbial fermentation in a continuous culture system. *Anim. Feed Sci. Technol.* **2005**, *123–124*, 597–613. [CrossRef]
4. Castillejos, L.; Calsamiglia, S.; Ferret, A.; Losa, R. Effects of a specific blend of essential oil compounds and the type of diet on rumen microbial fermentation and nutrient flow from a continuous culture system. *Anim. Feed Sci. Technol.* **2005**, *119*, 29–41. [CrossRef]
5. Cardozo, P.W.; Calsamiglia, S.; Ferret, A.; Kamel, C. Screening for the effects of natural plant extracts at different pH on *in vitro* rumen microbial fermentation of a high-concentrate diet for beef cattle. *J. Anim. Sci.* **2005**, *83*, 2572–2579. [CrossRef]
6. Bassolé, I.H.N.; Juliani, H.R. Essential oils in combination and their antimicrobial properties. *Molecules.* **2012**, *17*, 3989–4006. [CrossRef] [PubMed]
7. Hyldgaard, M.; Mygind, T.; Meyer, R.L. Essential oils in food preservation: Mode of action, synergies, and interactions with food matrix components. *Front. Microbiol.* **2012**, *3*, 1–24. [CrossRef]
8. Langeveld, W.T.; Veldhuizen, E.J.A.; Burt, S.A. Synergy between essential oil components and antibiotics: A review. *Crit. Rev. Microbiol.* **2013**, *40*, 76–94. [CrossRef] [PubMed]
9. NRC. *Nutrient Requirements of Beef Cattle*, 7th rev. ed; Nat. Acad. Press: AcadWashington DC, USA, 1996.
10. McDougall, E.I. Studies on ruminant saliva. 1. The composition and output of sheep's saliva. *Biochem. J.* **1948**, *43*, 99–109. [CrossRef] [PubMed]
11. Chaney, A.L.; Marbach, E.P. Modified reagents for determination of urea and ammonia. *Clin. Chem.* **1962**, *8*, 130–132. [PubMed]
12. Jouany, J.P. Volatile fatty acid and alcohol determination in digestive contents, silage juices, bacterial cultures and anaerobic fermentor contents. *Sci. Aliment.* **1982**, *2*, 131–144.
13. Benchaar, C.; Duynisveld, J.L.; Charmley, E. Effects of monensin and increasing dose levels of a mixture of essential oil compounds on intake, digestion and growth performance of beef cattle. *Can. J. Anim. Sci.* **2006**, *86*, 91–96. [CrossRef]
14. Nanon, A.; Suksombat, W.; Yang, W.Z. Effects of essential oils supplementation on *in vitro* and *in situ* feed digestion in beef cattle. *Anim. Feed Sci. Technol.* **2014**, *196*, 50–59. [CrossRef]
15. Wallace, R.J.; Czerkawski, J.W.; Breckenridge, G. Effect of monensin on the fermentation of basal rations in the Rumen Simulation Technique (Rusitec). *Br. J. Nutr.* **1981**, *46*, 131–148. [CrossRef] [PubMed]
16. Chalupa, W.; Corbett, W.; Brethour, J.R. Effects of monensin and amicloral on rumen fermentation. *J. Anim. Sci.* **1980**, *51*, 170–179. [CrossRef] [PubMed]
17. Castillejos, L.; Calsamiglia, S.; Ferret, A. Effect of essential oil active compounds on rumen microbial fermentation and nutrient flow in in vitro systems. *J. Dairy Sci.* **2006**, *89*, 2649–2658. [CrossRef]
18. Dorman, H.J.D.; Deans, S.G. Antimicrobial agents from plants: Antibacterial activity of plant volatile oils. *J. Appl. Microbiol.* **2000**, *88*, 308–316. [CrossRef] [PubMed]
19. Walsh, S.E.; Maillard, J.Y.; Russell, A.D.; Catrenich, C.E.; Charbonneau, D.L.; Bartolo, R.G. Activity and mechanisms of action of selected biocidal agents on Gram-positive and -negative bacteria. *J. Appl. Microbiol.* **2003**, *94*, 240–247. [CrossRef]
20. Lambert, R.J.W.; Skandamis, P.N.; Coote, P.J.; Nychas, G.J.E. A study of the minimum inhibitory concentration and mode of action of oregano essential oil, thymol and carvacrol. *J. Appl. Microbiol.* **2001**, *91*, 453–462. [CrossRef]

21. Juven, B.J.; Kanner, J.; Schved, F.; Weisslowicz, H. Factors that interact with the antibacterial action of thyme essential oil and its active constituents. *J. Appl. Bacteriol.* **1994**, *76*, 626–631. [CrossRef]
22. Castillejos, L.; Calsamiglia, S.; Martín-Tereso, J.; Ter Wijlen, H. *In vitro* evaluation of effects of ten essential oils at three doses on ruminal fermentation of high concentrate feedlot-type diets. *Anim. Feed Sci. Technol.* **2008**, *145*, 259–270. [CrossRef]
23. Macheboeuf, D.; Morgavi, D.P.; Papon, Y.; Mousset, J.L.; Arturo-Schaan, M. Dose–response effects of essential oils on *in vitro* fermentation activity of the rumen microbial population. *Anim. Feed Sci. Technol.* **2008**, *145*, 335–350. [CrossRef]
24. Davidson, P.M.; Naidu, A.S. Phyto-phenols. In *Natural Food Antimicrobial Systems*; CRC Press: Boca Raton, FL, USA, 2000; pp. 265–293.
25. Hristov, A.N.; Ropp, J.K.; Zaman, S.; Melgar, A. Effects of essential oils on *in vitro* ruminal fermentation and ammonia release. *Anim. Feed Sci. Technol.* **2008**, *144*, 55–64. [CrossRef]
26. Gunal, M.; Ishlak, A.; AbuGhazaleh, A.A.; Khattab, W. Essential oils effect on rumen fermentation and biohydrogenation under *in vitro* conditions. *Czech., J. Anim. Sci.* **2014**, *59*, 450–459. [CrossRef]
27. Busquet, M.; Calsamiglia, S.; Ferret, A.; Kamel, C. Plant Extracts Affect *In Vitro* Rumen Microbial Fermentation. *J. Dairy Sci.* **2006**, *89*, 761–771. [CrossRef]
28. Devant, M.; Ferret, A.; Gasa, J.; Calsamiglia, S.; Casals, R. Effects of protein concentration and degradability on performance, ruminal fermentation, and nitrogen metabolism in rapidly growing heifers fed high-concentrate diets from 100 to 230 kg body weight. *J. Anim. Sci.* **2000**, *78*, 1667–1676. [CrossRef]

Article

Combining Orchardgrass and Alfalfa: Effects of Forage Ratios on In Vitro Rumen Degradation and Fermentation Characteristics of Silage Compared with Hay

Zhulin Xue [1], Nan Liu [1], Yanlu Wang [2], Hongjian Yang [2], Yuqi Wei [1], Philipe Moriel [3], Elizabeth Palmer [3] and Yingjun Zhang [1,*]

[1] Key Laboratory of Grasslands Management and Utilization, Ministry of Agriculture and Rural Affairs, College of Grassland Science and Technology, China Agricultural University, Beijing 100193, China; xzl2007.ok@163.com (Z.X.); liunan@cau.edu.cn (N.L.); weiyuqi91@126.com (Y.W.)

[2] State Key Laboratory of Animal Nutrition, College of Animal Science and Technology, China Agricultural University, Beijing 100193, China; yanluwang@yeah.net (Y.W.); yang_hongjian@sina.com (H.Y.)

[3] IFAS-Range Cattle Research and Education Center, University of Florida, Gainesville, FL 32611, USA; pmoriel@ufl.edu (P.M.); e.palmer@ufl.edu (E.P.)

* Correspondence: zhangyj@cau.edu.cn

Received: 9 December 2019; Accepted: 23 December 2019; Published: 28 December 2019

Simple Summary: Forages are an essential portion of ruminant rations to maintain rumen function. Exploring how orchardgrass and alfalfa interact in the rumen is necessary to better understand their feed use potential as both hay and silage. This study evaluated in vitro rumen degradation, fermentation characteristics, and methane production responses to different forage ratios of alfalfa and orchardgrass. The results indicate that dry matter and organic matter degradability and methane production were greater for mixed silages compared to mixed hays. A forage ratio of 50:50 for orchardgrass and alfalfa favor the growth of rumen microorganisms without compromising nutrient digestion and rumen fermentation.

Abstract: This study aimed to investigate the effects of different forage ratios of orchardgrass (*Dactylis glomerata*) and alfalfa (*Medicago sativa*) on in vitro rumen degradation and fermentation characteristics. Orchardgrass and alfalfa were harvested separately and prepared as hay and silage mixtures at ratios of 100:0, 75:25, 50:50, 25:75, and 0:100 (w/w on a dry matter basis) and anaerobically incubated for 48 h with rumen fluid obtained from lactating dairy cows. Fermented residues and cultured fluids were used to determine nutrient degradability, fermentation parameters, and associative effect indices. Increasing the proportion of alfalfa in hay and silage mixtures quadratically increased in vitro organic matter disappearance (IVOMD, up +5.14%) and marginally decreased in vitro neutral detergent fiber disappearance (NDFD, down −1.79%). Meanwhile, increasing the proportion of alfalfa accelerated the rumen fermentation process (e.g., gas production) and remarkably enhanced the growth of rumen microbes as indicated by microbial protein production (MCP, 13.4% increase). Increments of rumen degradability and methane production were more pronounced in silage mixtures than hay mixtures. In combination, a forage ratio of 50:50 for orchardgrass and alfalfa is recommended for both hay and silage in order to improve the feed use potential in ruminants.

Keywords: forage quality; gas production; methane; ruminant

---

1. Introduction

Orchardgrass (*Dactylis glomerata*) is one of the most productive cool-season grasses, fairly drought resistant, and is widely distributed in North America, Europe, and East Asia [1]. Previous studies noted that orchardgrass had greater nutritive value than other forages commonly used as feedstuff for livestock, such as bromegrass, tall fescue, and reed canarygrass [2,3]. However, dietary inclusion of orchardgrass hay and silage in steers and sheep should be around 60–75% to avoid negative effects on rumen fermentation [4,5]. Additionally, polyphenol oxidase activity in orchardgrass was able to decrease protein degradation and lipolysis during aerobic ensiling and rumen fermentation [6]. Alfalfa (*Medicago sativa*) has been shown to be rich in crude protein (CP) as well as rapidly degradable protein in the rumen and high inclusion in the diet usually reduces nitrogen (N) utilization and increases urine N excretion [7,8]. However, it is unclear whether a mixture of orchardgrass and alfalfa has the potential to optimize ruminal digestibility and fermentation characteristics through complementary effects of nutrients in the forages.

Digestive interactions may occur in the rumen when ruminants are fed grass-legume mixtures [9,10] and these interactions are due to the diverse impacts on rumen digestion, fermentation, and metabolism efficiency [11]. Positive or negative interactions indicate that the response of ruminants to grass–legume mixtures is either greater or less than the arithmetical calculation obtained from using the responses for grass or legume alone. It would be helpful to determine the proportion of alfalfa at which maximal benefits can be reached in terms of rumen digestibility and fermentation efficiency in order to more precisely formulate rations [12].

The two most common preservation methods of forages, haymaking or ensiling, result in different nutritional compositions, which coincide with changes in ruminal digestibility and fermentation characteristics. For example, alfalfa preserved as hay has less effective protein degradability and digestibility than alfalfa silage, but replacement of alfalfa hay with alfalfa silage did not change ruminal volatile fatty acids (VFAs) or ammonia N [13,14]. Although some studies have separately evaluated the digestibility and fermentation characteristics of silage or hay mixtures [15,16], digestion parameters of silage in comparison with hay mixtures have not been assessed.

The in vitro digestion and gas production method has been widely used to routinely evaluate the associative effects of basal and supplementary feeds with many combinations and is less expensive and time-consuming than in vivo trials [17]. We hypothesized that mixtures of orchardgrass and alfalfa at suitable ratios would increase rumen degradability and fermentation efficiency, while decreasing methane (CH_4) production compared to alfalfa or orchardgrass alone. Therefore, the objectives of this study were to (1) investigate the associative effects of orchardgrass and alfalfa mixtures on rumen degradation and fermentation characteristics and (2) compare ruminal degradation and fermentation characteristics of hay and silage mixtures.

2. Materials and Methods

2.1. Experimental Site for Forage Planting

The field component of our experiment was conducted in Dongchengfang County (39°37′ N, 115°51′ E) in Zhuozhou City of Hebei Province of China. In September 2014, orchardgrass (cv. Aba) and alfalfa (cv. WL534) were established as monocultures in 2 field plots (21 × 75 m), each with three replicates. Sowing rate was 15 kg/ha and row spacing was 15 cm for alfalfa and orchardgrass monocultures. Monocultures were grown under the same management practices, with weed-free conditions and no insect pests. The sward soil was a sandy loam with organic matter (OM) content of 13.8 g/kg, total N content of 0.850 g/kg, plant-available phosphorus (P) content of 15.42 mg/kg, available potassium (K) content of 80.12 mg/kg and a pH of 7.80. A total of 75 kg/ha N fertilizer was applied annually in the swards.

2.2. Hay and Silage Mixture Preparation

Whole orchardgrass and alfalfa plants were harvested (second cutting) with a reaping hook from the field plots at the jointing and early bloom stages. Ten square areas (1 × 1 m) in each field plot were randomly selected and plants were harvested at a stubble height of 5 cm above ground. The harvested fresh forage was weighed separately and recorded. Representative forage samples (approximately 1 kg) from each square harvest were oven-dried at 65 °C for 48 h to determine the dry matter (DM) yield and chemical composition. Based on the forage DM obtained from four annual harvests, the annual DM yields of orchardgrass and alfalfa averaged 1065 g/m^2 and 1480 g/m^2, respectively. The chemical composition of orchardgrass and alfalfa prior to hay making and ensiling was shown in Table 1. After the forage was harvested, representative samples (approximately 20 kg) of each preservation method were divided equally into two subsamples. The first subsample, of both orchardgrass and alfalfa, was air-dried in the sun for 48 h, oven-dried at 65 °C for 48 h, thoroughly mixed to obtain forage ratios of 100:0, 75:25, 50:50, 25:75, and 0:100 (orchardgrass: alfalfa, w/w), and five representative hay samples of each forage ratio were prepared. All of these hay samples were ground in a mill (FW100; Tianjin Taisite Instrument Co., Ltd., Tianjin, China) to pass through a 1.0 mm sieve and stored in plastic bags for chemical analysis and in vitro batch culturing.

The second subsample, of both orchardgrass and alfalfa, was sun-wilted for 2 h until the DM content increased to 400 g/kg and was chopped into 2.5 cm pieces. A microwave oven (M1-L213B, Midea Group Co., Ltd., Foshan, China) was used for the rapid determination of DM content at regular intervals. The wilted forage samples were mixed at ratios of 100:0, 75:25, 50:50, 25:75, and 0:100 (orchardgrass: alfalfa, w/w) based on DM content and were packed into polyethylene silos (1 L capacity) until the total forage weight reached approximately 750 g. Five silo replicates of each forage ratio were prepared, and all silos were sealed with screw lids and kept in the dark at approximately 25 °C. After 40 days of ensiling, a total of 20 g of each silage mixture was homogenized in a blender with 180 mL of distilled water for 1 min. The samples were filtered through a four-layered cheesecloth to prepare the silage extract for determination of ensiling characteristics such as pH, ammonia N, and organic acids, including lactic acid, acetic acid, propionic acid, and butyric acid. Silage mixtures were then dried in a forced-air oven at 65 °C for 48 h and ground in a mill to pass through a 1.0 mm sieve for subsequent analysis.

Table 1. Chemical composition of orchardgrass and alfalfa prior to hay making and ensiling based on dry matter (g/kg).

Item	Orchardgrass	Alfalfa
Dry matter	260	268
Crude protein	167	214
Neutral detergent fiber	562	484
Acid detergent fiber	424	385
Acid detergent lignin	37.0	72.2
Ash	111	92.0
Ether extract	26.8	24.0
Calcium	7.04	14.1
Phosphorus	3.42	3.36
Potassium	33.6	27.8
Magnesium	3.72	3.16

2.3. Rumen Fluid Collection

All experimental procedures used in this study were conducted in accordance with the Institutional Animal Care Committee of China Agricultural University (CAU20171014-1). Rumen fluid was collected from five rumen-cannulated (Type 2c; Bar Diamond Inc., Parma, ID, USA) lactating Holstein dairy cows with a live weight of 530 ± 31.2 kg. Cows were uniformly fed equal rations twice daily at 7:00 a.m. and 7:00 p.m. and provided free access to water. The ration contained 4.0 kg/day of alfalfa hay, 3.5 kg/day of maize silage and 5.5 kg/day of commercial concentrate with a net energy for lactation of 6.69 MJ/kg DM and a CP content of 160 g/kg. Fresh rumen contents (mix of liquid and solid) from different sites inside the rumen were withdrawn from all dairy cows 1 h before the morning feeding with a pair of rubber fabric tongs inserted through the rumen cannula to form one composite sample. The composite sample was squeezed through four layers of surgical gauze to obtain the rumen fluid. Rumen fluid was kept in a thermos flask pre-warmed at 39 °C and immediately transferred to the laboratory. Rumen fluid from the five dairy cows was mixed in equal proportions kept at 39 °C and filled with carbon dioxide (CO_2) prior to in vitro inoculation.

2.4. In Vitro Batch Culturing and Sample Collection

Samples (500 mg) of each forage ratio, of both the hay and silage mixtures, were transferred into 120 mL glass bottles with Hungate stoppers and screw caps, resulting in a total of 50 samples in glass bottles (5 forage ratios × 2 preservation methods × 5 fermentation bottles). Fifty mL of fresh buffer solution at a pH of 6.85 [18] and 25 mL of filtered rumen fluid were mixed in each bottle, which were all purged with nitrogen gas (N_2) to maintain anaerobic conditions. Thereafter, each bottle was connected to an automated trace gas recording instrument (AGRS-III) [19] to continuously record the cumulative gas production (GP). Five bottles without forage samples were incubated as blanks. Simultaneously, an additional 30 glass bottles (5 forage ratios × 2 preservation methods × 3 fermentation bottles) prepared in the same manner were separately connected to pre-emptied air bags that collected all of the fermentation gases for subsequent determination and calculation. Three fermentations without samples were also included as blanks. All bottles were incubated at 39 °C for 48 h, and these batch cultures were repeated in 3 experimental runs. Two hundred and forty bottles (5 forage ratios × 2 preservation methods × 5 fermentation bottles × 3 experimental replicates + 5 forage ratios × 2 preservation methods × 3 fermentation bottles × 3 experimental replicates) plus 8 bottles in each replicate that were used as blanks (rumen fluid + buffer) were examined in this experiment.

After incubation, the remaining contents in each bottle were separately filtered through preweighed nylon bags (8 × 12 cm, 42 μm pore size). The filtrate was sampled for the determination of ammonia N, microbial protein production (MCP) and volatile fatty acids (VFAs). The nylon bags were thoroughly rinsed and dried at 65 °C for 48 h to a constant weight (the difference between two consecutive weights was no more than 0.20 mg per g) to determine in vitro dry matter disappearance (IVDMD), in vitro organic matter disappearance (IVOMD), in vitro neutral detergent fiber disappearance (NDFD) and in vitro acid detergent fiber disappearance (ADFD) after the chemical analysis of residues remaining in the bags.

2.5. Chemical Analysis

Determination of DM, OM, ether extract (EE), CP, and calcium (Ca), K, and magnesium (Mg) followed the methods of the Association of Official Analytic Chemists (AOAC) [20]. Following the method of Van Soest et al. [21], the contents of neutral detergent fiber (NDF), acid detergent fiber (ADF), and acid detergent lignin (ADL) were determined using a Fiber Analyzer (ANKOM Technology Corporation, Fairport, NY, USA) and expressed including residual ash. The concentrations of ammonia N and organic acids, including lactic acid, acetic acid, propionic acid, and butyric acid, in the silage extract were determined according to the method described by Li and Meng [22].

For each cultured fluid, the concentrations of ammonia N [23] and MCP [24] were determined. The determination of VFAs was conducted according to the method described by Cui et al. [25]. Briefly, the filtrate of each cultured fluid kept in polypropylene tubes (10 mL) was centrifuged at 3500× *g* for 15 min at 4 °C and 1.0 mL of the supernatant sample was mixed with 0.3 mL of 250 g/L orthophosphoric acid solution into a polypropylene microtube for 30 min. Thereafter, the mixture was cooled at 4 °C for 2 h and centrifuged at 15,000× *g* for 10 min at 4 °C. The supernatants were kept at −20 °C for later VFAs analysis. Methane, CO_2, and hydrogen gas (H_2) produced per kg OM digested were measured with a gas chromatograph according to the method described by Cui et al. [25].

2.6. Calculations

Flieg's score was calculated according to the method of Kilic [26], as shown in Equation (1).

$$\text{Flieg's score} = 220 + (2 \times \% \, DM - 15) - 40 \times pH, \tag{1}$$

The values at 81–100, 61–80, 41–60, 21–40, and 0–20 represented a very good, good, medium, low and poor silage quality, respectively.

Relative forage quality (RFQ) was calculated according to the method of Moore and Undersander [27], as shown in Equation (2).

$$RFQ = DMI \times TDN \, / \, 1.32, \tag{2}$$

where DMI is dry matter intake (% of body weight) and TDN is the total digestible nutrients (% of DM). For the alfalfa and orchardgrass–alfalfa mixtures, TDN and DMI were determined according to the methods described by the NRC [28] and Mertens [29], respectively, with NDFD adjusted according to Oba and Allen [30]. For orchardgrass, TDN and DMI were calculated following the methods of Moore and Undersander [27] and Moore and Kunkle [31], respectively.

Cumulative GP data [$GP_{(t)}$, mL/g DM] at time (t) were exported from the AGRS-III instrument into a Microsoft Excel file and fitted to a nonlinear model [32] by iterative regression analysis, as shown in Equation (3).

$$GP_t = A \times \left[1 - e^{-c \times (t - \text{Lag})}\right], \tag{3}$$

where A represents the asymptotic GP generated at a constant fractional rate (c) per unit time, t is the time of the gas recording, *e* is the base of the natural logarithm, and Lag represents the lag time before GP commences. The average GP rate (AGPR) at half of A was calculated according to the equation of García-Martínez et al. [33], as shown in Equation (4).

$$AGPR = \frac{A \times c}{2 \times (\text{Ln2} + c \times \text{Lag})}, \tag{4}$$

Single-factor associative effects index (SFAEI, %) and multiple-factor associative effects index (MFAEI, %) were calculated according to the methods described by Niderkorn et al. [34] and Wang [35], as shown in Equations (5)–(7).

$$SFAEI = \frac{\text{observed value} - \text{calculated value}}{\text{calculated value}} \times 100 \tag{5}$$

$$\begin{aligned} \text{Calculated value} = &\text{ the measured value of sole orchardgrass} \times \text{the proportion of orchardgrass} \\ &\text{in mixtures} + \text{the measured value of sole alfalfa} \times \text{the proportion of alfalfa in mixtures} \end{aligned} \tag{6}$$

$$MFAEI = \text{the sum of each single} - \text{factor associative effect value}, \tag{7}$$

In our current study, MFAEI is the sum of the six SFAEI of RFQ, OMD, NDFD, GP_{48}, c, and MCP.

2.7. Statistical Analysis

Data were analyzed using a randomized complete block design with two preservation methods (hay vs. silage) and five orchardgrass to alfalfa ratios (100:0, 75:25, 50:50, 25:75, and 0:100) for each block. One-way ANOVA was used to determine the linear and quadratic responses to the proportion of alfalfa inclusion on chemical composition and ensiling characteristics of the hay and the silage mixtures, respectively. The in vitro experimental data consisted of five fermentation replicates per forage mixture sample in each of the three experimental runs, for a total of 150 observations of the mixed ratios and preservation methods treatments. After averaging fermentation replicates within an experimental run, the data of 30 observations were subjected to ANOVA using a general linear model in which the fixed effects of preservation methods and mixed ratios of orchardgrass to alfalfa were considered. All statistical analyses were performed using R software (version 3.2.3). The least squares means and standard error (SEM) were calculated, and orthogonal contrasts were performed to examine the linear and quadratic effects of the proportion of alfalfa in hay and silage mixtures. Significance was declared at $p < 0.05$ unless otherwise noted.

3. Results

3.1. Chemical Composition of Hay and Silage Mixtures

Increasing the proportion of alfalfa increased the contents of CP, ADL, and Ca and decreased NDF, ADF, ash, and Mg in both hay and silage mixtures (Figure 1). Crude protein, NDF, ADF, and ADL were greater in the hay vs. silage mixtures. The TDN, DMI, and RFQ showed a quadratic increase with increasing alfalfa proportions, and these values were reduced in the hay mixtures compared to the silage mixtures (Figure 1).

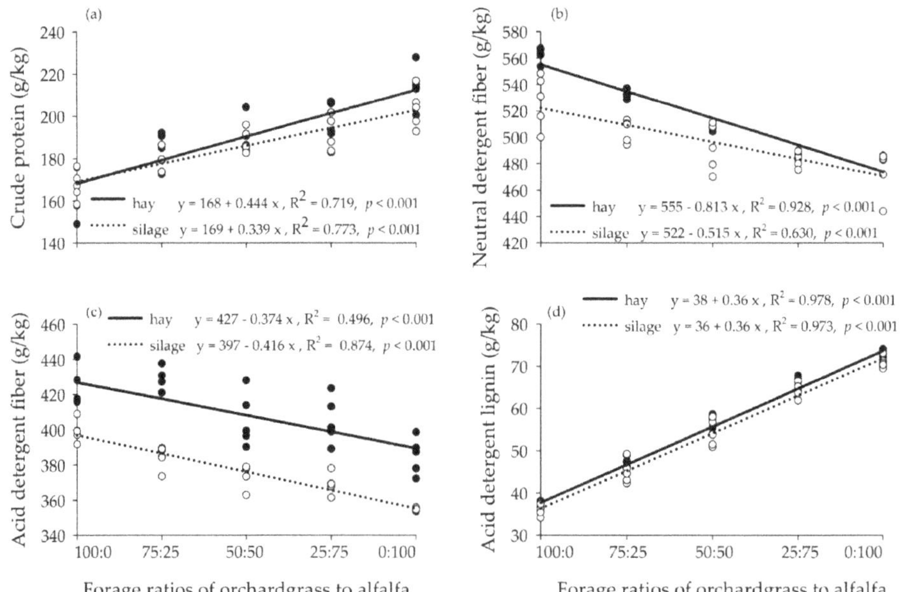

Figure 1. *Cont.*

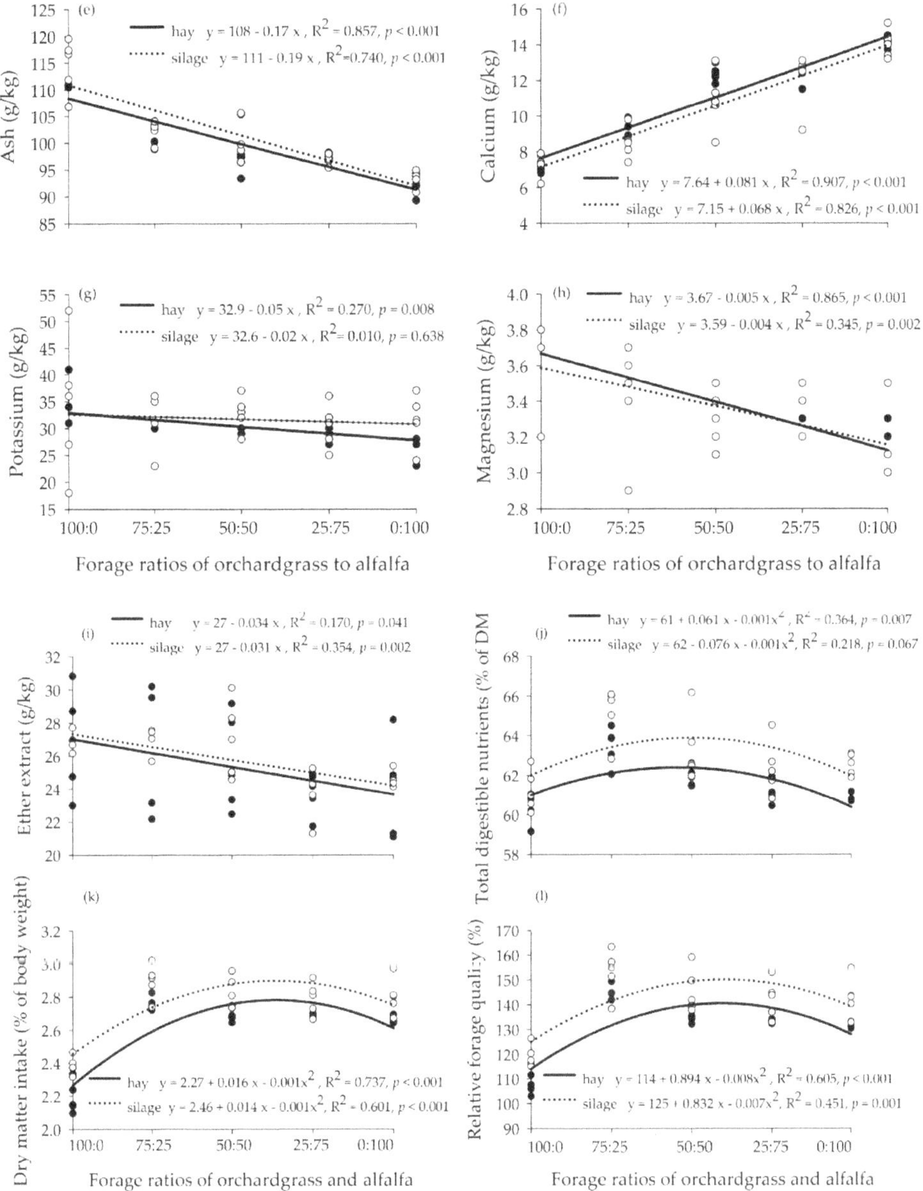

Figure 1. Chemical composition of orchardgrass and alfalfa mixtures preserved as hay and silage: (**a**) crude protein; (**b**) neutral detergent fiber; (**c**) acid detergent fiber; (**d**) acid detergent lignin; (**e**) ash; (**f**) calcium; (**g**) potassium; (**h**) magnesium; (**i**) ether extract; (**j**) total digestible nutrients; (**k**) dry matter intake; (**l**) relative forage quality.

3.2. Ensiling Characteristics of Orchardgrass and Alfalfa Silage Mixtures

The silage pH and the concentrations of ammonia N and acetic acid increased, whereas the concentration of lactic acid decreased as the proportion of alfalfa increased in silage mixtures (Table 2). The concentration of ammonia N in alfalfa was approximately twice that of orchardgrass (11.1 mmol/L and 5.36 mmol/L, respectively). However, the concentration of lactic acid in alfalfa was 355 mmol/L lower than that in orchardgrass (Table 2). According to the Flieg's scores for silage quality, mixtures of orchardgrass and alfalfa produced excellent quality silages when alfalfa proportion was ≤ 50% (Table 2). The silage mixtures of orchardgrass and alfalfa resulted in higher SFAEI of lactic acid and lower SFAEI of pH, compared with the sole orchardgrass and alfalfa (Table 2).

Table 2. Ensiling characteristics of orchardgrass and alfalfa silage mixtures.

Item	Forage Ratios of Orchardgrass to Alfalfa					SEM [2]	p [3]	
	100:0	75:25	50:50	25:75	0:100		Linear	Quadratic
pH	4.50	4.83	4.75	5.03	4.96	0.049	<0.001	<0.001
Ammonia N (mmol/L)	5.36	7.96	9.96	11.2	11.1	0.585	<0.001	<0.001
Lactic acid (mmol/L)	963	998	969	754	608	50.7	0.004	0.005
Acetic acid (mmol/L)	1.58	2.55	7.54	7.88	8.63	0.860	<0.001	<0.001
Propionic acid (mmol/L)	0.38	0.30	0.29	0.31	0.14	0.023	0.002	0.005
Butyric acid (mmol/L)	0.12	0.10	0.13	0.14	0.15	0.008	0.081	0.159
Flieg's score [1]	95.6	89.2	87.6	79.3	76.8	1.67	<0.001	<0.001
Single-factor associative effects indices (%) of ensiling characteristics [4]								
pH	-	4.6	0.5	3.8	-	0.76		
Ammonia N (mmol/L)	-	17.0	20.6	15.4	-	4.09		
Lactic acid (mmol/L)	-	14.2	23.4	8.2	-	5.76		
Acetic acid (mmol/L)	-	−23.7	12.8	14.8	-	10.2		
Propionic acid (mmol/L)	-	−6.4	11.0	51.1	-	11.93		

[1] Flieg's score was calculated by means of the pH values and DM contents of the silages: Flieg's score = 220 + (2 × % DM − 15) − 40 × pH. Flieg's score: < 20, very bad; 21–40, bad; 41–60, medium; 61–80, good; 81–100, very good. Since there was no difference on DM contents (ranging from 376 g/kg to 383 g/kg) among silage mixtures, the data were not shown. [2] SEM, standard error of the mean. [3] Linear and quadratic represent the effects of alfalfa inclusion on different silage mixtures. [4] Single-factor associative effects indices (SFAEI, %) = 100 × [(observed value − calculated value)/calculated value]. Calculated value was determined based on the proportional contribution of sole forage (orchardgrass vs. alfalfa) in the silage mixture. Positive or negative values indicated positive or negative associative effects of orchardgrass and alfalfa mixtures.

3.3. Degradation Characteristics and Gas Production Kinetics of Hay and Silage Mixtures

The IVDMD increased linearly (Figure 2a, $p < 0.001$), whereas ADFD decreased linearly (Figure 2d, $p < 0.001$) with increasing proportions of alfalfa in hay and silage mixtures. Positive associative effects were observed on IVOMD (Figure 2b, $p < 0.001$), cumulative GP_{48} (Table 3, $p < 0.05$), and asymptotic GP (A) (Table 3, $p < 0.05$); furthermore, the greatest values of these measurements were observed in the 50:50 orchardgrass-alfalfa mixtures. Negative associative effects on NDFD (Figure 2c, $p < 0.05$) were observed when the proportion of alfalfa increased in hay and silage mixtures. The IVDMD (Figure 2a, $p < 0.001$) and IVOMD (Figure 2b, $p < 0.05$) were greater in the silage mixtures than in the hay mixtures, but no effects of preservation method were observed for NDFD (Figure 2c), ADFD (Figure 2d), GP_{48} (Table 3) or other parameters of GP kinetics (Table 3). Positive associative effects were also observed for the fractional GP rate (c) and AGPR as the proportion of alfalfa increased in hay and silage mixtures ($p < 0.05$). Regarding fermentation end-product gases, no changes were observed in CO_2 or H_2 production; however, increasing the proportion of alfalfa linearly increased CH_4 production ($p < 0.05$). Greater CH_4 production was observed in the silage mixtures than in the hay mixtures ($p < 0.05$; Table 3). No interactions between forage ratios and preservation method were observed on degradation or gas production parameters ($p > 0.05$; Table 3; Figure 2).

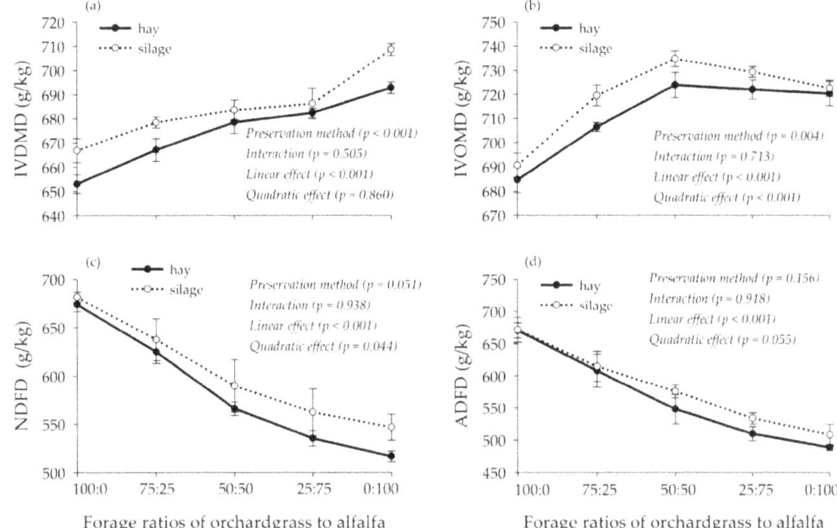

Figure 2. In vitro degradation characteristics of orchardgrass and alfalfa mixtures preserved as hay and silage: (**a**) IVDMD, in vitro dry matter disappearance; (**b**) IVOMD, in vitro organic matter disappearance; (**c**) NDFD, in vitro neutral detergent fiber disappearance; (**d**) ADFD, in vitro acid detergent fiber disappearance. The preservation method indicates whether the forage was preserved as hay or silage, and the interaction indicates the interaction between the preservation method and the ratio of orchardgrass to alfalfa. Orthogonal contrasts were used to examine the linear and quadratic effects of alfalfa inclusion on different hay and silage mixtures. Bars show the mean ± standard error.

Table 3. Gas production kinetics of orchardgrass and alfalfa mixtures preserved as hay and silage.

Item [1]	Preservation Method	Forage Ratios of Orchardgrass to Alfalfa					SEM [2]	p [3]		
		100:0	75:25	50:50	25:75	0:100		Preservation Method	Linear	Quadratic
GP_{48} (mL/g DM)	Hay	122	128	133	134	131	1.1	0.847	<0.001	0.026
	Silage	124	124	136	134	134				
GP kinetics										
A (mL/g DM)	Hay	123	128	134	135	131	1.1	0.837	<0.001	0.025
	Silage	124	124	136	134	134				
c	Hay	0.13	0.14	0.16	0.16	0.15	0.002	0.165	0.011	0.004
	Silage	0.13	0.14	0.16	0.14	0.14				
Half-time (h)	Hay	2.71	2.69	2.55	2.60	2.60	0.012	0.339	0.005	0.005
	Silage	2.72	2.66	2.57	2.65	2.65				
AGPR (mL/h)	Hay	23.4	25.1	28.4	29.2	29.8	0.37	0.319	<0.001	0.025
	Silage	24.2	24.1	28.4	28.5	29.1				
Fermentation gas pattern (mL/g OM digested)										
CO_2	Hay	174	176	178	180	176	0.8	0.081	0.150	0.675
	Silage	176	166	178	175	177				
CH_4	Hay	31.6	34.5	36.2	36.4	37.1	0.62	0.045	0.006	0.287
	Silage	34.6	37.5	37.8	39.0	39.2				
H_2	Hay	0.59	0.36	0.82	0.76	1.51	0.121	0.476	0.513	0.164
	Silage	2.00	0.84	0.94	0.66	0.40				

[1] GP_{48}, cumulative gas production at 48 h; A, asymptotic gas production; c, fractional rate of the gas production of 'A'; half-time, the time when 'A' reached its half-maximal value; AGPR, average gas production rate when half of 'A' occurred. [2] SEM, standard error of the mean. [3] The preservation method indicates whether the forage was preserved as hay or silage. Linear and quadratic represent the effects of alfalfa inclusion in the different hay and silage mixtures. Since there were no interactions between the preservation method and the ratios of orchardgrass to alfalfa, we did not present p-values on interaction in the Table 3.

3.4. Fermentation Characteristics of Hay and Silage Mixtures

Concentration of ammonia N increased linearly ($p < 0.05$) when the proportion of alfalfa increased in hay and silage mixtures (Table 4). A positive associative effect on MCP was detected as the proportion of alfalfa increased ($p < 0.05$). No effects of alfalfa inclusion on the molar proportions of acetate, propionate, butyrate, and isobutyrate were observed ($p > 0.05$). However, the concentrations of total VFAs, valerate and isovalerate increased linearly with increasing alfalfa proportion ($p < 0.05$). No effects of preservation method were observed on these fermentation parameters ($p > 0.05$; Table 4). There were no interactions between forage ratios and preservation method on fermentation parameters ($p > 0.05$; Table 4).

Table 4. Fermentation characteristics of orchardgrass and alfalfa mixtures preserved as hay and silage.

Item [1]	Preservation Method	Forage Ratios of Orchardgrass to Alfalfa					SEM [2]	p [3]		
		100:0	75:25	50:50	25:75	0:100		Preservation Method	Linear	Quadratic
Ammonia	Hay	20.0	20.0	20.4	21.5	21.5	0.31	0.635	0.029	0.996
N (mg/dL)	Silage	20.0	20.2	21.2	21.7	21.8				
MCP	Hay	270	276	329	345	338	5.2	0.555	<0.001	0.034
(µg/mL)	Silage	282	296	315	338	303				
Total VFAs	Hay	110	116	118	117	120	1.2	0.208	0.049	0.688
(mmol/L)	Silage	113	110	115	110	119				
VFA pattern (mmol/L)										
Acetate	Hay	71.2	76.6	78.4	77.4	77.8	1.10	0.085	0.309	0.809
	Silage	74.5	70.1	72.6	68.8	76.2				
Propionate	Hay	18.8	17.6	17.4	18.9	19.0	0.17	0.297	0.136	0.631
	Silage	17.2	17.9	19.5	17.4	18.4				
Butyrate	Hay	9.09	9.76	9.51	9.33	10.2	0.31	0.620	0.300	0.887
	Silage	9.45	9.56	9.82	10.3	10.4				
Isobutyrate	Hay	2.14	2.63	2.56	2.14	2.20	0.084	0.257	0.900	0.811
	Silage	2.36	2.34	2.64	2.63	2.83				
Valerate	Hay	2.61	3.02	3.10	2.81	3.50	0.093	0.801	0.015	0.710
	Silage	2.80	2.88	2.94	3.25	3.40				
Isovalerate	Hay	6.04	6.20	6.68	6.54	7.16	0.185	0.097	0.024	0.704
	Silage	6.60	6.76	7.18	7.23	7.99				

[1] Ammonia N, ammonia nitrogen; MCP, microbial protein production; VFAs, volatile fatty acids. [2] SEM, standard error of the mean. [3] The preservation method indicates whether the forage was preserved as hay or silage. Linear and quadratic represent the effects of alfalfa inclusion in different hay and silage mixtures. Since there were no interactions between the preservation method and the ratios of orchardgrass to alfalfa, we did not present p-values on interaction in the Table 4.

3.5. SFAEI and MFAEI of Hay and Silage Mixtures

The SFAEI of RFQ, IVOMD, c, MCP, and GP_{48} was positive in both hay and silage mixtures of orchardgrass and alfalfa (Figure 3). The SFAEI of RFQ was greatest at 75:25 orchardgrass to alfalfa hay and silage mixtures. Increasing the proportion of alfalfa increased the SFAEI of IVOMD, GP_{48}, c, and MCP by +5.14%, +5.52%, +16.1%, +13.4%, respectively, and decreased the SFAEI of NDFD (−1.79%) when compared with the calculated value. Positive MFAEI was observed in all mixtures, and the highest MFAEI (up +40%) was observed in 50:50 of orchardgrass to alfalfa hay and silage mixtures.

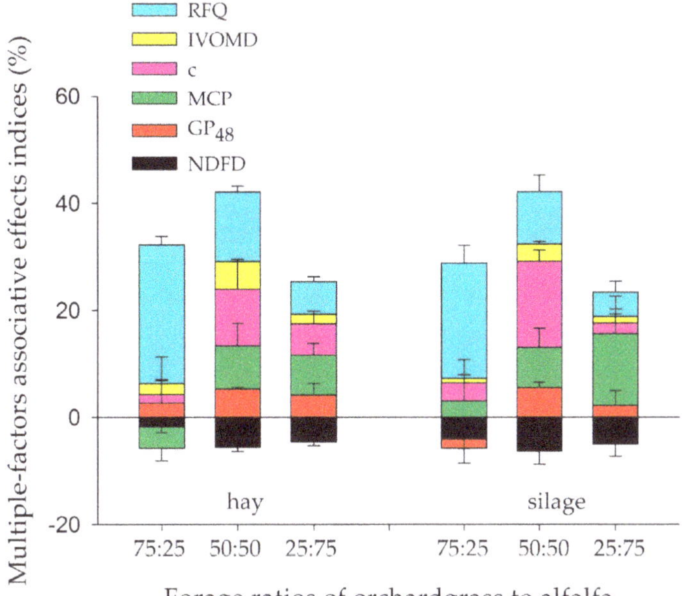

Figure 3. Single-factor associative effects indices (SFAEI, %) and multiple-factors associative effects indices (MFAEI, %) of hay and silage mixtures. RFQ, relative forage quality; IVOMD, in vitro dry matter disappearance; NDFD, in vitro neutral detergent fiber disappearance; c, gas production rate; GP_{48}, gas production at 48 h; MCP, microbial protein production. Bars show the mean ± standard error. Single-factor associative effects indices (SFAEI, %) = 100 × [(observed value − calculated value)/calculated value]. Calculated value was determined based on the proportional contribution of sole forage (orchardgrass vs. alfalfa) in the mixture. Positive or negative values indicated positive or negative associative effects of orchardgrass and alfalfa mixtures.

4. Discussions

As hypothesized, combining orchardgrass with alfalfa optimized the ruminal degradation and fermentation efficiency through complementary effects of nutrients in forage mixtures and the greatest MFAEI (up about +40%) was obtained at 50:50 of orchardgrass to alfalfa in both hay and silage mixtures. Although mixtures of orchardgrass and alfalfa decreased NDFD, this antagonistic effect was quite little for the total MFAEI, which presumably has been compensated for by synergistic effects of other nutrients. Considering the observed associative effects of mixing orchardgrass and alfalfa, it is possible to optimize forage use efficiency in ruminants.

4.1. Chemical Composition of Hay and Silage Mixtures

Greater CP and reduced NDF and ADF contents are usually observed in legumes compared to grasses, due to N fixation from the atmosphere and less cell wall material in legumes [36,37], similar to the current study. A reduction in CP, NDF, and ADF contents were observed in the silage mixtures than in the hay mixtures, probably because a smaller fraction of the protein was broken down into peptides, amino acids, and ammonia and some of the cellulose and hemicellulose were hydrolyzed during the ensiling process [13,38]. Greater values of RFQ were observed in alfalfa-containing mixtures compared with the orchardgrass alone, suggesting that alfalfa inclusion has the potential to improve the overall forage quality of mixtures and the subsequent animal performance [39,40].

4.2. Ensiling Characteristics of Orchardgrass and Alfalfa Silage Mixtures

Legumes are usually more difficult to preserve as silage than are grasses due to a reduction in water-soluble carbohydrates and a greater buffering capacity [7]. The results from the current study indicate that alfalfa silage had increased pH and ammonia N concentrations and decreased lactic acid concentrations compared to orchardgrass silage and silage mixtures of orchardgrass and alfalfa showed suitable ensiling characteristics and high associative effects indices. In accordance with previous studies [41,42], silage mixtures could be a feasible way to preserve nutrition and improve fermentation quality through the role of dominated Lactobacillus, compared with the sole silage crops.

4.3. Associative Effects on the Degradation Characteristics and Gas Production Kinetics of Hay and Silage Mixtures

The IVDMD linearly increased in response to increasing proportions of alfalfa in the present study due to its high CP and low fiber contents [43]. The IVDMD of silage mixtures was greater than that of hay mixtures, which agrees with previous results that the apparent digestibility of DM and OM in alfalfa silage was greater than that in alfalfa hay [44].

Negative associative effects on NDFD were detected as the proportion of alfalfa increased in mixtures, which is in agreement with other studies [10,45]. These results may be attributed to the antagonistic effects that occur in the fermentation extent and degradation rate of fiber fractions. Similarly, these antagonistic effects were more evident when the proportion of alfalfa was greater than 25% in alfalfa–Caucasian bluestem hay mixtures [46]. The dominant fibrolytic bacteria, that are not proteolytic, require NH_3 and branched-chain VFAs that are usually derived from active proteolytic bacteria and amino acid fermentation bacteria that do not ferment carbohydrates [47,48]. The degradable and rapidly digested N from legumes compensated for the slow release of N from grasses and coincided with associations and shifts in the ruminal microbial community [15,49] to better degrade the fiber fractions in grass-legume mixtures [46,50].

However, the NDF of legumes is more resistant to degrade than that of grasses because legumes have higher proportions of indigestible NDFs [51] with more complicated structures (e.g., cuticle sheets and vascular bundles) and physicochemical properties (e.g., alkali solubility and ether links), which limit the attachment of fibrolytic bacteria and the enzymatic hydrolysis of forage polysaccharides [37] and decrease the potential for fiber digestion [52]. In the current study, it appears that the positive effect of alfalfa inclusion on NDFD by supplying a rapidly digested N source for fibrolytic bacteria was offset by the negative effect of increased indigestible fiber fractions, subsequently decelerating the fermentation extent and degradation rate of NDF. High levels of fat, starch, and sugars that are rapidly fermented to VFAs can also reduce rumen pH and hence depress the activity of fibrolytic bacteria, decreasing NDFD [53–55].

The volume of GP was positively correlated with OM degradability, which is indirectly reflected in the potential of microbes to degrade feed and the extent of microbial fermentation [56]. The current results indicate positive associative effects on cumulative GP and IVOMD when the proportion of alfalfa increased in hay and silage mixtures, implying that complementary digestible nutrients in orchardgrass-alfalfa mixtures provided favorable substrates for microbes, resulting in improved fermentation. The presence of positive associative effects on the asymptotic fractional GP rate (c) but negative associative effects on the time to reach half of the asymptotic GP (half-time) suggests that the inclusion of alfalfa in mixtures accelerated microbial fermentation of digestible components and/or resulted in a shift in the site of digestion of cell contents, including N [9]. In summary, no differences were observed in IVDMD and IVOMD for either hay or silage mixtures containing 50, 75, or 100% alfalfa, however, all mixtures at these proportions had greater IVDMD and IVOMD and reduced NDFD and ADFD values than did mixtures with 25% or 0% alfalfa. If the DM and OM degradability of a mixture needs to be increased, at least 50% alfalfa is required based on the results obtained in this study.

There was no effect on CO_2 and H_2 with increasing alfalfa proportions, this could be explained by similar responses in acetate, propionate, and butyrate [57]. The amount of CH_4 produced per unit of

OM digested increased linearly with increasing alfalfa proportion, and the higher nitrogen-free extract (NFE) (44.07% vs. 43.75%) and lower crude fiber (CF) (21.0% vs. 33.93%) in alfalfa compared with orchardgrass could explain the comparatively high CH_4 production with alfalfa [58,59]. Feedstuffs with higher GP and IVDMD tended to have greater CH_4 production [60]; this trend is in accordance with the current study where silage mixtures, which had greater IVDMD and IVOMD, had increased CH_4 production per gram of OM digested than did the hay mixtures.

4.4. Associative Effects on the Fermentation Characteristics of Hay and Silage Mixtures

Ammonia N showed a linear increase in cultured fluids as the proportion of alfalfa increased, which was partially due to the proteolysis in alfalfa that occurred during in vitro incubation [61]. Rumen fermentation and microbial protein synthesis proceeded at a lower than maximal rate when the concentrations of ammonia N in the rumen fluid were less than 23.5 mg/dL and 5 mg/dL, respectively [62,63]. In the current study, the concentration of ammonia N in the cultured fluid was above the requirements to maximize MCP synthesis (>5 mg/dL) and thus the sufficient energy should match the synthetic abilities of microorganisms to effectively utilize these N resources [64]. Positive associative effects on MCP were observed as the proportion of alfalfa increased in our study, implying that mixtures of orchardgrass and alfalfa contributed more to the balance of readily degradable carbohydrates and readily degradable protein than did the monocultures, subsequently improving microbial protein synthesis and increasing bacterial N flow and microbial growth.

The concentration of total VFAs linearly increased with increasing proportions of alfalfa, and similarly, increased VFAs production was also observed in sweet sorghum-alfalfa mixtures [15]. In our study, alfalfa inclusion did not change the molar proportions of acetate, propionate, and butyrate, probably because the ruminal microbial community and metabolic pathways by which specific microbes utilize substrates to produce VFAs end-products were not altered [65]. The concentrations of valerate and isovalerate increased linearly with increasing alfalfa proportions, which was in accordance with a previous study [17] showing that the CP level was significantly correlated with valerate and isovalerate production.

5. Conclusions

The forage ratio of 50:50 for orchardgrass and alfalfa hay and silage mixtures favored the growth of rumen microorganisms and improved nutrient digestion and ruminal fermentability. Increments of rumen degradability and methane production were more pronounced in silage mixtures than hay mixtures. This study provided potential evidence on how to maximize forage use efficiency in ruminant systems, and further research could be performed to assess the synergistic effects of this practice on agronomic characteristics and ruminant production performance.

Author Contributions: Conceptualization, Z.X. and Y.Z.; Methodology, Z.X., Y.Z. and H.Y.; Formal analysis, Z.X. and Y.Z.; Investigation, Y.W. (Yanlu Wang) and Y.W. (Yuqi Wei). Data curation, Z.X. and Y.Z.; Writing—original draft preparation, Z.X.; Writing—review and editing, N.L., H.Y., P.M., E.P. and Y.Z.; Supervision, Y.Z.; Project administration, Y.Z.; Funding acquisition, Y.Z. All authors have read and agreed to the published version of the manuscript.

Funding: This study was funded by the China Forage and Grass Research System (CARS-34).

Conflicts of Interest: The authors declare no conflict of interest.

References

1. Stewart, A.V.; Ellison, N.W. Dactylis. In *Wild Crop Relatives: Genomic and Breeding Resources*; Kole, C., Ed.; Springer: Berlin/Heidelberg, Germany, 2011; pp. 73–87.
2. Turner, L.; Donaghy, D.; Lane, P.; Rawnsley, R. A comparison of the establishment, productivity, and feed quality of four cocksfoot (*Dactylis glomerata* L.) and four brome (*Bromus* spp.) cultivars, under leaf stage based defoliation management. *Aust. J. Agric. Res.* **2007**, *58*, 900–906. [CrossRef]

3. Butkutė, B.; Lemežienė, N.; Kanapeckas, J.; Navickas, K.; Dabkevičius, Z.; Venslauskas, K. Cocksfoot, tall fescue and reed canary grass: Dry matter yield, chemical composition and biomass convertibility to methane. *Biomass Bioenergy* **2014**, *66*, 1–11. [CrossRef]

4. Bourquin, L.D.; Titgemeyer, E.C.; Merchen, N.R.; Fahey, G.C. Forage level and particle size effects on orchardgrass digestion by steers.1. Site and extent of organic matter, nitrogen, and cell wall digestion. *J. Anim. Sci.* **1994**, *72*, 746. [CrossRef] [PubMed]

5. Niderkorn, V.; Martin, C.; Rochette, Y.; Julien, S.; Baumont, R. Associative effects between orchardgrass and red clover silages on voluntary intake and digestion in sheep: Evidence of a synergy on digestible dry matter intake. *J. Anim. Sci.* **2015**, *93*, 4967–4976. [CrossRef] [PubMed]

6. Lee, M.R.; Olmos Colmenero, J.D.J.; Winters, A.L.; Scollan, N.D.; Minchin, F.R. Polyphenol oxidase activity in grass and its effect on plant-mediated lipolysis and proteolysis of Dactylis glomerata (cocksfoot) in a simulated rumen environment. *J. Sci. Food Agric.* **2006**, *86*, 1503–1511. [CrossRef]

7. Phelan, P.; Moloney, A.P.; McGeough, E.J.; Humphreys, J.; Bertilsson, J.; O'Riordan, E.G.; O'Kiely, P. Forage Legumes for Grazing and Conserving in Ruminant Production Systems. *Crit. Rev. Plant Sci.* **2015**, *34*, 281–326. [CrossRef]

8. Peyraud, J.; Astigarraga, L. Review of the effect of nitrogen fertilization on the chemical composition, intake, digestion and nutritive value of fresh herbage: Consequences on animal nutrition and N balance. *Anim. Feed Sci. Technol.* **1998**, *72*, 235–259. [CrossRef]

9. Patra, A. Effects of supplementing low-quality roughages with tree foliages on digestibility, nitrogen utilization and rumen characteristics in sheep: A meta-analysis. *J. Anim. Physiol. Anim. Nutr.* **2010**, *94*, 338–353. [CrossRef]

10. Johansen, M.; Søegaard, K.; Lund, P.; Weisbjerg, M.R. Digestibility and clover proportion determine milk production when silages of different grass and clover species are fed to dairy cows. *J. Dairy Sci.* **2017**, *100*, 8861–8880. [CrossRef]

11. INRA. *INRA Feeding System for Ruminants*; Wageningen Academic Publishers: Wageningen, The Netherlands, 2018; pp. 44–59.

12. Niderkorn, V.; Martin, C.; Bernard, M.; Le Morvan, A.; Rochette, Y.; Baumont, R. Effect of increasing the proportion of chicory in forage-based diets on intake and digestion by sheep. *Animal* **2019**, *13*, 718–726. [CrossRef]

13. Hristov, A.N.; Sandev, S.G. Proteolysis and rumen degradability of protein in alfalfa preserved as silage, wilted silage or hay. *Anim. Feed Sci. Technol.* **1998**, *72*, 175–181. [CrossRef]

14. Plaizier, J.C. Replacing chopped alfalfa hay with alfalfa silage in barley grain and alfalfa-based total mixed rations for lactating dairy cows. *J. Dairy Sci.* **2004**, *87*, 2495–2505. [CrossRef]

15. Chen, L.; Dong, Z.; Li, J.; Shao, T. Ensiling characteristics, in vitro rumen fermentation, microbial communities and aerobic stability of low-dry matter silages produced with sweet sorghum and alfalfa mixtures. *J. Sci. Food Agric.* **2019**, *99*, 2140–2151. [CrossRef]

16. Dal Pizzol, J.; Ribeiro-Filho, H.; Quereuil, A.; Le Morvan, A.; Niderkorn, V. Complementarities between grasses and forage legumes from temperate and subtropical areas on in vitro rumen fermentation characteristics. *Anim. Feed Sci. Technol.* **2017**, *228*, 178–185. [CrossRef]

17. Getachew, G.; Robinson, P.; DePeters, E.; Taylor, S. Relationships between chemical composition, dry matter degradation and in vitro gas production of several ruminant feeds. *Anim. Feed Sci. Technol.* **2004**, *111*, 57–71. [CrossRef]

18. Menke, K.H.; Steingass, H. Estimation of the energetic feed value obtained from chemical analysis and in vitro gas production using rumen fluid. *Anim. Res. Dev.* **1988**, *28*, 7–55.

19. Zhang, D.; Yang, H. In vitro ruminal methanogenesis of a hay-rich substrate in response to different combination supplements of nitrocompounds; pyromellitic diimide and 2-bromoethanesulphonate. *Anim. Feed Sci. Technol.* **2011**, *163*, 20–32. [CrossRef]

20. AOAC. *Official Methods of Analysis of AOAC International*, 17th ed.; Association of Official Analytical Chemists: Gaithersburg, MD, USA, 2005.

21. Van Soest, P.J.; Robertson, J.B.; Lewis, B.A. Methods for dietary fiber, neutral detergent fiber, and nonstarch polysaccharides in relation to animal nutrition. *J. Dairy Sci.* **1991**, *74*, 3583–3597. [CrossRef]

22. Li, Y.; Meng, Q. Effect of different types of fibre supplemented with sunflower oil on ruminal fermentation and production of conjugated linoleic acids in vitro. *Arch. Anim. Nutr.* **2006**, *60*, 402–411. [CrossRef]

23.   Verdouw, H.; Vanechteld, C.J.A.; Dekkers, E.M.J. Ammonia determination based on indophenol formation with sodium salicylate. *Water Res.* **1978**, *12*, 399–402. [CrossRef]

24.   Makkar, H.; Sharma, O.; Dawra, R.; Negi, S. Simple determination of microbial protein in rumen liquor. *J. Dairy Sci.* **1982**, *65*, 2170–2173. [CrossRef]

25.   Cui, J.; Yang, H.; Yu, C.; Bai, S.; Wu, T.; Song, S.; Sun, W.; Shao, X.; Jiang, L. Effect of urea fertilization on biomass yield, chemical composition, in vitro rumen digestibility and fermentation characteristics of straw of highland barley planted in Tibet. *J. Agric. Sci.* **2016**, *154*, 151–164. [CrossRef]

26.   Kilic, A. *Silo Feed (Instruction, Education and Application Proposals)*; Bilgehan Press: Izmir, Turkey, 1986.

27.   Moore, J.E.; Undersander, D.J. Relative forage quality: An alternative to relative feed value and quality index. In Proceedings of the 13th Annual Florida Ruminant Nutrition Symposium, Gainesville, FL, USA, 10–11 January 2002.

28.   National Research Council. *Nutrient Requirements of Dairy Cattle*; National Academies Press: Washington, DC, USA, 2001.

29.   Mertens, D. Predicting intake and digestibility using mathematical models of ruminal function. *J. Anim. Sci.* **1987**, *64*, 1548–1558. [CrossRef] [PubMed]

30.   Oba, M.; Allen, M. Evaluation of the importance of the digestibility of neutral detergent fiber from forage: Effects on dry matter intake and milk yield of dairy cows. *J. Dairy Sci.* **1999**, *82*, 589–596. [CrossRef]

31.   Moore, J.; Kunkle, W. Evaluation of equations for estimating voluntary intake of forages and forage-based diets. *J. Anim. Sci.* **1999**, *1*, 204.

32.   France, J.; Dijkstra, J.; Dhanoa, M.; Lopez, S.; Bannink, A. Estimating the extent of degradation of ruminant feeds from a description of their gas production profiles observed in vitro: Derivation of models and other mathematical considerations. *Br. J. Nutr.* **2000**, *83*, 143–150. [CrossRef] [PubMed]

33.   García-Martínez, R.; Ranilla, M.; Tejido, M.; Carro, M. Effects of disodium fumarate on in vitro rumen microbial growth, methane production and fermentation of diets differing in their forage: Concentrate ratio. *Br. J. Nutr.* **2005**, *94*, 71–77. [CrossRef]

34.   Niderkorn, V.; Baumont, R.; Le Morvan, A.; Macheboeuf, D. Occurrence of associative effects between grasses and legumes in binary mixtures on in vitro rumen fermentation characteristics. *J. Anim. Sci.* **2011**, *89*, 1138–1145. [CrossRef]

35.   Wang, J. *Methods in Ruminant Nutrition Research*; Modern Education Press: Beijing, China, 2011. (In Chinese)

36.   Krawutschke, M.; Kleen, J.; Weiher, N.; Loges, R.; Taube, F.; Gierus, M. Changes in crude protein fractions of forage legumes during the spring growth and summer regrowth period. *J. Agric. Sci.* **2013**, *151*, 72–90. [CrossRef]

37.   Buxton, D.R.; Redfearn, D.D. Plant limitations to fiber digestion and utilization. *J. Nutr.* **1997**, *127*, 814S–818S. [CrossRef]

38.   Kondo, M.; Shimizu, K.; Jayanegara, A.; Mishima, T.; Matsui, H.; Karita, S.; Goto, M.; Fujihara, T. Changes in nutrient composition and in vitro ruminal fermentation of total mixed ration silage stored at different temperatures and periods. *J. Sci. Food Agric.* **2016**, *96*, 1175–1180. [CrossRef] [PubMed]

39.   Salama, H.S.A.; Zeid, M.M.K. Hay quality evaluation of summer grass and legume forage monocultures and mixtures grown under irrigated conditions. *Aust. J. Crop Sci.* **2016**, *11*, 1543. [CrossRef]

40.   Favre, J.R.; Castiblanco, T.M.; Combs, D.K.; Wattiaux, M.A.; Picasso, V.D. Forage nutritive value and predicted fiber digestibility of Kernza intermediate wheatgrass in monoculture and in mixture with red clover during the first production year. *Anim. Feed Sci. Technol.* **2019**, *258*, 114298. [CrossRef]

41.   Wang, M.; Wang, L.; Yu, Z. Fermentation dynamics and bacterial diversity of mixed lucerne and sweet corn stalk silage ensiled at six ratios. *Grass Forage Sci.* **2019**, *74*, 264–273. [CrossRef]

42.   Ni, K.; Zhao, J.; Zhu, B.; Su, R.; Pan, Y.; Ma, J.; Zhou, G.; Tao, Y.; Liu, X.; Zhong, J. Assessing the fermentation quality and microbial community of the mixed silage of forage soybean with crop corn or sorghum. *Bioresour. Technol.* **2018**, *265*, 563–567. [CrossRef] [PubMed]

43.   Jung, H.G.; Allen, M.S. Characteristics of plant-cell walls affecting intake and digestibility of forages by ruminants. *J. Anim. Sci.* **1995**, *73*, 2774–2790. [CrossRef]

44.   Hristov, A.N.; Broderick, G.A. Synthesis of microbial protein in ruminally cannulated cows fed alfalfa silage, alfalfa hay, or corn silage. *J. Dairy Sci.* **1996**, *79*, 1627–1637. [CrossRef]

45.   Reid, R.L.; Templeton, W.C., Jr.; Ranney, T.S.; Thayne, W.V. Digestibility, intake and mineral utilization of combinations of grasses and legumes by lambs. *J. Anim. Sci.* **1987**, *64*, 1725–1734. [CrossRef]

46. Bowman, J.; Asplund, J. Evaluation of mixed lucerne and caucasian bluestem hay diets fed to sheep. *Anim. Feed Sci. Technol.* **1988**, *20*, 19–31. [CrossRef]

47. Hobson, P.N.; Stewart, C.S. *The Rumen Microbial Ecosystem*, 2nd ed.; Blackie Academic & Professional: London, UK, 1997.

48. Millen, D.D.; Arrigoni, M.D.B.; Pacheco, R.D.L. *Rumenology*, 1st ed.; Springer International Publishing: Cham, Switzerland, 2016.

49. Liu, J.; Zhang, M.; Xue, C.; Zhu, W.; Mao, S. Characterization and comparison of the temporal dynamics of ruminal bacterial microbiota colonizing rice straw and alfalfa hay within ruminants. *J. Dairy Sci.* **2016**, *99*, 9668–9681. [CrossRef]

50. Kammes, K.L.; Allen, M.S. Nutrient demand interacts with forage family to affect digestion responses in dairy cows. *J. Dairy Sci.* **2012**, *95*, 3269–3287. [CrossRef] [PubMed]

51. Bhatti, S.A.; Bowman, J.G.P.; Firkins, J.L.; Grove, A.V.; Hunt, C.W. Effect of intake level and alfalfa substitution for grass hay on ruminal kinetics of fiber digestion and particle passage in beef cattle. *J. Anim. Sci.* **2008**, *86*, 134–145. [CrossRef] [PubMed]

52. Van Soest, P.J. *Nutritional Ecology of the Ruminant*, 2nd ed.; Cornell University press: Ithaca, NY, USA, 1994.

53. Chen, K.J.; Jan, D.F.; Chiou, P.W.S.; Yang, D.W. Effects of dietary heat extruded soybean meal and protected fat supplement on the production, blood and ruminal characteristics of Holstein cows. *Asian-Australas. J. Anim. Sci.* **2002**, *15*, 821–827. [CrossRef]

54. Brask, M.; Lund, P.; Hellwing, A.L.F.; Poulsen, M.; Weisbjerg, M.R. Enteric methane production, digestibility and rumen fermentation in dairy cows fed different forages with and without rapeseed fat supplementation. *Anim. Feed Sci. Technol.* **2013**, *184*, 67–79. [CrossRef]

55. Oba, M.; Mewis, J.L.; Zhining, Z. Effects of ruminal doses of sucrose, lactose, and corn starch on ruminal fermentation and expression of genes in ruminal epithelial cells. *J. Dairy Sci.* **2015**, *98*, 586–594. [CrossRef]

56. Menke, K.; Raab, L.; Salewski, A.; Steingass, H.; Fritz, D.; Schneider, W. The estimation of the digestibility and metabolizable energy content of ruminant feedingstuffs from the gas production when they are incubated with rumen liquor in vitro. *J. Agric. Sci.* **1979**, *93*, 217–222. [CrossRef]

57. Pen, B.; Sar, C.; Mwenya, B.; Kuwaki, K.; Morikawa, R.; Takahashi, J. Effects of Yucca schidigera and Quillaja saponaria extracts on in vitro ruminal fermentation and methane emission. *Anim. Feed Sci. Technol.* **2006**, *129*, 175–186. [CrossRef]

58. Lee, H.; Lee, S.; Kim, J.; Oh, Y.; Kim, B.; Kim, C.; Kim, K. Methane production potential of feed ingredients as measured by in vitro gas test. *Asian Australas. J. Anim. Sci.* **2003**, *16*, 1143–1150. [CrossRef]

59. Cronje, P.B. *Ruminant Physiology: Digestion, Metabolism, Growth and Reproduction*; CABI Publishing: Wallingford, UK, 2000.

60. Jayanegara, A.; Wina, E.; Soliva, C.; Marquardt, S.; Kreuzer, M.; Leiber, F. Dependence of forage quality and methanogenic potential of tropical plants on their phenolic fractions as determined by principal component analysis. *Anim. Feed Sci. Technol.* **2011**, *163*, 231–243. [CrossRef]

61. Zhang, S.J.; Chaudhry, A.S.; Osman, A.; Shi, C.Q.; Edwards, G.R.; Dewhurst, R.J.; Cheng, L. Associative effects of ensiling mixtures of sweet sorghum and alfalfa on nutritive value, fermentation and methane characteristics. *Anim. Feed Sci. Technol.* **2015**, *206*, 29–38. [CrossRef]

62. Mehrez, A.; Ørskov, E.; McDonald, I. Rates of rumen fermentation in relation to ammonia concentration. *Br. J. Nutr.* **1977**, *38*, 437–443. [CrossRef] [PubMed]

63. Satter, L.; Slyter, L. Effect of ammonia concentration on rumen microbial protein production in vitro. *Br. J Nutr.* **1974**, *32*, 199–208. [CrossRef] [PubMed]

64. Wang, D.; Fang, J.; Xing, F.; Yang, L. Alfalfa as a supplement of dried cornstalk diets: Associative effects on intake, digestibility, nitrogen metabolisation, rumen environment and hematological parameters in sheep. *Livest. Sci.* **2008**, *113*, 87–97. [CrossRef]

65. Eun, J.S.; Beauchemin, K.; Schulze, H. Use of exogenous fibrolytic enzymes to enhance in vitro fermentation of alfalfa hay and corn silage. *J. Dairy Sci.* **2007**, *90*, 1440–1451. [CrossRef]

 *animals*

Article

In Vitro Digestibility, In Situ Degradability, Rumen Fermentation and N Metabolism of Camelina Co-Products for Beef Cattle Studied with a Dual Flow Continuous Culture System

Hèctor Salas, Lorena Castillejos *, Montserrat López-Suárez and Alfred Ferret

Animal Nutrition and Welfare Service (SNIBA), Department of Animal and Food Science, Universitat Autonòma de Barcelona, 08193 Bellaterra, Spain; hectorsalasolive@gmail.com (H.S.); Montserrat.Lopez.Suarez@uab.cat (M.L.-S.); Alfred.Ferret@uab.cat (A.F.)

* Correspondence: Lorena.Castillejos@uab.cat; Tel.: +34-581-15-56

Received: 6 November 2019; Accepted: 29 November 2019; Published: 3 December 2019

Simple Summary: Currently, vegetable protein sources such as soybean meal and rapeseed meal are expensive and with volatile prices. These economic circumstances are driving the research of potential new protein resources for beef cattle diets that can reduce the ration cost without compromising animal productive yields. As possible candidates, camelina meal and camelina expeller have been studied; they are co-products with a high protein percentage, obtained after oil extraction from the oil seeds of *Camelina sativa*. The objectives of this study were to characterize these camelina co-products and ascertain if they could be useful ingredients for beef cattle diets. The results indicate that the diets formulated with camelina meal and camelina expeller do not show differences in the efficiency of microbial protein synthesis compared to the current reference proteins, camelina meal diet being the most similar to soybean meal and rapeseed meal diets, and camelina expeller the diet with the highest fermentation potential. The results of soybean meal as an individual ingredient reveal more differences with camelina co-products. In vivo studies are necessary to draw conclusions, but in vitro results obtained suggest that camelina meal and camelina expeller are potential substitutes for rapeseed meal in beef cattle diets.

Abstract: Camelina meal (CM) and camelina expeller (CE) were compared with soybean meal (SM) and rapeseed meal (RM). Trial 1 consisted of a modified Tilley and Terry in vitro technique. Trial 2 was an in situ technique performed by incubating nylon bags within cannulated cows. Trial 3 consisted in dual-flow continuous culture fermenters. In Trial 1, CM, CE and RM showed similar DM digestibility and OM digestibility, and SM was the most digestible ingredient ($p < 0.05$). Trial 2 showed that CE had the numerically highest DM degradability, but CP degradability was similar to RM. Camelina meal had a DM degradability similar to SM and RM and had an intermediate coefficient of CP degradability. In Trial 3, CE diet tended to present a higher true OM digestibility than SM diet ($p = 0.06$). Total volatile fatty acids (VFA) was higher in CE and CM diets than in SM diet ($p = 0.009$). Crude protein degradation tended to be higher ($p = 0.07$), and dietary nitrogen flow tended to be lower ($p = 0.06$) in CE diet than in CM diet. The efficiency of microbial protein synthesis was not affected by treatment ($p > 0.05$). In conclusion, CE and CM as protein sources differ in CP coefficient of degradability but their results were similar to RM. More differences were detected with regard to SM.

Keywords: beef cattle; protein sources; camelina co-products; rumen microbial fermentation

---

1. Introduction

Meeting protein requirements in ruminant nutrition can be costly. The main reasons are the high and unstable prices of protein sources, such as soybean meal, and their availability, which is affected by global trade [1]. This situation makes it necessary to search for new alternatives to replace totally or partially the protein sources currently used in ruminant diets. In this line of research, *Camelina sativa* co-products could be one such alternative. *Camelina sativa* or false flax is an oilseed crop of the Brassica family, which originates from the Mediterranean and Central Asia. It is an annual or overwintering herb with low agronomic requirements [2] and is more tolerant to frost, heat, and drought than other plants of the same family [3], such as rapeseed meal. The biofuel industry's growing interest in its cultivation is attributable to the 40% oil content of the seed, which is used to produce biodiesel [4–8]. When oil is extracted from the seed, camelina expeller (CE) and camelina meal (CM) are produced, the former being obtained after mechanical oil extraction of the seed and the latter after mechanical and subsequent chemical oil extraction. According to Zubr [9], the resulting meal after oil extraction contains 30–40% CP and 12% fiber. However, the main limitation of using CM and CE is related to the presence of anti-nutritional components. Camelina co-products contain glucosinolates and erucic acid [10], which, according to Tripathi and Mishra [11] affect the thyroid and the cardiovascular system. That said, CM has been used in beef steers without effect on growth performance or thyroid function [12,13]. Our hypothesis was that given the high protein content of both co-products [14], CE and CM could be alternative protein sources in ruminant nutrition. The aims of this study were to characterize these camelina co-products and to compare in vitro beef cattle diets made either with them or with more commonly used protein sources like soybean meal and 00-rapeseed meal.

2. Materials and Methods

Animal procedures were approved by the Institutional Animal Care and Use Committee (reference CEEAH 1585) of the Universitat Autònoma de Barcelona (Spain), in accordance with the European directive 2010/63/EU.

Three different trials were conducted to achieve the objectives. Trial 1 was performed to determine the in vitro digestibility of four protein ingredients commonly used in cattle diets (Table 1). Two of them were sources supplied by the feed industry and frequently used for beef cattle: Soybean meal 44% CP (SM) and 00-rapeseed meal (RM), while the other two were CE and CM, supplied by Camelina Company, S.L. (Madrid, Spain). The CE and CM samples were selected according to their chemical composition similarity to others studies in the case of CE [8,15] and to RM in the case of CM. To select the SBM and RM samples, we performed chemical analyses of different SBM and RM varieties and selected those most similar to the most commonly used varieties in beef diets in our commercial conditions. In Trial 2, the in situ technique was used to estimate the dry matter (DM) and the crude protein (CP) effective degradability of these protein ingredients. Finally, in Trial 3 a dual-flow continuous culture system was used to study the true digestibility, rumen fermentation and nitrogen metabolism of four isoenergetic and isonitrogenous diets (Table 2). Each one was formulated with one of these protein sources and for a targeted gain of 1.4 kg/day according to Fundación Española para el Desarrollo Animal (FEDNA) [16] recommendations for beef cattle. All diets were designed with a 90:10 concentrate to barley straw ratio.

Table 1. Chemical composition of protein sources [1] (DM basis).

Item	SM	CM	CE	RM
Chemical composition (g/kg)				
DM	881	915	928	883
OM	824	858	880	814
CP	467	395	351	398
EE	26.1	12.9	135.7	15.5
NDF	107	375	327	358
ADF	120	174	144	214
Lignin	5.3	40.3	25.8	88.2
NDICP	29.3	118.8	107.1	86.4
ADICP	13.5	62.7	63.0	58.2
Gross energy (Mcal/Kg)	4.38	4.35	4.96	4.29
Antinutritional factors Allyl isothiocyanate (mg/g)	-	0.09	0.09	0.13
Erucic acid (g/100g fat)	-	0.05	0.04	<0.01

[1] SM, Soybean meal 44% CP; CM, Camelina meal; CE, Camelina expeller; RM, Rapeseed meal 00.

Table 2. Ingredients and chemical composition of treatment diets [1] tested with the dual-flow continuous culture system.

Item	SMD [1]	CMD	CED	RMD
Ingredients (g/kg DM)				
Corn grain	376	379	343	364
Barley grain	318	272	308	286
Soybean meal 44%	114	-	-	-
Camelina meal	-	152	-	-
Camelina expeller	-	-	163	-
Rapeseed 00 meal	-	-	-	149
Barley straw	100	100	100	100
Palm oil	10	27	10	28
Soybean hulls	69	56	62	60
Calcium carbonate	9	10	10	9
Vitamin and mineral premix [2]	4	4	4	4
Chemical composition (g/kg DM)				
DM	889	894	893	889
OM	852	860	858	854
CP	125	132	132	130
EE	35.8	43.1	43.3	45.9
NDF	217	224	242	221
ADF	112	117	124	124
Lignin	12.2	104	20.4	21.7
Gross energy (Mcal/Kg)	3.97	4.07	4.08	4.07

[1] SMD, diet with soybean meal 44% CP; CMD, diet with camelina meal; CED, diet with camelina expeller; RMD, diet with rapeseed meal 00. [2] Vitamin and mineral premix: Vitamin A 8400 IU/Kg; Vitamin D3 1680 IU/Kg; Zinc oxide 85 mg/Kg; Iron carbonate 39 mg/Kg; Manganese oxide 30 mg/Kg.

2.1. Chemical Composition

Chemical composition of protein sources used in Trial 1 and Trial 2, and composition of ingredients and chemical composition of diets in Trial 3 are described in Tables 1 and 2, respectively. Samples, of either ingredients or diets, were ground to pass through a 1-mm sieve and analyzed in duplicate. The dry matter (DM) content was determined by drying samples for 24 h in a 103 °C forced air oven, and organic matter (OM) was determined after ignition of a sample in a muffle furnace at 550 °C overnight according to AOAC ID 950.05 [17]. Nitrogen content was determined by the Kjeldahl procedure AOAC ID 976.05 [17]. Ether extract (EE) was performed according to AOAC ID 920.30 [17].

The neutral detergent fiber (NDF) and acid detergent fiber (ADF) contents were determined sequentially by using an Ankom Fiber Analyzer (Ankom Technology, Fairport, NY, USA) in accordance with the methodology provided by the company. This is based on the procedure of Van Soest et al. [18] using a thermostable α–amylase and sodium sulfite. Ash determination was performed at the end of the sequential process to express fiber results on an ash-free basis. In the case of both camelina co-products, the concentration of the α–amylase and sodium sulfite were doubled compared to the original procedure to achieve acceptable repeatability of the determination. The lignin content was determined after fiber procedures using sulfuric acid 72%. Neutral detergent insoluble crude protein (NDICP) and acid detergent insoluble crude protein (ADICP) were determined by the aforementioned Kjeldahl procedure in residues obtained after fiber determinations. Gross energy content was measured by completely burning a sample of the ingredients to measure the heat produced using a bomb calorimeter (Parr 6300, Parr Instrument Company, Moline, IL, USA). The allyl isothiocyanate level was determined by a destilation-volumetry procedure according to the European Directive 71/250/EEC. The erucic acid content was analyzed by chromatography (Model 6890, Hewlett Packard, Palo Alto, CA, USA), according to American Oil Chemists' society (AOCS) method CE 2-66 [19].

2.2. Trial 1: In Vitro Digestibility

In vitro DM and OM digestibility of samples were obtained following the Tilley and Terry [20] procedure modified by Stern and Endres [21]. The experiment was performed in two periods with three replicates per period. The rumen liquid obtained from two different cows fed a 60:40 forage to concentrate ratio diet was diluted with the buffer solution proposed by McDougall [22] at a ratio of 1:4 (rumen fluid:buffer solution). A sample of 0.5 g was incubated in each tube with 50 mL of mixed solution in continuous agitation and anaerobiosis, which was achieved with the introduction of CO_2 at 39 °C. After 48 h of incubation, 0.2 g of pepsin and 2 mL of HCl were added in each tube. Tubes were maintained in continuous agitation at 39 °C for 24 h. The content of each tube was filtered using filtration crucibles, and the DM and OM of the residue obtained were determined as mentioned previously.

2.3. Trial 2: In Situ Degradability

Two cannulated cows fed ad libitum with a 60:40 forage to concentrate ratio diet were used. The DM and CP in situ degradability analysis was carried out by incubating nylon bags (Ankom Technology Corporation, Fairport, NY, USA) in the rumen, which were 5 cm × 10 cm with a pore size of 50 μm containing 1 g of sample that was ground to pass through a 1.5-mm sieve. In three repeated periods, one bag per sample, animal and hour was inserted into the rumen at 8.30 am directly before feeding. Incubation periods were 2, 4, 6, 8, 12, 24, 48 and 72 h. After rumen incubation, bags were subjected to a washing procedure consisting of three washing cycles of 5 min, and immediately frozen at −18 °C for further analyses. The 0-h nylon bags were treated with the same methodology but without passing through the rumen. Degradation of DM and CP parameters was calculated using the equation of Ørskov and McDonald [23]:

$$D = a + b\,(1 - e^{-ct}) \tag{1}$$

where D is the disappearance of either DM or CP at time t; a is an intercept representing the DM or CP soluble fraction, b is the fraction of insoluble but degradable DM or CP, c is the rate of disappearance of fraction b, and t is the time of incubation. The non-linear parameters a, b and c were estimated by an iterative least-squares procedure of SAS (v. 9.1; SAS Inst. Inc., Cary, NC, USA). The effective degradability (ED) of DM and CP were calculated using the equation:

$$ED = a + [bc/(c + k)] \tag{2}$$

where *a*, *b* and *c* are the same parameters as described earlier and k is the estimated rate of passage of 0.06/h according to Institut National de la Recherche Agronomique (INRA) [24].

2.4. Trial 3: Dual-Flow Continuous Culture System

2.4.1. Fermenters

Eight 1320 mL dual-flow continuous culture fermenters developed by Hoover et al. [25] were used in two replicated periods. Each experimental period consisted of 5 days for adaptation and 3 days for sampling. Fermenters were inoculated with ruminal liquid from two dairy cows fed a 60:40 forage to concentrate ratio diet. Fermentation conditions were maintained constant with a temperature of 39 °C, and pH at 6.1 ± 0.05 by infusion of 3 N HCl or 5 N NaOH, monitored and controlled by a computer. Liquid and solid constant dilution rates (0.1 and 0.05 h-1, respectively) were obtained by a continuous infusion of artificial saliva [26]. Finally, anaerobic conditions were maintained by infusion of N2 gas at a rate of 40 mL/min. Fermenters were daily fed 90 g of diet (DM basis) in three equidistant doses of 30 g.

2.4.2. Sample Collection

On the sampling days, collection vessels were maintained at 4 °C to prevent microbial activity. Solid and liquid effluents were mixed and homogenized for 1 min, and a 600 mL sample was removed by aspiration. Upon completion of each period, effluents from each fermenter collected over the three sampling days were composited and mixed and homogenized for 1 min. Subsamples were taken for total N, ammonia N, and VFA analyses. The remainder of the sample was lyophilized. Dry samples were analyzed for DM, ash, NDF, ADF, and purine contents.

Bacteria were isolated from the fermenter contents on the last day of each experimental period. Solid and liquid associated bacteria were isolated using a combination of several detachment procedures, which were selected to obtain the maximum detachment without affecting cell integrity [27]. To remove attached bacteria, 100 mL of a 2 g/L methylcellulose solution and small marbles (thirty measuring 2 mm in diameter, and fifteen of 4 mm) were added to each fermenter, incubated in the same fermenter contents at 39 °C, and mixed for 1 h. After incubation, fermenter contents were refrigerated for 24 h at 4 °C, and subsequently agitated for 1 h to dislodge loosely attached bacteria. Finally, the fermenter contents were filtered through cheesecloth and washed with saline solution (8.5 g/L NaCl). Bacterial cells were isolated within 4 h by differential centrifugation at 1000× *g* for 10 min, to obtain a supernatant without feed particles, which was then centrifuged at 20,000× *g* for 20 min to isolate bacterial cells. Pellets were rinsed twice with saline solution and recentrifuged at 20,000× *g* for 20 min. The final pellet was recovered with distilled water to prevent contamination of bacteria with ash. Bacterial cells were lyophilized and analyzed for DM, ash, N, and purine contents. Digestion of DM, OM, NDF, ADF and CP, and flows of total, non-ammonia, microbial, and dietary N were calculated as described by Stern and Hoover [28].

2.4.3. Chemical Analyses

Effluent DM was determined by lyophilizing 200 mL aliquots in triplicate. The DM, OM, total N, NDF, ADF, lignin and EE content of the lyophilized effluents, bacterial samples and diets were determined as described in Section 2.1. Effluent ammonia N was analyzed by colorimetry as described by Chaney and Marbach [29], where 4 mL of a 0.2 N HCl solution were added to 4 mL of filtered rumen fluid and frozen at −20 °C until later analysis. Samples were centrifuged at 15,000× *g* for 15 min, and the supernatant was used to determine ammonia N. Effluent samples for VFA determination were prepared as described by Jouany [30] and analyzed by gas chromatography: 1 mL of a solution made up of a 2 g/L solution of mercuric chloride, 0.002 mg/L of 4-methylvaleric acid as an internal standard, and 2 g/L orthophosphoric acid was added to 4 mL of filtered rumen fluid and frozen at −20 °C until later analysis. Samples were centrifuged at 3000× *g* for 30 min, and the supernatant analyzed by gas chromatography (Model 6890, Hewlett Packard, Palo Alto, CA, USA) using a polyethylene glycol TPA treated capillary column (BP21, SGE, Europe Ltd., Buckinghamshire, UK). The dimensions of the column were 30 m × 0.25 mm ID, with 0.25 μm film thickness. The injector was set at 275 °C with a split ratio of 4:1. Helium was used as the carrier gas. The initial temperature was 85 °C for 1 min, and increased by 3C/min until a final temperature of 220 °C was reached, and held for a further 2 min. The detector temperature was 275 °C. Samples of lyophilized effluent and bacterial cells were analyzed for adenine and guanine content by HPLC as described by Balcells et al. [31], using allopurinol as internal standard.

2.5. Statistical Analysis

Data from the in vitro digestibility experiment were analyzed using the MIXED procedure of SAS (v. 9.1; SAS Inst. Inc., Cary, NC, USA). The model contained the fixed effect of treatment, and the random effect of period. Statistical analysis of the dual-flow continuous culture system data was performed using the MIXED procedure, and multiple comparisons were performed by LSMEANS adjusted with the Tukey test using SAS (v. 9.1; SAS Inst. Inc., Cary, NC, USA). The model contained treatment as the fixed effect and period as the random effect. Statistical significance was declared at $p < 0.05$ and differences among means with $0.05 < p < 0.10$ were considered tendencies.

3. Results

3.1. In Vitro Digestibility with the Two-Stage Tilley and Terry Technique

In vitro DM digestibility (DMD) and OM digestibility (OMD) values are presented in Table 3. Soybean meal was the protein source with the highest digestibility ($p < 0.05$), both DMD and OMD. There were no significant differences ($p > 0.05$) among CM, CE and RM in DMD, and OMD values.

Table 3. In vitro digestibility of protein sources [1].

Item	SM	CM	CE	RM	SEM [2]	*p*-Value
Coefficients of digestibility						
Dry matter	0.74 [a3]	0.65 [b]	0.64 [b]	0.62 [b]	0.160	<0.001
Organic matter	0.87 [a]	0.72 [b]	0.70 [b]	0.72 [b]	0.131	<0.001

[1] SM, soybean meal 44% CP; CM, camelina meal; CE, camelina expeller; RM, rapeseed meal 00. [2] SEM, standard error of mean. [3] Means with different superscript differ statistically ($p < 0.05$).

3.2. In Situ Rumen Degradation

Nonlinear parameter estimates and ED values are presented in Table 4, and ruminal degradation kinetics of DM and CP are shown in Figures 1 and 2, respectively. The greatest DM soluble fraction was recorded with CE, while the remaining protein sources showed similar values. In contrast, the greatest fraction *b* was obtained in SM, being intermediate in CM and RM, and lowest in CE. The DM rate of the disappearance of fraction *b* was similar in CM and CE, and in an intermediate position between SM, with the lowest value, and RM, with the greatest. The ED of DM was similar in SM, CM and RM, and the greatest value was recorded in CE. In descending order, the CP *a* fraction values came out as follows: CE, RM, CM and SM. Soybean meal showed the highest *b* value, followed by CM, RM and CE. The lowest rate of CP degradation was found in SM, followed by CM and RM, and the greatest was recorded in CE. The lowest ED of CP was obtained in SM, the greatest in CE and RM, and the value was intermediate for CM.

Table 4. In situ nonlinear estimates [1] and effective degradability values of dry matter (ED) and crude protein (EDCP) of protein sources [2].

Item	SM	CM	CE	RM
Dry matter				
a	0.32 ± 0.012 [3]	0.35 ± 0.022	0.50 ± 0.013	0.37 ± 0.025
b	0.57 ± 0.015	0.44 ± 0.025	0.33 ± 0.015	0.40 ± 0.028
c (/h)	0.077 ± 0.001	0.135 ± 0.017	0.132 ± 0.014	0.180 ± 0.027
ED	0.64	0.66	0.72	0.67
Crude Protein				
a	0.24 ± 0.017	0.29 ± 0.027	0.47 ± 0.030	0.38 ± 0.019
b	0.79 ± 0.022	0.66 ± 0.031	0.48 ± 0.034	0.57 ± 0.017
c (/h)	0.064 ± 0.001	0.160 ± 0.017	0.249 ± 0.038	0.199 ± 0.013
ED	0.64	0.77	0.85	0.82

[1] *a*: Soluble fraction; *b*: Insoluble but degradable fraction; *c*: The rate (/h) of disappearance of *b* fraction; [2] SM, soybean meal 44% CP; CM, camelina meal; CE, camelina expeller; RM, rapeseed meal 00; [3] Mean ± standard error.

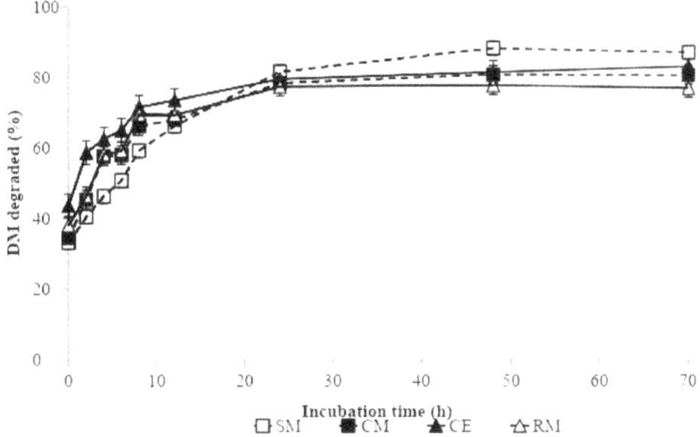

Figure 1. In situ dry matter degradation curve of protein sources: SM, Soybean meal 44% CP; CM, camelina meal; CE, camelina expeller; RM, rapeseed meal 00.

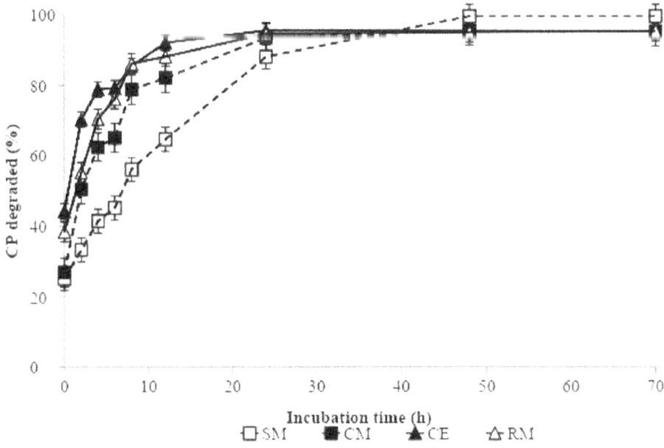

Figure 2. In situ crude protein degradation curve of protein sources: SM, Soybean meal 44% CP; CM, camelina meal; CE, camelina expeller; RM, rapeseed meal 00.

3.3. Dual-Flow Continuous Culture System

Results of dual-flow continuous culture system are presented in Tables 5–7.

Table 5. Effect of treatment diet on apparent digestibility and true digestibility in the dual-flow continuous culture system.

Item	Diet [1]				SEM [2]	*p*-Value
	SMD	CMD	CED	RMD		
Apparent digestibility (g/kg)						
DM	442	431	413	412	34.6	0.73
OM	346	352	372	360	22.2	0.58
NDF	380	357	408	359	67.2	0.83
ADF	370	385	421	345	77.8	0.76
True digestibility (g/kg)						
DM	548	531	563	521	37.8	0.66
OM	480 [b3]	486 [ab]	574 [a]	494 [ab]	37.1	0.06

[1] SMD, diet with soybean meal; CMD, diet with camelina meal; CED diet with camelina expeller; RMD, diet with rapeseed meal. [2] SEM, standard error of the mean. [3] Means with different superscript tended to differ ($p < 0.10$).

Table 6. Effect of treatment diet on volatile fatty acids (VFA) concentration and profile in the dual-flow continuous culture system.

Item	Diet [1]				SEM [2]	*p*-Value
	SMD	CMD	CED	RMD		
Total VFA (mM)	119.6 [b3]	135.6 [a]	138.8 [a]	121.0 [ab]	7.06	0.009
VFA (mol/100 mol)						
Acetate	45.2	46.0	46.2	44.0	2.44	0.777
Propionate	32.1	35.9	34.6	33.6	2.01	0.236
Butyrate	16.4 [a]	12.1 [b]	13.1 [ab]	15.0 [ab]	1.46	0.016
Iso-butyrate	0.39	0.41	0.45	0.43	0.04	0.357
Valerate	5.23	4.42	4.39	5.89	1.18	0.494
Iso-valerate	0.74	1.14	1.25	1.11	0.36	0.412
BCVFA [4] (mM)	1.14	1.55	1.70	1.54	0.40	0.399
Acetate:propionate ratio	1.42	1.31	1.35	1.32	0.12	0.773

[1] SMD, diet with soybean meal 44% CP; CMD, diet with camelina meal; CED, diet with camelina expeller; RMD, diet with rapeseed meal 00. [2] SEM, standard error of mean. [3] Means with different superscripts differ statistically ($p < 0.05$). [4] BCVFA, Branch-chained VFA.

Table 7. The effect of treatment diet on N metabolism in the dual-flow continuous culture system.

Item	Diet [1]				SEM [2]	*p*-Value
	SMD	CMD	CED	RMD		
NH3-N (mg/100 mL)	1.60	1.37	1.80	2.23	0.51	0.42
N flow (g/d)						
Total	2.37	2.49	2.46	2.46	0.07	0.34
Ammonia	0.05	0.04	0.06	0.07	0.02	0.41
Non-ammonia	2.32	2.44	2.36	2.39	0.09	0.51
Bacterial	0.86	0.83	1.23	0.87	0.19	0.13
Dietary	1.46 [ab3]	1.61 [a]	1.13 [b]	1.52 [ab]	0.18	0.06
Crude protein degradation (g/kg)	229 [ab]	196 [b]	423 [a]	236 [ab]	90.62	0.07
EMPS [4]	35.27	35.61	37.06	34.12	4.19	0.89

[1] SMD, diet with soybean meal 44% CP; CMD, diet with camelina meal; CED, diet with camelina expeller; RMD, diet with rapeseed meal 00. [2] SEM, standard error of mean. [3] Means with different superscript tended to differ statistically ($p < 0.10$). [4] The efficiency of microbial protein synthesis (g bacterial N/Kg organic matter truly digested).

Apparent digestibility of DM, NDF and ADF, and true DM digestibility were not different among diets (Table 5). In contrast, true OM digestibility tended to be affected by diet, with the highest value in CED and the lowest in SMD. Total VFA was higher in CED and CMD than in SMD ($p = 0.009$), but there was no difference between these diets and RMD (Table 6). Butyrate proportion was lower in CMD than in SMD ($p = 0.016$), but the proportion detected in CED and RMD was not different from that detected in CMD and SMD. There were no differences among diets in acetate, propionate, iso-butyrate, valerate, and iso-valerate proportions, the BCVFA, and the acetate to propionate ratio. Table 7 shows the results on nitrogen metabolism. There were no diet effects on total NH3-N, and flows of total, ammonia, non-ammonia and bacterial N, and EMPS. However, dietary N flow tended to be lower in CED than in CMD ($p = 0.06$), and CP degradation tended to be higher in CED than in CMD ($p = 0.07$).

4. Discussion

Chemical composition of feedstuffs commonly used in beef cattle diets, like SM and RM, was in accordance with published values [24,32]. Comparing both *Camelina sativa* co-products, CM contained higher values of CP, NDF, ADF and lignin compared to CE, but lower EE content. The CP content of the CM used in the present study was similar to the values referenced in the literature [8,33]. On the contrary, the EE content of these references was closer to the value found in CE. In reference to the anti-nutritional factors, the content of allyl isothiocyanate, as a major metabolite of glucosinolates of CM and CE, was below that of RM, an ingredient that is considered to have a negligible amount of glucosinolates. Tripathi and Mishra [11] obtained values between 0.3 mg/g and 2.1 mg/g of allyl isothiocyanate in different varieties of RM obtained in diverse oil extraction processes. Values of erucic acid presented by CM and CE were below 1% of the fat fraction that is considered to be the threshold of a rapeseed meal zero erucic acid variety [16]. Therefore, the comparison between camelina co-products and RM suggests that their use would not represent a nutritional problem for beef cattle nutrition.

In vitro DM and OM digestibility were not different in both camelina co-products. These digestibility values were similar to those recorded with RM but lower than in SM. The higher CP content and the lower fiber content of SM would explain these higher digestibility coefficients. Yong-Gang Liu et al. [33] and Moss and Givens [34], also using in vitro studies, reported equivalent OMD coefficients for SM and RM (0.89 and 0.74, respectively) to those reported in the present study.

Effective DM degradation was higher in CE than in CM. This difference would be related with the *a* and *b* fractions, which were higher and lower, respectively, in CE than in CM, but without differences in the rate of DM disappearance. Dry matter ED of CM was close to the values recorded in RM and SM. However, this result was obtained with different kinetic parameters because although the *a* fraction was similar among these protein sources, the *b* fraction was higher in SM and lower in CE, while rate of DM disappearance was higher in RM and lower in SM. In the comparison between SM and RM, our results agree with Heendeniya et al. [35] and Wulf and Südekum [36], who presented similar DM kinetic parameters between SM and RM, with a higher insoluble but degradable fraction and slower degradation rate in SM compared with RM. In contrast, Prestløkken [37] (1999) did not report differences between SM and RM, and Maxin et al. [38] found a greater DM degradation rate and ED in SM than in RM. The highest *b* fraction recorded in SM could be explained by its chemical composition: High CP content, low fiber content, and low content of NDICP and ADICP, resulting in the highest DMD and OMD in comparison with the remaining protein sources.

Effective CP degradation of CE was higher than that of CM, with a higher *a* fraction and rate of CP disappearance and a lower *b* fraction. Lawrence and Anderson [39] studied the kinetic parameters of a CM and recorded CP degradability very similar to the result obtained in the present study. However, the chemical composition of their CM was closer to our CE, because their EE content was 143 g/kg DM, similar to the 135 g/kg DM of our CE and different from the 13 g/kg DM of our CM. Camelina expeller and RM showed the same ED of CP, but the *a* fraction and rate of disappearance were lower and the *b* fraction higher in RM than in CE. The lowest ED of CP was found in SM, in comparison with the remaining protein sources. The values obtained with SM were in accordance with the literature [36,37,40]. Conversely, RM soluble fraction was very high in comparison with Prestløkken [37], Wulf and Südekum [36] and Heendeniya et al. [35], and close to the result obtained by Maxin et al. [38].

Chemical composition of the diets tested in the dual-flow continuous culture system confirmed that they were formulated to be isonergetic and isonitrogenous, with only a slight decrease in the CP content in the SMD diet. No differences among diets in apparent digestibilities and true DMD were observed. In contrast, Brandao et al. [41], in a dual-flow continuous culture system trial, observed lower NDF digestibility in diets with either 50% or 100% of CE instead of RM, meaning that RM was included at 10.3% and 0% (on DM basis), respectively, compared to the control diet with RM, in which RM was included at 20.6% as the main protein source. The lack of differences in our Trial between SMD and RMD digestibilities is in accordance with Paula et al. [42] who also compared, in a dual-flow continuous culture system, a SM diet with RM diets with different rumen-undegradable protein (RUP) content. However, true OMD tended to be affected by diet. The trend detected in the highest true OMD value in CED compared to SMD, although not statistically different from CMD and RMD, could be related with the highest fermentation activity observed in this diet, where high values of total VFA were recorded in comparison with SMD. With similar true OMD to SMD but also CED, CMD and RMD did not differ from CED in total VFA concentration. While there were no differences among diets in acetate and propionate proportion, butyrate proportion in CMD was lower than in SMD. During the synthesis of acetate and butyrate, there is also a methane generation that reduces its energy efficiency compared with the synthesis of propionate, in the formation of which there is no loss of any carbon during the reaction [43]. Therefore, it seems that CMD could be more efficient in energy usage. This could be related with the higher total VFA concentration detected in CMD and CED than in SMD. Although there was no different BCVFA content among diets, the numerically higher value found in CED diet could be related with the high CP degradability recorded in this diet, because BCVFA are produced when rumen microbes degrade protein [44]. The highest CP degradation detected in the CED would coincide with the highest CP degradation recorded in the in situ trial. In contrast, the same did not occur in the case of SMD and RMD when comparing in situ and fermenter results. In addition we obtained a lower CP degradation in CMD than in CED, and the same result between SMD and RMD. In the in vitro experiment of Brandao et al. [41], there were also no differences between RMD and CED treatments. In the in situ trial, in accordance with Huhtanen et al. [45] and most feed tables [24,46], a greater RUP content for SM compared with RM was recorded. However, Brito et al. [47] and Brito and Broderick [48] reported no differences in omasal RUP flows when isonitrogenous RM and SM diets were compared, in accordance with our fermenter results. Higher RUP of SMD, RMD and CMD could explain the effects of diet on dietary N flow, in which a tendency was found for the lowest flow in CED, highest in CMD, and intermediate in SMD and RMD diets. This difference in the RUP content between CED and CMD could be due to the second oil-extraction chemical treatment performed in CM, which is not applied in CE. This second extraction could modify the rumen digestibility of protein by decreasing the accessibility of rumen microorganisms to protein. Chemical and physical extractions are strategies commonly used to reduce ruminal CP degradability and increase RUP [45,49]. Moreover, this could also explain the numerical results observed in bacterial N flows, which were numerically higher in CED than in the remaining diets. When comparing a high RUP diet with a basal diet, Ipharraguerre and Clark [50] reported that a significant decrease in the flow of microbial N to the small intestine occurred with RUP supplements. Hoover and Stokes [51] concluded that when rumen degradable protein is replaced by RUP, the microbial growth in the rumen can decrease. The high OMD, the high CP degradation and the low RUP of CED could have promoted the microbial growth. This would be in agreement with Santos et al. [49], who concluded that high RUP diets resulted in decreased microbial protein synthesis. However, there was no significant difference between treatments in EMPS. It is important to remark that CED and CMD presented equivalent results in EMPS than diets formulated with SM and RM, the most common protein sources used in beef cattle. The lack of differences between treatments could be related with the low NH3-N concentration recorded in all effluents, which did not attain the 5 mg/dL, concentration usually recommended to ensure maximum microbial growth [52]. However, Russell et al. [53] reported no difference in microbial growth when NH3-N concentration was below 5 mg/dL or greater than 16 mg/dL. Owens and Bergen [54] concluded that the minimum amount of NH3-N to maximize bacterial growth was 2.5 mg/dL. Other studies have also reported NH3-N concentrations below 5 mg/ dL without reporting differences in microbial yield [55]. Finally, considering that barley protein is more degradable than corn protein [56,57], the differences in CP degradation detected in the present study could also be explained by the different barley to corn ratio of the diets, this ratio being 0.90, 0.85, 0.79, and 0.72 for CED, SMD, RMD, and CMD, respectively.

The results for digestibility, rumen fermentation and nitrogen flow recorded in CMD were similar to those found in diets formulated with standard proteins such as SMD and RMD, but with an increase in total VFA concentration. These similar results could be related to the close chemical composition of the main protein sources of these diets, and especially the very similar CP digestibility detected for these diets compared with CED. As has been argued previously in the specific case of the comparison between CMD and CED, the chemical oil extraction process could decrease the protein availability for the rumen microorganisms.

5. Conclusions

In conclusion, although both camelina co-products showed some differences between them in CP degradability and N flows, when used as a main protein source in isocaloric and isonitrogenous diets, neither of them differed from rapeseed meal in coefficients of digestibility, rumen fermentation and nitrogen metabolism. Thus, camelina co-products could be an alternative ingredient for beef cattle diets. Further in vivo research is necessary to confirm these results and to ascertain the suitable level of inclusion of these co-products.

Author Contributions: Conceptualization, L.C., A.F.; methodology, H.S, M.L.-S., L.C., A.F.; formal analysis, H.S, M.L.-S.; investigation, H.S, L.C., A.F.; data curation, H.S, L.C., A.F.; writing—original draft preparation, H.S.; L.C. and A.F. critically reviewed the manuscript. All authors revised and approved the final manuscript.

Funding: This research was funded by Spanish Ministry of Economy and Competitiveness and the European Regional Development Fund (Research Project RTC-2015-3265-5).

Acknowledgments: This manuscript had been proofread by Phil Grayston, a native English-speaking instructor.

Conflicts of Interest: The authors declare no conflict of interest.

References

1.   Zagorakis, K.; Liamadis, D.; Milis, C.; Dotas, V.; Dotas, D. Nutrient digestibility and in situ degradability of alternatives to soybean meal protein sources for sheep. *Small Rumin. Res.* **2015**, *124*, 38–44. [CrossRef]
2.   Putnam, D.H.; Budin, J.T.; Filed, L.A.; Breene, W.M. A promising low input oilseed. *New Crops. J.* **1993**, 314–322.
3.   Wittkop, B.; Snowdon, R.J.; Friedt, W. Status and perspectives of breeding for enhanced yield and quality of oilseed crops for Europe. *Euphytica* **2009**, *170*, 131–140. [CrossRef]
4.   Fröhlich, A.; Rice, B. Evaluation of *Camelina sativa* oil as a feedstock for biodiesel production. *Ind. Crop. Prod.* **2005**, *21*, 25–31. [CrossRef]
5.   Moser, B.R. Camelina (*Camelina sativa* L.) oil as a biofuels feedstock: Golden opportunity or false hope? *Lipid Technol.* **2010**, *22*, 270–273. [CrossRef]
6.   Shonnard, D.R.; Williams, L.; Kalnes, T.N. Camelina-derived jet fuel and diesel: Sustainable advanced biofuels. *Environ. Prog. Sustain. Energy* **2010**, *29*, 382–392. [CrossRef]
7.   Waraich, E.A.; Ahmed, Z.; Ahmad, R.; Yasin Ashraf, M.; Saifullah Naeem, M.S.; Rengel, Z. *Camelina sativa*, a climate proof crop, has high nutritive value and multiple-uses: A review. *Aust. J. Crop Sci.* **2013**, *7*, 1551–1559.
8.   Hurtaud, C.; Peyraud, J.L. Effects of Feeding Camelina (Seeds or Meal) on Milk Fatty Acid Composition and Butter Spreadability. *J. Dairy Sci.* **2007**, *90*, 5134–5145. [CrossRef]
9.   Zubr, J. Oil-seed crop: *Camelina sativa*. *Ind. Crop. Prod.* **1997**, *6*, 113–119. [CrossRef]

10. Betz, J.M.; Fox, W.D. High-performance liquid chromatographic determinations of glucosinolates in *Brassica* vegetables. Food Phytochemicals I: Fruits and vegetables. In *Food Phytochemicals for Cancer Prevention I*; American Chemical Society: Washington, DC, USA, 1994; pp. 181–196.

11. Tripathi, M.K.; Mishra, A.S. Glucosinolates in animal nutrition: A review. *Anim. Feed Sci. Technol.* **2007**, *132*, 1–27. [CrossRef]

12. Moriel, P.; Nayigihugu, V.; Cappellozza, B.I.; Gonçalves, E.P.; Krall, J.M.; Foulke, T.; Hess, B.W. Camelina meal and crude glycerin as feed supplements for developing replacement beef heifers. *J. Anim. Sci.* **2011**, *89*, 4314–4324. [CrossRef] [PubMed]

13. Cappellozza, B.I.; Cooke, R.F.; Bohnert, D.W.; Cherian, G.; Carroll, J.A. Effects of camelina meal supplementation on ruminal forage degradability, performance, and physiological responses of beef cattle. *J. Anim. Sci.* **2012**, *90*, 4042–4054. [CrossRef] [PubMed]

14. Bonjean, A.; Le Goffic, F. La caméline—*Camelina sativa* (L.) Crantz: Une opportunité pour l'agriculture et l'industrie Européennes. *Oléagineux Corps Gras Lipides* **1999**, *6*, 28–34.

15. Lawrence, R.D.; Anderson, J.L.; Clapper, J.A. Evaluation of camelina meal as a feedstuff for growing dairy heifers. *J. Dairy Sci.* **2016**, *99*, 6215–6228. [CrossRef] [PubMed]

16. Ferret, A.; Calsamiglia, S.; Bach, A.; Devant, M.; Fernández, C.; García-Rebollar, P. *Necesidades Nutricionales Para Rumiantes de Cebo. Normas FEDNA*; Ed Fundación Española para el Desarrollo de la Alimentación Animal: Madrid, Spain, 2008.

17. AOAC. *Official Methods of Analysis*, 15th ed.; Association of Official Analytical Chemists: Arlington, VA, USA, 1990.

18. Van Soest, J.P.; Robertson, J.B.; Lewis, B.A. Methods of dietary fiber, neutral detergent fiber, and nonstarch polysaccharides in relation to animal nutrition. *J. Dairy Sci.* **1991**, *71*, 1587–1597. [CrossRef]

19. American Oil Chemists' Society. *Official Methods and Recommended Practices of the AOCS*; American Oil Chemists' Society: Champaign, IL, USA, 1998.

20. Tilley, J.M.A.; Terry, R.A. A two-stage technique for the in vitro digestion of forage crops. *J. Br. Grassl. Soc.* **1963**, *18*, 104–111. [CrossRef]

21. Stern, M.D.; Endres, M.I. *Laboratory Manual*; Department of Animal Science, University of Minnesota: Heights, MN, USA, 1991; pp. 90–92.

22. McDougall, E.I. Studies on ruminant saliva. 1. The composition and output of sheep's saliva. *Biochem. J.* **1948**, *43*, 99–109. [CrossRef]

23. Orskov, E.R.; Mcdonald, I. The estimation of protein degradability in the rumen from incubation measurements weighted according to rate of passage. *J. Agric. Sci.* **1979**, *92*, 499–503. [CrossRef]

24. INRA. *Nutrition of Cattle, Sheep and Goats: Animal Needs-Values of Feeds*; Quae Editions: Paris, France, 2002.

25. Hoover, W.H.; Crooker, B.A.; Sniffen, C.J. Effects of Differential Solid-Liquid Removal Rates on Protozoa Numbers in Continous Cultures of Rumen Contents. *J. Anim. Sci.* **1976**, *43*, 528–534. [CrossRef]

26. Weller, R.A.; Pilgrim, A.F. Passage of Portozoa and volatile fatty acids from the rumen of the sheep and from a continuos in vitro fermentation system. *Br. J. Nutr.* **1974**, *32*, 341. [CrossRef]

27. Whitehouse, N.L.; Olson, V.M.; Schwab, C.G.; Chesbro, W.R.; Cunningham, K.D.; Lykos, T. Improved techniques for dissociating particle-associated mixed ruminal microorganisms from ruminal digesta solids. *J. Anim. Sci.* **1994**, *72*, 1335–1343. [CrossRef] [PubMed]

28. Stern, M.D.; Hoover, W.H. The dual flow continuous culture system. In Proceedings of the Continuous Culture Fermenters: Frustation or fermentation, Northeast ADSA-ASAS Regional meeting, Chazy, NY, USA, 8 July 1990; pp. 17–32.

29. Chaney, A.L.; Marbach, E.P. Modified reagents for determination of urea and ammonia. *Clin. Chem.* **1962**, *8*, 130–132. [PubMed]

30. Jouany, J.P. Volatile fatty acids and alcohol determination in digestive contents, silage juice, bacterial cultures and anaerobic fermentor contents. *Sci. Aliment.* **1982**, *2*, 131–144.

31. Balcells, J.; Guada, J.; Peiró, J.M.; Parker, D.S. Simultaneous determination of allantoin and oxypurines in biological fluids by high performance liquid chromatography. *J. Chromatogr.* **1992**, *575*, 153–157. [CrossRef]

32. NRC. *Nutrient Requirements of Dairy Cattle*, 8th ed.; National Academy of Sciences: Washington, DC, USA, 2016.

33. Liu, Y.G.; Steg, A.; Hindle, V.A. Rumen degradation and intestinal digestion of crambe and other oilseed by-products in dairy cows. *Anim. Feed. Sci. Technol.* **1994**, *45*, 397–409. [CrossRef]

34. Moss, A.R.; Givens, D.I. The Chemical-Composition, Digestibility, Metabolizable Energy Content and Nitrogen Degradability of Some Protein-Concentrates. *Anim. Feed. Sci. Technol.* **1994**, *47*, 335–351. [CrossRef]

35. Heendeniya, R.G.; Christensen, D.A.; Maenz, D.D.; McKinnon, J.J.; Yu, P. Protein fractionation byproduct from canola meal for dairy cattle. *J. Dairy Sci.* **2012**, *95*, 4488–4500. [CrossRef]

36. Wulf, M.; Südekum, K.H. Effects of chemically treated soybeans and expeller rapeseed meal on in vivo and in situ crude fat and crude protein disappearance from the rumen. *Anim. Feed. Sci. Technol.* **2005**, *118*, 215–227. [CrossRef]

37. Prestløkken, E. In situ ruminal degradation and intestinal digestibility of dry matter and protein in expanded feedstuffs. *Anim. Feed. Sci. Technol.* **1999**, *77*, 1–23. [CrossRef]

38. Maxin, G.; Ouellet, D.R.; Lapierre, H. Ruminal degradability of dry matter, crude protein, and amino acids in soybean meal, canola meal, corn, and wheat dried distillers grains. *J. Dairy Sci.* **2013**, *96*, 5151–5160. [CrossRef]

39. Lawrence, R.D.; Anderson, J.L. Ruminal degradation and intestinal digestibility of camelina meal and carinata meal compared with other protein sources. *Prof. Anim. Sci.* **2018**, *34*, 10–18. [CrossRef]

40. Batajoo, K.K.; Shaver, R.D. In situ dry matter, crude protein, and starch degradabilities of selected grains and by-product feeds. *Anim. Feed Sci. Technol.* **1998**, *71*, 165–176. [CrossRef]

41. Brandao, V.L.N.; Silva, L.G.; Paula, E.M.; Monteiro, H.F.; Dai, X.; Lelis, A.L.J.; Faccenda, A.; Poulson, S.R.; Faciola, A.P. Effect of replacing canola meal with solvent-extracted camelina meal on microbial fermentation in a dual-flow continuous culture system. *J. Dairy Sci.* **2018**, *101*, 9028–9040. [CrossRef] [PubMed]

42. Paula, E.M.; Monteiro, H.F.; Silva, L.G.; Benedeti, P.D.B.; Daniel, J.L.P.; Shenkoru, T.; Faciola, A.P. Effects of replacing soybean meal with canola meal differing in rumen-undegradable protein content on ruminal fermentation and gas production kinetics using 2 in vitro systems. *J. Dairy Sci.* **2017**, *100*, 5281–5292. [CrossRef] [PubMed]

43. McDonald, P. *Animal Nutrition*, 6th ed.; Pearson education: Edinburg, UK, 2002; pp. 84–215.

44. Bryant, M.P. Nutritional Requirements of the Predominant Rumen Cellulolytic Bacteria. *Fed. Proc.* **1973**, *32*, 1809–1813. [PubMed]

45. Huhtanen, P.; Hetta, M.; Swensson, C. Evaluation of canola meal as a protein supplement for dairy cows: A review and a meta-analysis. *Can. J. Anim. Sci.* **2011**, *91*, 529–543. [CrossRef]

46. NRC. *Nutrient Requirements of Dairy Cattle*, 7th ed.; National Academy of Sciences: Washington, DC, USA, 2001.

47. Brito, A.F.; Broderick, G.A. Effects of Different Protein Supplements on Milk Production and Nutrient Utilization in Lactating Dairy Cows. *J. Dairy Sci.* **2007**, *90*, 1816–1827. [CrossRef]

48. Brito, A.F.; Broderick, G.A.; Reynal, S.M. Effects of Different Protein Supplements on Omasal Nutrient Flow and Microbial Protein Synthesis in Lactating Dairy Cows. *J. Dairy Sci.* **2007**, *90*, 1828–1841. [CrossRef]

49. Santos, F.A.P.; Santos, J.E.P.; Theurer, C.B.; Huber, J.T. Effects of rumen-undegradable protein on dairy cow performance: A 12-year literature review. *J. Dairy Sci.* **1998**, *81*, 3182–3213. [CrossRef]

50. Ipharraguerre, I.R.; Clark, J.H. Impacts of the Source and Amount of Crude Protein on the Intestinal Supply of Nitrogen Fractions and Performance of Dairy Cows. *J. Dairy Sci.* **2005**, *88*, E22–E37. [CrossRef]

51. Hoover, W.H.; Stokes, S.R. Balancing carbohydrates and proteins for optimum rumen microbial yield. *J. Dairy Sci.* **1991**, *74*, 3630–3644. [CrossRef]

52. Satter, L.D.; Slyter, L.L. Effect of ammonia concentration on rumen microbial protein production in vitro. *Br. J. Nutr.* **1974**, *32*, 199. [CrossRef] [PubMed]

53. Russell, J.B.; Sniffen, C.J.; Van Soest, J. Effect of carbohydrate limitation on degradation and utilization of casein by mixed rumen bacteria. *J. Dairy Sci.* **1983**, *66*, 763–775. [CrossRef]

54. Owens, F.N.; Bergen, W.G. Nitrogen metabolsim of ruminant animals: Historical Perspective, Current Understanding and Future Implications. *J. Anim. Sci.* **1983**, *57*, 498–518. [PubMed]

55. Bach, A.; Stern, M.D. Effects of different levels of methionine and ruminally undegradable protein on the amino acid profile of effluet from continuous culture feremeters. *J. Anim. Sci.* **1999**, *77*, 3377–3384. [CrossRef]

56. Spicer, L.A.; Theurer, C.B.; Sowe, J.; Noon, T.H. Ruminal and post-ruminal utilization of nitrogen and starch from sorghum grain -corn- and barley-based diets by beef steers. *J. Anim. Sci.* **1986**, *62*, 521. [CrossRef]

57. Herrera-Saldana, R.E.; Huber, J.T.; Poore, M.H. Dry Matter, Crude Protein, and Starch Degradability of Five Cereal Grains. *J. Dairy Sci.* **1990**, *73*, 2386–2393. [CrossRef]

Article

Relations of Ruminal Fermentation Parameters and Microbial Matters to Odd- and Branched-Chain Fatty Acids in Rumen Fluid of Dairy Cows at Different Milk Stages

Keyuan Liu [1,2,†], Yang Li [2,†], Guobin Luo [2,3], Hangshu Xin [2], Yonggen Zhang [2,*] and Guangyu Li [1,*]

[1] Institute of Special Economic Animal and Plant Science, Chinese Academy of Agricultural Sciences, Changchun 130112, China; liukeyuan0212@163.com
[2] Department of Animal Science and Technology, Northeast Agricultural University, Harbin 150030, China; liyang1405053@sina.com (Y.L.); guobinluo@126.com (G.L.); laura_liuky@foxmail.com (H.X.)
[3] Zhejiang NHU Company Ltd., Shaoxing 312500, China
* Correspondence: zhangyonggen@sina.com (Y.Z.); liguangyu@caas.cn (G.L.)
† These authors contributed equally.

Received: 1 November 2019; Accepted: 18 November 2019; Published: 22 November 2019

Simple Summary: The objective of this study was to determine the relationships between milk odd- and branched-chain fatty acids (OBCFAs) and ruminal fermentation parameters, microbial populations, and base contents. Significant relationships existed between the concentrations of C11:0, *iso*-C15:0, *anteiso*-C15:0, C15:0, and *anteiso*-C17:0 in rumen and milk. The total OBCFA content in milk was positively related to the acetate molar proportion but negatively correlated with isoacid levels. The adenine/N ratio was negatively related to milk OBCFA content but positively associated with the *iso*-C15:0/*iso*-C17:0 ratio.

Abstract: The purpose of this research was to evaluate whether relationships exist between odd- and branched-chain fatty acids (OBCFAs) originating from milk fat and the corresponding data of ruminal fermentation parameters, microbial populations, and base contents that were used to mark microbial protein in rumen. Nine lactating Holstein dairy cows with similar body weights and parity were selected in this study, and the samples of rumen and milk were collected at the early, middle, and late stages, respectively. The rumen and milk samples were collected over three consecutive days from each cow, and the ruminal and milk OBCFA profiles, ruminal fermentation parameters, bacterial populations, and base contents were measured. The results showed that the concentrations of OBCFAs, with the exception of C11:0 and C15:0, were significantly different between milk and rumen ($p < 0.05$). The concentrations of *anteiso*-fatty acids in milk were higher than those in rumen, and the contents of linear odd-chain fatty acids were higher than those of branched-chain fatty acids in both milk and rumen. Significant relationships that existed between the concentrations of C11:0, *iso*-C15:0, *anteiso*-C15:0, C15:0, and *anteiso*-C17:0 in rumen and milk ($p < 0.05$). The total OBCFA content in milk was positively related to the acetate molar proportion but negatively correlated with isoacid contents ($p < 0.05$). The populations of *Ruminococcus albus*, *R. flavefaciens*, and *Eubacterium ruminantium* were significantly related to milk C13:0 contents ($p < 0.05$). The adenine/N ratio was negatively related to milk OBCFA content ($p < 0.05$) but positively associated with the *iso*-C15:0/*iso*-C17:0 ratio ($p < 0.05$). Milk OBCFAs were significantly correlated with ruminal fermentation parameters, ruminal bacterial populations, and base contents. Milk OBCFAs had the potential to predict microbial nitrogen flow, and the prediction equations for ruminal microbial nitrogen flow were established for OBCFAs in dairy milk.

Keywords: fermentation parameters; microbial populations; microbial bases; odd- and branched-chain fatty acids; lactation stages

---

1. Introduction

There is growing awareness that milk fat content can respond to physiological and metabolic health situations [1,2]. The milk fatty acid profile is a dynamic pattern influenced by many factors, such as lactational stage, season, and dietary composition [3–5]. The microbial processes of the rumen confer the ability to convert feeds into available nutrients for the ruminant animal [6]. Odd- and branched-chain fatty acids (OBCFAs) of ruminant milk generally originate from rumen bacteria [7] and are mainly present in bacterial membrane lipids [8]. Some studies have discussed the potential of OBCFAs as markers of rumen fermentation and ruminal bacteria [9–12].

Lactating dairy cow digestion is strongly determined by the microbial population in the rumen. In the rumen, microbial fermentation of feedstuffs produces volatile fatty acids (VFAs), which are the main energy supply substances in ruminants. There is a significant relationship between ruminal pH and the profile of VFAs available for absorption [13]. Hence, the composition and amount of milk fatty acids are determined by the proportions and the total amounts of fermentation end-products in ruminants [14]. Many studies have examined the effects of microbial protein synthesis and microbial nucleic acid composition in the rumen on protein nutrition [15]. The composition of the ruminal microbial ecosystem in the forestomach of ruminants is known to be affected by the type and quantity of the ration, feeding intervals, specific additives (e.g., antibiotics), and the host animal itself [16]. To identify the relationships between bacterial populations and milk OBCFA concentrations, seven kinds of bacteria species (cellulolytic or amylolytic bacteria) were selected for the current study. *Ruminococcus albus*, *R. flavefaciens*, *Fibrobacter succinogenes* [17], and *Eubacterium ruminantium* [18] are the predominant ruminal cellulolytic bacteria. The genus *Butyrivibrio fibrisolvens* is a heterogeneous bacterial taxon [19]. *Selenomonas ruminantium* [20] and *Streptococcus bovis* [21] are important for the degradation of starch and lactate, which are abundant in high-grain diets.

However, there are few studies on the relationships between milk OBCFAs during different lactation stages and ruminal bacterial populations. The objective of this research was to estimate the potential use of OBCFAs in milk to predict ruminal fermentation parameters and rumen microbial matter. First, we investigated whether there were relationships between the contents of OBCFA in milk and rumen, and the fermentation parameters, bacterial populations, and bases of milk during different stages of lactation. Second, we developed equations for fermentation parameters, bacterial populations, and bases using the independent datasets of OBCFAs and identified the best OBCFA combination.

2. Materials and Methods

2.1. Animals and Basal Diets

Nine lactating Holstein dairy cows of similar body weights (650 ± 33 kg body weight) were examined at the same fetal time, and samples of rumen and milk were collected at the early, middle, and late lactation stages. The milk yields were 35.44 ± 2.63, 37.62 ± 2.85, and 26.98 ± 2.79 kg/d in the early, middle, and late stages, respectively. The cows were fed total mixed rations (TMRs), whose composition and nutrition levels are shown in Table 1, according to the dairy nutrient requirements of the NRC (national research council) (2001) [22].

Table 1. Feed ingredients and chemical composition (g/kg DM) of the rations.

Ingredients	Content
Alfalfa hay	73
Chinese wildrye	43
Corn silage	334
Corn	220
Soybean meal	41
DDGS	214
Cottonseed meal	58
Molasses	5.0
NaCl	3.0
CaHPO$_4$	1.6
Limestone	5.4
Premix *	3.0
Nutrient levels	
NEL(MJ/kg) †	8.71
CP	162
NDF	312
ADF	183
Ca	8.7
TP	4.6

* The premix provided the following per kg of diets: Vitamin A 330,000 IU, Vitamin D 60,000 IU, Vitamin E 1000 IU, Zn 2100 mg, Mn 1500 mg, Cu 535 mg, Se 12 mg, I 45 mg. † NEL was an estimated value [22]. DM, dry matter; DDGS, Corn distillers dried grains; CP, crude protein; NDF, neutral detergent fiber; ADF, acid detergent fiber; TP, total phosphorus.

2.2. Samples Collection and Analysis Method

All the TMR samples were analyzed for DM, nitrogen (N) (AOAC 968.06), calcium (Ca) (AOAC 927.02), and total phosphorus (TP) (AOAC 965.17) according to the procedures of the AOAC (Association of Official Analytical Chemists) (1990) after air-drying at 60 ± 5 °C [23]. The content of crude protein (CP) was calculated as N × 6.25. The acid detergent fiber (ADF) and neutral detergent fiber (NDF) concentrations were analyzed based on the procedures described by Van Soest et al. [24] using the Ankom system (Ankom 220 fiber analyzer; Ankom, New York, USA) with a heat-stable α-amylase and expressed exclusive of residual ash. Net energy for lactation (NE$_L$) at a production level was calculated using an NRC summative approach from the dairy nutrient requirement [22].

The rumen contents were evacuated via the gastric canal over 3 consecutive days. One part of the rumen contents was filtered through four layer of cheesecloth and the filtrates were preserved at −20 °C for the analysis of the concentrations of VFAs and ammonia nitrogen (NH$_3$-N), after determining the pH that was obtained from samples by a pH meter (Sartorius Basic pH Meter, Gottingen, Germany). The filtered samples were treated according to the description of Li and Meng [25]. Then, the contents of VFAs were analyzed by a gas chromatography (GC 2010, Shimadzu, Tokyo, Japan) with an FFAP capillary column (HP-INNOWAX, 30 m × 0.25 mm × 0.2 μm, Agilent, California, USA), and using an ammonia-sensing electrode (Expandable Ion Analyzer EA 940, Orion, Massachusetts, USA) to determine the concentration of NH$_3$-N.

The other part of the rumen contents were squeezed through two layers of cheesecloth, and about 5 mL was preserved at −80 °C for the extraction of DNA [26] and nearly 100 mL was preserved at −20 °C for the analysis of OBCFA, microbial bases, and the total nitrogen (N). The samples of OBCFA were treated according to the description of Zhang et al. [27] and analyzed by a gas chromatography (GC 2010, Shimadzu, Tokyo, Japan) with an SP-2560^{TM} column for fatty acid methyl esters (100 m × 0.25 mm × 0.2 μm, Supelco, Pennsylvania, USA). The carrier gas was highly pure, and the injector pressure was held constant at 266.9 Kpa. The initial oven temperature was held at 170 °C for 30 min, and increased at 1.5 °C/min to 200 °C and held for 20 min, and then increased by 5 °C/min to 230 °C and held for 5 min. The rumen bases were extracted from freeze-dried samples using perchloric acid, as described by Vlaeminck et al. [9]. The standards of individual bases (≥99.5%, Aladdin, Shanghai, China) were formulated to a concentration of 50 mg/L. The mixed solution mixed by the standard base solutions with the same volume was serially diluted into 5 gradients and subsequently analyzed by HPLC (high performance liquid chromatography) using a C18 column (5 μ, 250 × 4.6 mm, Diamonsil, Guangzhou, China). The buffer solution (20 mM $NH_4H_2PO_4$) was run isocratically at 1 mL/min, and the effluent was monitored at 254 nm.

DNA was extracted from the rumen contents by the bead-beating procedure described previously by Reilly and Attwood [28]. In detail, 1.5 mL of rumen liquid was placed into a 2-mL centrifuge tube and later centrifuged for 5 min at 12,000 g/min, and the supernatant was subsequently removed. Phosphate buffer saline (PBS; 1.5 mL; pH = 8.0) was then added, and the sample was mixed and centrifuged for 5 min at 12,000 g/min. The supernatant was then removed. CTAB (hexadecyl trimethyl ammonium bromide) buffer (800 μL; sterilized solution containing 4 g of CTAB, 16.364 g of NaCl, 20 mL of 1 mol/L Tris-HCl, pH = 8.0, and 8 mL of 0.5 mol/L EDTA (Ethylene Diamine Tetraacetic Acid) to a final volume of 200 mL) was then added, and the sample was mixed, cultured for 20 min in 70 °C, and centrifuged for 10 min at 10,000 g/min. The supernatant was then transferred to a new centrifuge tube, and 5 μL of RNA enzymes (10 mg/mL) were added. The sample was mixed and then incubated for 30 min at 37 °C. An equal volume of phenol/chloroform/isoamyl alcohol 25:24:1 solution was then added, and the sample was mixed for 15 to 30 s until a white emulsion appeared and then centrifuged for 10 min at 13,000 rpm. The supernatant was added to a new centrifuge tube, and the last step was then repeated until the interface became clear. After a clear interface was obtained, an 0.8-fold volume of the isopropyl alcohol mixture was added into the tube, and the sample was gently mixed, placed at room temperature for 5 min, and then placed at −20 °C overnight. The tube was then centrifuged for 15 to 20 min at 13,000 g/min, and the supernatant was carefully decanted to leave a white DNA precipitation in the bottom of the tube. Ice-cold 70% ethanol (750 μL) was then added to the tube to gently resuspend the DNA precipitation, and the sample was then centrifuged for 10 to 15 min at 12,000 g/min. The supernatant was decanted, and the DNA was allowed to air dry. The DNA was then dissolved with 100 μL (depending on the precipitation volume) of TE (Tris-EDTA) buffer and incubated at 70 °C for 5 min. The DNA solution was then stored at −20 °C until use. The air-dried DNA pellet was redissolved in TE buffer (10 mmol/L Tris-HCl, 1 mmol/L EDTA, pH 8.0) and diluted to concentrations of 10 ng/μL and stored at 4 °C until real-time PCR amplification. The bacterial 16SrRNA genes were amplified using absolute quantification PCR (qPCR). The standard DNA in real-time PCR used plasmid DNA containing the respective target gene sequence, which was obtained by PCR cloning using the species-specific primer set. The specific method was described by Singha et al. [29]. The primer sequences for the 16SrRNA genes and specific amplifications of the correct size are shown in Table 2. The qPCR protocol was performed with ABI 7500 system software (ABI 7500, Massachusetts, USA) using TOYOBO (Osaka, Japan) DNA Master SYBR Green II.

Table 2. Primers for real time-PCR.

Bacteria species	Primer Sequence (5′-3′)	Product Size (bp)	Reference
Fibrobacter succinogenes	F-GGCGGGATTGAATGTACCTTGAGA R-TCCGCCTGCCCCTGAACTATC	204	Yang (2007) [30]
Ruminococcus albus	F-GTTTTAGGATTGTAAACCTCTGTCTT R-CCTAATATCTACGCATTTCACCGC	273	Yang (2007) [30]
Ruminococcus flavefaciens	F-GATGCCGCGTGGAGGAAGAAG R-CATTTCACCGCTACACCAGGAA	278	Yang (2007) [30]
Butyrivibrio fibrisolvens	F-TAACATGAGTTTGATCCTGGCTC R-CGTTACTCACCCGTCCGC	113	Yang (2007) [30]
Eubacterium ruminantium	F-CTCCCGAGACTGAGGAAGCTTG R-GTCCATCTCACACCACCGGA	184	Stevenson, et al. (2007) [31]
Streptococcus bovis	F-TTCCTAGAGATAGGAAGTTTCTTCGG R-ATGATGGCAACTAACAATAGGGGT	127	Stevenson, et al. (2007) [31]
Selenomonas ruminantium	F-CAATAAGCATTCCGCCTGGG R-TTCACTCAATGTCAAGCCCTGG	138	Stevenson, et al. (2007) [31]

Nearly 100-mL milk samples were collected and preserved at −20 °C, and treated the samples that referred to a description of Vlaeminck et al. [28]. The OBCFA compositions were analyzed by a gas chromatography (GC 2010, Tokyo, Shimadzu, Japan) and the analysis program was the same as the rumen samples.

2.3. Statistical Analyses

All data statistical analyses were performed using SAS 9.2 (SAS Institute Inc.,Cary, NC, USA).

The data of variable analysis was done by the MIXED procedure. The MIXED model procedure with the inclusion of the random effect of the study was as described by St-Pierre [32]. The effect of different dietary F:C ratios on OBCFAs were estimated following:

$$Y_{ijk} = \mu + T_i + P_j + C_k + \epsilon_{ijk},$$ (1)

where Y_{ijk} is the individual observation, μ is the overall mean, T_i is the effect of the dietary treatment ($i = 3$; F:C = 30:70, 50:50, and 70:30), P_j is the effect of the experimental period, C_k is the effect of the cow, and ϵ_{ijk} is the residual error. The effect of the cow was treated as a random effect. For all statistical analyses, significance was declared at $p < 0.05$.

The correlation between the milk OBCFA profile, and rumen OBCFA concentrations, and data, which were obtained from the VFAs, NH$_3$-N, and pH, were analyzed by CORR PROC using the Pearson correlation method. The correlations were determined to be significant at $p \leq 0.05$.

The data of milk OBCFA were considered as an independent data set and the data of VFAs, NH$_3$-N, and pH were a dependent dataset. Multiple regression was applied using the STEPWISE method of the REG procedure. The SLENTRY and START values were all 0.05. The equations were determined by least squares estimation ($p \leq 0.05$). The regression equations were evaluated based on the root mean square error (RMSE) and coefficient of multiple determinations (R^2) of the regression model:

$$\text{RMSE} = \sqrt{\frac{1}{n} \times \sum_{i=1}^{n} (y_i - \hat{y}_i)^2},$$ (2)

where n is the number of observations, and y_i and $\hat{y}_i$ are the observed and predicted values, respectively.

3. Results

3.1. Changes during Different Milk Stages in OBCFA Production in Rumen Fluid and Milk

The changes in OBCFAs in rumen fluid and milk fat are shown in Table 3. The C11:0 and C13:0 contents were significantly higher in the late milk stage than in the early and middle stages, while the concentrations of C15:0 and *anteiso*-C17:0 were lower in the late stage than in the other stages in the rumen fluid ($p < 0.05$). The contents of *anteiso*-C15:0 and *iso*-C16:0 were most abundant in the late milk stage, and *iso*-C17:0 concentrations were lowest in milk fat ($p < 0.05$).

The total OBCFA and odd *anteiso*-chain fatty acids contents contained in milk were higher than those in rumen, and the linear odd-chain fatty acids were more abundant both in milk and rumen (Figure 1). Except for C15:0, the concentrations of other OBCFAs were significantly different in milk and rumen. Fatty acids with 15 carbon atoms were the major kinds in rumen, but the *anteiso*-chain fatty acids were much more abundant than other kinds in milk, especially concentrations of *anteiso*-C17:0.

Table 3. Changes of different milk stages on OBCFA production in rumen fluid and milk (g/100 g fatty acids).

OBCFA Profile	Rumen Fluid					Milk				
	Early Stage	Middle Stage	Late Stage	SEM	p	Early Stage	Middle Stage	Late Stage	SEM	p
C11:0	0.03 [b]	0.04 [b]	0.05 [a]	0.003	0.001	0.03	0.03	0.04	0.01	0.17
C13:0	0.06 [b]	0.06 [b]	0.08 [a]	0.004	0.004	0.01 [b]	0.02 [a]	0.01 [b]	0.002	0.04
Iso-C15:0	0.39	0.33	0.36	0.04	0.61	0.14	0.15	0.15	0.01	0.12
Anteiso-C15:0	0.58	0.55	0.56	0.04	0.79	0.38	0.39	0.38	0.03	0.93
C15:0	1.37 [a]	1.01 [b]	1.04 [b]	0.08	0.004	0.84	0.84	0.78	0.04	0.44
Iso-C16:0	0.30	0.30	0.33	0.02	0.62	0.25 [a]	0.19 [b]	0.16 [b]	0.02	0.02
Iso-C17:0	0.27	0.26	0.29	0.02	0.33	0.44 [b]	0.52 [a]	0.55 [a]	0.03	0.04
Anteiso-C17:0	0.07 [b]	0.09 [a]	0.07 [b]	0.004	0.002	0.83 [b]	1.13 [a]	1.01 [b]	0.07	0.02
C17:0	0.25	0.24	0.23	0.02	0.51	0.57	0.43	0.47	0.06	0.23
TOBCFA	3.32	2.87	3.00	0.14	0.08	3.54	3.57	3.69	0.09	0.42

OBCFA, odd- and branched-chain fatty acids; TOBCFA, the total odd- and branched-chain fatty acids. a, b, c, means with different letters within the same line and the same item differ significantly ($p < 0.05$). SEM = standard error of mean.

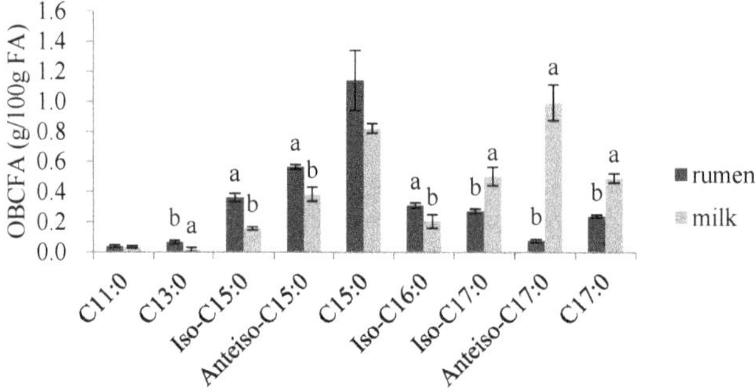

Figure 1. Comparison of rumen fluid and milk OBCFA of different milk stages. The data are means and deviations of three milk stages and error bars the show standard error of the mean. Bars without a common letter (a and b) differ ($p < 0.05$). OBCFA, odd- and branched-chain fatty acids.

3.2. Correlation between Rumen and Milk OBCFA during Different Milk Stages

The contents of OBCFA in rumen fluid were significantly related to that in milk fat (Table 4). Significant relationships of concentrations of C11:0, *iso*-C15:0, *anteiso*-C15:0, C15:0, and *anteiso*-C17:0 were found between rumen fluid and milk fat ($r = 0.39 \sim 0.66$, $p = 0.0002$–0.04). The concentrations of C11:0 in milk were significantly related to the concentrations of *anteiso*-C15:0 in rumen ($r = 0.39$, $p = 0.04$). As for the milk, C13:0 contents were significantly correlated with the ruminal *anteiso*-C15:0, C15:0, and total OBCFA concentrations ($r = 0.48 \sim 0.52$, $p = 0.005$–0.01). The concentrations of *iso*-C15:0 in milk were positively associated with the contents of C11:0, C13:0, and *anteiso*-C15:0 in rumen fluid ($r = 0.39 \sim 0.41$, $p = 0.03$–0.04). The contents of *anteiso*-C15:0 in milk were significantly related to the contents of C11:0, C13:0, and total OBCFA in rumen fluid ($r = 0.46$–0.65, $p = 0.0002$–0.02). The milk C15:0 contents were positively correlated with the concentrations of *anteiso*-C15:0 and total OBCFA in rumen ($r = 0.58$–0.62, $p = 0.001$–0.002). The contents of *iso*-C16:0 in milk were positively correlated with the ruminal C11:0 contents ($r = 0.44$, $p = 0.02$); however, they were negatively related to the ruminal *anteiso*-C17:0 contents ($r = -0.39$, $p = 0.04$). There was significantly relationship between the concentrations of *iso*-C17:0 in milk and the concentrations of C11:0 in rumen ($r = -0.59$, $p = 0.001$). Besides, the contents of *anteiso*-C17:0 in milk were negatively linked with the contents of *iso*-C15:0, C15:0, and total OBCFA in rumen ($r = -0.39$ to -0.49, $p = 0.01$–0.046). A negative relationship that existed between the milk C17:0 contents and ruminal C11:0 contents ($r = -0.44$, $p = 0.02$).

Table 4. Correlation between rumen and milk OBCFA of different milk stages.

Rumen Fluid	Milk Fat																				
	C11:0		C13:0		Iso-C15:0		Anteiso-C15:0		C15:0		Iso-C16:0		Iso-C17:0		Anteiso-C17:0		C17:0		TOBCFA	
	r	p	r	p	r	p	r	p	r	p	r	p	r	p	r	p	r	p	r	p
C11:0	0.40	0.04	0.13	0.53	0.39	0.04	0.50	0.01	0.16	0.43	0.44	0.02	−0.59	0.001	−0.32	0.11	−0.44	0.02	−0.03	0.87
C13:0	0.26	0.19	0.14	0.47	0.41	0.04	0.65	0.0002	−0.01	0.96	0.37	0.06	−0.36	0.07	−0.25	0.21	−0.23	0.25	0.03	0.89
Iso-C15:0	−0.02	0.93	0.34	0.08	0.42	0.03	0.25	0.20	0.36	0.07	−0.06	0.76	−0.15	0.46	−0.39	0.046	0.16	0.43	−0.03	0.88
Anteiso-C15:0	0.39	0.04	0.48	0.01	0.43	0.03	0.60	0.001	0.58	0.002	−0.24	0.23	−0.004	0.98	−0.30	0.13	0.09	0.65	0.31	0.11
C15:0	0.20	0.33	0.52	0.005	0.20	0.31	0.12	0.54	0.66	0.0002	−0.14	0.50	0.13	0.51	−0.44	0.02	0.22	0.27	0.06	0.76
Iso-C16:0	0.08	0.68	−0.07	0.72	0.17	0.39	0.18	0.38	−0.03	0.90	0.38	0.052	−0.19	0.34	−0.14	0.48	−0.05	0.81	−0.02	0.90
Iso-C17:0	0.29	0.14	0.18	0.36	0.27	0.18	0.37	0.06	0.45	0.02	−0.07	0.74	0.003	0.99	−0.22	0.26	−0.14	0.47	0.13	0.52
Anteiso-C17:0	−0.09	0.65	−0.12	0.57	−0.13	0.52	0.06	0.78	−0.02	0.93	−0.39	0.04	0.31	0.12	0.39	0.04	0.14	0.49	0.34	0.08
C17:0	−0.04	0.83	0.31	0.12	0.02	0.90	−0.03	0.89	−0.01	0.96	−0.29	0.15	0.07	0.72	−0.16	0.44	0.17	0.40	−0.16	0.43
TOBCFA	0.29	0.14	0.49	0.01	0.36	0.07	0.46	0.02	0.62	0.001	−0.06	0.77	−0.07	0.73	−0.49	0.01	0.11	0.57	0.09	0.64

OBCFA, odd- and branched-chain fatty acids; TOBCFA, the total odd- and branched-chain fatty acids. r = correlation coefficient.

3.3. Simple Statistics of Experimental Data Used for Model Development

Simple statistical data were gathered from of the present experimental conditions shown in Table 5. According to the CV%, the *Butyrivibrio flarisolvens* population changed less than the other bacterial populations. There was a wide range of rumen fermentation parameters in this research. With regard to rumen VFAs, larger differences were observed in isoacids than in linear-chain VFAs in the rumen. Compared to the variation of NH_3-N, the pH was more stable. With regard to the rumen bacterial bases, the variations were similar to each other.

Table 5. Simple statistics of experimental data used for model development.

Variables	N	Mean	SD	Minimum	Maximum	CV%
Milk odd and branched-chain fatty acids (g/100 g fatty acids)						
C11:0	27	0.03	0.02	0.01	0.08	50.53
C13:0	27	0.02	0.01	0.01	0.04	46.25
Iso-C15:0	27	0.16	0.03	0.09	0.21	18.39
Anteiso-C15:0	27	0.38	0.07	0.23	0.58	19.33
C15:0	27	0.82	0.12	0.64	1.02	14.26
Iso-C16:0	27	0.20	0.07	0.10	0.37	34.76
Iso-C17:0	27	0.50	0.10	0.23	0.62	19.46
Anteiso-C17:0	27	0.99	0.23	0.51	1.44	23.35
C17:0	27	0.49	0.18	0.34	1.08	36.52
TOBCFA	27	3.61	0.25	3.01	4.03	6.94
Rumen fermentation parameters						
Acetate (mmol/mol)	27	523.81	25.31	474.55	577.33	4.83
Propionate (mmol/mol)	27	294.02	18.24	255.62	327.79	6.21
Isobutyrate (mmol/mol)	27	14.84	2.20	10.06	19.35	14.81
Butyrate (mmol/mol)	27	120.65	4.81	112.00	127.81	3.99
Isovalerate (mmol/mol)	27	20.84	2.37	15.89	25.26	11.36
Valerate (mmol/mol)	27	25.84	2.32	21.25	30.63	8.98
TVFA (mmol/l)	27	63.12	13.50	48.76	110.34	21.39
NH3-N (mmol/l)	27	11.69	3.67	4.84	19.95	31.39
pH	27	6.35	0.24	5.87	6.70	3.77
Ruminal bacterial populations ($\log_{10}$ copies/mL)						
Fibrobacter succinogenes	27	7.99	0.32	7.33	8.56	4.00
Ruminococcus albus	27	7.03	0.30	6.55	7.55	4.29
Ruminococcus flavafaciens	27	7.73	0.14	7.42	8.02	1.85
Butyrivibro flarisolvens	27	8.91	0.07	8.79	9.08	0.80
Eubacterium ruminantium	27	6.62	0.28	5.88	7.23	4.19
Streptococcus bovis	27	5.27	0.30	4.45	5.74	5.61
Selenomonas ruminantium	27	7.35	0.19	6.97	7.62	2.61
Ruminal bacterial bases (g/kg DM)						
Cytosine	27	1.21	0.11	0.98	1.38	8.94
Uracil	27	0.98	0.14	0.63	1.16	14.10
Guanine	27	2.37	0.29	1.60	2.80	12.20
Adenine	27	1.86	0.16	1.43	2.16	8.57
Ruminal bacterial bases/N (g/100 g N)						
Cytosine/N	27	3.11	0.24	2.75	3.66	0.08
Uracil/N	27	2.52	0.33	1.80	3.01	0.13
Guanine/N	27	6.09	0.62	4.52	7.16	0.10
Adenine/N	27	4.80	0.44	4.13	6.11	0.09

TVFA, the total contents of acetate, propionate, isobutyrate, butyrate, isovalerate, and valerate in rumen. OBCFA, odd- and branched-chain fatty acids; TOBCFA, the total odd- and branched-chain fatty acids. SD = standard deviation; CV = coefficient of variation.

3.4. Relationship of Milk OBCFA Pattern during Different Milk Stages to Ruminal Fermentation Parameters, Bacteria Populations, and Microbial Bases

Relationships existed between the fermentation parameters in rumen fluid and the OBCFA patterns (Table 6). Although there no significant correlations that were found between the molar proportions of acetate and individual OBCFA concentrations, the molar proportions of acetate were positively related to total OBCFA concentrations ($r = 0.40$, $p = 0.04$). A negative correlation existed between the propionate molar proportions and *iso*-C17:0 concentrations ($r = -0.39$, $p = 0.045$). The total OBCFA concentrations were negatively correlated with the molar proportions' isobutytate ($r = -0.49$, $p = 0.01$) while the concentrations of *anteiso*-C15:0 were positively associated with the molar proportions' butyrate ($r = 0.47$, $p = 0.01$). There was a negative relationship between the milk C15:0 concentration and isovalelate proportion in rumen ($r = -0.41$, $p = 0.03$). The total OBCFA concentrations were negatively related to isovalelate and valelate proportions ($r = -0.48$ to -0.49, $p = 0.01$) but were positively correlated with total VFA contents ($r = 0.51$, $p = 0.007$) and the *iso*-C15:0/*iso*-C17:0 ratio and *anteiso*-C15:0/anteiso-C17:0 ratio ($r = -0.56$ to -0.60, $p = 0.001$–0.002). The concentrations of NH_3-N were negatively related to *iso*-C15:0 and *anteiso*-C15:0 concentrations ($r = -0.50$ to -0.56, $p = 0.003$–0.008), but positively correlated with *anteiso*-C17:0 concentrations ($r = 0.47$, $p = 0.01$). However, there were no apparent relationships between pH and OBCFA concentrations.

The relationships between bacterial populations and OBCFA concentrations are shown in Table 7. The copy numbers of *Fibrobacter succinogenes* were not related to single OBCFAs concentrations, but positively linked with the *iso*-C15:0/*iso*-C17:0 ratio ($r = 0.47$, $p = 0.01$). The concentrations of C11:0 and C13:0 were positively correlated with the populations of *Ruminococcus albus* ($r = 0.53$–0.57, $p = 0.002$–0.005). The copies of *Ruminococcus flavefaciens* were significantly related to the C13:0 and total OBCFA concentrations ($r = 0.38$–0.50, $p = 0.009$–0.049). A positive correlation was found between C13:0 concentrations and *Eubacterium ruminantium* copy numbers ($r = 0.43$, $p = 0.03$). The ratio of *anteiso*-C15:0 to *anteiso*-C17:0 was positively correlated with *Streptococcus bovis* copies ($r = 0.48$, $p = 0.01$); however, no significant relationships that were found between *Selenomonas ruminantium* copies and individual OBCFA concentrations. The populations of *Selenomonas ruminantium* were positively associated with the total OBCFA and branched-chain fatty acid contents ($r = 0.39$–0.57, $p = 0.002$–0.04).

Several relationships were found between the microbial bases and OBCFA patterns (Table 8). Cytosine concentrations were negatively related to *iso*-C17:0 and total OBCFA concentrations ($r = -0.39$ to -0.50, $p = 0.007$–0.04). The concentrations of uracil were significantly correlated with the *iso*-C17:0 content, the sum of *iso*-C15:0 and *iso*-C17:0 contents ($r = -0.51$ to -0.54, $p = 0.003$–0.004). The concentrations of guanine were negatively related to the concentrations of *iso*-C17:0, the total OBCFA contents, and the sum of *iso*-C15:0 and *iso*-C17:0 contents ($r = -0.41$ to -0.54, $p = 0.004$–0.03). A negative relation was presented by the concentrations of adenine and C17:0 ($r = -0.39$, $p = 0.04$), while the contents of *anteiso*-C15:0 and the C15:0/C17:0 ratio were positively linked with the adenine concentrations ($r = 0.44$–0.49, $p = 0.01$–0.02). The correlations of microbial protein marked by different microbial bases/N ratio with the OBCFAs are also shown in Table 8. The cytosine/N ratio was negatively related to total OBCFA concentrations in milk ($r = -0.40$, $p = 0.04$). In addition, some negative relationships were found between the guanine/N ratio and some milk OBCFAs, which were *iso*-C17:0 and the sum of *iso*-C15:0 and *iso*-C17:0 ($r = -0.49$ to -0.55, $p = 0.003$–0.01). The adenine/N ratio was negatively connected with *iso*-C17:0, *anteiso*-C17:0, the sum of *iso*-C15:0 and *iso*-C17:0, the sum of *anteiso*-C15:0 and *anteiso*-C17:0, the total *iso*-branched chain fatty acids, the total *anteiso*-branched chain fatty acids, and the total branched chain fatty acids in milk ($r = -0.42$ to -0.59, $p = 0.001$–0.03). However, a positive relationship was found between the adenine/N ratio and the ratio of *iso*-C15:0 and *iso*-C17:0 in milk ($r = 0.50$, $p = 0.01$).

Table 6. Correlation between ruminal fermentation parameters and milk OBCFA pattern.

Item	Acetate		Propionate		Isobutyrate		Butyrate		Isovalerate		Valerate		TVFA		NH$_3$-N		pH	
	r	p	r	p	r	p	r	p	r	p	r	p	r	p	r	p	r	p
C11:0	−0.22	0.28	0.12	0.56	0.36	0.06	0.17	0.41	0.19	0.33	0.26	0.18	−0.31	0.11	0.06	0.75	0.22	0.26
C13:0	0.14	0.48	−0.22	0.28	0.08	0.69	−0.21	0.30	0.16	0.44	−0.05	0.82	−0.20	0.33	0.19	0.34	0.29	0.14
Iso-C15:0	−0.07	0.73	0.19	0.35	−0.08	0.69	0.05	0.79	−0.10	0.63	−0.07	0.74	0.18	0.37	−0.50	0.008	−0.05	0.81
Anteiso-C15:0	−0.34	0.08	0.31	0.11	0.32	0.11	0.47	0.01	0.11	0.60	0.28	0.16	−0.22	0.27	−0.56	0.003	0.21	0.29
C15:0	0.18	0.37	−0.11	0.60	−0.31	0.11	−0.05	0.81	−0.41	0.03	−0.33	0.097	0.37	0.06	−0.28	0.16	0.20	0.32
Iso-C16:0	−0.12	0.54	0.20	0.32	0.19	0.35	0.03	0.90	0.02	0.93	0.22	0.28	−0.06	0.76	−0.21	0.29	−0.31	0.12
Iso-C17:0	0.36	0.06	−0.39	0.045	−0.37	0.06	−0.24	0.22	−0.25	0.20	−0.35	0.08	0.11	0.59	0.22	0.27	0.12	0.56
Anteiso-C17:0	0.26	0.19	−0.35	0.07	−0.19	0.35	−0.12	0.55	−0.17	0.39	−0.21	0.28	0.17	0.40	0.47	0.01	0.18	0.37
C17:0	0.21	0.28	−0.06	0.77	−0.37	0.06	−0.32	0.0996	−0.28	0.16	−0.38	0.05	0.23	0.26	0.12	0.55	−0.05	0.81
TOBCFA	0.40	0.04	−0.33	0.09	−0.49	0.01	−0.26	0.18	−0.48	0.01	−0.49	0.01	0.51	0.007	0.06	0.76	0.08	0.71
C15:0 + C17:0	0.12	0.57	0.03	0.89	−0.29	0.14	−0.11	0.58	−0.37	0.06	−0.32	0.10	0.37	0.06	−0.25	0.22	0.07	0.74
Iso-C15:0 + *iso*-C17:0	0.30	0.13	−0.36	0.06	−0.32	0.10	−0.12	0.55	−0.23	0.26	−0.27	0.17	0.23	0.26	0.02	0.90	0.17	0.39
Anteiso-C15:0 + *anteiso*-C17:0	0.14	0.48	−0.25	0.21	−0.07	0.72	0.04	0.83	−0.11	0.58	−0.09	0.65	0.09	0.65	0.30	0.12	0.22	0.27
Total *iso*-fatty acids	0.32	0.0997	−0.34	0.08	−0.28	0.16	−0.24	0.23	−0.18	0.38	−0.19	0.35	0.22	0.27	−0.02	0.94	−0.23	0.24
Total *anteiso*-fatty acids	0.14	0.48	−0.25	0.21	−0.07	0.71	0.05	0.82	−0.11	0.57	−0.09	0.64	0.09	0.64	0.31	0.12	0.22	0.27
Total odd-chain fatty acids	0.09	0.67	0.05	0.81	−0.26	0.19	−0.09	0.65	−0.33	0.096	−0.30	0.13	0.34	0.08	−0.26	0.20	0.10	0.63
Total branched-chian fatty acids	0.16	0.44	−0.25	0.22	−0.08	0.68	−0.03	0.90	−0.19	0.35	−0.09	0.66	0.12	0.55	0.22	0.28	0.15	0.47
C15:0/C17:0	−0.13	0.51	0.04	0.85	0.17	0.40	0.30	0.13	−0.03	0.90	0.18	0.37	−0.04	0.83	−0.31	0.11	0.27	0.17
Iso-C15:0/*iso*-C17:0	−0.25	0.21	0.35	0.08	0.14	0.49	0.22	0.28	0.05	0.80	0.16	0.42	0.09	0.65	−0.56	0.002	−0.13	0.52
Anteiso-C15:0/*anteiso*-C17:0	−0.32	0.10	0.36	0.06	0.31	0.12	0.30	0.12	0.14	0.50	0.30	0.13	0.07	0.74	−0.60	0.001	0.04	0.84

TVFA, the total contents of acetate, propionate, isobutyrate, butyrate, isovalerate, and valerate in rumen. OBCFA, odd- and branched-chain fatty acids; TOBCFA, the total odd- and branched-chain fatty acids. *r* = correlation coefficient.

Table 7. Correlation between bacteria populations and milk OBCFA pattern.

Item	Fibrobacter succinogenes		Ruminococcus albus		Ruminococcus flavafaciens		Butyrivibro flarisolvens		Eubacterium ruminantium		Streptococcus bovis		Selenomonas ruminantium	
	r	p	r	p	r	p	r	p	r	p	r	p	r	p
C11:0	0.17	0.38	0.57	0.002	0.18	0.37	0.12	0.54	0.36	0.06	0.19	0.34	0.01	0.95
C13:0	0.11	0.60	0.53	0.005	0.50	0.009	0.15	0.45	0.43	0.03	0.22	0.27	−0.01	0.97
Iso-C15:0	0.37	0.055	−0.03	0.87	0.35	0.07	−0.09	0.65	0.05	0.79	0.02	0.94	−0.05	0.79
Anteiso-C15:0	0.18	0.37	0.35	0.07	−0.04	0.83	0.14	0.47	0.22	0.26	0.33	0.09	−0.25	0.20
C15:0	0.21	0.29	0.25	0.21	0.17	0.39	−0.28	0.16	0.05	0.81	0.22	0.28	0.27	0.18
Iso-C16:0	0.14	0.48	−0.17	0.39	−0.18	0.36	0.14	0.48	−0.07	0.71	−0.21	0.29	−0.09	0.65
Iso-C17:0	−0.25	0.21	−0.03	0.88	0.13	0.53	0.05	0.81	−0.01	0.94	−0.02	0.92	0.33	0.09
Anteiso-C17:0	−0.14	0.49	0.01	0.98	0.13	0.51	−0.19	0.34	0.03	0.88	−0.34	0.09	0.36	0.06
C17:0	−0.22	0.27	−0.19	0.34	0.22	0.28	0.18	0.38	−0.09	0.65	−0.01	0.96	0.37	0.06
TOBCFA	0.05	0.81	0.03	0.89	0.38	0.049	−0.03	0.89	0.17	0.40	−0.18	0.38	0.57	0.002
C15:0 + C17:0	0.16	0.42	0.06	0.77	0.16	0.44	−0.04	0.83	−0.03	0.90	0.24	0.23	0.32	0.098
Iso-C15:0 + iso-C17:0	−0.06	0.77	0.13	0.53	0.28	0.16	0.01	0.95	0.11	0.59	0.09	0.67	0.38	0.051
Anteiso-C15:0 + anteiso-C17:0	−0.005	0.98	0.09	0.66	0.17	0.39	−0.22	0.27	0.06	0.78	−0.30	0.13	0.27	0.18
Total iso-fatty acids	0.02	0.93	−0.03	0.89	0.23	0.25	0.13	0.52	0.06	0.76	−0.10	0.61	0.31	0.12
Total anteiso-fatty acids	−0.01	0.96	0.08	0.69	0.16	0.41	−0.22	0.27	0.05	0.79	−0.30	0.13	0.26	0.18
Total odd-chain fatty acids	0.18	0.38	0.11	0.59	0.19	0.35	−0.04	0.84	0.01	0.97	0.24	0.22	0.31	0.12
Total branched-chian fatty acids	0.04	0.86	0.11	0.59	0.18	0.37	−0.15	0.45	0.08	0.69	−0.28	0.16	0.39	0.04
C15:0/C17:0	0.32	0.104	0.33	0.09	−0.05	0.81	−0.32	0.11	0.10	0.63	0.17	0.41	−0.10	0.61
Iso-C15:0/iso-C17:0	0.47	0.01	−0.05	0.82	0.13	0.51	−0.10	0.63	−0.04	0.86	0.07	0.75	−0.21	0.28
Anteiso-C15:0/anteiso-C17:0	0.23	0.24	0.29	0.15	−0.12	0.54	0.21	0.28	0.17	0.39	0.48	0.01	−0.31	0.12

OBCFA, odd- and branched-chain fatty acids; TOBCFA, the total odd- and branched-chain fatty acids. r = correlation coefficient.

Table 8. Correlation between microbial protein marked by nucleic acids, marker:N ratio in rumen (when using nucleic acids as the microbial marker), and milk OBCFA pattern (*n* = 27).

Item	Cytosine		Uracil		Guanine		Adenine		Cytosine/N		Uracil/N		Guanine/N		Adenine/N	
	r	*p*	*r*	*p*	*r*	*p*	*r*	*p*	*r*	*p*	*r*	*p*	*r*	*p*	*r*	*p*
C11:0	0.05	0.81	0.20	0.33	0.07	0.74	0.16	0.42	0.02	0.91	0.10	0.62	0.19	0.35	0.08	0.68
C13:0	0.12	0.54	−0.09	0.65	0.13	0.51	0.21	0.28	0.27	0.17	0.00	0.99	−0.23	0.24	0.13	0.50
Iso-C15:0	−0.11	0.58	−0.18	0.36	−0.06	0.78	0.18	0.36	−0.13	0.52	−0.05	0.79	−0.14	0.48	0.20	0.32
Anteiso-C15:0	−0.06	0.76	0.12	0.56	0.07	0.74	0.49	0.01	−0.07	0.72	−0.17	0.39	0.16	0.44	0.08	0.68
C15:0	−0.19	0.35	−0.32	0.11	−0.28	0.16	0.30	0.13	−0.25	0.21	0.04	0.86	−0.14	0.48	0.10	0.63
Iso-C16:0	0.09	0.67	0.38	0.051	0.31	0.12	−0.09	0.66	0.35	0.07	−0.07	0.72	0.26	0.19	0.20	0.31
Iso-C17:0	−0.39	0.04	−0.51	0.006	−0.54	0.004	−0.21	0.29	−0.17	0.40	−0.24	0.22	−0.49	0.01	−0.59	0.001
Anteiso-C17:0	−0.18	0.36	−0.12	0.55	−0.25	0.20	−0.13	0.52	−0.22	0.27	−0.09	0.65	−0.17	0.40	−0.47	0.01
C17:0	−0.30	0.13	0.11	0.59	−0.23	0.24	−0.39	0.04	−0.12	0.54	−0.13	0.51	0.19	0.35	0.20	0.32
TOBCFA	−0.50	0.007	−0.28	0.16	−0.41	0.03	0.03	0.90	−0.40	0.04	−0.33	0.09	−0.17	0.40	−0.37	0.06
C15:0 + C17:0	−0.28	0.16	−0.04	0.84	−0.19	0.34	0.00	0.99	−0.24	0.23	−0.09	0.65	0.08	0.69	0.22	0.28
Iso-C15:0 + *iso*-C17:0	−0.35	0.08	−0.54	0.003	−0.53	0.005	−0.09	0.66	−0.21	0.28	−0.27	0.17	−0.55	0.003	−0.55	0.003
Anteiso-C15:0 + *anteiso*-C17:0	−0.18	0.38	−0.04	0.83	−0.21	0.28	−0.001	1.00	−0.26	0.19	−0.16	0.42	−0.13	0.52	−0.49	0.01
Total *iso*-fatty acids	−0.28	0.16	−0.33	0.09	−0.31	0.11	−0.17	0.39	0.05	0.81	−0.34	0.08	−0.37	0.06	−0.42	0.03
Total *anteiso*-fatty acids	−0.18	0.37	−0.05	0.82	−0.22	0.27	−0.01	0.97	−0.26	0.19	−0.16	0.42	−0.13	0.52	−0.49	0.01
Total odd-chain fatty acids	−0.24	0.23	−0.03	0.89	−0.15	0.45	0.04	0.83	−0.23	0.24	−0.08	0.67	0.09	0.66	0.23	0.24
Total branched-chian fatty acids	−0.23	0.26	−0.11	0.60	−0.26	0.19	−0.04	0.86	−0.21	0.30	−0.26	0.19	−0.24	0.22	−0.56	0.002
C15:0/C17:0	0.16	0.42	0.00	0.99	−0.05	0.81	0.44	0.02	0.01	0.96	0.23	0.24	−0.03	0.87	0.06	0.78
Iso-C15:0/*iso*-C17:0	0.10	0.60	0.23	0.25	0.23	0.25	0.21	0.28	0.04	0.83	0.10	0.61	0.25	0.21	0.50	0.01
Anteiso-C15:0/*anteiso*-C17:0	0.06	0.77	0.22	0.27	0.17	0.39	0.38	0.053	0.02	0.91	−0.05	0.79	0.24	0.22	0.34	0.08

OBCFA, odd- and branched-chain fatty acids; TOBCFA, the total odd- and branched-chain fatty acids. *r* = correlation coefficient.

The quadratic, ratio and reciprocal of milk OBCFAs were used in the predicted models shown in Tables 9–11, respectively. The equations of fermentation parameters, with the exception of that for valerate, were all significant. According to the R^2 value, the isobutyrate and butyrate models were more accurate than those for other VFAs. The regression equations showed that the ruminal acetate, propionate, and isobutyrate molar proportions and pH had inverse proportional relationships to milk C11:0 concentrations. In addition, inverse proportional relationships existed between the ruminal isobutyrate proportion and milk C17:0 concentrations, between the ruminal butyrate proportion and milk *anteiso*-C15:0 concentrations, and between ruminal NH_3-N and milk *anteiso*-C15:0 concentrations. The accuracy of the prediction equations for different bacterial populations in the rumen was variable. Regression analysis showed that milk C11:0 concentrations had inverse proportional relationships with *Eubacterium ruminantium*, *Ruminococcus albus*, and *Streptococcus bovis* populations in the rumen. Inverse proportional relationships were found between milk *iso*-C15:0 content and ruminal *Fibrobacter succinogenes* populations as well as between milk C17:0 content and ruminal *Selenomonas ruminantium* populations. Significant regression equations for ruminal microbial markers were obtained using milk OBCFAs as an independent data set. Additionally, the compositions of these equations were complex.

Table 9. Predicted equations of ruminal fermentation parameters from milk OBCFA.

Variables	Predicted Equation	RMSE	R^2	P
Acetate (mmol/mol)	$Y = 55.73 - 1.60/(C11:0)$	2.38	0.15	0.049
Propionate (mmol/mol)	$Y = 26.99 + 1.15/(C11:0)$	1.75	0.12	0.0496
Isobutyrate (mmol/mol)	$Y = -0.32 + 0.48/(C17:0) + 20.23 \times C11:0 \times anteiso - C15:0 + 1.04/(C11:0)$	0.17	0.46	0.002
Butyrate (mmol/mol)	$Y = 14.37 - 0.37/(anteiso - C15:0) - 4.08 \times C11:0 \times C17:0 - 102.26 \times C13:0 \times iso - C16:0$	0.41	0.35	0.02
Isovalerate (mmol/mol)	$Y = 1.13 - 0.88 \times C15:0 - 0.68 \times anteiso - C17:0 \times C17:0$	0.21	0.28	0.02
Valerate (mmol/mol)	$Y = 3.21 - 1.36 \times C11:0 \times C17:0 - 0.38 \times C15:0 \times anteiso - C17:0$	0.21	0.27	0.08
TVFA (mmol/L)	$Y = 24.56 + 155.95 \times C11:0 \times C15:0 + 38.78 \times anteiso - C17:0 \times C17:0$	12.03	0.27	0.02
NH3-N (mmol/L)	$Y = 5.46 + 0.11/(iso - C16:0) + 3.71 \times C11:0 \times iso - C16:0$	2.95	0.40	0.002
pH	$Y = 6.03 + 0.15/(C11:0)$	0.16	0.22	0.04

R^2 = coefficient of determination; RMSE = root mean square error. TVFA, the total contents of acetate, propionate, isobutyrate, butyrate, isovalerate, and valerate in rumen.

Table 10. Predicted equations of ruminal bacterial populations (log_{10} copies/mL) from milk OBCFA.

Variables	Predicted Equation	RMSE	R^2	P
Fibrobacter succinogenes	$Y = 8.48 - 0.15 \times (iso - C17:0)/(iso - C15:0)$	0.20	0.29	0.02
Ruminococcus albus	$Y = 6.87 - 0.007/(C11:0)$	0.26	0.18	0.03
Ruminococcus flavafaciens	$Y = 7.04 - 0.01/(C11:0) + 35.41 \times C11:0 \times anteiso - C15:0$	0.24	0.41	0.002
Butyrivibro flarisolvens	$Y = 7.30 + 73.80 \times C11:0 \times iso - C15:0 + 0.53 \times anteiso - C17:0 \times C17:0$	0.12	0.38	0.09
Eubacterium ruminantium	$Y = 5.86 - 0.22/(C11:0)$	0.28	0.15	0.04
Streptococcus bovis	$Y = 7.08 - 0.22/C17:0 + 1.20 \times C11:0 \times anteiso - C17:0 + 0.65 \times C15:0 \times anteiso - C17:0$	0.14	0.53	0.0005

R^2 = coefficient of determination; RMSE = root mean square error.

Table 11. Predicted equations of ruminal bacterial bases from milk OBCFA.

Variables	Predicted Equation	R^2	RMSE	P
Ruminal bases (g/kg DM)				
Cytosine	$Y = 1.54 - 0.79 \times C11:0 \times C17:0 - 0.93 \times iso - C17:0 \times anteiso - C17:0$	0.32	0.09	0.01
Uracil	$Y = 1.18 - 4.11 \times iso - C15:0 \times iso - C17:0 + 1.59 \times anteiso - C15:0 \times iso - C16:0$	0.50	0.10	0.0002
Guanine	$Y = 2.23 + 0.01/(C11:0) + 1.17 \times C11:0 \times anteiso - C17:0 + 23.00 \times C13:0 \times anteiso - C17:0 - 1.48 \times iso - C17:0 \times anteiso - C17:0$	0.58	0.20	0.001
Adenine	$Y = 1.96 + 2.33 \times iso - C16:0 \times iso - C16:0 + 52.67 \times C11:0 \times anteiso - C15:0 - 91.61 \times C13:0 \times iso - C16:0 - 0.5 \times anteiso - C17:0 \times C17:0$	0.49	0.12	0.004
Ruminal bases /N (g/100 g N)				
Cytosine:N	$Y = 1.72 + 0.01/(C13:0) + 0.20/(iso - C16:0) - 1.90 \times iso - C16:0 \times anteiso - C17:0$	0.58	0.18	0.002
Uracil:N	$Y = 2.93 + 0.01/(C13:0) - 2.88 \times iso - C17:0 \times anteiso - C17:0 + 273.13 \times C11:0 \times C13:0 - 1.44 \times (iso - C15:0)/(iso - C17:0)$	0.60	0.23	0.0004
Guanine:N	$Y = 7.17 - 2.12 \times iso - C17:0 \times anteiso - C17:0$	0.37	0.50	0.001
Adenine:N	$Y = 3.27 - 2.53 \times iso - C17:0 \times C17:0 + 3.81 \times C15:0 \times iso - C16:0 + 0.73 \times (anteiso - C15:0)/(iso - C16:0)$	0.62	0.28	<0.0001

N, the total contents of nitrogen in rumen. R^2 = coefficient of determination; RMSE = root mean square error.

4. Discussion

The OBCFA in milk are mainly derived from the ruminal bacterial cell membrane, which was already recognized half a century ago [7]. For reasons previously described by Fievez et al. [33], the milk OBCFA profile was not completely similar to the rumen OBCFA profile. Different proportions of OBCFAs in the rumen and milk were also found in this research. In this study, the fatty acids with 15-carbon atoms were more plentiful than other components in the rumen, but the contents of these fatty acids and *iso*-C16:0 were much lower than those in milk. The summary statistics by Vlaeminck et al. [34] suggested that the linear odd-chain fatty acids accounted for the major proportion of milk OBCFAs. The linear odd-chain fatty acid contents were also much higher in the rumen than in milk in our study. Previous studies have indicated that the mammary gland tissue can de novo synthesize odd-chain fatty acids and their *anteiso*-isomers, through the incorporation of propionyl-CoA instead of acetyl-CoA [35–37]. Milk secretion of linear odd-chain fatty acids is higher than that of these fatty acids in duodenal flow, which further demonstrates endogenous chain elongation in the mammary gland [34]. This research also found that the contents of odd-chain fatty acids, except those of C15:0, were higher in milk than in the rumen. Scheerlinck et al. [38] summarized a representative subset of 92 milk samples adopted from a sample database of different dietary treatments, and they reported that the highest concentrations of *iso*-C17:0, *anteiso*-C17:0, and some mono-unsaturated fatty acids were observed in milk. Additionally, Vlaeminck et al. [39] reported that two-carbon elongation of branched-chain FAs occurs postruminal. Increased contents of *anteiso*-C17:0 in milk were found in this research.

At the beginning of lactation, cows are in a negative energy balance, causing the mobilization of adipose fatty acids and the synthesis of these long-chain fatty acids in milk fat [40]. The OBCFAs with 14- and 15-carbon atoms, such as *iso*-C14:0, *iso*-C15:0, C15:0, and *anteiso*-C15:0, were increased in early lactation, which changed with the lactation curves of the short- and medium-chain fatty acids in milk [41]. Increased levels of ruminal C15:0 contents were also found in early lactation in this study. In contrast, levels of OBCFAs with 17-carbon atoms decreased during early lactation, which showed a similar pattern to that of long-chain fatty acids in milk [41]. In the present study, however, the concentrations of *iso*-C17:0 in milk were more abundant in early lactation, which may be caused by the synthesis of these fatty acids. A previous study found that long-chain fatty acids are de novo synthesized from some fatty acids, which causes rumen acidosis in milk at an early lactation stage [40].

In the mammary gland, fatty acids with 4- to 14-carbon atoms mainly come from endogenous synthesis, and the long-chain fatty acids greater than 16-carbon atoms are mostly derived from exogenous transformation [42]. In addition, Verbeke et al. [43] found that the mammary gland had the ability to synthesize fatty acids from 2-methylbutyryl-CoA, isovaleryl-CoA, and isobutyryl-CoA. The same types of fatty acids with less than 16-carbon atoms collected from rumen fluid and milk were significantly or nearly related to each other in this research. These results may be due to the minor amount of fatty acids synthesized in the mammary gland. Odd-chain fatty acids are further metabolized under the action of dehydrogenase in the mammary gland, but the transformation of C17:0 to C17:1 was statistically significant [44]. Hence, negative correlations existed between the milk fatty acids with 17-carbon atoms and other ruminal fatty acids in this study. The $C15:0_{milk}/C15:0_{duodenum}$ and $C17:0 + C17:1_{milk}/C17:0_{duodenum}$ values changed from 1.5 to 2.25 [34,45,46], which suggested that the metabolic and transport processes of the two types of fatty acids were different in the mammary gland.

Increasing the acetate supply has a positive effect on milk yield and milk fat content [47]. In this study, a positive correlation existed between acetate content in the rumen and the total concentrations of OBCFAs in milk. Vlaeminck et al. [10] found that milk *iso*-C14:0 and *iso*-C15:0 contents are positively correlated with ruminal proportions of acetate. We also found a positive relationship between the concentrations of *iso*-C17:0 in milk and the molar proportions of ruminal acetate, but this relationship was not significant. Moreover, milk *iso*-C14:0 and *iso*-C15:0 contents are negatively related to ruminal propionate proportions according to Vlaeminck et al. [10]. In this work, the contents of ruminal propionate were negatively related to *iso*-fatty acids, especially milk *iso*-C17:0. Negative relationships were found between ruminal *iso*-fatty acids and the contents of total OBCFAs in milk. Isoacids are also used as primers for branched-chain fatty acids [34], and their contents are low in the rumen [48,49]. Compared to linear and *iso*-fatty acids, *anteiso*-fatty acids are more susceptible to effects on bacterial membrane properties [50]. Cabrita et al. [51] found a negative correlation between the concentrations of ruminal NH_3 and milk *anteiso*-C17:0. In this study, there was a negative relationship between NH_3-N concentrations in the rumen and *anteiso*-C15:0 concentrations in milk. However, milk *anteiso*-C17:0 concentrations were significantly positively related to ruminal NH_3 concentrations. In addition, the ruminal NH_3-N concentrations were positively related to *iso*-C15:0/*iso*-C17:0 or *anteiso*-C15:0/*anteiso*-C17:0 values.

Variation in the OBCFA profile leaving the rumen has been suggested to reflect changes in the relative abundance of specific ruminal bacterial populations [33]. Any deviation in the rumen's inner circumstance may influence the microbial population and its fermentation products [52]. Vlaeminck et al. [34] summarized that *Fibrobacter succinogenes* abundantly produces linear odd-chain fatty acids and fatty acids with 15-carbon atoms. In the present study, *Fibrobacter succinogenes* populations in the rumen were not significantly related to separate OBCFA concentrations, but they were positively correlated with the *iso*-C15:0/*iso*-C17:0 ratio in milk. Hence, the different results from the cultivable ruminal bacteria might be due to de novo fatty acid synthesis and elongation in the mammary gland. Some previous studies have reported that cellulolytic bacteria contain a high amount of *iso*-fatty acids, with differences between different species [19,53–55]. Specifically, *Ruminococcus flavefaciens* has a high level of odd-chain *iso*-fatty acids, and *Ruminococcus albus* has enriched even-chain *iso*-fatty acids [34]. Although the contribution of OBCFA endogenous synthesis in milk from dairy cows was negligible [56], linear odd-chain fatty acids, or their *anteiso*-isomers, can be de novo synthesized in the mammary gland [35]. Ifkovits and Ragheb [53] found that a pure culture of *Eubacterium ruminantium* contains high levels of C15:0 and *iso*-C15:0, but Minato et al. [54] reported that *Eubacterium ruminantium* is enriched in *anteiso*-C13:0. These studies suggest that different isolated strains of *Eubacterium ruminantium* have variable OBCFA profiles. In the present work, *Ruminococcus albus*, *Ruminococcus flavefaciens*, and *Eubacterium ruminantium* abundance were correlated with C13:0 contents in milk, which suggested that milk C13:0 might reflect the population of cellulolytic bacteria. The copy number of *Fibrobacter succinogenes* was positively correlated with the ratio of *iso*-C15:0 to *iso*-C17:0, which indicated that the transformation of *iso*-C15:0 to *iso*-C17:0 may occur in *Fibrobacter succinogenes*. Various bacterial strains of *Butyrivibrio fibrisolvens* form a heterogeneous group of bacteria that ferment a wide array of substrates [19], including fiber, starch, and fatty acids [52], and produce large amounts of lactate and acetate [57], which can increase both the growth rate [58] and butyrate production rate [59]. These previous studies showed that *Butyrivibrio fibrisolvens* is a key link between long-chain and short-chain fatty acids, which may be attributed to the relationship between the ruminal microbial population and milk OBCFA concentrations. Previous studies have suggested that amylolytic bacteria are relatively enriched in linear odd-chain fatty acids [34]. The populations of *Selenomonas ruminantium* in the rumen are highly correlated with total OBCFAs and branched-chain fatty acid concentrations in milk, suggesting that OBCFA concentrations in milk are involved in a complex interaction with ruminal microorganisms.

Vlaeminck et al. [9] first suggested that milk OBCFAs can be used as markers for the duodenal flow of microbial matter and found that changes in OBCFAs are closely associated with changes in uracil and purine bases in the rumen, thus confirming the potential of OBCFAs to predict microbial matter in the rumen. Several significant correlations existed between ruminal microbial base concentrations and milk OBCFA contents in the present study and in previous studies. Cabrita et al. [51] found that milk C17:0, *iso*-C17:0, and *anteiso*-C17:0 contents are significantly and negatively associated with dietary crude protein content, and they suggested that C17:0 is a marker of protein deficiency. This study implied that fatty acids with 17-carbon atoms in milk are associated with protein degradability in the rumen. Additionally, protein degradability is one of the factors that may affect rumen microbial growth [60]. The odd-chain fatty acid synthesis process is different from that of branched-chain fatty acids. [34]. Wongtangtintharn et al. [61] showed that all branched-chain fatty acids equally inhibit fatty acid synthesis. In this study, changes in the relationships between milk branched-chain fatty acids and ruminal microbial base contents were found. Moreover, the elongation of *iso*-C15:0 and *anteiso*-C15:0 adds to the existing *iso*-C17:0 and *anteiso*-C17:0 content [33], which may explain the lower *iso*-C15:0/*iso*-C17:0 and *anteiso*-C15:0/*anteiso*-C17:0 ratios in milk compared to the ratios found in rumen bacteria or duodenal content. The *iso*-C15:0/*iso*-C17:0 ratio in milk was positively correlated with the adenine/N ratio in the rumen, which suggested that the *iso*-C15:0/*iso*-C17:0 ratio in milk reflects the abundance of ruminal microbial.

Vlaeminck et al. [10] developed equations based on milk OBCFAs to predict molar proportions of individual VFAs in the rumen. Bhagwat et al. [12] improved the prediction accuracy of VFA proportions from measured milk OBCFA concentrations through the development and application of quadratic terms and interactions as well as the ratio of linear expressions. Bhagwat et al. [12] indicated that the more complex methods provide better predictions with fewer OBCFAs. Hence, we established an equation with mixed variables containing the first and second degree of dependent data as well as the ratio of any two datasets and the reciprocal of each dependent datasets in this research. Using this method, we generated a prediction model for major bacterial populations and microbial protein markers. In this model, there were some inversely proportional and quadratic relationships from the predicted equations, which suggested that there were some nonlinear relationships between milk OBCFAs and ruminal fermentation parameters, ruminal bacterial populations, and base contents.

5. Conclusions

OBCFAs originating from milk were significantly correlated with ruminal fermentation parameters, ruminal bacterial populations, and base contents. The results suggested that milk odd-chain fatty acids have the potential to be used as a noninvasive technique to assess rumen function in terms of microbial populations, substrates, and interactions. To increase the accuracy of the predicted equations for ruminal parameters, ruminal bacterial populations, and base contents established based on milk OBCFAs, a large number of experiments are required.

Author Contributions: Conceptualization, K.L.; Data curation, K.L., Y.L. and G.L.; Funding acquisition, Y.L., H.X., Y.Z., and G.L.; Writing—original draft, K.L.; Writing—review and editing, K.L., Y.Z.

Funding: The research was supported by the Innovation Program of Agricultural Science and Technology in CAAS (CAAS-ASTIP-2017-ISAPS) and the China Agriculture Research System (CARS-36), the Central Public-Interest Scientific Institution Basal Research Fund, the Postdoctoral Foundation in Heilongjiang Province (LBH-Z17035), Young Talents "Project of Northeast Agricultural University" (18QC35) and Technology Research Project from Education Department of Heilongjiang Province (No. 12511036).

Conflicts of Interest: The authors declare no conflict of interest. The funders had no role in the design of the study; in the collection, analyses, or interpretation of data; in the writing of the manuscript, or in the decision to publish the results.

References

1. Bauman, D.E.; Griinari, J.M. Nutritional regulation of milk fat synthesis. *Annu. Rev. Nutr.* **2003**, *23*, 203–227. [CrossRef] [PubMed]

2. Bainbridge, M.L.; Cersosimo, L.M.; Wright, A.D.G.; Kraft, J. Content and composition of branched-chain fatty acids in bovine milk are affected by lactation stage and breed of dairy cow. *PLoS ONE* **2016**, *11*, e0150386. [CrossRef]

3. Gross, J.; Van Dorland, H.A.; Bruckmaier, R.M.; Schwarz, F.J. Milk fatty acid profile related to energy balance in dairy cows. *J. Dairy Res.* **2011**, *78*, 479–488. [CrossRef] [PubMed]

4. Stoop, W.M.; Bovenhuis, H.; Heck, J.M.L.; van Arendonk, J.A.M. Effect of lactation stage and energy status on milk fat composition of Holstein-Friesian cows. *J. Dairy Sci.* **2009**, *92*, 1469–1478. [CrossRef] [PubMed]

5. Garnsworthy, P.C.; Masson, L.L.; Lock, A.L.; Mottram, T.T. Variation of milk citrate with stage of lactation and de novo fatty acid synthesis in dairy cows. *J. Dairy Sci.* **2006**, *89*, 1604–1612. [CrossRef]

6. Dewhurst, R.J.; Davies, D.R.; Merry, R.J. Microbial protein supply from the rumen. *Anim. Feed Sci. Technol.* **2000**, *85*, 1–21. [CrossRef]

7. Keeney, M.; Katz, I.; Allison, M.J. On the probable origin of some milk fat acids in rumen microbial lipids. *J. Am. Oil Chem. Soc.* **1962**, *39*, 198–201. [CrossRef]

8. Kaneda, T. Iso- and anteiso-fatty acids in bacteria: Biosynthesis, function, and taxonomic significance. *Microbiol. Mol. Biol. Rev.* **1991**, *55*, 288–302.

9. Vlaeminck, B.; Dufour, C.; van Vuuren, A.M.; Cabrita, A.R.J.; Dewhurst, R.J.; Demeyer, D.; Fievez, V. Use of odd and branched-chain fatty acids in rumen contents and milk as a potential microbial marker. *J. Dairy Sci.* **2005**, *88*, 1031–1042. [CrossRef]

10. Vlaeminck, B.; Fievez, V.; Tamminga, S.; Dewhurst, R.J.; Van Vuuren, A.; De Brabander, D.; Demeyer, D. Milk odd- and branched-chain fatty acids in relation to the rumen fermentation pattern. *J. Dairy Sci.* **2006**, *89*, 3954–3964. [CrossRef]

11. Vlaeminck, B.; Fievez, V.; Van Laar, H.; Demeyer, D. Rumen odd and branched chain fatty acids in relation to in vitro rumen volatile fatty acid productions and dietary characteristics of incubated substrates. *J. Anim. Physiol. Anim. Nutr.* **2004**, *88*, 401–411. [CrossRef] [PubMed]

12. Bhagwat, A.M.; De Baets, B.; Steen, A.; Vlaeminck, B.; Fievez, V. Prediction of ruminal volatile fatty acid proportions of lactating dairy cows based on milk odd- and branched-chain fatty acid profiles: New models, better predictions. *J. Dairy Sci.* **2012**, *95*, 3926–3937. [CrossRef] [PubMed]

13. Dijkstra, J.; Ellis, J.L.; Kebreab, E.; Strathe, A.B.; López, S.; France, J.; Bannink, A. Ruminal pH regulation and nutritional consequences of low pH. *Anim. Feed Sci. Technol.* **2012**, *172*, 22–33. [CrossRef]

14. Sutton, J.D. Digestion and Absorption of Energy Substrates in the Lactating Cow. *J. Dairy Sci.* **1985**, *68*, 3376–3393. [CrossRef]

15. Fujihara, T.; Shem, M.N. Metabolism of microbial nitrogen in ruminants with special reference to nucleic acids. *Anim. Sci. J.* **2011**, *82*, 198–208. [CrossRef]

16. Bryant, M.P.; Robinson, I.M. Effects of diet, time after feeding, and position sampled on numbers of viable bacteria in the bovine rumen. *J. Dairy Sci.* **1968**, *51*, 1950–1955. [CrossRef]

17. Shi, Y.; Odt, C.L.; Weimer, P.J. Competition for cellulose among three predominant ruminal cellulolytic bacteria under substrate-excess and substrate-limited conditions. *Appl. Environ. Microb.* **1997**, *63*, 734–742.

18. Bryant, M.P. Bacterial species of the rumen. *Microbiol. Mol. Biol. Rew.* **1959**, *23*, 125–153.

19. Miyagawa, E. Cellular fatty-acid and fatty alaehyde composition rumen bacteria. *J. Gen. Appl. Microbiol.* **1982**, *28*, 389–408. [CrossRef]

20. Fernando, S.C.; Purvis, H.T., II; Najar, F.Z.; Sukharnikov, L.O.; Krehbiel, C.R.; Nagaraja, T.G.; Roe, B.A.; DeSilva, U. Rumen microbial population dynamics during adaptation to a high-grain diet. *Appl. Environ. Microb.* **2010**, *76*, 7482–7490. [CrossRef]

21. Klieve, A.V.; Hennessy, D.; Ouwerkerk, D.; Forster, R.J.; Mackie, R.I.; Attwood, G.T. Establishing populations of *Megasphaera elsdenii* YE 34 and *Butyrivibrio fibrisolvens* YE 44 in the rumen of cattle fed high grain diets. *J. Appl. Microbiol.* **2003**, *95*, 621–630. [CrossRef] [PubMed]

22. NRC. *Nutrient Requirements of Dairy Cattle*, 7th ed.; National Academy of Science: Washington, DC, USA, 2001.

23. AOAC. *Offical Methods of Analysis of the AOAC*; Association of Official Analytical Chemists Inc.: Arlington, VA, USA, 1990; Volume 2.

24. Van Soest, P.J.; Roberson, J.B.; Lewis, B.A. Methods for dietary fiber, neutral detergent fiber, and nonstarch polysaccharides in relation to animal nutrition. *J. Dairy Sci.* **1991**, *74*, 3583–3597. [CrossRef]

25. Li, Y.; Meng, Q. Effect of different types of fibre supplemented with sunflower oil on ruminal fermentation and production of conjugated linoleic acids in vitro. *Arch. Anim. Nutr.* **2006**, *60*, 402–411. [CrossRef] [PubMed]

26. Tajima, K.; Aminov, R.I.; Nagamine, T.; Matsui, H.; Nakamura, M.; Benno, Y. Diet-dependent shifts in the bacterial population of the rumen revealed with real-time PCR. *Appl. Environ. Microb.* **2001**, *67*, 2766–2774. [CrossRef] [PubMed]

27. Zhang, Y.; Liu, K.; Hao, X.; Xin, H. The relationships between odd- and branched-chain fatty acids to ruminal fermentation parameters and bacterial populations with different dietary ratios of forage and concentrate. *J. Anim. Physiol. Anim. Nutr.* **2016**, *101*, 1103–1114. [CrossRef]

28. Reilly, K.; Attwood, G.T. Detection of Clostridium proteoclasticum and closely related strains in the rumen by competitive PCR. *Appl. Environ. Microb.* **1998**, *64*, 907–913.

29. Singha, K.M.; Pandyab, P.R.; Tripathia, A.K.; Patelb, G.R.; Parnerkarb, S.; Kotharic, R.K.; Joshia, C.G. Study of rumen metagenome community using qPCR under different diets. *Meta Gene* **2014**, *2*, 191–199. [CrossRef]

30. Yang, S. Effect of Soybean Oil and Linseed Oil Supplementation on Population of Ruminal Bacteria and Fermentation Parameters in Dairy Cows. Ph.D. Thesis, Chinese Academy of Agricultural Sciences, Beijing, China, 2007.

31. Stevenson, D.M.; Weimer, P.J. Dominance of *Prevolla* and low abundance of classical ruminal bacterial species in the bovine rumen revealed by relative quantification real-time PCR. *Appl. Microb. Biotechnol.* **2007**, *75*, 165–174. [CrossRef]

32. St-Pierre, N.R. Invited review integrating quantitative findings from multiple studies using mixed model methodology. *J. Dairy Sci.* **2001**, *84*, 741–755. [CrossRef]

33. Fievez, V.; Colman, E.; Castro-Montoya, J.M.; Stefanov, I.; Vlaeminck, B. Milk odd- and branched-chain fatty acids as biomarkers of rumen function. *Anim. Feed Sci. Technol.* **2012**, *172*, 51–65. [CrossRef]

34. Vlaeminck, B.; Fievez, V.; Cabrita, A.R.J.; Fonseca, A.J.M.; Dewhurst, R.J. Factors affecting odd- and branched-chain fatty acids in milk: A review. *Anim. Feed Sci. Technol.* **2006**, *131*, 389–417. [CrossRef]

35. Smith, S. The animal fatty acid synthase: One gene, one polypeptide, seven enzymes. *FASEB J.* **1994**, *8*, 1248–1259. [CrossRef] [PubMed]

36. Dodds, P.F.; Guzman, M.G.; Chalberg, S.C.; Anderson, G.J.; Kumar, S. Acetoacetyl-CoA reductase activity of lactating bovine mammary fatty acid synthase. *J. Biol. Chem.* **1981**, *256*, 6282–6290. [PubMed]

37. Massart-Leèn, A.M.; Roets, E.; Peeters, G.; Verbeke, R. Propionate for fatty acid synthesis by the mammary gland of the lactating goat. *J. Dairy Sci.* **1983**, *7*, 1445–1454. [CrossRef]

38. Scheerlinck, K.; De Baets, B.; Stefanov, I.; Fievez, V. Subset selection from multi-experiment data sets with application to milk fatty acid profiles. *Comput. Electron. Agric.* **2010**, *73*, 200–212. [CrossRef]

39. Vlaeminck, B.; Gervais, R.; Rahman, M.M.; Gadeyne, F.; Gorniak, M.; Doreau, M.; Fievez, V. Postruminal synthesis modifies the odd- and branched-chain fatty acid profile from the duodenum to milk. *J. Dairy Sci.* **2015**, *98*, 4829–4840. [CrossRef]

40. Belyea, R.L.; Adams, M.W. Energy and nitrogen utilization of high versus low producing dairy cows. *J. Dairy Sci.* **1990**, *73*, 1023–1030. [CrossRef]

41. Craninx, M.; Steen, A.; Laar, H.V.; Nespen, T.V.; Martín-Tereso, J.; Baets, B.D.; Fievez, V. Effect of Lactation Stage on the Odd- and Branched-Chain Milk Fatty Acids of Dairy Cattle Under Grazing and Indoor Conditions. *J. Dairy Sci.* **2008**, *91*, 2662–2677. [CrossRef]

42. Bauman, D.E.; Mellenberger, R.W.; Ingle, D.L. Metabolic adaptations in fatty acid and lactose biosynthesis by sheep mammary tissue during cessation of lactation. *J. Dairy Sci.* **1974**, *57*, 719–723. [CrossRef]

43. Verbeke, R.; Lauryssens, M.; Peeters, G.; James, A.T. Incorporation of DL-[1-14C] leucine and [1-14C] isovaleric acid into milk constituents by the perfused cow's udder. *Biochem. J.* **1959**, *73*, 24–29.

44. Fievez, V.; Vlaeminck, B.; Dhanoa, M.S.; Dewhurst, R.J. Use of principal component analysis to investigate the origin of heptadecenoic and conjugated linoleic acids in milk. *J. Dairy Sci.* **2003**, *86*, 4047–4053. [CrossRef]

45. Dewhurst, R.J.; Moorby, J.M.; Vlaeminck, B.; Fievez, V. Apparent recovery of duodenal odd- and branched-chain fatty acids in milk of dairy cows. *J. Dairy Sci.* **2007**, *90*, 1775–1780. [CrossRef] [PubMed]

46. Gervais, R.; Vlaeminck, B.; Fanchone, A.; Nozière, P.; Doreau, M.; Fievez, V. Odd-and branched-chain fatty acids duedenal flows and milk yield in response to N underfeeding and enegy source in dairy cows. *J. Dairy Sci.* **2011**, *94*, 125–126.

47. Dijkstra, J. Production and absorption of volatile fatty acids in the rumen. *Livest. Prod. Sci.* **1994**, *39*, 61–69. [CrossRef]

48. Misra, A.K.; Thakur, S.S. Influence of isobutyric acid supplementation on nutrient intake, its utilization and growth performance of crossbred calves fed low protein and urea containing rations. *Indian J. Anim. Sci.* **2002**, *72*, 476–479.

49. Felix, A.; Cook, R.M.; Huber, J.T. Effect of feeding isoacids with urea on growth and nutrient utilization by lactating cows. *J. Dairy Sci.* **1980**, *63*, 1943–1946. [CrossRef]

50. Kaneda, T. Fatty acids of the genus Bacillus: An example of branched-chain preference. *Microbiol. Mol. Biol. Rev.* **1977**, *41*, 391–418.

51. Cabrita, A.R.J.; Fonseca, A.J.M.; Dewhurst, R.J.; Gomes, E. Nitrogen supplementation of corn silages. 2. Assessing rumen function using fatty acid profiles of bovine milk. *J. Dairy Sci.* **2003**, *86*, 4020–4032. [CrossRef]

52. Buccioni, A.; Decandia, M.; Minieri, S.; Molle, G.; Cabiddu, A. Lipid metabolism in the rumen: New insights on lipolysis and biohydrogenation with an emphasis on the role of endogenous plant factors. *Anim. Feed Sci. Technol.* **2012**, *174*, 1–25. [CrossRef]

53. Ifkovits, R.W.; Ragheb, H.S. Cellular Fatty Acid Composition and Identification of Rumen Bacteria. *Appl. Environ. Microb.* **1968**, *16*, 1406–1413.

54. Minato, H.; Ishibashi, S.; Hamaoka, T. Cellular fatty-acid and suger composition of reppesentative strains of rumen bacteria. *J. Gen. Appl. Microbiol.* **1988**, *34*, 303–319. [CrossRef]

55. Saluzzi, L.; Smith, A.; Stewart, C.S. Analysis of bacterial phospholipid markers and plant monosaccharides during forage degradation by Ruminococcus flavefaciens and Fibrobacter succinogenes in co-culture. *J. Gen. Microbiol.* **1993**, *139*, 2865–2873. [CrossRef] [PubMed]

56. Croom, W.J.; Bauman, D.E.; Davis, C.L. Methylmalonic acid in low-fat milk syndrome. *J. Dairy Sci.* **1981**, *64*, 649–654. [CrossRef]

57. Diez-Gonzalez, F.; Bond, D.R.; Jennings, E.; Russell, J.B. Alternative schemes of butyrate production in Butyrivibrio fibrisolvens and their relationship to acetate utilization, lactate production, and phylogeny. *Arch. Microbiol.* **1999**, *171*, 324–330. [CrossRef]

58. Roché, C.; Albertyn, H.; Van Gylswyk, N.O.; Kistner, A. The growth response of cellulolytic acetate-utilizing and acetate-producing *Butyrivibrios* to volatile fatty acids and other nutrients. *J. Gen. Microbiol.* **1973**, *78*, 253–260. [CrossRef]

59. Latham, M.J.; Legakis, N.J. Cultural factors influencing the utilization or production of acetate by Butyrivibrio fibrisolvens. *J. Gen. Microbiol.* **1976**, *94*, 380–388. [CrossRef]

60. Stern, M.D.; Hoover, W.H. Methods for determining and factors affecting rumen microbial protein synthesis: A review. *J. Anim. Sci.* **1979**, *49*, 1590–1603. [CrossRef]

61. Wongtangtintharn, S.; Oku, H.; Iwasaki, H.; Toda, T. Effect of branched-chain fatty acids on fatty acid biosynthesis of human breast cancer cells. *J. Nutr. Sci. Vitaminol.* **2004**, *50*, 137–143. [CrossRef]

Article

Comparison of Faecal versus Rumen Inocula for the Estimation of NDF Digestibility

Maria Chiaravalli, Luca Rapetti, Andrea Rota Graziosi, Gianluca Galassi, Gianni Matteo Crovetto and Stefania Colombini *

Dipartimento di Scienze Agrarie e Ambientali, Università degli Studi di Milano, 20133 Milano, Italy; maria.chiaravalli@unimi.it (M.C.); luca.rapetti@unimi.it (L.R.); andrea.rota@unimi.it (A.R.G.); gianluca.galassi@unimi.it (G.G.); matteo.crovetto@unimi.it (G.M.C.)
* Correspondence: stefania.colombini@unimi.it

Received: 22 October 2019; Accepted: 5 November 2019; Published: 7 November 2019

Simple Summary: The evaluation of fibre digestibility is very important for the formulation of ruminant diets. Fibre digestibility is usually determined in lab with rumen inoculum obtained from cannulated cows. The research of alternative and less invasive inoculum sources is a critical issue that should be addressed. The present study evaluated the potential of faecal inocula, obtained from cows fed different diets, to assess fibre digestibility of different substrates at different incubation times (48, 240 and 360 h). At short incubation time, fibre digestibility obtained with rumen fluid was always higher than those obtained with faecal inocula, confirming a lower activity of the faecal inocula compared with rumen fluid. However, the type of diets fed to the donor animals had a significant effect on fibre digestibility, with a more active faecal inoculum for cows fed a diet based on maize silage. Despite the differences obtained at the short incubation time, the digestibility values at longer intervals showed that faecal inoculum could replace rumen inoculum. As a consequence, faeces may replace rumen fluid as inoculum for end-point measures, avoiding the use of cannulated animals and decreasing the analytical costs.

Abstract: Cow faeces have been investigated as alternative inoculum to replace rumen fluid to determine neutral detergent fibre (NDF) digestibility (NDFD). Aims of this study were to estimate: (1) the NDFD (48 h) of feed ingredients using a rumen inoculum in comparison with faecal inocula from cows fed diets with different forage basis; (2) the undigestible NDF (uNDF) at 240 and 360 h with ruminal fluid and faecal inocula from lactating cows fed two different diets. At 48 h incubation, the NDFD was affected both by feed and type of inoculum ($p < 0.01$) and by their interaction ($p = 0.03$). Overall, the mean NDFD was higher for rumen inoculum than for faecal inocula (585 vs. 389 g/kg NDF, $p < 0.05$), and faecal inoculum obtained from cows fed hay-based diets gave lower NDFD than those from cows fed maize silage (367 vs. 440 g/kg, $p < 0.05$). At long incubation times, the average uNDF was affected by substrate, inoculum and incubation time ($p < 0.01$), but not by their interactions. For each inoculum, significantly lower values were obtained at 360 than at 240 h. Regressions between uNDF with rumen and with the tested faecal inocula resulted in $r^2 \geq 0.98$. Despite the differences at 48 h, the uNDF showed that faecal inoculum could replace rumen fluid at longer incubation times.

Keywords: NDF digestibility; faecal inoculum; diet composition; in vitro

---

1. Introduction

Neutral detergent fibre (NDF) concentration of forages varies from 30% to 80% of DM (Dry Matter) with a wide variation for its digestibility [1]: from less than 40% for highly lignified mature legumes to greater than 90% for unlignified immature grass [2]. The accurate estimation of the NDF digestibility (NDFD) is important because ruminant nutritionists and forage plant breeders use in vitro measures of NDFD to assess forage quality, predict diet digestibility, and select plant genotypes for breeding [2]. For these purposes, NDFD is generally determined at short incubation times (for example: 24, 30, 48 h). However, NDFD can also be measured at longer incubation time to estimate the undegradable NDF (uNDF) of feedstuff and, consequently, to predict the potentially digestible NDF (pdNDF) and to estimate the extent of NDF digestion. Moreover, the uNDF content of faeces and total mixed ration (TMR) is used as a marker to estimate in vivo NDF digestibility with a field application at farm scale.

The in vitro techniques developed to determine NDFD (for example: [3,4]) involve the use of ruminal inoculum. However, in recent time fresh faeces from ruminants have been investigated as alternative inoculum to replace rumen fluid [5–7]. Using a fresh faecal inoculum would have some advantages: it is easier to obtain faeces than rumen liquor and rumen cannulated animals are not needed. Faecal inocula from cows have been used to evaluate the in vitro organic matter digestibility (OMD) of forages [6,8] and gas production (GP) of different feeds [5,7]. For example, Akhter et al. [8] obtained significant regression equations between in vitro OMD of eight forages determined with sheep rumen liquor and cow faeces. Another study [6] showed that the OMD of different forages estimated with faecal inoculum was comparable to the OMD determined with cow rumen fluid after 48 h of incubation. Regarding GP, a study [7] showed that in vitro total GP was greater for feeds incubated with cow rumen inoculum as compared to cow faecal inoculum. Mauricio et al. [5] confirmed these results: they observed that cow faecal inoculum gave lower GP volumes and longer lag time than rumen inoculum; however, potential GP was highly correlated between rumen liquor and faecal inoculum. Overall, all the studies indicate that faeces have the potential to be used as inoculum, but some limitations have been identified; one of the most important is the lower enzymatic activity of faecal inocula compared to rumen liquor [5,8]. In this regard, the type of diets fed to the donor animals can change the microbial population within the digestive tract, and therefore in the faeces of animals. For example, Kim et al. [9] observed that the community structure of cow faecal microbiota is greatly affected by diet and it is particularly associated to the dietary forage and concentrate ratio. These changes should be considered in using faeces as microbial inoculum and should be better evaluated. To the best of our knowledge, the effects of diets fed to cows on faecal inoculum activity were never evaluated. Similarly, as far as we know, there are no studies in literature that determined in vitro NDFD with cow faecal inoculum. Overall further research seems needed to fully understand and develop in vitro digestibility techniques using faecal inoculum.

The aims of this study were: (1) to estimate the NDFD (48 h) of several feed ingredients using a rumen inoculum in comparison with faecal inocula obtained both from dry and lactating cows fed diets characterised by a different forage basis; (2) to estimate the uNDF concentrations of different substrates (faeces and feeds) at two incubation times (240 and 360 h) with ruminal fluid and faecal inocula obtained from lactating cows fed two different diets.

2. Materials and Methods

The study was conducted at Cascina Baciocca, the experimental farm of the University of Milan, with the authorisation of the Ministry of Health, authorisation n. 904/2016-PR.

2.1. In Vitro Incubations with Faecal and Rumen Inocula at 48 h

The NDFD analyses were conducted on four feeds and using six inocula (five different faecal inocula and one rumen inoculum). The feeds were selected in order to have different NDF contents, as follows: grass hay (533 aNDF g/kg DM), wheat bran (497 aNDF g/kg DM), maize distiller (387 aNDF g/kg DM), and maize silage (373 aNDF g/kg DM). The feeds were dried at 60 °C for 48 h in a forced-air oven and ground to pass a 1-mm Fritsch mill (Fritsch Pulverisette, Idar-Oberstein, Germany). Each sample was weighed (0.250 g) in duplicate in Ankom F57 bags (Ankom Technology, Macedon, NY, USA); two blank Ankom bags (i.e., bags without sample) were also incubated for each inoculum source. Each bag was placed into a pre-warmed 100 mL Erlenmeyer flask closed by a rubber stopper with a Bunsen valve for gas release and maintained at 39 °C in a water bath with constant agitation.

Different faeces were tested as inoculum for a total of five: faeces collected from two cannulated Holstein dry cows fed a diet composed by (g/kg DM) grass hay (700), lucerne hay (130), maize meal (135), soybean meal (30) and vitamin mineral supplement (5), faeces collected from two Holstein lactating cows fed different diets based on the following forages: maize silage (FL-MS), ryegrass and lucerne silages (FL-GLS), wheat and lucerne silages (FL-WLS), and ryegrass and lucerne hays (FL-GLH). Two different donor cows fed the same diet were used for each incubation run. The composition of the lactating cow diets is in Table 1.

Table 1. Composition of the faecal inoculum donor lactating cow diets (on a dry matter basis)[1].

Ingredient	MS	GLS	WLS	GLH
(g DM/kg total DM)				
maize silage	493			
maize high moisture ear		286	291	
ryegrass hay	172			253
lucerne silage		268	104	
ryegrass silage		191		
wheat silage			200	
lucerne hay			106	253
grass hay				
soybean meal 44% CP			127	
soybean meal 48% CP	157	81.3		89.8
maize grain ground fine	121	114	127	228
maize grain flaked				86.0
maize gluten feed dry				43.9
wheat bran				11.5
molasses cane	30.3	34.6	19.1	21.5
limestone ground	9.59	8.97	9.11	4.74
sodium bicarbonate	6.15	5.75	5.84	3.16
sodium chloride	3.64	3.40	3.46	2.39
magnesium oxide	3.13	2.93	2.97	1.04
dicalcium phosphate	2.10	1.97	2.00	0.27
smartamine M	0.36	0.30	0.34	0.44
minvit suppl.[2]	1.22	1.14	1.16	1.29

[1] MS: maize silage diet; GLS: ryegrass and lucerne silage diet; WLS: wheat and lucerne silage diet; GLH: ryegrass and lucerne hay diet; [2] minvit composition: vitamin A 400,000 UI/kg, vitamin D3 60,000 UI/kg, vitamin E 1000 mg/kg, vitamin B1 60 mg/kg, vitamin PP 6000 mg/kg, biotin 40 mg/kg, copper sulfate pentahydrate 1900 mg/kg, cupric chelate of amino acids hydrate 4900 mg/kg, calcium iodate anhydrous 40 mg/kg, ferrous carbonate 1800 mg/kg, manganese chelate of amino acids hydrate 9000 mg/kg, sodium selenite 31 mg/kg, zinc oxide 1900 mg/kg, zinc chelate of amino acids hydrate 600 mg/kg, propyl gallate 12 mg/kg, butyl hydroxytoluene 34 mg/kg.

All faecal samples were collected immediately after defecation from the floor of the pen and transported in pre-warmed sealed containers to the laboratory for processing. The time lap in faecal collection and processing was within the range of 20 min. Rumen inoculum was collected from the two rumen cannulated donor cows. The rumen fluid was collected before the morning meal and was immediately strained through four layers of cheesecloth into a pre-warmed (39 °C) flask with CO_2 and mixed with the buffer solution [10] in a 1:2 ratio. Faecal samples were first mixed with the same buffer solution (at the same 1:2 ratio) and maintained for thirty minutes at 39 °C with CO_2, and then strained through four layers of cheesecloth [5,6,11].

Following the method described by Spanghero et al. [12], 90 mL of each inoculum was dispensed into the Erlenmeyer flasks under anaerobic conditions, flushing the flask with CO_2. Two runs of incubations were made in a shaking water bath at 39 °C for 48 h. At the end of incubation, the F57 bags were rinsed with cold water until the water ran clear and then placed in a 60 °C forced-air oven to dry. Subsequently, NDF concentration was determined for each bag using the fibre analyser (Ankom Technology, Macedon, NY, USA) and in vitro NDFD was calculated.

2.2. In Vitro Incubations with Faecal and Rumen Inocula at 240 and 360 h

The uNDF content was determined on seven substrates using three different inocula (one rumen and two faeces) and two incubation times at 240 and 360 h.

The substrates were: barley meal (293 aNDF g/kg DM), maize silage (505 aNDF g/kg DM), lucerne hay (538 aNDF g/kg DM), lactating cow TMR (321 aNDF g/kg DM) and the faeces (595 aNDF g/kg DM) produced by a lactating Holstein cow fed the same TMR, grass hay (662 aNDF g/kg DM) and the faeces (753 aNDF g/kg DM) produced by a dry Holstein cow fed the same grass hay. Each substrate was dried at 60 °C for 48 h in a forced-air oven and then ground to pass a 1-mm Fritsch mill (Fritsch Pulverisette, Idar-Oberstein, Germany) Incubations were conducted weighting (0.250 g) duplicate samples in F57 bags and using the DaisyII incubator jars (Ankom Technology, Macedon, NY, USA). Each jar contained four F57 bags for each substrate (two bags for each incubation time) and four F57 blank bags (i.e., bags without sample; two blanks for each incubation time).

The inocula RD-GH, FL-MS and FL-GLH previously described were tested. Faeces and rumen liquor were collected and treated as described above. The buffer was composed by two solutions as reported by Ankom protocol [13]. Faeces were mixed with the buffer in a ratio of 450 g/L and processed as decribed by Hughes et al. [6], while rumen fluid was added at a dose of 400 mL/jar, using a 1:3 ratio with the buffer [13].

Each inoculum was poured into one pre-warmed jar and two fermentation runs were carried out. Two F57 bags for each substrate and blank were removed at two times: 240 and 360 h. The inocula were renewed at 120 and 240 h replacing the content of jars with a new inoculum. Upon completion of the incubation, jars were emptied and the F57 bags were treated as described above.

2.3. Statistical Analysis

All data were analysed by the mixed procedure of SAS (version 9.4) considering the main effects of inoculum, run, substrate and time and their interactions. Data are reported as LS-MEANS.

Linear regressions between the rumen and the faecal inocula at each incubation time were performed by PROC REG procedure of SAS (version 9.4, SAS Institute Inc., Cary, NC, USA).

3. Results

3.1. NDFD at 48 h of Incubation Using Rumen and Faecal Inocula

The results of NDFD for rumen and faecal inocula at 48 h are presented in Table 2.

Table 2. Effect of inocula[1] on neutral detergent fibre (NDF) digestibility (NDFD, g/kg NDF) of different feeds at 48 h of incubation.

Sample	RD-GH	FL-MS	FL-GLS	FL-WLS	FL-GLH	FD-GH	P	SE
grass hay	578[a]	379[b]	405[b]	411[b]	361[b]	446[b]	0.03	32.4
wheat bran	521[a]	431[b]	425[b]	415[b]	408[b]	407[b]	0.01	18.0
maize distiller	741[a]	534[b]	467[b,c]	441[b,c]	428[b,c]	383[c]	<0.01	34.9
maize silage	498[a]	418[a,b]	288[b,c]	315[b,c]	214[c]	286[b,c]	0.04	50.4
all substrates	585[a]	440[b]	396[b,c]	396[b,c]	353[c]	381[c]	<0.01	14.9

Mean values in a row with different superscripts (a, b, c) differ significantly ($p < 0.05$); [1]RD-GH: rumen, dry cow—grass hay diet; FL-MS: faeces, lactating cow—maize silage diet; FL-GLS: faeces, lactating cow—ryegrass and lucerne silage diet; FL-WLS: faeces, lactating cow—wheat and lucerne silage diet; FL-GLH: faeces, lactating cow—ryegrass and lucerne hay diet; FD-GH: faeces, dry cow—grass hay diet.

The values of NDFD were affected by inoculum ($p < 0.01$) and by the interaction between inoculum and feed ($p = 0.03$).

Overall, the mean NDFD value was higher for RD-GH (585 g/kg NDF) than for the faecal inocula (389 g/kg NDF, on average) ($p < 0.05$) but there were differences among faecal inocula. Particularly, faecal inocula of cows fed diets based on hays (FD-GH and FL-GLH) resulted, on average, in lower NDFD (367 g/kg NDF) than FL-MS (440 g/kg NDF) ($p < 0.05$). The values obtained using FL-WLS and FL-GLS were intermediate.

Considering each individual feed sample, all feed samples had higher NDFD values using RD-GH treatment ($p < 0.05$), except for maize silage which was not significantly different between RD-GH (498 g/kg) and FL-MS (418 g/kg). The NDFD of maize distillers obtained with FD-GH inoculum was also significantly lower ($p < 0.05$) than the value obtained using FL-MS (383 vs. 534 for FD-GH and FL-MS, respectively).

3.2. Determination of uNDF Using Rumen and Faecal Inocula

The results of uNDF with faecal and rumen inocula at both incubation times (240 and 360 h) are presented in Table 3.

Table 3. Effect of inocula[1] and incubation time on undigestible NDF (uNDF, g/kg DM) of different substrates.

Sample	240 h			360 h			P		SE
	RD-GH	FL-MS	FL-GLH	RD-GH	FL-MS	FL-GLH	inoculum	time	
grass hay	223	287	237	198	233	225	0.16	0.15	22.2
lucerne hay	294	309	298	293	295	303	0.62	0.69	9.16
barley meal	79.7	92.9	81.3	58.7	69.8	67.1	0.43	0.03	8.72
maize silage	155	217	175	136	168	160	0.11	0.12	18.3
lactating cow TMR	78.6	104	84.5	74.4	74.9	83.6	0.33	0.13	8.09
faeces dry cow	400	454	434	372	385	403	0.16	0.02	17.0
faeces lactating cow	322	367	334	282	319	289	0.20	0.04	20.6
all substrates	222[b]	262[a]	235[b]	202[c]	221[b]	219[b,c]	<0.01	<0.01	6.06

Mean values in a row with different superscripts (a, b, c) differ significantly ($p < 0.05$); TMR: total mixed ration; [1] RD-GH: rumen, dry cow—grass hay; FL-MS: faeces, lactating cow—maize silage diet; FL-GLH: faeces, lactating cow—ryegrass and lucerne hay diet.

The average uNDF value was affected by inoculum and incubation time ($p < 0.01$), but not by the interactions between substrate, inoculum and time. Considering the effect of inoculum, the uNDF was not affected by the type of inoculum for RD-GH and FL-GLH; on the contrary, FL-MS had a higher value than RD-GH both at 240 and 360 h, and a higher value at 240 h in comparison with FL-GLH (Table 3). For each inoculum, the time effect was significant with lower values obtained at 360 h (on average, 240 vs. 214 g uNDF/kg DM for 240 and 360 h of incubation, respectively, $p < 0.01$). This time effect has to be ascribed particularly to the faeces samples. Moreover, uNDF decreased more for FL-MS treatment (−15.6%) than for RD-GH (−9.0%) and FL-GLH (−6.8%) from 240 to 360 h of incubation.

Considering the values for each substrate (Table 3), the dry cow faeces had the highest ($p < 0.05$) uNDF concentration (408 g/kg DM), while the lowest concentrations were obtained for barley meal and lactating cow TMR (74.9 g/kg DM and 83.4 g/kg DM, respectively). Intermediate values ($p < 0.05$) were registered for maize silage (169 g/kg DM), grass hay (234 g/kg DM), lucerne hay (298 g/kg DM) and lactating cows faeces (319 g/kg DM).

All linear regressions (Figures 1 and 2) between uNDF determined with rumen and faecal inocula at 240 and 360 h resulted in high r^2 values ($r^2 = 0.98$ for RD-GH vs. FL-MS at 240 h; $r^2 = 0.996$ for RD-GH vs. FL-GLH at 240 h; $r^2 = 0.980$ for RD-GH vs. FL-MS at 360 h; $r^2 = 0.994$ for RD-GH vs. FL-GLH at 360 h) and slopes within the range 0.92–0.97 showing a high relationship between data.

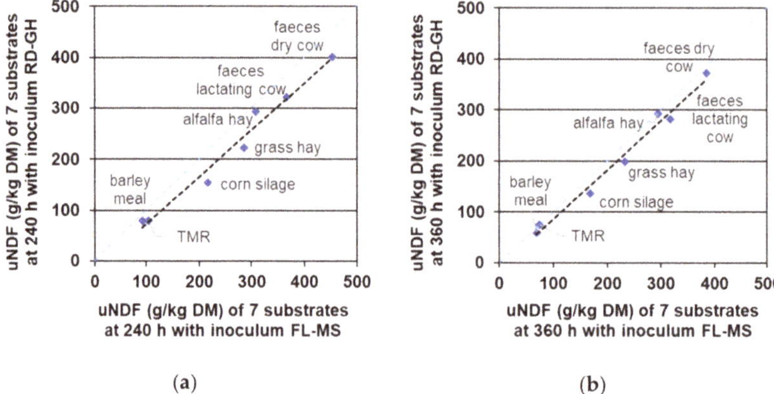

Figure 1. Linear regression between uNDF (g/kg DM) of the tested substrates with the inocula RD-GH (rumen, dry cow—grass hay diet) and FL-MS (faeces, lactating cow—maize silage diet). (**a**) At 240 h, $Y = 0.92X$ ($p < 0.01$) $- 19.5$ ($p = 0.35$); $r^2 = 0.98$, MSE = 443; (**b**) At 360 h, $Y = 0.97X$ ($p < 0.01$) $- 11.4$ ($p = 0.44$); $r^2 = 0.98$, MSE = 276.

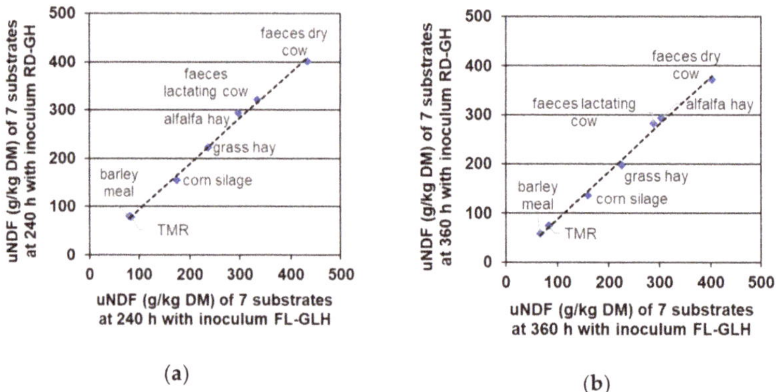

Figure 2. Linear regression between uNDF (g/kg DM) of the tested substrates with the inocula RD-GH (rumen, dry cow—grass hay) and FL-GLH (faeces, lactating cow—ryegrass and lucerne hay diet). (**a**) At 240 h, $Y = 0.94X$ ($p < 0.01$) $+ 0.03$ ($p = 1.00$); $r^2 = 0.995$, MSE = 823; (**b**) At 360 h, $Y = 0.96X$ ($p < 0.01$) $- 8.68$ ($p = 0.33$); $r^2 = 0.99$, MSE = 97.0.

4. Discussion

The objective of this study was to evaluate different faecal inocula (obtained from donor cows fed different diets) to determine NDFD at different incubation times (48, 240 and 360 h). It is well known that in ruminants the intestinal microbes can ferment fibre not digested in the rumen, hence there is the interest to evaluate the potential of faecal inoculum to digest NDF. The evaluation was done on a different set of samples including both forages and fibrous by-products, characterised by a wide variation in NDF digestibility and with an NDF content in the range between 373 and 533 g/kg DM in the first experiment.

Cow faeces were used in other studies as inoculum source, however, other parameters such as GP [5,7] or in vitro OM digestibility [6,8] were evaluated. The results of these studies underlined that faeces can be used as source of inoculum although the digestibility and the GP determined with faecal inocula were always lower than those found with rumen liquor. Several studies [7,14] showed a consistently longer lag phase with faecal inocula rather than rumen inocula, and this determined lower GP values with faecal inoculum. Similarly, Jiao et al. [15] evaluated fibre digestion in different segments of the post-ruminal tract demonstrating that detectable changes in hydrolysis of NDF occurred after 24 h of in vitro incubation.

In agreement with the cited studies, the results of the present study showed lower NDFD values at 48 h for cow faecal inocula in comparison with rumen inoculum. The lower NDFD values for faecal inocula were expected since the populations of cellulolytic bacteria are more numerous in rumen inoculum as compared to faecal inoculum [16]. In the rumen, the existence of a greater number of microbial populations is due to the medium stability and various diet compositions, whilst, on the other hand, in the gut nutrients are poor and the nutrient absorption rate is quick [17]. However, Van Vliet et al. [18] showed that diet composition can affect faecal bacterial biomass concentration with highest values associated with high energy diets due to a higher microbial growth in the rumen. Similarly, Lettat et al. [19] showed that increasing the maize silage proportion in the diet reduced the ruminal richness and diversity of the bacterial community but increased the number of total bacteria, which in turn should be reflected in the faecal microbial biomass and hence in the enzymatic activity of inoculum. As far as we know, no studies were conducted to evaluate NDFD using faecal inocula collected from cows fed different diets, although it is known that diet greatly influences the faecal microbiota of cattle [9,20,21]. In the present study FL-MS inoculum significantly affected maize silage NDFD in comparison with the other faecal inocula; it has to be underlined that maize silage was the main ingredient of the diet fed to donor cows of FL-MS inoculum. Although microbial communities were not determined in the present study, it can be speculated that there were some differences in rumen microbial community depending on diet, which in turn affects the faecal microbial community. Lengowski et al. [22] found higher abundance of the cellulolytic species *Fibrobacter succinogenes* during in vitro incubation of maize silage compared to grass silage. Similarly, another study [23] found a larger population of *F. succinogenes* in maize stems compared to other cellulolytic bacteria in in vitro conditions, while Lettat et al. [19] observed higher numbers of *F. succinogenes* in the rumen of dairy cows fed diets high in maize silage compared to those fed diets high in lucerne silage. Maize silage is a C_4 plant; C_4 plants usually have a lower nutritive value than C_3 plants due to a higher proportion of thick-walled cells associated with indigestible fibre in their leaves [24]; therefore, as suggested by Lengowski et al. [25], it may be that *F. succinogenes* has an advantage in degrading maize silage cell walls. The enzymatic system of *F. succinogenes* is more effective at degrading cellulose than the mechanisms used by the other cellulolytic organisms that occupy the same environment [26]. Moreover, *Fibrobacter spp.* have been detected in herbivorous species including the bovine rumen and caecum and in the faeces of different animal species [26]. Overall, it can be speculated that the higher NDFD for FL-MS inoculum can be due both to a higher bacterial concentration and to a possible higher presence of *F. succinogenes* due to the high concentration of maize silage in the diet [19].

Despite the difference between rumen and faecal inocula in the NDFD values at 48 h, the incubation conducted at 240 and 360 h showed promising results for faecal inocula to determine uNDF, especially for the inoculum FL-GLH. For all inocula and considering all the substrates, the average uNDF was affected by incubation time with lower values at 360 than 240 h. The uNDF fraction can be estimated by long-term (240 h) in vitro fermentations [27] or by incubating the samples in bags placed in the rumen for 288 h [28]. In the present study, an unconventional incubation time of 360 h (15 days) was also tested in order to compensate for the longer expected lag time of faecal inocula [5]. Unexpectedly, with the in vitro method used in the present study, the incubation time was significant also for the rumen inoculum; with significantly lower uNDF values at 360 than 240 h. Despite the wide application of uNDF, few documented recommendations of the method exist [28]. Recently, Raffrenato et al. [29] indicated that 288 h (12 days) of fermentation were necessary to reach the maximum extent of NDF digestion for the Daisy incubator; hence the incubation time of 240 h used in the present study seems not adequate to evaluate uNDF for some of the incubated substrates (faeces), confirming the results of Raffrenato et al. [29].

The effect of the diet fed to donor cows was also significant at long incubation times; however, differently from 48 h incubation, in the long incubation runs the FL-GLH inoculum gave results similar to RD-GH. As suggested by Aiple et al. [30], to obtain an active faecal inoculum, microbial populations in the hind gut should be supplied with fermentable substrate from a varied diet. Although knowledge of the nutrient requirements and supply of the hindgut microbes is limited, it is likely that their requirements are similar to those of the ruminal bacteria [31]. In the present study, cows fed GLH diet (inoculum FL-GLH) had a higher DMI and lower total tract organic matter digestibility (data not presented) than cows fed MS diet. As a consequence, a higher quantity of undigested organic matter was present in FL-GLH inoculum which could have better supported the microbial growth in the long time incubations and positively affecting NDFD. However, it has to be underlined that overall both faecal inocula gave results close to the rumen inocula, as confirmed by the slopes obtained with the linear regression analysis. The long time incubation results are promising since in feed evaluation systems the uNDF is normally computed from the ADL content multiplied by 2.4. However, Raffrenato et al. [32] showed that the ratio between ADL and uNDF is not constant, and, as reported also by Colombini et al. [33], a wrong evaluation of uNDF results in a bias in the prediction of fibre digestibility. Therefore, the use of an appropriate faecal inoculum seems a valuable alternative method to estimate the uNDF avoiding the use of cannulated cows.

5. Conclusions

The NDFD results at 48 h obtained with rumen fluid were always higher than those obtained with faecal inocula confirming a lower activity of the faecal inocula compared with rumen fluid. However, the type of diets fed to the donor animals had a significant effect on NDFD values determined by faecal inocula. This should be considered in using faeces as microbial inoculum source. As a prospective view to improve the results at 48 h, the use of fibrolytic enzymes added to the faecal inoculum should be evaluated.

Despite the differences obtained at 48 h, the uNDF results showed that faecal inoculum could replace rumen fluid at longer incubation times. As a consequence, in agreement with the review of Mould et al. [34], faeces may replace rumen fluid as an inoculum for end-point measures. Within the in vitro method applied in the present study, the incubation time of 240 h seems to be not enough to measure uNDF for both faecal and rumen inocula. In the present study, an unconventional time (360 h) was used; hence, further studies seem to be necessary to better define for different substrates (faeces, feeds and TMR) and inoculum the optimum incubation length (in the interval between 240 and 360 h) to evaluate uNDF.

Author Contributions: Conceptualization, S.C. and L.R.; methodology, M.C., A.R.G. and G.G.; software, M.C. and L.R.; validation, M.C. and A.R.G.; formal analysis, M.C.; and A.R.G.; investigation, S.C.; resources, S.C.; data curation, M.C. and S.C.; writing—original draft preparation, M.C.; writing—review and editing, S.C., G.M.C., L.R. and G.G.; visualization, S.C.; supervision, S.C.; project administration, S.C.; funding acquisition, S.C.

Funding: This research was funded by the Italian Ministry of Education, University and Research (PRIN 2015 2015FP39B9-LS9). PRIN Project: Metodologie innovative per studi di fermentazione ruminale in vitro senza l'impiego di animali da esperimento (Innovative methodologies for in vitro rumen fermentation studies conducted without test animals).

Conflicts of Interest: The authors declare no conflict of interest. The funders had no role in the design of the study; in the collection, analyses, or interpretation of data; in the writing of the manuscript, or in the decision to publish the results.

References

1. Bender, R.W.; Cook, D.E.; Combs, D.K. Comparison of in situ versus in vitro methods of fiber digestion at 120 and 288 hours to quantify the indigestible neutral detergent fiber fraction of corn silage samples. *J. Dairy Sci.* **2016**, *99*, 5394–5400. [CrossRef] [PubMed]
2. Goeser, J.P.; Combs, D.K. An alternative method to assess 24-h ruminal in vitro neutral detergent fiber digestibility. *J. Dairy Sci.* **2009**, *92*, 3833–3841. [CrossRef] [PubMed]

3. Hall, M.B.; Mertens, D.R. In vitro fermentation vessel type and method alter fiber digestibility estimates. *J. Dairy Sci.* **2008**, *91*, 301–307. [CrossRef] [PubMed]

4. Lopes, F.; Ruh, K.; Combs, D.K. Validation of an approach to predict total-tract fiber digestibility using a standardized in vitro technique for different diets fed to high-producing dairy cows. *J. Dairy Sci.* **2015**, *98*, 2596–2602. [CrossRef] [PubMed]

5. Mauricio, R.M.; Owen, E.; Mould, F.L.; Givens, I.; Theodorou, M.K.; France, J.; Davis, D.R.; Dhanoa, M.S. Comparison of bovine rumen liquor and bovine faeces as inoculum for an in vitro gas production technique for evaluating forages. *Anim. Feed Sci. Technol.* **2001**, *89*, 33–48. [CrossRef]

6. Hughes, M.; Mlambo, V.; Lallo, C.H.O.; Jennings, P.G.A. Potency of microbial inocula from bovine faeces and rumen fluid for in vitro digestion of different tropical forage substrates. *Grass Forage Sci.* **2012**, *67*, 263–273. [CrossRef]

7. Ramin, M.; Lerose, D.; Tagliapietra, F.; Huhtanen, P. Comparison of rumen fluid inoculum vs. faecal inoculum on predicted methane production using a fully automated in vitro gas production system. *Livest. Sci.* **2015**, *181*, 65–71. [CrossRef]

8. Akhter, S.; Owen, E.; Theodorou, M.K.; Butler, E.A.; Minson, D.J. Bovine faeces as a source of micro-organisms for the in vitro digestibility assay of forages. *Grass Forage Sci.* **1999**, *54*, 219–226. [CrossRef]

9. Kim, M.; Kim, J.; Kuehn, L.A.; Bono, J.L.; Berry, E.D.; Kalchayanand, N.; Freetly, H.C.; Benson, A.K.; Wells, J.E. Investigation of bacterial diversity in the feces of cattle fed different diets. *J. Anim. Sci.* **2014**, *92*, 683–694. [CrossRef]

10. Menke, K.H.; Steingass, H. Estimation of the energetic feed value obtained from chemical analysis and in vitro gas production using rumen fluid. *Anim. Res. Dev.* **1988**, *28*, 7–55.

11. Zicarelli, F.; Calabro, S.; Cutrignelli, M.I.; Infascelli, F.; Tudisco, R.; Bovera, F.; Piccolo, V. In vitro fermentation characteristics of diets with different forage/concentrate ratios: Comparison of rumen and faecal inocula. *J. Sci. Food Agric.* **2011**, *91*, 1213–1221. [CrossRef] [PubMed]

12. Spanghero, M.; Chiaravalli, M.; Colombini, S.; Fabro, C.; Froldi, F.; Mason, F.; Moschini, M.; Sarnataro, C.; Schiavon, S.; Tagliapietra, F. Rumen inoculum collected from cows at slaughter or from a continuous fermenter and preserved in warm, refrigerated, chilled or freeze-dried environments for in vitro tests. *Animals* **2019**, *9*, 815. [CrossRef] [PubMed]

13. Ankom Technology. In Vitro True Digestibility Using the DAISY Incubator. 2005. Available online: http://www.ankom.com/media/documents/IVDMD_0805_D200.pdf (accessed on 13 September 2010).

14. Pandian, C.S.; Reddy, T.J.; Sivaiah, K.; Blümmel, M.; Ramana Reddy, Y.R. Faecal matter as inoculum for in vitro gas production technique. *Anim. Nutr. Feed Technol.* **2016**, *16*, 271. [CrossRef]

15. Jiao, J.; Lu, Q.; Tan, Z.; Guan, L.; Zhou, C.; Tang, S.; Han, X. In vitro evaluation of effects of gut region and fiber structure on the intestinal dominant bacterial diversity and functional bacterial species. *Anaerobe* **2014**, *28*, 168–177. [CrossRef]

16. El-Meadaway, A.; Mir, Z.; Mir, P.S.; Zaman, M.S.; Yanke, L.J. Relative efficacy of inocula from rumen fluid or faecal solution for determining in vitro digestibility and gas production. *Can. J. Anim. Sci.* **1998**, *78*, 673–679. [CrossRef]

17. Guzmán, M.L.; Sager, R.L. Ruminant Fecal Inolucum for In Vitro Feed Digestibility Analysis. 2016. Available online: https://unsl.academia.edu/RicardoSager (accessed on 6 November 2019).

18. Van Vliet, P.C.J.; Reijs, J.W.; Bloem, J.; Dijkstra, J.; De Goede, R.G.M. Effects of cow diet on the microbial community and organic matter and nitrogen content of feces. *J. Dairy Sci.* **2007**, *90*, 5146–5158. [CrossRef]

19. Lettat, A.; Hassanat, F.; Benchaar, C. Corn silage in dairy cow diets to reduce ruminal methanogenesis: Effects on the rumen metabolically active microbial communities. *J. Dairy Sci.* **2013**, *96*, 5237–5248. [CrossRef]

20. Callaway, T.R.; Dowd, S.E.; Edrington, T.S.; Anderson, R.C.; Krueger, N.; Bauer, N.; Kononoff, P.J.; Nisbet, D.J. Evaluation of bacterial diversity in the rumen and feces of cattle fed different levels of dried distillers grains plus solubles using bacterial tag-encoded FLX amplicon pyrosequencing. *J. Anim. Sci.* **2010**, *88*, 3977–3983. [CrossRef]

21. Shanks, O.C.; Kelty, C.A.; Archibeque, S.; Jenkins, M.; Newton, R.J.; McLellan, S.L.; Huse, S.M.; Sogin, M.L. Community structures of fecal bacteria in cattle from different animal feeding operations. *Appl. Environ. Microbiol.* **2011**, *77*, 2992–3001. [CrossRef]

22. Lengowski, M.B.; Zuber, K.H.; Witzig, M.; Möhring, J.; Boguhn, J.; Rodehutscord, M. Changes in rumen microbial community composition during adaption to an in vitro system and the impact of different forages. *PLoS ONE* **2016**, *11*, e0150115. [CrossRef]

23. Kozakai, K.; Nakamura, T.; Kobayashi, Y.; Tanigawa, T.; Osaka, I.; Kawamoto, S.; Hara, S. Effect of mechanical processing of corn silage on in vitro ruminal fermentation, and in situ bacterial colonization and dry matter degradation. *Can. J. Anim. Sci.* **2007**, *87*, 259–267. [CrossRef]

24. Wilson, J.R. Cell wall characteristics in relation to forage digestion byruminants. *J. Agric. Sci.* **1994**, *122*, 173–182. [CrossRef]

25. Lengowski, M.B.; Witzig, M.; Möhring, J.; Seyfang, G.M.; Rodehutscord, M. Effects of corn silage and grass silage in ruminant rations on diurnal changes of microbial populations in the rumen of dairy cows. *Anaerobe* **2016**, *42*, 6–16. [CrossRef]

26. McDonald, J.E.; Rooks, D.J.; McCarthy, A.J. Methods for the isolation of cellulose-degrading microorganisms. *Methods Enzymol.* **2012**, *510*, 349–374. [PubMed]

27. Palmonari, A.; Gallo, A.; Fustini, M.; Canestrari, G.; Masoero, F.; Sniffen, C.J.; Formigoni, A. Estimation of the indigestible fiber in different forage types. *J. Anim. Sci.* **2016**, *94*, 248–254. [CrossRef] [PubMed]

28. Krizsan, S.J.; Huhtanen, P. Effect of diet composition and incubation time on feed indigestible neutral detergent fiber concentration in dairy cows. *J. Dairy Sci.* **2013**, *96*, 1715–1726. [CrossRef]

29. Raffrenato, E.; Ross, D.A.; Van Amburgh, M.E. Development of an in vitro method to determine rumen undigested aNDFom for use in feed evaluation. *J. Dairy Sci.* **2018**, *101*, 9888–9900. [CrossRef]

30. Aiple, K.P.; Steingass, H.; Menke, K.H. Suitability of a buffered faecal suspension as the inoculum in the Hohenheim gas test: 1. Modification of the method and its ability in the prediction of organic matter digestibility and metabolizable energy content of ruminant feeds compared with rumen fluid as inoculum. *J. Anim. Physiol. Anim. Nutr.* **1992**, *67*, 57–66.

31. Frey, J.C.; Pell, A.N.; Berthiaume, R.; Lapierre, H.; Lee, S.; Ha, J.K.; Angert, E.R. Comparative studies of microbial populations in the rumen, duodenum, ileum and faeces of lactating dairy cows. *J. Appl. Microbiol.* **2010**, *108*, 1982–1993. [CrossRef]

32. Raffrenato, E.; Erasmus, L.J. Variability of indigestible NDF in C3 and C4 forages and implications on the resulting feed energy values and potential microbial protein synthesis in dairy cattle. *S. Afr. J. Anim. Sci.* **2013**, *43*, 93–97.

33. Colombini, S.; Galassi, G.; Crovetto, G.M.; Rapetti, L. Milk production, nitrogen balance, and fiber digestibility prediction of corn, whole plant grain sorghum, and forage sorghum silages in the dairy cow. *J. Dairy Sci.* **2012**, *95*, 4457–4467. [CrossRef] [PubMed]

34. Mould, F.L.; Kliem, K.E.; Morgan, R.; Mauricio, R.M. In vitro microbial inoculum: A review of its function and properties. *Anim. Feed Sci. Technol.* **2005**, *123*, 31–50. [CrossRef]

Article

Rumen Inoculum Collected from Cows at Slaughter or from a Continuous Fermenter and Preserved in Warm, Refrigerated, Chilled or Freeze-Dried Environments for In Vitro Tests

Mauro Spanghero [1,*], Maria Chiaravalli [2], Stefania Colombini [2], Carla Fabro [1], Federico Froldi [3], Federico Mason [4], Maurizio Moschini [3], Chiara Sarnataro [1], Stefano Schiavon [5] and Franco Tagliapietra [5]

[1] Dipartimento di Scienze Agroalimentari, Ambientali e Animali, University of Udine, 33100 Udine, Italy; carla.fabro@uniud.it (C.F.); sarnataro.chiara@spes.uniud.it (C.S.)

[2] Dipartimento di Scienze Agrarie e Ambientali, University of Milan, 20122 Milano, Italy; maria.chiaravalli@unimi.it (M.C.); stefania.colombini@unimi.it (S.C.)

[3] Dipartimento di Scienze Animali, della Nutrizione e degli Alimenti, Università Cattolica del Sacro Cuore of Piacenza, 29122 Piacenza, Italy; federico.froldi@unicatt.it (F.F.); maurizio.moschini@unicatt.it (M.M.)

[4] Department of Biodiversity Protection, Institute of Animal Reproduction and Food Research of Polish Academy of Sciences (IARFR PAS), 10-748 Olsztyn, Poland; f.mason@pan.olsztyn.pl

[5] Dipartimento di Agronomia, Animali, Alimenti, Risorse naturali e Ambiente, University of Padova, 35122 Padova, Italy; stefano.schiavon@unipd.it (S.S.); franco.tagliapietra@unipd.it (F.T.)

[*] Correspondence: mauro.spanghero@uniud.it; Tel.: +39-432-558193 or +39-432-558199

Received: 7 September 2019; Accepted: 8 October 2019; Published: 16 October 2019

Simple Summary: The utilization of animal donors of rumen fluid for laboratory experiments can raise ethical concerns due to invasive methods of collection (rumen cannulated or intubated animals). Societies are strongly oriented to support cruelty free experiments and alternatives to the collection of rumen fluids from live animals are urgently requested from the scientific community. Thus, in order to attenuate the dependence of laboratories on animal donors, this study compared the rumen inoculum collected at slaughter with the fermentation liquid from a rumen continuous fermenter and both rumen inoculum were used fresh or preserved (by refrigeration, chilling and freeze-drying). The results support the possibility of using continuous fermenters to generate inoculum for in vitro purposes, and short-term refrigeration is confirmed to be a valuable storage system to facilitate transfer inoculum from the collection sites. These findings should attenuate the need for laboratories' frequent collections from animals while continuing research in ruminant nutrition.

Abstract: The utilization of animal donors of rumen fluid for laboratory experiments can raise ethical concerns, and alternatives to the collection of rumen fluids from live animals are urgently requested. The aim of this study was to compare the fresh rumen fluid (collected at slaughter, W) with that obtained from a continuous fermenter (RCF) and three methods of rumen fluid preservation (refrigeration, R, chilling, C, and freeze-drying, FD). The fermentability of different inoculum was evaluated by three in vitro tests (neutral detergent fiber (NDF) and crude protein (CP) degradability and gas production, NDFd, RDP and GP, respectively) using six feeds as substrates. Despite the two types of inoculum differed in terms of metabolites and microbiota concentration, the differences in vitro fermentability between the two liquids were less pronounced than expected (-15 and 20% for NDFd and GP when the liquid of fermenter was used and no differences for RDP). Within each in vitro test, the data obtained from rumen and from fermenter liquids were highly correlated for the six feeds, as well as between W and R (r: 0.837–0.985; $p < 0.01$). The low fermentative capacity was found for C and, particularly, FD for liquids. RCF could be used to generate inoculum for in vitro purposes and short-term refrigeration is a valuable practice to manage inoculum.

Keywords: rumen liquid; in vitro fermentation; rumen degradability; gas production

---

1. Introduction

Rumen fluid sampled from live animals is used in laboratory experiments (e.g., in vitro rumen fermentation) to evaluate the nutritive value and gaseous emissions of ruminant feeds [1,2] or to inoculate continuous fermenters for studies of rumen fermentation [3].

In general, the utilization of animals as donors of rumen fluid can raise some ethical concerns, because the collection of rumen fluid is an invasive practice, which requires donor animals that are surgically modified (e.g., rumen cannulated), or immobilised and intubated with esophageal probes. An alternative is to collect rumen fluid from animals slaughtered for production purposes in commercial slaughterhouses, but it is difficult to monitor feeding before slaughter. Overall, societies are strongly oriented to support cruelty free experiments and alternatives to collection of rumen fluids from live animals are urgently requested of the scientific community to continue research activity in ruminant nutrition.

Rumen continuous fermenters (RCFs) are laboratory apparatus developed to simulate the rumen conditions for studies of rumen metabolism. They generate a fermenting fluid by starting from an initial rumen inoculum and by a continuous influx of artificial saliva, an output of fermentation products and a constant supply of nutrients (substrates). However, they could also be modified and adapted to be used as artificial generators of rumen fermentation fluid, which could be standardised with respect to several conditions (type and amount of fermentable substrate, pH, dilution, etc.). There are studies which have compared fermentation liquids from different RCFs or between fluids collected from rumens and fermenters [3–7]. The liquid from fermenters is less concentrated in terms of volatile fatty acids (VFA) and protozoa, and the cellulosolitic bacterial strains seem reduced in some types of RCF (e.g., Rusitec, [8]) while in other fermenters, the bacteria microbiota was comparable to that measured directly on the rumen inoculum collected in vivo [6,7]. However, no experiments have simultaneously compared rumen and RCFs fluids as inoculum in terms of results of different in vitro tests.

Overall, independently from the mode of inoculum provision, the possibility of preserving rumen fluid and to create stocks would allow the concentration of the collection in specialized centers to facilitate the transfer of the inoculum and reduce the need of laboratories to frequently collect liquids from live animals. The preservation techniques of rumen fluid have been investigated in several studies, which have mainly considered the usage of low temperatures (e.g., refrigeration and chilling at -20, at $-80\,°C$, or in liquid N), the addition of cryoprotectants, and also freeze-drying [9–13]. In general, there are encouraging findings, but also a scarce homogeneity among trials in terms of measurement of the maintenance of the fermentative capacity of liquid after preservation.

The present research has the general aim to attenuate the needs of direct collection of rumen fluid from animals by: (i) artificially generating rumen fermenting fluids; (ii) evaluating preservation methods for stock rumen fluid. These technologies should reduce the need for laboratories to frequently collect from animals, with the potential advantage of reducing inoculum variability. The specific aim is to compare the fresh rumen fluid (collected at slaughter) with that obtained from a stratified single-flow RCF system [14] and three methods of rumen fermenting fluid preservation (refrigeration, chilling and freeze-drying).

Unlike other research, the novelty of this study is to compare the different inoculum in terms of the results obtained by three widely utilised in vitro methods (degradability of neutral detergent fiber, NDFd, degradability of protein (RDP), and gas production, (GP) respectively). The chosen in vitro tests allow a wide evaluation of the fermentation potential of inoculum, because they quantify the fermentation of main dietary components of ruminant rations, such as fibers (NDFd), protein (RDP) and (mainly) non-fibrous carbohydrates (GP).

2. Materials and Methods

2.1. Trial Organization

The experiment was organised in two subsequent identical fermentation trials (runs) and was carried out by four Italian research groups from the University of Milano, Padova, Piacenza, and Udine (Labs 1, 2, 3, and 4, respectively).

For each trial, 4 dry multiparous Holstein Friesian cows were slaughtered for production purposes in a commercial slaughterhouse, after being fed for 3 weeks a basal diet (32.3% meadow hay, 34.0% corn silage, 27.3 % compound feed and 6.4 % soybean meal). The cows of both runs were in good health, housed in the same barn located 50 km from the slaughterhouse, fed the last feeding in the morning 2 h before being moved to the slaughterhouse, and had free access to fresh water until slaughter. After slaughter, the rumen liquors, collected in equal amounts from the 4 cows by Lab 4, were coarsely filtered, the liquids bulked together during continuous flushing with CO_2, and then divided into 2 main amounts (see Figure 1). The first amount represented the rumen inoculum for the in vitro tests (NDFd, RDP and GP test), while the second amount was used to inoculate the RCF system. Both inoculum were used as: warm at 39 °C (W), refrigerated at 4 °C (R), chilled at −80 °C (C), and freeze-dried (FD). After the collection at slaughter, W (kept inside pre-warmed thermic bottles flushed with CO_2) and R (kept inside bottles flushed with CO_2, immersed in ice water within a portable fridge to quickly lower the temperature to 4 °C) were divided in aliquots of 250–300 mL each and immediately delivered to Lab 1, 2, and 3 to start the in vitro tests within 6 h from slaughter. The amounts to be preserved as C, FD or to be used in the RCF fermenter were immediately brought to Lab 4 (maintained warm at 39 °C). Here, the C or FD, amounts were divided in aliquots of 250–300 mL each. The C aliquots were chilled at −80 °C, while the FD aliquots were processed according to Luchini et al. [15]. Briefly, after centrifugation at 5000 g for 30 min at 4 °C, the supernatant was discarded, and the residue obtained was chilled at −80 °C, and subsequently freeze-dried. The C and FD aliquots were delivered to Lab 1, 2, 3 after 30 d of conservation. Before being used as inoculum, the C liquids were thawed in a bain-marie at 39 °C within 2 h and kept at the same temperature for another 2 h, while the FD liquids were reconstituted to the original volume with the artificial saliva used for the in vitro tests and kept at 39 °C for 2 h.

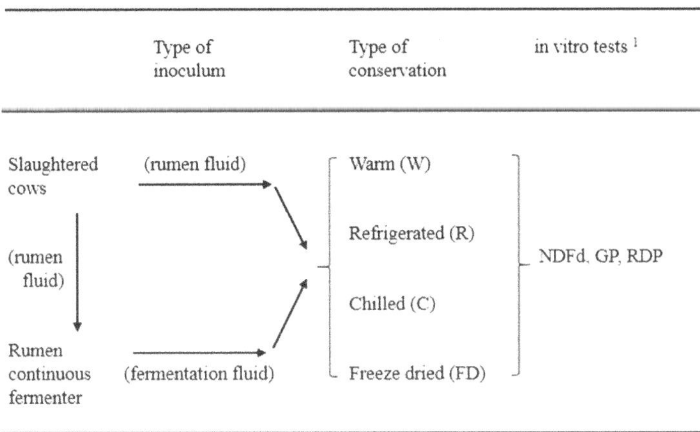

[1] NDFd : NDF degradability; GP : gas production; RDP : rumen degradable protein.

Figure 1. Schematic representation of the trial organization.

The remaining amount of rumen liquid was immediately used to inoculate the RCF fermenter in Lab 4. On the 9th day of incubation in the RCF, the fermentation fluid was collected, and divided in aliquots for each liquid type (W, R, C and FD). The W and R aliquots, prepared as previously described for the rumen inoculum, were immediately transferred to Lab 1, 2 and 3 to perform the in vitro tests within 6 h from the collection of the liquid from the RCF system. The C and FD aliquots were delivered to Labs 1, 2, and 3 after 30 d of conservation.

As incubations of liquids C and FD were delayed by 30 d with respect to W and R and the incubations of liquid from the fermenter were delayed by 9 d from the incubation of rumen liquids. There were in total 4 incubation sessions within each fermentation run.

2.2. Preparation of the Substrates for the In Vitro Tests

All the in vitro tests were performed on the following six ingredients: meadow hay, corn silage, wheat bran, distillers, soybean and barley meal. The samples of corn silage were dried (48 h at 60 °C) and all feed samples were milled through a 1 mm sieve and then analysed for chemical composition.

2.3. The Rumen Continuous Fermenter (RCF) System and the In Vitro Tests

The RCF system, used in Lab 4, was described in detail by Mason et al. [14]. In brief, the system consists of 8 × 2 L glass bottles, immersed in a water bath at 39 °C. A peristaltic pump supplies the buffer solution [16] from a reservoir to the fermenters (dilution rate of 5 %/h) and the outflow, located at the bottom of the bottles, allows stratification of the feeding material. The bottles were inoculated with 600 mL of strained rumen fluid and 800 mL of artificial saliva, and each bottle received a total of 15 g/d of dry matter (DM), in two equal doses, at 09:00–17:00, of the same diet used to feed the donor cows before slaughtering. Each fermentation lasted 9 d and on the last day, the fermentation fluids of the 8 bottles were collected, pooled, and processed to prepare the different inoculum, as previously described.

In all 3 in vitro tests, the inoculum was strained through 4 layers of cheesecloth into pre-warmed (39 °C) flasks with CO_2 and mixed with the buffer solutions ([17] for NDFd and GP and Van Soest buffer [18] for RDP, in a 1:2 and 1:4 ratio, respectively).

The NDFd was tested by Lab 1. Each feed sample was weighed (0.250 ± 0.005 g) in duplicate in Ankom F57 bags (Ankom Technology, Macedon, NY, USA). The bags were incubated in a pre-warmed 100 mL Erlenmeyer flask closed by a rubber stopper with a Bunsen valve for gas release and maintained at 39 °C in a water bath with shaking for 48 h. Each flask was inoculated with 90 mL of the solution under anaerobic conditions, flushing the flask with CO_2. At the end of the incubations, the bags were rinsed with cold water until the water ran clear and then placed for 3 h in a 105 °C forced-air oven to dry. Subsequently, the NDF concentration was determined for each bag using the fiber analyser (Ankom Technology, Macedon, NY, USA).

The RDP was tested by Lab 3 according to the rumen step of the Ross method [19]. Briefly, each sample was weighed (1.000 ± 0.020 g) into a 120 mL glass bottle equipped with a cap and placed into a 39 °C water bath 1 h before the in vitro fermentation. The neutral detergent residue from corn silage was used for microbial contamination correction of the post fermentation feed residues. Each bottle was inoculated with 100 mL solution under CO_2 flushing, closed with the cap, and incubated for 16 h at 39 °C in a shaking water bath. After the incubation, the bottle content was vacuum filtered (110 mm diameter Whatman Filter Papers 54) and the residue was analyzed for Kjeldahl nitrogen.

The GP was measured after 24 h incubation by Lab 2 by using a commercial GP apparatus (Ankom Technology, Macedon, NY, USA; [20]). The system consists of 50 bottles hermetically sealed, equipped with a wireless pressure sensor connected to a computer. Each bottle (317 mL) was filled with 0.500 ± 0.010 g of feed and 75 mL of fermenting solution, obtaining a headspace volume of 242 mL. The bottles were placed in a ventilated incubator at 39 ± 0.4 °C and automatically vented at a fixed pressure (6.8 kPa), to prevent overpressure.

All the tests were performed for each feed substrate and for each inoculum type in duplicate (8 inoculum types × 6 feeds in duplicate) and were replicated in a second fermentation run. The between-run determinations were considered as experimental repetitions. To account for the incubation session effect, a standard rumen fluid was included in each incubation session by each laboratory. Before conducting the whole experiment, Lab 4 prepared a rumen fluid to be used as the control by other labs. The liquid was collected at slaughter from 4 dairy cows (culled in good health) and delivered to Lab 4 within 30 min in airtight glass-bottles refluxed with carbon dioxide and maintained at 39 °C. The whole amount was divided in small aliquots (200 mL), frozen at −20 °C and distributed to Lab 1, 2 and 3. The frozen-thawed inoculum (in a bain-marie at 39 °C within 2 h and kept at the same temperature for another 2 h) were added by each Lab into each incubation run (two incubation bottles added with corn silage as substrate) to detect anomalies between runs.

2.4. Analysis

2.4.1. Inoculum Sample Preparation

The samples of the two inoculum for each preservation treatment were collected before performing the in vitro tests for the following analyses: pH, VFA, NH_3, and microbial population. The samples for the VFA analysis were acidified with H_2SO_4 0.1 N and stored at −20 °C while the samples for NH_3 were directly stored at −20 °C. The samples for bacterial DNA analysis were chilled in liquid nitrogen (N) and stored at −80 °C.

2.4.2. Chemical Analysis

The DM content of the feeds was determined by heating at 105 °C for 3 h (method 930.15; [21]), and the ash content was subsequently determined after incineration at 550 °C for 2 h (method 942.05; [21]). The neutral and acid detergent fiber (NDF and ADF, respectively) analysis was performed with a fiber analyzer (Ankom Technology, Macedon, NY, USA) following the procedure of Van Soest et al. [18] without correction for residual ash. The ether extract (EE) and the N contents were respectively determined by the solvent extraction and by the Kjeldahl methods (methods 954.02 and 976.05, [21]).

The pH and the NH_3 content of the inoculum were measured with a glass electrode pH meter (GLP 22, Crison Instruments, S.A. Barcelona, Spain) and an ammonia electrode (Ammonia Gas Sensing Combination Electrode, ©Hach Company, Loveland, CO, USA, 2001).

For the VFA analysis, the aliquots of the inoculum were centrifuged at 20,000 g for 30 min at 20 °C and the supernatant was then filtered using polypore 0.45 μm filters (Alltech Italia, Milan, Italy). The filtrate was injected into a high-performance liquid chromatography instrument (Perkin-Elmer, Norwalk, CN, USA), set to 220 nm according to the method described by Martillotti and Puppo [22].

2.4.3. Microbial Analysis

Genomic DNA was extracted from about 700 μL of inoculum samples using the PowerSoil DNA extraction kit (MoBio Laboratories Inc., Carlsbad, CA, USA) with some modifications as described by Kong et al. [23] (2010). DNA concentration, eluted in 50 μL, was determined by NanoDrop One (Thermo Fisher Scientific Inc., Wilmington, DE, USA).

The quantitative polymerase chain reaction analysis was performed by CFX96 Real Time System (Bio-Rad Technologies Inc, Hercules, CA, USA) using iQ SYBR Green Supermix (Bio-Rad Technologies Inc, Hercules, CA, USA) mixed with 0.3 μL of each forward and reverse primer (0.3 μM), 8.4 μL of sterile water and 1 μL of gDNA to obtain a reaction volume of 20 μL. The amplification program included the denaturation step at 98 °C for 3 min followed by 40 cycles at 98 °C for 15 s, annealing at 60 °C for 30 s and elongation at 72 °C for 30 s.

To determine the specificity of the amplification of each primer, the melting curve was performed. The amplification efficiency was calculated using the formula: $E = 10^{(-1/\text{slope})} - 1$. The relative abundance of the target bacterial groups or species was expressed in a proportion of the total bacteria 16S rRNA gene and was calculated using the following formula: $2^{-\Delta CT}$.

2.4.4. Statistical Analyses

The data from the in vitro tests (NDFd, RDP and GP, within each test and within each feed substrate) and the chemical and microbial composition of fermentation fluids differing for origin and preservation (pH values, VFA and ammonia contents and bacteria abundance) were statistically analysed as a factorial 2×4 completely randomised block (2 fermentation runs as blocks) design:

$$Y_{ijk} = \mu + \alpha_i + \beta_j + \gamma_k + (\beta\gamma)_{jk} + \varepsilon_{ijk},$$

where: y_{ijk} is the experimental data; μ is the overall mean; α_i is the random effect (block) of the fermentation run ($i = 1, 2$); β_j is the fixed effect of the type of inoculum ($j = 1, 2$); γ_k is the fixed effect of the inoculum preservation method ($k = 1,4$); and ε_{ijk} is the random error.

Within each in vitro test data, the Pearson correlation coefficient (r) was calculated for the 6 feeds tested in 2 runs with fermentation liquids, differing for origin (liquids from fermenter versus liquids from rumen, 12 data points) or for the type of conservation (W versus R, W versus C and W versus FD liquids, 12 data points for each correlation).

For all statistical analyses, the probability significance levels (p) were 0.05 and 0.01 ($p < 0.05$ and $p \leq 0.01$, respectively).

3. Results

The chemical composition of feeds used as substrates for the in vitro tests and the characteristics of inocula are reported in Tables 1 and 2, respectively.

Table 1. Chemical composition of feeds used as substrates in the in vitro tests.

Feeds	DM	Ash	CP	EE	NDF	ADF
	%	% DM	% DM	% DM	% DM	% DM
Corn silage [1]	91.88	4.16	7.67	3.10	36.79	20.74
Wheat bran	89.71	5.98	16.74	3.47	47.80	13.40
Meadow hay	95.54	11.30	7.34	1.45	58.38	34.82
Distillers	89.15	5.87	34.32	7.79	42.87	11.86
Soya meal, extr.	88.88	6.95	46.09	1.19	21.11	8.88
Barley	89.79	3.08	10.65	1.55	31.89	8.30

[1] pre-dried samples at 60 °C.

Among the parameters reported in Table 2, the concentration of ammonia and the relative abundance of *Fibrobacter succinogenes* in fermentation fluids showed a significant interaction ($p < 0.05$) between the origin of the liquid and the preservation method. However, for both parameters the impact of the interaction in terms of contribution to the mean square of the model was much lower than that of the main effects as well the level of significance ($p < 0.05$ vs. $p \leq 0.01$). Therefore, this study decided to show and to discuss the main effects of the model.

The liquid from the fermenter had a lower total VFA content, acetic percentage on total VFA, acetic:propionic ratio and ammonia concentration compared with the liquid from rumen (34.9 versus 122.1 mMol, 57.8 versus 63.3%, 2.9 versus 5.7 and 15.4 versus 26.1 mg/dL, respectively, $p \leq 0.01$). On the contrary, the valeric and isovaleric acid percentages on total VFA were higher in fluid from the fermenter than the rumen liquid (1.8 versus 1.1% and 2.7 versus 1.4%, respectively, $p \leq 0.01$). The FD preservation differed significantly ($p \leq 0.01$) from the others for the lowest proportion of acetic, propionic and isovaleric acids and for the very high proportion of isobutyric acid. The FD also had a very low content of ammonia in comparison with W and R (6.4 versus 21.8–22.3 mg/dL, $p \leq 0.01$), while the C liquid showed the highest concentration (32.6 mg/dL, $p \leq 0.01$).

Table 2. Volatile fatty acid (VFA) content, pH, ammonia (NH_3) content and bacteria abundance of fermentation liquids from inoculum differing for origin and for conservation.

Items	Type of Inoculum (TI)		Type of Conservation [1] (TC)				TI [2]	TC [2]	TI × TC [2]	RMSE [2]
	From Rumen	From Fermenter	Warm (W)	Refrigerated (R)	Chilled (C)	Freeze—Dried (FD)				
Samples, *n*	8	8	4	4	4	4				
Total VFA (mMol)	122.1	34.9	76.4	74.1	100.3	63.4	**	ns	ns	25.98
VFA composition (% of total VFA)										
Acetic (A)	63.3	57.8	67.3 [A]	66.0 [A]	65.7 [A]	43.4 [B]	*	**	ns	3.72
Propionic (P)	12.5	20.2	18.2 [A]	18.5 [A]	16.9 [A]	11.9 [B]	**	**	ns	1.84
Isobutyric	10.9	9.6	0.4 [B]	0.4 [B]	4.2 [B]	36.0 [A]	ns	**	ns	5.77
Butyric	10.5	8.0	10.1	10.8	9.5	6.5	ns	ns	ns	2.35
Isovaleric	1.4	2.7	2.1 [A]	2.4 [A]	2.3 [A]	1.2 [B]	**	**	ns	0.34
Valeric	1.1	1.8	1.6	1.6	1.5	1.2	*	ns	ns	0.40
A:P	5.7	2.9	3.9	3.8	4.0	5.4	**	ns	ns	1.27
pH	6.4	6.7	6.5	6.5	6.5	6.8	ns	ns	ns	0.51
NH_3 (mg/dL)	26.1	15.4	21.8 [B]	22.3 [B]	32.6 [A]	6.4 [C]	**	**	*	3.28
Relative abundance (% of total bacteria)										
Genus *Prevotella*	45.1	19.7	25.8	31.7	34.6	37.5	**	ns	ns	8.90
Fibrobacter succinogenes	0.39	2.04	2.36 [A]	2.10 [A]	0.39 [B]	0.01 [B]	**	**	*	0.63
Ruminococcus albus group	0.017	0.003	0.013	0.010	0.011	0.006	**	ns	ns	0.0056

[1] Means of the type of conservation with different superscripts (A, B, C) are statistically different. [2] *: $p < 0.05$; **: $p \leq 0.01$; ns: not significant; RMSE: root mean square error.

The relative abundances of the genus *Prevotella* and *Ruminococcus albus* group were lower ($p < 0.01$) in the liquid from the fermenter compared to the liquid from rumen, while the preservation method did not show effects. On the contrary, the *Fibrobacter succinogenes* was higher ($p < 0.01$) in the liquid from the fermenter and its relative abundance in W and R liquids was higher ($p < 0.01$) than the others.

The results of the in vitro tests on the six feeds and the correlation coefficients calculated on the six feeds by using the sets of two inocula differing for origin (liquid from rumen versus liquid from fermenter) or preservation (W versus R, W versus C and W versus FD liquids) are reported in Tables 3 and 4, respectively. The in vitro tests used different inoculum and were performed in eight subsequent incubation sessions by each Lab. To correct any possible effect of the incubations session, a standard rumen fluid was included in each fermentation, which was used on the same substrate (corn silage) as the control. The data from the standard allowed the exclusion of any appreciable effect of the incubation session and the in vitro results from the different incubation sessions did not require corrections.

The NDFd data of the corn silage samples for the FD mode of preservation were removed from the analysis, having an abnormal fermentation due to technical problems.

The utilization of the liquid from the fermenter determined a significant NDFd reduction for three feeds (corn silage, wheat bran, and distillers) and the correlation analysis indicated a close relationship for NDFd between the liquid from rumen and diluted liquids from the fermenter (r = 0.960, $p \leq 0.01$). Considering the preservation of the inoculum, the W and R liquids were not different for all the tested feeds and the correlation indicated a close correspondence (r = 0.985, $p \leq 0.01$). The C and FD liquids depressed ($p \leq 0.01$) the NDFd for two and three feeds, respectively. However, the correlation W versus C was high (r = 0.905, $p \leq 0.01$), while the W versus FD was non-significant.

The RDP results indicated a good correspondence between fluid from the fermenter and liquid from the rumen liquids and the correlation between the liquid from rumen and the liquid from the fermenter data was high (r = 0.837, $p \leq 0.01$). The R and C liquids gave similar RDP results to W liquid for all feeds and the regression W versus R showed a good correspondence of values (r = 0.892), while the W versus C was not significant. The RDP obtained with the FD liquid were quite variable, with the values for two feeds being higher than those obtained with W, while for the soya, the opposite was true, and the correlation W versus FD was not significant.

Unlike NDFd and RDP, the gas production showed a significant interaction—inoculum $\times$ preservation ($p \leq 0.01$)—for some feeds (corn silage, wheat bran, and distillers). However, the impact of the interaction in terms of the contribution to the mean square of the model was much lower than that of the main effects as well the level of significance ($p < 0.05$ versus $p \leq 0.01$). Therefore, it was decided to show the results and to discuss the main effects of the model. For all the feeds, the liquid from rumen gave higher gas production than the liquid from the fermenter ($p \leq 0.01$ and, only for soya, $p < 0.05$).

Moreover, the preservation method had a statistical effect on the gas production ($p \leq 0.01$ and, only for soya and barley, $p < 0.05$). In general, the gas production of C was not statistically different from W and R liquids, except for wheat bran and barley where C gas production was not statistically different from W. The FD liquid generated the lowest yields of gas and all the ingredients differed significantly from those of the W liquid. All correlations of preserved liquids with W were statistically significant (r = 0.797–0.921, $p \leq 0.01$).

Table 3. In vitro degradability of NDF (NDFd), CP (RDP), and gas production (at 24 h of fermentation, GP) of six substrates using rumen inoculum differing for origin and for conservation method.

Items	Type of Inoculum (TI)		Type of Conservation [1] (TC)				TI [2]	TC [2]	TI × TC [2]	RMSE [2]
	From Rumen	From Fermenter	Warm (W)	Refrigerated (R)	Chilled (C)	Freeze-Dried (FD)				
NDFd (%)										
Corn silage	37.5	30.1	39.0 ^{A}	40.0 ^{A}	22.5 ^{B}	-	*	**	ns	6.34
Wheat bran	45.9	38.2	46.7 ^{A}	46.2 ^{A}	39.3 ^{B}	36.0 ^{B}	**	**	ns	3.40
Meadow hay	44.5	40.7	51.8 ^{A}	47.9 ^{A,B}	39.0 ^{B,C}	31.8 ^{C}	ns	**	ns	5.36
Distillers	52.5	41.3	49.4	51.7	47.9	38.5	*	ns	ns	6.85
Soya meal, extr.	81.4	78.5	92.2 ^{A}	93.7 ^{A}	85.5 ^{A}	48.5 ^{B}	ns	**	ns	6.26
Barley	59.4	54.3	58.0	58.8	55.7	55.1	ns	ns	ns	5.39
RDP (%)										
Corn silage	66.8	66.2	57.0 ^{b}	49.8 ^{b}	66.0 ^{b}	93.3 ^{a}	ns	*	ns	15.21
Wheat bran	59.9	61.9	58.1	58.9	60.4	66.2	ns	ns	ns	10.20
Meadow hay	39.0	45.6	32.5 ^{b}	27.8 ^{b}	44.3 ^{b}	64.7 ^{a}	ns	*	ns	11.35
Distillers	33.8	34.5	34.8	37.5	33.8	30.6	ns	ns	ns	5.00
Soya meal, extr.	45.1	53.4	52.8 ^{A}	56.8 ^{A}	47.8 ^{A,B}	39.7 ^{B}	*	*	ns	5.10
Barley	59.0	57.4	51.8	49.7	60.4	71.0	ns	ns	ns	17.38
GP (mL)										
Corn silage	223	193	242 ^{A}	234 ^{A}	184 ^{B}	172 ^{B}	**	**	**	26.13
Wheat bran	183	146	176 ^{A,B}	188 ^{A}	157 ^{B,C}	136 ^{C}	**	**	**	18.40
Meadow hay	162	119	177 ^{A}	198 ^{A}	124 ^{B}	64 ^{C}	**	**	ns	24.81
Distillers	193	153	197 ^{A}	207 ^{A}	156 ^{B}	131 ^{B}	**	**	**	50.44
Soya meal, extr.	178	134	176 ^{B}	213 ^{A}	136 ^{C}	100 ^{D}	*	*	ns	23.31
Barley	266	226	263 ^{A,B}	264 ^{A}	231 ^{A,B}	226 ^{B}	**	*	ns	27.85

[1] Means of the type of conservation with different superscripts (A, B, C and a, b) are statistically different. [2],*: $p < 0.05$; **: $p \leq 0.01$; ns: not significant; RMSE: root mean square error.

Table 4. Pearson correlation coefficients (r) of in vitro data obtained for the six feeds used as substrates in two fermentation run (12 observations) by using inocula differing for origin and for conservation.

| | Type of Inoculum (TI) | | Type of Conservation (TC) | | | | | |
| | Rumen vs. Fermenter | | Warm vs. Refrigerated | | Warm vs. Chilled | | Warm vs. Freeze-Dried | |
	r	p [1]	r	p [1]	r	p [1]	r	p [1]
NDFd [2]	0.960	**	0.985	**	0.905	**	0.554	ns
RDP [3]	0.837	**	0.892	**	0.345	ns	0.431	ns
GP [4]	0.939	**	0.921	**	0.797	**	0.850	**

[1] statistical significance of r: **: $p \leq 0.01$; ns: not significant. [2] NDFd: NDF degradability; [3] RDP: CP degradability; [4] GP: gas production (at 24 h of fermentation).

4. Discussion

4.1. Type of Inoculum

The present work used a rumen fluid collected from cows immediately after slaughter in controlled conditions (animals fed a known diet, not slaughtered in emergency, in good health status, transported from farms located near the slaughterhouse, rumen fluid sampled within 20 min of slaughter) as a reference in vivo rumen liquor. To our knowledge, there are no specific comparisons between rumen fluid collected via cannula, through the esophageal tube and at slaughtering. The authors reason that rumen fluid collected at slaughter in controlled conditions has a limited difference from that sampled by other methods. This is based on our previous work [24] where a close relationship between NDF degradability measured in vitro (average of rumen collected by different methods) and in situ was found. Our recent data [25] showed comparable values of in vitro NDF degradability using rumen fluids from slaughtered cows and from cows with rumen cannula fed similar diets. Moreover, the collection at slaughter is an accepted method of sampling rumen fluid for microbiota studies by the Rumen Microbial Genomics Network [26] and it is mentioned as an alternative to sampling via cannula [2].

A first aim of this study was to compare the fresh rumen inoculum with that obtained from a continuous culture system (RCF, see Figure 1). The data from the literature [4,5] showed that fermentation liquids from continuous cultures are less concentrated than rumen fluids collected directly from the rumen in terms of fermentation metabolites. This is confirmed also by data of this experiment, where a particularly low VFA concentration was found in the continuous fermenter liquids, being approximately 30% of the VFA concentration of rumen fluid. The concentration of VFAs measured in the rumen liquids were comparable to those found in the literature, while those of the RCF liquids were slightly lower than those obtained previously with the same in vitro system [14,27]. In addition, the ammonia reduction in the fermenter liquids was less marked than VFA, being approximately 60% of the ammonia in the rumen fluid. Moreover, a higher proportion of valeric and isovaleric acids was found in RCF liquids. These latter acids mainly originated from the metabolism of amino acids and this agrees with the relatively high ammonia concentration mentioned above.

The microbial population was represented in this study by only some bacterial strains and was affected by the type of inoculum. After the incubation period in the RCF, *Prevotella* genus and *Ruminococcus albus* group decreased the relative abundance, while *Fibrobacter succinogenes* increased, compared to the rumen liquids. *F. succinogenes*, both in the liquid from rumen and in the RCF liquids, was present at higher percentages than the *R. albus* group, as reported also from by Soto et al. [6,7] and by Koike et al. [28]. Our limited observations on the bacterial rumen microbiota confirm that the RCF in vitro environment shifts bacteria composition, because it depresses some strains and is favourable to others.

Despite the previously discussed differences between the two types of inoculum in terms of metabolites and microbiota concentration, the effects on the in vitro tests were less pronounced than expected, at least for NDF and crude protein degradability. On average, compared to the rumen fluid, the utilization of fermenter liquid inoculum determined a small reduction of NDFd (−12%), while the degradability of protein was slightly increased (+5%). Furthermore, the correspondence of NDFd data obtained on the six ingredients by the two inoculum was very close (r: 0.960, $p \leq 0.01$). On the contrary, gas production was depressed by 20% with the usage of RCF, but also in this case, there was a close correlation between the values obtained with the two types of inoculum (r: 0.939, $p \leq 0.01$). Therefore, despite the differences described above in metabolite and microbiota composition, the overall in vitro fermentability and the ranking of feeds in term of in vitro rumen fermentation do not seem to be greatly affected using the RCF fluid in comparison to rumen inoculum. This surprising result could be in part explained because the in vitro tests work at a high ratio of fermentation liquid to substrate (20, 50 and 180 mL/g of DM substrate for RDP, GP and NDFd, respectively). In such conditions, there is an abundance of degrading capacity and it could be speculated that the RCF fermenting fluid is not so diluted to greatly reduce the fermentability measured by the in vitro tests, at least for the fiber and protein fractions.

4.2. Method of Inoculum Preservation

A second aim of this study was to compare the fresh rumen inocula with those preserved by refrigeration, chilling and freeze-drying (Figure 1). The rumen fluid maintained at 39 °C for 5–6 h prior to incubation represented the reference rumen fluid because Robinson et al. [29] demonstrated that the storage in such conditions had no detrimental effects on in vitro NDF fermentation. The results of the present study show that rumen fluid refrigerated for 5–6 h had fermentative parameters very close to that of the fluid preserved at the animal body temperature. On the contrary, freezing at −80 °C determined a high ammonia increment and the freeze-drying process caused pronounced variations in comparison with fresh rumen fluid because both acetic and propionic acids proportions were reduced, and the iso-butyric increased abnormally and the ammonia was severely reduced.

The NDFd values obtained from the fluid preserved by refrigeration were very similar to those attained from the inoculum maintained at 39 °C, as previously found by Robinson et al. [29]. A similar situation was found also for the RDP values, apart from the two forage samples. The same close relationship between refrigerated and fluid from rumen was also the case for gas production and this result confirms those found by Hervás et al. [30], who suggested that the preservation of rumen fluid for up to 6 h in crushed ice is a practical alternative to the use of freshly collected rumen fluid as inoculum for in vitro ruminal fermentation studies. Therefore, refrigerated inoculum can be a more practical system of storing rumen inoculum during transport from the site of collection to the laboratory of utilization within a 4 to 6 h period.

The possibility of storing the inoculum for long periods was evaluated in terms of chilling at −80 °C and by freeze-drying. The chilled inoculum determined a limited reduction of NDF degradation (−14%) in comparison with the warm inoculum and there was a satisfactory relationship between feeds (r: 0.905).

Furthermore, the degradation of protein was, on average, not depressed and this agrees with the findings of Luchini et al. [31] who suggested that rumen inoculum preserved frozen might be used for in vitro rumen protein degradation experiments. However, the regression between feeds showed an insufficient degree of relation.

Finally, less gas was produced by the chilled inocula (−20%) with a satisfactory relationship among feeds (r: 0.797). The protection of the microbiota from freezing damage can be obtained by adding organic chemicals with cryoprotectant properties but these compounds can interfere with the subsequent in vitro tests because they are fermentable substrates (e.g., glycerol, [12]) or can have toxic effects (e.g., dimethyl sulfoxide). A possible improvement of our chilling procedure [9,12] could be to avoid freezing large amounts of inoculum and to divide it into small amounts (e.g., 30–50 mL) because the high surface/volume accelerates the freezing.

The freeze-drying process produced the lowest fermentation activity and a very poor relation with fresh liquid among feeds, confirming similar findings obtained by Belanche et al. [9]. Several methodological factors can explain these poor results, such as the absence of cryoprotectants, the procedure of pellet preparation in terms of centrifugation intensity, and an improper process of rehydration. Systemic experimental work would be necessary to evaluate the role of specific conditions on the microbiota survival throughout whole process. Finally, it is worth noting that both preservation methods based on low temperatures, determined a drastic drop in the relative abundance of the *Fibrobacter succinogenes*, which plays a main role in fiber degradation. This variation could be associated with a depression of degradability and, consequently, with the low NDFD and GP values measured.

5. Conclusions

The results from the present trial indicate that fermentation liquid from rumen continuous fermenters can be used to generate inoculum for in vitro purposes. This result has relevant implications, not only in terms of artificially generating rumen fermenting fluids by RCF systems, but also from the perspective of the standardization of the fermentation liquids by adopting controlled fermentation conditions (e.g., substrate, dilution rate of saliva, fermentation, and pH). Short-term refrigeration is confirmed to be a valuable storage system which has practical advantages in managing rumen fluid, particularly in the case of laboratories distant from collection sites. Finally, low fermentative capacity was found for chilled and, particularly, for freeze-dried liquids, and the procedures adopted to obtain such preserved inoculum probably require substantial improvements.

Author Contributions: Conceptualization, M.S., S.C., F.M., M.M. and F.T.; Data curation, M.S.; Formal analysis, M.C., C.F., F.F., F.M., C.S., S.S. and F.T.; Funding acquisition, M.S.; Project administration, M.S.; Writing—original draft, M.S.; Writing—review & editing, S.C., F.M., M.M., S.S. and F.T.. All Authors participated in critical revision of the manuscript and all have approved the final draft submitted.

Funding: This work was financed by the Italian Ministry of Education, University and Research (PRIN 2015 2015FP39B9-LS9).

Conflicts of Interest: The authors declare no conflict of interest. The funders had no role in the design of the study; in the collection, analyses, or interpretation of data; in the writing of the manuscript, or in the decision to publish the results.

References

1. López, S. In vitro and in situ techniques for estimating digestibility. In *Quantitative Aspects of Ruminant Digestion and Metabolism*, 2nd ed.; CABI Publishing: Wallingford, UK, 2005; pp. 87–121.
2. Yáñez-Ruiz, D.R.; Bannink, A.; Dijkstra, J.; Kebreab, E.; Morgavi, D.P.; O'Kiely, P.; Reynolds, C.K.; Schwarm, A.; Shingfield, K.J.; Yu, Z.; et al. Design, implementation and interpretation of in vitro batch culture experiments to assess enteric methane mitigation in ruminants—A review. *Anim. Feed Sci. Technol.* **2016**, *216*, 1–18. [CrossRef]
3. Hristov, A.N.; Lee, C.; Hristova, R.; Huhtanen, P.; Firkins, J.L. A meta-analysis of variability in continuous-culture ruminal fermentation and digestibility data. *J. Dairy Sci.* **2012**, *95*, 5299–5307. [CrossRef] [PubMed]
4. Carro, M.D.; Ranilla, M.J.; Martin-García, A.I.; Molina-Alcaide, E. Comparison of microbial fermentation of high- and low-forage diets in Rusitec, single-flow continuous-culture fermenters and sheep rumen. *Animal* **2009**, *3*, 527–534. [CrossRef] [PubMed]
5. Muetzel, S.; Lawrence, P.; Hoffmann, E.M.; Becker, K. Evaluation of a stratified continuous rumen incubation system. *Anim. Feed Sci. Technol.* **2009**, *151*, 32–43. [CrossRef]
6. Soto, E.C.; Yánez-Ruiz, D.R.; Cantalapiedra-Hijar, G.; Vivas, A.; Molina-Alcaide, E. Changes in ruminal microbiota due to rumen content processing and incubation in single-flow continuous culture fermenters. *Anim. Prod. Sci.* **2012**, *52*, 813–822. [CrossRef]
7. Soto, E.C.; Molina-Alcaide, E.; Khelil, H.; Yáñez-Ruiz, D.R. Ruminal microbiota developing in different in vitro simulation systems inoculated with goats' rumen liquor. *Anim. Feed Sci. Technol.* **2013**, *185*, 9–18. [CrossRef]

8.	Martínez, M.E.; Ranilla, M.J.; Tejido, M.L.; Saro, C.; Carro, M.D. Comparison of fermentation of diets of variable composition and microbial populations in the rumen of sheep and Rusitec fermenters. II. Protozoa population and diversity of bacterial communities. *J. Dairy Sci.* **2010**, *93*, 3699–3712. [CrossRef]

9.	Belanche, A.; Palma-Hidalgo, J.M.; Nejjam, I.; Serrano, R.; Jiménez, E.; Martín-García, I.; Yáñez-Ruiz, D.R. In vitro assessment of the factors that determine the activity of the rumen microbiota for further applications as inoculum. *J. Sci. Food Agric.* **2018**, *99*, 163–172. [CrossRef]

10.	Chaudhry, A.S.; Mohamed, R.A.I. Fresh or frozen rumen contents from slaughtered cattle to estimate in vitro degradation of two contrasting feeds. *Czech J. Anim. Sci.* **2012**, *57*, 265–273. [CrossRef]

11.	Denek, N.; Can, A.; Avci, M. Frozen rumen fluid as microbial inoculum in the two-stage in vitro digestibility assay of ruminant feeds. *S. Afr. J. Anim. Sci.* **2010**, *40*, 251–256. [CrossRef]

12.	Prates, A.; de Oliveira, J.A.; Abecia, L.; Fondevila, M. Effects of preservation procedures of rumen inoculum on in vitro microbial diversity and fermentation. *Anim. Feed Sci. Technol.* **2010**, *155*, 186–193. [CrossRef]

13.	Zeigler, L.D.; Schlegel, M.L.; Edwards, M.S. Development of a rumen fluid preservation technique and application to an in vitro dry matter digestibility assay. In *Proceedings of the Fifth Conference on Zoo and Wildlife Nutrition, Chester, PA, USA, 5–8 October 2003*; Ward, A., Brooks, M., Maslanka, M., Eds.; AZA Nutrition Advisory Group: Minneapolis, MN, USA, 2003.

14.	Mason, F.; Zanfi, C.; Spanghero, M. Testing a stratified continuous rumen fermenter system. *Anim. Feed Sci. Technol.* **2015**, *201*, 104–109. [CrossRef]

15.	Luchini, N.D.; Broderick, G.A.; Combs, D.K. Preservation of ruminal microorganisms for in vitro determination of ruminal protein degradation. *J. Anim. Sci.* **1996**, *74*, 1134–1143. [CrossRef]

16.	Slyter, L.L.; Bryant, M.P.; Wolin, M.J. Effect of pH on population and fermentation in a continuously cultured rumen ecosystem. *J. Appl. Microbiol.* **1966**, *14*, 573–578.

17.	Menke, K.H.; Steingass, H. Estimation of the energetic feed value obtained from chemical analysis and gas production using rumen fluid. *Anim. Res. Dev.* **1988**, *28*, 7–55.

18.	Van Soest, P.J.; Robertson, J.B.; Lewis, B.A. Methods for dietary fiber, neutral detergent fiber, and nonstarch polysaccharides in relation to animal nutrition. *J. Dairy Sci.* **1991**, *74*, 3583–3597. [CrossRef]

19.	Ross, D.A.; Gutierrez-Botero, M.; van Amburgh, M.E. Development of an in-vitro intestinal digestibility assay for ruminant feeds. In Proceedings of the Cornell Nutrition Conference, Syracuse, NY, USA, 22–24 October 2013; pp. 190–202.

20.	Tagliapietra, F.; Cattani, M.; Bailoni, L.; Schiavon, S. In vitro rumen fermentation: Effect of headspace pressure on the gas production kinetics of cornmeal and meadow hay. *Anim. Feed. Sci. Technol.* **2010**, *158*, 197–201. [CrossRef]

21.	AOAC International. *Official Methods of Analysis of AOAC International*, 17th ed.; AOAC International: Gaithersburg, MD, USA, 2000.

22.	Martillotti, F.; Puppo, S. Liquid chromatographic determination of organic acids in silages and rumen fluids. *Ann. dell'Istituto Sper. Zootec.* **1985**, *18*, 1–10.

23.	Kong, Y.; Teather, R.; Forster, R. Composition, spatial distribution, and diversity of the bacterial communities in the rumen of cows fed different forages. *FEMS Microbiol. Ecol.* **2010**, *74*, 612–622. [CrossRef]

24.	Spanghero, M.; Berzaghi, P.; Fortina, R.; Masoero, F.; Rapetti, L.; Zanfi, C.; Tassone, S.; Gallo, A.; Colombini, S.; Ferlito, J.C. Technical note: Precision and accuracy of in vitro digestion of neutral detergent fiber and predicted net energy of lactation content of fibrous feeds. *J. Dairy Sci.* **2010**, *93*, 4855–4859. [CrossRef]

25.	Spanghero, M. (Dipartimento di Scienze Agroalimentari, Ambientali e Animali, University of Udine, Udine, Italy). Unpublished work. 2019.

26.	Rumen Microbial Genomics Network. A Report in Support of the Rumen Microbial Genomics (RMG) Network Describing Standard Guidelines and Protocols for Data Acquisition, Analysis and Storage. Available online: http://www.rmgnetwork.org/user/file/37/RMG%20Network%20Report%20standard%20guidelines.pdf (accessed on 14 October 2019).

27.	Spanghero, M.; Mason, F.; Zanfi, C.; Nikulina, A. Effect of diets differing in protein concentration (low *vs* medium) and nitrogen source (urea vs soybean meal) on in vitro rumen fermentation and on performance of finishing Italian Simmental bulls. *Livest. Sci.* **2017**, *196*, 14–21. [CrossRef]

28.	Koike, S.; Pan, J.; Kobayashi, Y.; Tanaka, K. Kinetics of in sacco fiber-attachment of representative ruminal cellulolytic bacteria monitored by competitive PCR. *J. Dairy Sci.* **2003**, *86*, 1429–1435. [CrossRef]

29. Robinson, P.H.; Campbell Mathews, M.; Fadel, J.G. Influence of storage time and temperature on in vitro digestion of neutral detergent fibre at 48 h, and comparison to 48 h in sacco neutral detergent fibre digestion. *Anim. Feed Sci. Technol.* **1999**, *80*, 257–266. [CrossRef]

30. Hervás, G.; Frutos, P.; Javier Giráldez, F.; Mora, M.J.; Fernández, B.; Mantecón, A.R. Effect of preservation on fermentative activity of rumen fluid inoculum for in vitro gas production techniques. *Anim. Feed Sci. Technol.* **2005**, *123*, 107–118. [CrossRef]

31. Luchini, N.D.; Broderick, G.A.; Combs, D.K. In vitro determination of ruminal protein degradation using freeze-stored ruminal microorganisms. *J. Anim. Sci.* **1996**, *74*, 2488–2499. [CrossRef] [PubMed]

Article

Variability and Potential of Seaweeds as Ingredients of Ruminant Diets: An In Vitro Study

Ana de la Moneda [1], Maria Dolores Carro [2], Martin R. Weisbjerg [3], Michael Y. Roleda [4,5], Vibeke Lind [4], Margarita Novoa-Garrido [4,6] and Eduarda Molina-Alcaide [1,*]

[1] Estación Experimental del Zaidin (Consejo Superior de Investigaciones Cientificas), Profesor Albareda, 1, 18008 Granada, Spain; anamonedarod@gmail.com

[2] Departamento de Producción Agraria. Escuela Técnica Superior de Ingeniería Agronómica, Alimentaria y de Biosistemas, Universidad Politécnica de Madrid, Ciudad Universitaria, 28040 Madrid, Spain; mariadolores.carro@upm.es

[3] Aarhus University, AU Foulum, Blichers Allé 20, 8830 Tjele, Denmark; martin.weisbjerg@anis.au.dk

[4] Norwegian Institute of Bioeconomy Research (NIBIO), PB 115, 1431 Ås, Norway; Michael.Roleda@nibio.no (M.Y.R.); vibeke.lind@nibio.no (V.L.); margarita.novoa-garrido@nord.no (M.N.-G.)

[5] The Marine Science Institute, College of Science, University of the Philippines, Diliman, Quezon City 1101, Philippines

[6] Faculty of Biosciences and Aquaculture, Nord University, 8049 Bodø, Nordland, Norway

* Correspondence: molina@eez.csic.es; Tel.: +34-95-857-2757 (ext. 351)

Received: 26 September 2019; Accepted: 17 October 2019; Published: 22 October 2019

Simple Summary: The use of seaweeds as ingredients of ruminant diets can be an alternative to conventional feedstuffs, but it is necessary to assess their nutritive value. The aim of this study was to analyze the chemical composition and in vitro rumen fermentation of eight brown, red and green seaweed species collected in Norway during both spring and autumn. The in vitro ruminal fermentation characteristics of 17 diets composed of oat hay:concentrate in a 1:1 ratio, with the concentrate containing no seaweed or including one of the 16 seaweed samples, was also studied. Species and season determined differences in chemical composition and in vitro fermentation of seaweeds. Most of the tested seaweeds can be included in the diet (up to 200 g/kg concentrate) without negative effects on in vitro ruminal fermentation.

Abstract: This study was designed to analyze the chemical composition and in vitro rumen fermentation of eight seaweed species (Brown: *Alaria esculenta*, *Laminaria digitata*, *Pelvetia canaliculata*, *Saccharina latissima*, Red: *Mastocarpus stellatus*, *Palmaria palmata* and *Porphyra* sp.; Green: *Cladophora rupestris*) collected in Norway during spring and autumn. Moreover, the in vitro ruminal fermentation of seventeen diets composed of 1:1 oat hay:concentrate, without (control diet) or including seaweeds was studied. The ash and N contents were greater ($p < 0.001$) in seaweeds collected during spring than in autumn, but autumn-seaweeds had greater total extractable polyphenols. Nitrogen in red and green seaweeds was greater than 2.20 and in brown seaweeds, it was lower than 1.92 g/kg DM. Degradability after 24 h of fermentation was greater in spring seaweeds than in autumn, with *Palmaria palmata* showing the greatest value and *Pelvetia canaliculata* the lowest. Seaweeds differed in their fermentation pattern, and autumn *Alaria esculenta*, *Laminaria digitata*, *Saccharina latissima* and *Palmaria palmata* were similar to high-starch feeds. The inclusion of seaweeds in the concentrate of a diet up to 200 g/kg concentrate produced only subtle effects on in vitro ruminal fermentation.

Keywords: seaweeds; chemical composition; in vitro rumen fermentation; goats; methane

1. Introduction

The expected growth in the human population and the demand for animal products in the forthcoming years have increased the need for searching for alternative sources of nutrients for livestock feeding [1]. Seaweeds had been proposed as alternative feeds that might also have potential benefits on the health of the animals and the consumers of animal products due to their content in bioactive compounds [2,3]. Moreover, seaweeds offer additional advantages, as their cultivation does not compete with terrestrial agriculture, do not need fresh water, and the aquatic photosynthesis contribute to reduce CO_2 levels. The use of seaweeds in animal feeding could also help to alleviate the environmental pollution caused by management of seaweeds in coastal zones. On the other hand, seaweed farming is known to render environmental benefits by recycling nutrients and preventing eutrophication [4].

Although there are studies [5] reporting the traditional use of seaweeds for feeding sheep in the Artic coastal areas and deers in Scotland and Alaska, their widespread use in ruminants is still limited, partly due to the lack of information on the species-specific variability in their the nutritional value and consistency in their chemical composition that may exhibit spatial (site-specific or regional) and temporal (i.e., seasonal and interannual) variations [6–9]. A characteristic common to all seaweeds is their high water content, which may be an important limitation to their direct use in livestock feeding. Another possible limitation is their high salt content [10]. In addition, the presence of compounds that can be a challenge for the digestive system of terrestrial animals may also limit the use of seaweeds in animal feeding [2]. Some recent studies have shown that seaweeds can contain bioactive compounds with antimetanogenic activity, and therefore, they could contribute to reducing the enteric CH_4 emission from ruminants [11–14].

The use of seaweeds as ingredients of ruminant diets requires the assessment of their nutritive value. The first objective of this study was to investigate the chemical composition and in vitro ruminal fermentation of eight different species of seaweeds (three brown, four red and one green) harvested in Norway during spring and autumn. The second objective was to compare the in vitro ruminal fermentation of diets containing these seaweeds with a control diet not including seaweed that was formulated for goat feeding. The gas production technique was used for this study, as it is a relatively cheap and rapid technique that has being widely used in recent years for nutritive evaluation of different ruminant feeds, including seaweeds [8].

2. Materials and Methods

2.1. Seaweeds

The seaweeds used in the present study were chosen based on their biomass availability, potential for cultivation and traditional use for feeding livestock in the Artic areas where they were collected [15]. Eight different seaweed species were collected manually both in spring (March–April) and autumn (October–November) of 2015 in Bodø (northern Norway, 67°19′00″ N, 14°28′60″ E) during low tide. The tested seaweed species corresponded to three groups (Phyla) of seaweeds: the brown (Ochrophyta: *Alaria esculenta*, *Laminaria digitata*, *Pelvetia canaliculata* and *Saccharina latissima*), the red (Rhodophyta: *Mastocarpus stellatus*, *Palmaria palmata* and *Porphyra* sp.), and the green (Chlorophyta: *Cladophora rupestris*). The collected biomass was cleaned in a seawater bath to remove the remains of sand and associated fauna. Then, they were washed with a 30:70 mixture of seawater:freshwater, and finally, in fresh water to reduce the surface salt. The excess of surface water was manually drained and the seaweeds were frozen at −20 °C until their subsequent lyophilization. Once lyophilized, they were ground through a 1 mm sieve in a ZM 200 mill (Retsch GmbH, Haan, Germany).

2.2. Experimental Diets

Seventeen diets based on oat hay and concentrate in a 1:1 ratio were studied. The concentrate in the control diet was high in cereals (633 g/kg fresh matter) to be representative of those fed to goats in the practice and did not include any seaweed. Concentrates in the other 16 experimental diets included seaweeds (Table 1) replacing different amounts of feed ingredients (corn, wheat, soyabean meal, sunflower meal, palm soap and salts) present in the control concentrate.

Table 1. Ingredient composition (g/kg fresh matter) of the experimental concentrates [1].

Ingredient	Control	AS	AA	LS	LA	PS	PA	SS	SA	MS	MA	PPS	PPA	POS	POA	CS	CA
Wheat bran	250	165	165	166	166	186	186	250	250	250	250	195	195	250	250	250	250
Corn	250	250	250	245	245	164	164	100	100	159	159	206	206	250	250	250	250
Wheat	133	18	18	22	22	133	133	133	133	133	133	133	133	133	133	133	133
Soybean meal	124	124	124	124	124	124	124	124	124	104	104	93	93	100	100	96	96
Sunflower meal	84	84	84	84	84	84	84	84	84	45	45	84	84	24	24	28	28
Soybean husk	100	100	100	100	100	100	100	100	100	100	100	100	100	100	100	100	100
Others [2]	59	59	59	59	59	59	59	59	59	59	59	59	59	59	59	59	59
Seaweed	-	200	200	200	200	150	150	150	150	150	150	130	130	84	84	84	84

[1] Each experimental concentrate contained one seaweed (A: Alaria esculenta; L: Laminaria digitata; P: Pelvetia canaliculata; S: Saccharina latissima; M: Mastocarpus stellatus; PP: Palmaria palmata; PO: Porphyra sp.; C: Cladophora rupestris harvested either in spring (AS, LS, PS, SS, MS, PPS, POS and CS concentrates) or in autumn (AA, LA, PA, SA, MA, PPA, POA and CA concentrates). [2] All the concentrates included: 10 g of calcium carbonate, 10 g of sugarbeet molasses, 10 g of sepiolite, 14 g of palm soap, 5 g of NaCl, 5 g of Na_2CO_3, and 5 g of mineral-vitamin mixture per kg.

2.3. Donor Animals and Feeding

Four rumen-cannulated Murciano-Granadina goats with an average body weight of 43.8 ± 3.95 kg were used as donors of ruminal content for the in vitro incubations. The animals were fed a diet composed of oat hay and a commercial concentrate in a 50:50 ratio and were housed in pens in pairs with free access to drinking water. The level of intake was that of energy maintenance requirements [16] and the diet was supplied twice a day in equal amounts. The care and handling of the goats were carried out by trained personnel in accordance with the Spanish guidelines for the protection of animals used for experimentation or other purposes, and the experimental procedures were approved by the Animal Welfare Committee at the Zaidín Experimental Station of the Spanish National Research Council (Approval number: 05/24/2016/091).

2.4. In Vitro Trials

In vitro incubations were conducted using the seaweed samples alone and the 17 experimental diets (oat hay and concentrate 1:1) as substrates. Incubations were carried out in batch cultures of ruminal microorganisms using 120-mL glass bottles and ruminal fluid from goats as inoculum. The ruminal content was obtained from each of the four goats before the morning feeding, mixed, and immediately transported to the laboratory in thermal flasks pre-warmed at 39 °C. The ruminal content was filtered through four layers of surgical gauze and mixed with a buffer solution in a 1:4 ratio [17]; no trypticase added and under a continuous CO_2 flow. A total of six incubation runs were carried out. In the first three incubation runs, seaweeds were used as substrate and three feeds commonly used in goat feeding (oat hay, barley straw and a commercial concentrate) were also included for comparative purposes. In the last three incubation runs, the substrates were the 17 experimental diets. In all the incubation runs, four bottles per substrate were used, and four blanks (bottles without substrate) were included.

Five hundred mg of each substrate were carefully weighed in each bottle and 60 mL of the mixture of ruminal fluid and buffer solution were added under a continuous flow of CO_2. Bottles were sealed with butyl rubber stoppers and aluminum caps and placed in a water bath at 39 °C. The pressure inside the bottles and the volume of gas produced in two bottles per substrate and two blanks were measured at 2, 4, 6, 8, 12, 24, 48, 72, 96, 120 and 144 h of incubation using a pressure gauge scope (Sper Scientific LTD, Scottsdale, AZ, USA) and a calibrated glass syringe (Ruthe®, Normax Marinha Grande, Portugal). Additionally, in the incubations using seaweeds as substrate, the content of each bottle at the end of the 144 h of incubation was weighed, frozen at −20 °C and analyzed for neutral detergent fibre (NDF) content to estimate the true dry matter (DM) digestibility ($TDMD_{144}$) as described by Van Soest et al. [18].

In each incubation run, the other two bottles for each substrate (either seaweeds or the experimental diets) and blanks were incubated for 24 h. Gas production measurement was done as described above and a gas sample (5 mL) was stored in a vacuum tube (Terumo Europe NV, Leuven, Belgium) for analysis of CH_4. The fermentation was stopped by chilling on ice. The content of the bottles was homogenized and the following samples were taken: 2 mL were added to 2 mL of a deproteinizing solution (20 g of metaphosphoric acid and 0.6 g of crotonic acid per liter) for the analysis of volatile fatty acids (VFA), and 1 mL was mixed with 1 mL of 0.5 M HCl for the analysis of NH_3-N. Additionally, in the incubations using seaweeds as substrate, the content of each bottle was weighed (before sampling), frozen at −20 °C, and analyzed for neutral detergent fibre (NDF) content to estimate the true dry matter (DM) digestibility ($TDMD_{24}$) as described by Van Soest et al. [18].

2.5. Chemical Analyses

The DM content of the seaweeds and experimental concentrates was determined by lyophilization and subsequent drying of the lyophilized material in an oven at 103 °C for 24 h [19]. Ash content in seaweed (ID 048.13) and ether extract (ID 945.16) were determined according the AOAC procedures [19].

The total N content was analyzed by the Kjeldahl method. The NDF content in the in vitro incubation residues was determined following the procedure of Goering and van Soest [17] using a FibertecTMM6 system (Foss Analytical, Hillerød, Denmark). In the NDF analysis of concentrates, heat-stable amylase was added [20], and all results are expressed as ash-free. The content in total extractable polyphenols (TEP) was analyzed following the procedure of Julkunen-Tiito [21]. The concentrations of individual VFA in the content of the bottles and CH_4 in the gas produced were analyzed by gas chromatography using a HP Hewlett 5890 Packard Series II gas chromatograph (Waldbronn, Germany) equipped with a flame ionization detector (FID) and an HPINNOWAX cross linked polyethylene glycol column (25 m× 0.2 mm× 0.2 μm; Teknokroma, Madrid, Spain) as described by Molina-Alcaide et al. [8]. The concentration of $N\text{-}NH_3$ was determined following the colorimetric method of Weatherburn [22] using a spectrophotometer (Thermo Scientific, Genesys 10 uV Scanning, Madison, WI 53711 USA).

2.6. Calculations and Statistical Analysis

The gas production data were adjusted to the exponential model: gas = A $(1 - e^{(-c\,(t-lag))})$, where A is the asymptotic gas production, c is the gas production rate, lag is the delay at the start of gas production, and t is the time of gas measurement. Parameters A, c and lag were estimated using an iterative least-square procedure following the NLIN procedure of SAS (version 9.4, SAS Inst. Inc., Cary, NC, USA). The average gas production rate (AGPR, ml/h) is defined as the average rate of gas production between the start of incubation and the time at which half of A is reached, and was calculated as AGPR = A c/[2 (ln2 + c lag)]. The amount of VFA in each bottle after 24 h of incubation was corrected by the amount of VFA added with the ruminal fluid used as inoculum.

Data on the chemical composition of seaweed were analyzed by ANOVA using the PROC GLM of SAS (version 9.4, SAS Inst. Inc., Cary, NC, USA) in which the seaweed species and harvest season were the main effects. Fermentation data of seaweeds were analyzed using the PROC MIXED of SAS as a mixed model (version 9.4, SAS Inst. Inc., Cary, NC, USA), in which the seaweed species, harvest season and seaweed species x season interaction were considered as fixed effects, and the incubation run was considered random. The model for the analysis of data of experimental diets included the fixed effect of diet and the random effect of the incubation run. When a significant effect was detected ($p \leq 0.05$), the differences between the means were tested using Tukey's multiple comparison test.

3. Results

3.1. Chemical Composition and In Vitro Fermentation of Seaweeds

Both seaweed species and harvest season affected ($p < 0.001$) all chemical fractions analyzed (Table 2). Ash and N content was greater ($p < 0.001$) in seaweeds collected in spring than in those harvested in autumn (224 vs. 121 g/kg DM and 3.08 vs. 1.92 g/kg DM, respectively). Ash content ranged from 88.2 g/kg DM in *Porphyra* sp. to 225 g/kg DM in *Laminaria digitata* and *Saccharina latissima* (values averaged across seasons). There were also wide variations in total N content, with red and green seaweeds having values greater than 2.20 g/kg DM (values averaged for both collection seasons) and brown species showing values lower than 1.90 g/kg DM. The TEP content was greater ($p < 0.001$) in autumn than in spring (12.1 vs. 6.82 g/kg DM), and the greatest values corresponded to *Alaria esculenta* and *Pelvetia canaliculata*.

Table 2. Chemical composition (g/kg dry matter unless otherwise stated) of different seaweed species harvested in spring and autumn in northern Norway and of feeds commonly used in ruminant diets.

Species	Season	Dry Matter (g/100 g Fresh Matter)	Ash	Nitrogen	Total Extractable Polyphenols
Brown seaweeds					
Alaria esculenta	Spring	110	288	23.4	4.51
	Autumn	277	73.6	13.1	28.1
	Average	193 [f]	181 [d]	18.2 [d]	16.3 [e]
Laminaria digitata	Spring	115	311	23.0	1.44
	Autumn	189	138	6.77	6.08
	Average	152 [b]	225 [e]	14.9 [c]	3.76 [d]
Pelvetia canaliculata	Spring	237	199	16.3	26.9
	Autumn	237	174	6.88	40.4
	Average	237 [g]	187 [d]	11.6 [a]	33.7 [f]
Saccharina latissima	Spring	87.0	350	17.6	3.87
	Autumn	220	100	6.03	5.21
	Average	154 [c]	225 [e]	11.8 [a]	4.54 [d]
Red seaweeds					
Mastocarpus stellatus	Spring	261	183	26.4	4.36
	Autumn	245	194	18.1	3.57
	Average	253 [h]	189 [d]	22.2 [e]	3.97 [ab]
Palmaria palmata	Spring	121	213	43.0	3.86
	Autumn	191	103	14.6	1.93
	Average	156 [d]	158 [c]	28.8 [f]	2.89 [b]
Porphyra sp.	Spring	90.0	97.9	59.8	4.75
	Autumn	116	78.4	50.9	5.85
	Average	103 [a]	88.2 [a]	55.4 [h]	5.30 [c]
Green seaweeds					
Cladophora rupestris	Spring	191	149	37.1	4.88
	Autumn	181	105	37.0	5.39
	Average	186 [e]	127 [b]	37.1 [g]	5.14 [bc]
p value					
Species		<0.001	<0.001	<0.001	<0.001
Season		<0.001	<0.001	<0.001	<0.001
SEM		0.004	0.670	0.128	0.035
Feeds					
Oat hay		896	62.7	12.7	6.82
Barley straw		941	43.7	3.07	NA [1]
Commercial concentrate		933	77.4	23.0	NA [1]

[a–e] For each parameter, average values for each seaweed not sharing the same superscript differ ($p < 0.001$); [1] NA: not analysed.

As shown in Table 3, seaweed species x season interactions ($p < 0.001$) were detected for all the parameters of gas production (A, *c*, *lag* and AGPR) and $TDMD_{244}$. There were differences ($p < 0.001$) among seaweed species in all the parameters of gas production and in vitro digestibility values. *Palmaria palmata* had the greatest ($p < 0.05$) A and AGPR values (143 mL and 4.95 mL/g DM, respectively) with A values being similar to those in the three feedstuffs used as reference and AGPR values higher than those for feedstuffs. The lowest ($p < 0.05$) values were shown by *Pelvetia canaliculata* (8.2 mL and 1.38 mL/g DM, respectively, for A and AGPR) and were much lower than A for any of the feedstuffs and AGPR similar to this value in barley straw. The *lag* values were 0.00 for most seaweed samples, with the exception of *Alaria esculenta* in autumn, *Saccharina latissima* in spring and *Palmaria palmata*, but all the values were lower than 1 h except those for *Alaria esculenta* in autumn (2.58 h). The collecting season affected ($p < 0.001$) the values of A, *lag*, AGPR and $TDMD_{24}$. Compared with spring seaweeds, those collected in autumn had greater A (65.5 vs. 87.5 mL), lag (0.01 vs. 0.42 mL) and AGPR (2.14 vs. 2.93), but lower $TDMD_{144}$ values (87.9 vs. 83.0%).

Table 3. Parameters of gas production kinetics (A, c, *lag* and AGPR) and true dry matter (DM) digestibility (TDMD$_{144}$) after 144 h of in vitro incubation of different seaweed species harvested in spring and autumn in northern Norway and of feeds commonly used in ruminant diets [1].

Seaweed Species	Season	A (ml)	c (h^{-1})	*lag* (h)	AGPR (ml/h)	TDMD$_{144}$ (%)
Brown seaweeds						
Alaria esculenta	Spring	85.9	0.034	0.00	2.11	93.2
	Autumn	104.9	0.033	2.58	2.20	75.4
	Average	95.4 [e]	0.034 [a]	1.29 [c]	2.16 [b]	84.3 [c]
Laminaria digitata	Spring	85.2	0.027	0.00	1.68	98.3
	Autumn	107.4	0.034	0.00	2.59	79.1
	Average	96.3 [e]	0.031 [a]	0.00 [a]	2.14 [b]	88.7 [d]
Pelvetia canaliculata	Spring	6.3	0.351	0.00	1.58	67.8
	Autumn	10.0	0.162	0.00	1.17	68.4
	Average	8.15 [a]	0.257 [b]	0.00 [a]	1.38 [a]	68.1 [a]
Saccharina latissima	Spring	84.0	0.030	0.07	1.82	97.6
	Autumn	147.1	0.043	0.00	4.58	94.6
	Average	116 [f]	0.037 [a]	0.04 [a]	3.20 [c]	96.1 [e]
Red seaweeds						
Mastocarpus stellatus	Spring	31.0	0.068	0.00	1.52	89.3
	Autumn	20.6	0.078	0.00	1.16	91.0
	Average	25.8 [b]	0.073 [a]	0.00 [a]	1.34 [a]	90.2 [d]
Palmaria palmata	Spring	114.6	0.060	0.03	4.93	95.8
	Autumn	171.9	0.042	0.74	4.97	96.4
	Average	143 [g]	0.051 [a]	0.39 [b]	4.95 [d]	96.1 [e]
Porphyra sp.	Spring	54.8	0.063	0.00	2.51	87.3
	Autumn	64.7	0.071	0.00	3.31	90.0
	Average	59.8 [c]	0.067 [a]	0.00 [a]	2.91 [c]	88.7 [d]
Green seaweeds						
Cladophora rupestris	Spring	62.4	0.020	0.00	0.99	73.5
	Autumn	73.1	0.066	0.00	3.47	74.3
	Average	67.8 [d]	0.043 [a]	0.00 [a]	2.19 [b]	73.9 [b]
p value						
Species		<0.001	<0.001	<0.001	<0.001	<0.001
Season		<0.001	0.301	<0.001	<0.001	<0.001
Species x season		<0.001	<0.001	<0.001	<0.001	<0.001
SEM		0.296	0.008	0.029	0.057	0.347
Feeds						
Oat hay		129.1	0.037	0.00	3.43	79.7
Barley straw		124.6	0.017	0.41	1.53	56.9
Commercial concentrate		146.3	0.064	0.00	6.75	91.4

[a–e] For each parameter, the average values for each seaweed not sharing the same superscript differ ($P < 0.05$). [1] A: asymptotic gas production; c: rate of gas production; lag: lag time before fermentation starts; AGPR: average gas production rate; DMED$_{24}$: dry matter effective degradability calculated for a rumen passage rate of 0.041 per h. Data are expressed per 0.5 g DM fermented.

There were differences ($p < 0.001$ to 0.003) among seaweed species in total VFA production, VFA profile and acetate/propionate ratio (Table 4). *Pelvetia canaliculata* had the lowest ($p < 0.05$) VFA production, whereas *Alaria esculenta* and *Saccharina latissima* had the greatest production ($p < 0.05$). The VFA production was not affected ($p = 0.821$) by the harvesting season, and no seaweed species x season interaction ($p = 0.609$) was detected. In contrast, seaweed species x season interactions ($p < 0.001$) were detected for molar proportions of acetate, propionate, isobutyrate and isovalerate. *Palmaria palmata* had the lowest proportion of acetate and the greatest propionate proportion (58.5% and 30.1%, respectively), whereas *Mastocarpus stellatus* had the lowest proportion of propionate and the greatest of butyrate (15.1% and 9.50%). The production of minor VFA (isobutyrate, isovalerate and valerate) also differed among seaweed species, with *Porphyra* sp. having the greatest ($p < 0.05$) proportions of isobutyrate and isovalerate and *Pelvetia canaliculata* the greatest valerate proportions. Compared to seaweeds harvested in spring, autumn seaweeds had lower ($p < 0.001$ to 0.020) proportions of acetate (69.0% vs. 59.3%) and minor VFA, as well as greater propionate (18.4% vs. 27.1%) and butyrate (6.37% vs. 8.87%) proportions. The acetate/propionate ratio was highly variable, with values ranging from 1.47 mol/mol in *Alaria esculenta* to 4.76 mol/mol for *Mastocarpus stellatus* both collected in autumn. Spring seaweeds had greater ($p < 0.001$) acetate/propionate ratios than those collected in autumn (3.91 vs. 2.52).

Table 4. Fermentation parameters and true dry matter (DM) digestibility ($TDMD_{24}$) after 24 h of in vitro incubation of different seaweed species harvested in spring and autumn in northern Norway and of feeds commonly used in ruminant diets.

Seaweed Species	Season	VFA (mmol/g DM)	Molar Proportions (mol/100 mol)						Acetate/Propionate (mol/mol)	NH_3-N (mg/100 mL)	CH_4 (mL/g DM)	CH_4/VFA (mL/mmol)	$TDMD_{24}$ (%)
			Acetate	Propionate	Butyrate	Isobutyrate	Isovalerate	Valerate					
Brown Seaweeds													
Alaria esculenta	Spring	3.44	71.1	19.3	5.80	0.78	1.09	1.86	3.69	7.52	28.6	8.31	90.8
	Autumn	3.42	53.6	37.1	8.18	0.02	0.03	1.15	1.47	1.58	28.7	8.39	75.5
	Average	3.43 [cd]	62.4 [b]	28.2 [de]	7.0 [ab]	0.40 [a]	0.56 [a]	1.51 [a]	2.58 [b]	4.55 [a]	28.7 [b]	8.37 [b]	83.2 [c]
Laminaria digitata	Spring	3.16	76.1	16.2	4.00	0.80	1.17	1.75	4.73	10.7	17.0	5.38	92.6
	Autumn	3.24	54.8	32.7	10.0	0.66	0.50	1.32	1.70	0.80	38.5	11.9	75.7
	Average	3.20 [bc]	65.5 [cd]	24.4 [c]	7.00 [ab]	0.73 [b]	0.84 [ab]	1.54 [a]	3.22 [cd]	5.75 [bc]	27.8 [b]	8.69 [bc]	84.2 [c]
Pelvetia canaliculata	Spring	1.08	65.9	21.1	7.05	0.46	1.06	4.37	3.13	7.28	4.91	4.54	70.1
	Autumn	1.10	64.6	19.4	10.1	0.53	1.01	4.34	3.40	5.58	6.10	5.55	69.3
	Average	1.09 [a]	65.3 [cd]	20.3 [b]	8.60 [bc]	0.50 [a]	1.04 [b]	4.36 [c]	3.27 [cde]	6.43 [c]	5.50 [a]	5.05 [a]	69.7 [b]
Saccharina latissima	Spring	3.56	74.9	16.3	4.85	0.93	1.30	1.74	4.61	9.86	24.6	6.91	92.8
	Autumn	4.62	52.7	35.4	9.22	0.78	0.65	1.29	1.49	0.51	47.1	10.2	85.2
	Average	4.09 [d]	63.8 [bc]	25.8 [cd]	7.00 [ab]	0.86 [c]	0.98 [b]	1.52 [a]	3.05 [c]	5.18 [ab]	35.9 [b]	8.78 [b]	89.0 [de]
Red seaweeds													
Mastocarpus stellatus	Spring	2.64	67.6	15.8	9.30	1.38	2.81	3.05	4.28	15.1	18.0	6.82	88.3
	Autumn	1.44	68.3	14.3	9.66	1.49	2.99	3.22	4.76	13.5	11.1	7.71	88.8
	Average	2.04 [ab]	67.9 [e]	15.1 [a]	9.50 [c]	1.44 [e]	2.90 [c]	3.14 [b]	4.52 [f]	14.3 [e]	14.6 [a]	7.16 [a]	88.6 [d]
Palmaria palmata	Spring	7.18	61.1	26.4	6.65	1.33	1.71	2.76	2.31	22.1	64.6	9.00	94.9
	Autumn	6.56	55.8	33.7	7.89	0.70	0.55	1.35	1.67	0.78	60.2	9.18	88.7
	Average	6.87 [e]	58.5 [a]	30.1 [e]	7.30 [ab]	1.02 [d]	1.13 [b]	2.06 [a]	1.99 [a]	11.4 [d]	62.4 [c]	9.08 [bc]	91.8 [e]
Porphyra sp.	Spring	3.28	66.4	16.1	8.33	2.42	3.66	3.17	4.18	39.9	34.3	10.5	82.7
	Autumn	3.00	61.0	19.9	9.28	2.62	4.00	3.23	3.09	37.7	39.3	13.1	79.9
	Average	3.14 [bc]	63.7 [bc]	17.9 [b]	8.80 [bc]	2.52 [g]	3.83 [d]	3.20 [b]	3.64 [e]	38.8 [g]	36.8 [b]	11.7 [cd]	81.3 [c]
Green seaweeds													
Cladophorarupestris	Spring	2.38	69.0	16.0	4.97	2.07	3.66	4.25	4.32	29.8	31.9	13.4	63.9
	Autumn	3.00	63.3	24.4	6.59	1.40	1.85	2.43	2.60	19.5	35.9	12.0	62.0
	Average	2.70 [bc]	66.2 [de]	20.2 [b]	5.80 [a]	1.74 [f]	2.76 [c]	3.34 [b]	3.46 [de]	24.7 [f]	33.9 [b]	12.6 [e]	63.0 [a]
p value													
Species		<0.001	<0.001	<0.001	0.013	<0.001	<0.001	<0.001	<0.001	<0.001	<0.001	0.001	<0.001
Season		0.821	<0.001	<0.001	<0.001	<0.001	<0.001	0.021	<0.001	<0.001	<0.001	0.083	<0.001
Species x season		0.609	<0.001	<0.001	0.104	<0.001	<0.001	0.066	<0.001	<0.001	0.267	0.312	<0.001
SEM		0.154	0.310	0.330	0.249	0.015	0.041	0.086	0.047	0.164	1.588	0.394	0.355
Feeds													
Oat hay		3.84	68.6	19.5	8.10	0.75	0.86	2.13	3.52	6.09	28.2	7.34	64.0
Barley straw		2.68	69.3	20.4	7.03	0.70	0.91	1.66	3.40	2.97	21.7	8.10	38.0
Commercial concentrate		5.48	57.5	25.7	1.10	1.44	1.87	2.53	2.24	12.2	56.4	10.3	86.8

[a–g] For each parameter, the average values for each seaweed not sharing the same superscript differ ($p < 0.05$).

Both species and season affected ($p < 0.001$) NH_3-N concentrations, CH_4 production and $TDMD_{24}$, (Table 4), and seaweed species × season interactions were detected for NH_3-N concentrations and $TDMD_{24}$. *Alaria esculenta* and *Porphyra* sp. had the lowest and greatest NH_3-N concentrations, respectively, whereas *Pelvetia canaliculata* and *Palmaria palmata* had the lowest and greatest CH_4 productions, respectively. The values of $TDMD_{24}$ ranged from 63.0% to 91.8%, the lowest and greatest values corresponding to *Cladophora rupestris* and *Palmaria palmata*, respectively. Greater ($p < 0.001$) NH_3-N concentrations and $TDMD_{24}$ values and lower ($p < 0.001$) CH_4 production were observed for the samples collected in spring (17.8 mg/100 mL, 84.5% and 26.7 mL, respectively) compared to those collected in autumn (9.99 mg/100 mL, 78.1% and 33.4 mL).

3.2. Chemical Composition and In Vitro Fermentation of Experimental Diets

The chemical composition of the experimental diets is shown in Table 5. In general, ash content was greater in the diets containing seaweeds than in the control diet, whereas the opposite was observed for ether extract content. There were only small differences among diets in N content, which ranged from 17.6 to 19.8 g/kg DM, whereas TEP content varied from 4.75 to 7.98 g/kg DM.

Table 5. Chemical composition (g/kg dry matter (DM) unless otherwise stated) of diets containing 50% of oat hay and 50% of concentrate either including no seaweeds (control) or different seaweed species harvested in spring and autumn in northern Norway.

Seaweed Species	Harvesting Season	Concentrate	Dry Matter (g/kg Fresh Matter)	Ash	Nitrogen	Ether Extract	Total Extractable Polyphenols
-	-	Control	924	68.6	17.9	17.7	5.92
Alaria esculenta	Spring	AS	902	86.4	19.8	16.6	5.73
	Autumn	AA	901	72.1	18.4	17.8	6.29
Laminaria digitata	Spring	LS	910	81.9	19.4	13.8	5.64
	Autumn	LA	911	109	17.6	14.4	5.98
Pelvetiacanaliculata	Spring	PS	904	77.9	18.6	15.8	5.69
	Autumn	PA	903	85.9	17.6	15.8	6.10
Saccharinalatissima	Spring	SS	905	97.9	19.5	15.9	4.82
	Autumn	SA	905	79.5	17.8	16.9	6.58
Mastocarpusstellatus	Spring	MS	904	80.9	19.4	14.8	4.75
	Autumn	MA	905	79.7	18.1	15.5	5.83
Palmariapalmata	Spring	PPS	907	76.7	18.7	15.5	4.99
	Autumn	PPA	908	80.5	17.6	15.8	4.79
Porphyra sp.	Spring	POS	902	70.6	17.8	15.9	5.65
	Autumn	POA	901	64.1	18.1	15.8	5.96
Cladophorarupestris	Spring	CS	902	60.2	18.7	16.4	4.99
	Autumn	CA	900	70.8	17.8	16.6	5.62

As shown in Table 6, the diets including *Palmaria palmata* collected in autumn, and *Porphyra sp* and *Cladophora rupestris* collected in spring and autumn had greater ($p < 0.05$) potential gas production values (A) compared with the rest of the diets, including the control one. All the diets including seaweeds, except that with *Palmaria palmata* collected in autumn, had lower ($p < 0.05$) fractional rates of gas production and AGPR than the control.

Table 7 shows the in vitro fermentation parameters of the experimental diets. There were no differences ($p \geq 0.152$) in total VFA production, minor VFA molar proportions and NH_3-N concentrations. Compared with the control, diets including spring-harvested *Alaria esculenta*, *Saccharina latissima*, *Palmaria palmata*, *Laminaria digitata*, *Pelvetia canaliculata*, and *Mastocarpus stellatus* from both seasons had greater ($p < 0.05$) acetate proportions. All the diets except that including autumn-harvested *Alaria esculenta* had lowers ($p < 0.05$) propionate molar proportions than the control. Butyrate molar proportions were lowest for the diets with *Alaria esculenta*, *Laminaria digitata* and *Saccharina latissima* and greatest for the diets with *Porphyra sp.* and *Cladophora rupestris*, with the control diet having an intermediate value. Most diets including seaweeds had greater ($p < 0.05$) acetate/propionate ratios than the control diet, except those including autumn-harvested *Alaria esculenta, Laminaria digitata, Saccharina latissima* and *Palmaria palmata*. All the diets with autumn-harvested seaweeds had lower CH_4 production than the control diet.

Table 6. Parameters of gas production kinetics (A, c and AGPR) after 144 h of in vitro incubation of diets containing 50% of oat hay and 50% of concentrate either including no seaweeds (control) or different seaweed species harvested in spring and autumn in northern Norway [1].

Seaweed Species	Harvesting Season	Concentrate	A (ml)	c (h^{-1})	AGPR (ml/h)
-	-	Control	138 [a]	0.050 [b]	4.98 [b]
Alariaesculenta	Spring	AS	134 [a]	0.044 [a]	4.30 [a]
	Autumn	AA	138 [a]	0.043 [a]	4.28 [a]
Laminaria digitata	Spring	LS	131 [a]	0.040 [a]	3.78 [a]
	Autumn	LA	133 [a]	0.042 [a]	4.03 [a]
Pelvetiacanaliculata	Spring	PS	136 [a]	0.041 [a]	4.02 [a]
	Autumn	PA	129 [a]	0.041 [a]	3.82 [a]
Saccharinalatissima	Spring	SS	133 [a]	0.043 [a]	4.13 [a]
	Autumn	SA	137 [a]	0.043 [a]	4.25 [a]
Mastocarpusstellatus	Spring	MS	135 [a]	0.044 [a]	4.28 [a]
	Autumn	MA	131 [a]	0.043 [a]	4.03 [a]
Palmariapalmata	Spring	PPS	135 [a]	0.045 [a]	4.38 [a]
	Autumn	PPA	145 [b]	0.047 [ab]	4.92 [b]
Porphyra sp.	Spring	POS	147 [b]	0.041 [a]	4.35 [a]
	Autumn	POA	148 [b]	0.042 [a]	4.48 [a]
Cladophorarupestris	Spring	CS	146 [b]	0.041 [a]	4.35 [a]
	Autumn	CA	149 [b]	0.040 [a]	4.30 [a]
p value			<0.001	0.033	0.215
SEM			0.56	0.0014	0.098

[a-b] For each parameter, the mean values for each diet not sharing the same superscript differ ($p < 0.05$). [1] A: asymptotic gas production; c: rate of gas production; AGPR: average gas production rate. The values of *lag* were 0 for all samples. Data are expressed per 0.5 g DM fermented.

Table 7. Fermentation parameters after 24 h of in vitro incubation of diets containing 50% of oat hay and 50% of concentrate either including no seaweeds (control) or different seaweed species harvested in spring and autumn in northern Norway [1].

Seaweed Species	Harvesting Season	Concentrate	VFA (mmol/g DM)	Molar Proportions (mol/100 mol)						Acetate/Propionate (mol/mol)	NH$_3$-N (mg/100 mL)	CH$_4$ (mL/g DM)	CH$_4$/VFA (mL/mmol)
				Acetate	Propionate	Butyrate	Isobutyrate	Isovalerate	Valerate				
–	–	Control	5.06	42.8[a]	21.2[b]	11.9[b]	0.98	1.20	1.79	2.96[a]	10.2	64.6[ab]	12.8
Alaria esculenta	Spring	AS	4.88	45.3[b]	19.3[a]	11.3[a]	0.99	1.38	1.73	3.39[b]	9.88	65.4[b]	13.4
	Autumn	AA	5.02	43.8[a]	22.5[c]	10.4[a]	0.88	1.03	1.49	2.84[a]	7.09	63.2[ab]	12.4
Laminaria digitata	Spring	LS	4.86	45.8[b]	19.8[a]	10.5[a]	0.97	1.32	1.65	3.33[b]	9.84	64.0[ab]	13.2
	Autumn	LA	4.86	44.6[b]	21.8[a]	9.93[a]	0.91	1.14	1.63	2.96[a]	7.57	61.9[a]	12.7
Pelvetia canaliculata	Spring	PS	4.80	44.8[b]	19.4[a]	11.9[b]	0.95	1.31	1.65	3.34[b]	10.1	63.6[ab]	13.2
	Autumn	PA	5.06	45.0[b]	19.7[a]	11.0[b]	1.04	1.36	1.81	3.30[b]	8.10	66.6[b]	13.2
Saccharina latissima	Spring	SS	4.86	45.2[b]	20.3[a]	10.6[a]	0.98	1.31	1.65	3.22[b]	9.49	63.1[ab]	13.0
	Autumn	SA	5.22	43.6[a]	21.8[a]	10.6[a]	0.99	1.30	1.68	2.91[a]	7.83	65.0[b]	12.4
Mastocarpus stellatus	Spring	MS	4.90	44.6[b]	19.4[a]	12.0[b]	1.02	1.35	1.58	3.32[b]	10.3	65.5[b]	13.4
	Autumn	MA	4.66	44.7[b]	19.5[a]	11.9[b]	0.99	1.32	1.64	3.31[b]	9.51	61.6[a]	13.2
Palmaria palmata	Spring	PPS	5.26	44.3[b]	19.7[a]	12.0[b]	1.03	1.31	1.64	3.26[b]	10.4	68.4[b]	13.0
	Autumn	PPA	5.26	43.8[a]	20.6[a]	11.9[b]	0.96	1.22	1.51	3.10[ab]	8.87	67.7[b]	12.9
Porphyra sp.	Spring	POS	5.16	43.6[a]	19.9[a]	12.6[c]	0.95	1.30	1.62	3.19[b]	9.34	67.2[b]	13.0
	Autumn	POA	5.18	43.8[a]	19.5[a]	12.6[c]	1.03	1.36	1.66	3.27[b]	8.61	66.1[b]	12.8
Cladophora rupestris	Spring	CS	5.00	43.6[a]	20.0[a]	12.5[c]	0.95	1.26	1.67	3.19[b]	8.44	65.2[b]	13.0
	Autumn	CA	5.24	43.8[a]	19.6[a]	12.8[c]	0.98	1.30	1.53	3.26[b]	8.95	67.5[b]	13.2
p-value			0.152	<0.001	0.008	<0.001	0.767	0.861	0.636	0.033	0.960	0.049	0.569
SEM			0.042	0.088	0.129	0.063	0.011	0.025	0.023	0.023	0.349	1.59	0.52

[a–c] For each parameter, the mean values for each diet not sharing the same superscript differ ($p < 0.05$).

4. Discussion

4.1. Chemical Composition and In Vitro Fermentation of Seaweeds

The low DM and high ash content of seaweeds are frequently reported as the main limitations to their use in ruminant diets [7,8]. Both DM and ash contents were similar to those reported for the same seaweeds and others (*Ruppia maritima*, *Ulva lactuca* and *Chaetomorpha linum*) in previous studies [7,23]. In accordance with Tayyab et al. [7], the ash content of seaweeds was greater in spring than in autumn, and the values were greater than those found in conventional feeds used in ruminant nutrition (Table 2). As previously reported [7,8,24,25]. The N content was highly variable, and it was greater in spring-harvested seaweeds than in those collected in autumn. This has been attributed to high sunlight conditions that increase the photosynthesis and nutrient assimilation and to greater N concentration in water during spring compared with autumn [24]. Both *Porphyra sp* and *Cladophora rupestris* showed an N content greater than that in the commercial concentrate used as reference in our study (Table 2; 23.0 g/kg DM), but other seaweeds had an N content similar to that in the oat hay or even lower, especially those harvested in autumn. High-protein seaweeds may be used as an alternative to conventional high-protein feeds, such as soybean meal, and recent studies [25] showed that some amino acids in *Laminaria* and *Mastocarpus* species were protected against rumen degradation, making them potential sources of by-pass protein. In agreement with previous studies [8,9,26], brown seaweeds had, in general, a greater TEP content than both red and green seaweeds, and TEP content was lower in spring-harvested seaweeds than in those collected in autumn. Brown seaweeds are rich in phlorotannins [27], which seem to be different from the tannins in terrestrial plants, but their effect on ruminants is still unknown. Polyphenols have been reported to reduce protein degradation in the rumen, but they can also reduce the fibre degradation by decreasing the attachment of microbes to feed particles [3]. The negative relationships (n = 16) observed between the TEP content and $TDMD_{144}$ (r = 0.732; p = 0.001), $TDMD_{24}$ (r = 0.503; p = 0.047), and total VFA concentrations (r = 0.478; p = 0.061) indicates a negative effect of TEP on the in vitro rumen degradation of seaweeds. However, there were no correlations between TEP content and any of the gas production parameters, which supports the idea that gas measurement should be combined with measurements of feed degradability for a better interpretation of polyphenols effects, as pointed out by Makkar [3].

The high variability observed in the potential gas production values (A) of seaweeds reflects the differences in their potential degradation in the rumen. In fact, a positive relationship between A and $TDMD_{144}$ (r = 0.510; p = 0.044; n = 16) was detected. The lowest A and $TDMD_{144}$ values were observed for *Pelvetia canaliculata*, which agrees with the low DM degradability values reported for this seaweed by Tayyab et al. [7] using the in situ technique in dairy cows and by Molina-Alcaide et al. [8] in 24-h in vitro incubations with sheep ruminal fluid. The greatest A and $TDMD_{144}$ values were observed for *Palmaria palmata* and *Saccharina latissima*, which is in agreement with the high ruminal degradability observed in previous studies for both seaweeds [7,8].

A 24-h incubation period was chosen for the in vitro incubations in our study, as this rumen retention time can be found in goats and sheep fed at moderate levels of intake [28,29]. In agreement with the results of the gas production study, *Pelvetia canaliculata* promoted the lowest total VFA production, which was only 0.41 of that observed for barley straw, and *Palmaria palmata* and *Saccharina latissima* had the greatest values, which were 1.3 and 0.75 of those observed for the concentrate, respectively. Total VFA production for *Porphyra sp.* and *Cladophora rupestris* was similar to that for barley straw, whereas the fermentation of *Alaria esculenta* and *Laminaria digitata* promoted a VFA production only slightly lower than that from fermentation of medium-quality forage such as the oat hay used in our study. These results show that seaweeds can be fermented in the rumen to a variable extent. Although the collecting season had a marked influence on the chemical composition of seaweeds, no differences between seasons were observed in total VFA production. This agrees with the lack of differences between the two harvesting seasons in the ruminal degradability of the protein of nine seaweed species observed by Gaillard et al. [25], despite the marked differences detected in protein content.

There were pronounced differences among seaweed species with regard to VFA profile. *Alaria esculenta, Laminaria digitata, Saccharina latissima* and *Palmaria palmata* harvested in autumn had high propionate proportions (≥32.7%) and their acetate/propionate ratio (1.47:1.70) was similar to that observed in ruminants fed diets based on high-cereal concentrates [30,31]. Conversely, seaweeds harvested in spring, except *Palmaria palmata*, had acetate/propionate ratios (3.13:4.73) similar or even greater than those observed for the oat hay and barley straw used as reference, and the values were similar to those reported in forage-fed ruminants [32–34]. High variations between seaweed species in the in vitro VFA profile have also been previously observed [8,13,14].

The degradation of some amino acids produces branched-chain VFA, and therefore, they can be used as an index of protein degradation [35]. *Cladophora rupestris* and *Porphyra sp.* had the greatest N content (37.1 and 55.4 g/kg DM, respectively) and also the greatest proportions of minor VFA (calculated as the sum of isobyutyrate, isovalerate and valerate; 9.55% and 7.83%), whereas *Alaria esculenta* had the lowest proportions of minor VFA (2.47%) despite having an intermediate N content (18.2 g/kg DM). As pointed out by Hume [36], the interpretation of isoacids proportions is difficult because they are captured and used by the cellulolytic bacteria and the analyzed concentrations are the balance between the N produced from degradation and the N used by the bacteria to synthesize microbial protein in the rumen. Despite this, in our study the proportions of minor VFA were positively correlated with the N content of seaweeds (r = 0.730; p = 0.001; n = 16).The N content was also positively correlated with NH_3-N concentrations (r = 0.952; p < 0.001; n = 16), which reflects the balance between the NH_3-N produced by protein degradation and that captured by ruminal microorganisms. The NH_3-N concentrations for most of the seaweeds were above the level limiting in vitro ruminal microbial growth (5 mg/100 mL) [37], but concentrations for autumn-harvested *Alaria esculenta, Laminaria digitata, Saccharina latissima* and *Palmaria palmata* were clearly below this level (≤ 1.58 mg/100 mL), suggesting a possible limitation of microbial growth. These seaweeds had both low N content (ranging from 6.03 g/kg DM in *Saccharina latissima* to 14.6 g/kg DM in *Palmaria palmata*) and low proportions of minor VFA (1.20 in *Alaria esculenta* to 2.72% in *Saccharina latissima*), which would indicate low protein degradation. Interestingly, these seaweed samples promoted a high-propionate fermentation pattern (≥32.7% propionate), suggesting that the low NH_3-N concentrations could also have been due to a high NH_3-N capture by ruminal microorganisms, as was reported to occur in ruminants fed diets based on high-cereal concentrates [31,38].

The production of CH_4 from seaweed fermentation was highly variable, but the positive correlation observed between CH_4 and total VFA production (r = 0.881; p < 0.001; n = 16) suggests that the observed differences can be partly explained by the amount of substrate fermented, as both VFA and CH_4 derive from organic matter fermentation [8]. Several studies have investigated the possible antimethanogenic effect of marine seaweeds, with controversial results. Belanche et al. [12] observed no changes in in vitro CH_4 emissions when *Laminaria digitata* or *Ascophyllum nodosum* were included in the diet at 50 g/kg DM. However, Kinley et al. [14] and Machado et al. [39] observed an antimethanogenic effect of *Asparagopsis taxiformis* included in the diet at 20 g/kg, and Machado et al. [39] observed similar effects for a freshwater/brackish alga *Oedogonium* sp. at greater doses (>500 g/kg). The CH_4/total VFA ratio in the seaweeds (Table 4) was similar or slightly lower than that of the concentrate used as reference (10.3 mL/mmol), except for *Pelvetia canaliculata* (5.05 mL/mmol), *Porphyra sp.* (11.7 mL/mmol) and *Cladophora rupestris* (12.6 mL/mmol). The greater CH_4/VFA ratio observed in *Porphyra* sp. and *Cladophora rupestris* might be related to their high N content, as it has been shown that protein fermentation also contributes to CH_4 formation [40].

4.2. Chemical Composition and In Vitro Fermentation of Experimental Diets

The level of seaweed inclusion in the concentrates was chosen from its N content and degradability with the aim that all diets had a similar N content [7]. However, a maximum of 200 g of seaweed per kg concentrate was set up following the recommendations of Rjiba-Ktita et al. [23], who observed that inclusion levels of different seaweed species greater than 200 g/kg reduced the rate and extent of degradation of the mixture. In addition, a minimum of 93 g of soyabean meal per kg concentrate was fixed to guarantee the supply of essential amino acids (mainly lysine) for the host ruminant. *Alaria esculenta*, *Laminaria digitata*, *Pelvetia canaliculata* and *Saccharina latissima* were included as energy sources and therefore, they replaced different amounts of wheat bran, corn and wheat in the concentrate. *Mastocarpus stellatus* and *Palmaria palmata* were included as sources of both energy and protein, and therefore, they replaced different amounts of wheat bran, corn, soyabean meal and sunflower meal. Finally, *Porphyra* sp. and *Cladophora rupestris* were considered as protein sources and replaced both soyabean meal and sunflower meal.

The slightly lower N content observed in the diets including autumn-seaweed compared with those including spring-seaweed is consistent with the lower N content of the autumn seaweeds, as both spring and autumn samples of each seaweed were included in the same proportion in the diet (Table 1). The inclusion of seaweeds in the diet resulted in lower c and AGPR values than those in the control diet, which indicates that seaweeds were slower fermented than the conventional feeds (wheat, corn, soyabean meal, sunflower meal) they replaced in the concentrate. It has to be taken into account that the differences observed among diets in fermentation parameters are not only due to the inclusion of seaweeds, but also to the different proportions of each feed included in the corresponding concentrate. The diet including *Palmaria palmata* collected during autumn was the only seaweed that had c and AGPR values similar to those in the control diet, which was due to their rapid fermentation rate. As indicated by the values of the potential gas production (A), the inclusion of seaweeds in the concentrates at the level used in this study did not reduce the extent of fermentation, and in some cases (autumn-harvested *Palmaria palmata* and *Porphyra* sp. and *Cladophora* collected in both spring and autumn), even confirmed it.

The lack of negative effects of the seaweeds on the in vitro degradation of the diets was confirmed by the absence of differences among diets in total VFA production. In contrast, there were some differences among diets in the VFA profile, and acetate/propionate ratio was greater than that in the control diet for all seaweeds except *Alaria esculenta*, *Laminaria digitata*, *Saccharina latissima*, and *Palmaria palmata* collected in autumn. These results are in agreement with the low acetate/propionate ratios observed in the fermentation of these seaweeds, which was similar to those observed for ruminants fed high-cereal diets. The lack of differences among diets in NH_3-N concentrations and minor VFA proportions is in accordance with the similar N contents in all the diets and also indicates similar protein degradability in all the diets.

There were some differences among diets in CH_4 production, and the diets containing *Laminaria digitata* and *Mastocarpus stellatus* collected in autumn showed the lower values. The ratio CH_4/VFA can be used as an indicator of the efficiency of ruminal fermentation, as CH_4 is an energy loss to the host animal and VFA is used as an energy source and as substrates for the synthesis of other compounds [41]. The similar values of this ratio observed for all diets ($p = 0.569$) indicate that the observed differences in CH_4 production were mostly due to the amount of substrate fermented. The positive correlation observed between CH_4 and total VFA production ($r = 0.816$; $p < 0.001$; $n = 17$) supports this hypothesis. As discussed above, differences among diets in both CH_4 and VFA production are not only due to the inclusion of seaweeds, but also to the different feed ingredients in the concentrate. These results indicate that none of the tested seaweeds had a noticeable antimethanogenic effect.

5. Conclusions

The composition of the seaweeds was variable depending on both species and the harvesting season, with seaweeds collected in autumn having less N and ash and more polyphenols than spring-harvested seaweeds. The brown seaweeds studied are sources of energy, whereas *Porphyra sp.* and *Cladophora rupestris* are good protein sources and can be used as substitutes for conventional protein feeds. Seaweeds differed in their ruminal fermentation pattern and autumn-harvested *Alaria esculenta*,

Laminaria digitata, *Saccharina latissima* and *Palmaria palmata* were similar to conventional high-starch feeds used in ruminant feeding. The inclusion of variable levels of seaweeds in the concentrate of a diet (up to 200 g/kg concentrate) produced only subtle effects on in vitro ruminal fermentation.

Author Contributions: E.M.-A. and V.L. obtained the funding; E.M.-A. conceived the experiments; A.d.l.M. performed the in vitro trials, analyzed the samples and did data calculations; E.M.-A. and M.D.C. did the statistical analysis and wrote the draft; M.Y.R. and M.N.-G. and M.R.W. were responsible for collection and processing of seaweeds; All authors provided advice, revised progress of the manuscript and approved the final manuscript.

Funding: This research was funded by the Excellence Programme of Junta de Andalucia, Spain (grant number P12-AGR-587) and y the Research Council of Norway (grant number233682/E50).

Acknowledgments: Thanks to J. Fernandez for technical assistance.

Conflicts of Interest: The authors declare no conflict of interest.

References

1. Makkar, H.P.S. Review: Feed demand landscape and implications of food-not feed strategy for food security and climate change. *Animal* **2018**, *12*, 1744–1754. [CrossRef]
2. Makkar, H.P.S.; Tran, G.; Heuze, V.; Giger-Reverdin, S.; Lessire, M.; Lebas, F.; Ankers, P. Seaweeds for livestock diets: A review. *Anim. Feed Sci. Technol.* **2016**, *212*, 1–17. [CrossRef]
3. Makkar, H.P.S. Effects and fate of tannins in ruminant animals adaptation to tannins, and strategies to overcome detrimental effects of feeding tannin-rich diets. *Small Rumin. Res.* **2003**, *49*, 241–256. [CrossRef]
4. Roleda, M.Y.; Hurd, C.L. Seaweed nutrient physiology: Application of concepts to aquaculture and bioremediation. *Phycologia* **2019**, *58*, 552–562. [CrossRef]
5. Hanley, T.A.; Mckendrick, J.D. Potential nutritional limitations for black-tailed deer in a spruce hemlock forest southeastern Alaska. *J. Wildl. Manag.* **1985**, *49*, 103–114. [CrossRef]
6. Maehre, H.K.; Edvinsen, G.K.; Eilertsen, K.-E.; Elvevoll, E.O. Heat treatment increases the protein bioaccessibility in the red seaweed dulse (*Palmaria palmata*), but not in the brown seaweed winged kelp (*Alaria esculenta*). *J. Appl. Phycol.* **2016**, *28*, 581–590. [CrossRef]
7. Tayyab, U.; Novoa-Garrido, M.; Roleda, M.Y.; Lind, V.; Weisbjerg, M.R. Ruminal and intestinal protein degradability of various seaweed species measured in situ in dairy cows. *Anim. Feed Sci. Technol.* **2016**, *213*, 44–54. [CrossRef]
8. Molina Alcaide, E.; Carro, M.D.; Roleda, M.Y.; Weisbjerg, M.R.; Lind, V.; Novoa Garrido, M. In vitro ruminal fermentation and methane production of different seaweed species. *Anim. Feed Sci. Technol.* **2017**, *228*, 1–12. [CrossRef]
9. Roleda, M.Y.; Marfaing, H.; Desnica, N.; Jonsdottir, R.; Skjermo, J.; Rebours, C.; Nitschke, U. Variations in polyphenol and heavy meal contents of wild-harvested and cultivated seaweed bulk biomass: Health risk assessment and implication for food applications. *Food Control* **2019**, *95*, 121–134. [CrossRef]
10. Applegate, R.D.; Gray, P.B. Nutritional value of algae to ruminants. *Rangifer* **1995**, *15*, 15–18. [CrossRef]
11. Belanche, A.; Ramos-Morales, E.; Newbold, C.J. In vitro screening of natural feed additives from crustaceans, diatoms, algae and plant extracts to manipulate rumen fermentation. *J. Sci. Food Agric.* **2016**, *96*, 3069–3078. [CrossRef] [PubMed]
12. Belanche, A.; Jones, E.; Parveen, I.; Newbold, C.J. A metagenomics approach to evaluate the impact of dietary supplementation with ascophyllum nodosum or laminaria digitata on rumen function in rusitec fermenters. *Front. Microbiol.* **2016**, *7*, 299. [CrossRef] [PubMed]
13. Kinley, R.D.; Fredeen, A.H. In vitro evaluation of feeding North Atlantic stormtoss seaweeds on ruminal digestion. *J. Appl. Phycol.* **2015**, *27*, 282–289. [CrossRef]
14. Kinley, R.D.; De Nys, R.; Vucko, M.J.; Machado, L.; Tomkins, N.W. The red macroalgae Asparagopsis taxiformis is a potent natural antimethanogenic that reduces methane production during in vitro fermentation with rumen fluid. *Anim. Prod. Sci.* **2016**, *56*, 282–289. [CrossRef]
15. Bay-Larsen, I.; Risvoll, C.; Vestrum, I.; Bjorkhaug, H. Local protein sources in animal feed—Perceptions among arctic sheep farmers. *J. Rural Stud.* **2018**, *59*, 98–110. [CrossRef]
16. Prieto, C.; Aguilera, J.F.; Lara, L.; Fonolla, J. Protein and energy requirements for maintenance of indigenous Granadina goats. *Br. J. Nutr.* **1990**, *63*, 155–163. [CrossRef]

17. Goering, M.K.; Van Soest, P.J. Forage Fiber Analysis (Apparatus, Reagents, Procedures and Some Applications). In *Agricultural Handbook*; Agriculture Handbook No. 379; Agricultural Research Services: Washington, DC, USA, 1970.

18. Van Soest, P.J.; Win, R.; Moor, L. Estimation of the true digestibility of forages by the in vitro digestion of cell walls. In Proceedings of the 10th International Grassland Congress, Helsinki, Finland, 7–16 July 1966; pp. 438–441.

19. Association of Official Analytical Chemists (AOAC). *Official Methods of Analysis*, 18th ed.; AOAC International: Gaithersburg, MD, USA, 2005.

20. Mertens, D.R. Gravimetric determination of amylase-treated neutral detergent fiber in feeds with refluxing in beakers or crucibles: Collaborative study. *J. AOAC Int.* **2002**, *85*, 1217–1240. [CrossRef]

21. Julkunen-Tiito, R. Phenolics constituents in the leaves of northern willows: Methods for the analysis of certain phenolics. *J. Microbiol. Biotechnol. Food Sci.* **1985**, *33*, 213–217. [CrossRef]

22. Weatherburn, M.W. Phenol-hypochlorite reaction for determination of ammonia. *Anal. Chem.* **1967**, *39*, 971–974. [CrossRef]

23. Rjiba-Ktita, S.; Chermiti, A.; Bodas, R.; France, J.; Lopez, S. Aquatic plants and macroalgae as potential feed ingredients in ruminant diets. *J. Appl. Phycol.* **2017**, *29*, 449–458. [CrossRef]

24. Rodde, R.S.H.; Varum, K.M.; Larsen, B.A.; Myklestad, S.M. Seasonal and geographical variation in the chemical composition of the red alga *Palmaria palmata* (L.). Kuntze. *Bot. Mar.* **2004**, *47*, 125–133. [CrossRef]

25. Gaillard, C.; Bhatti, H.S.; Novoa-Garrido, M.; Lind, V.; Roleda, M.Y.; Weisbjerg, M.R. Amino acid profiles of nine seaweed species and their in situ degradability in dairy cows. *Anim. Feed Sci. Technol.* **2018**, *241*, 210–222. [CrossRef]

26. Mabeau, S.; Fleurence, J. Seaweed in food products: Biochemical and nutritional aspects. *Trends Food Sci. Technol.* **1993**, *4*, 103–107. [CrossRef]

27. Targett, N.M.; Arnold, T.M. Minireview—Predicting the effects of brown algal phlorotannins on marine.herbivores in tropical and temperate oceans. *J. Phycol.* **1998**, *34*, 195–205. [CrossRef]

28. Isac, M.D.; Garcia, M.A.; Aguilera, J.F.; Molina Alcaide, E. A comparative study of nutrient digestibility, kinetics of digestion and passage and rumen fermentation pattern in goats and sheep offered medium quality forages at the maintenance level of feeding. *Arch. Anim. Nutr.* **1994**, *46*, 37–50. [CrossRef] [PubMed]

29. Ranilla, M.J.; Lopez, S.; Giraldez, F.J.; Valdes, C.; Carro, M.D. Comparative digestibility and digesta flow kinetics in two breeds of sheep. *Anim. Sci.* **1998**, *66*, 389–396. [CrossRef]

30. Carro, M.D.; Ranilla, M.J.; Giraldez, F.J.; Mantecon, A.R. Effects of malate supplementation on feed intake, digestibility, microbial protein synthesis and plasma metabolites in lambs fed a high-concentrate diet. *J. Anim. Sci.* **2006**, *84*, 405–410. [CrossRef]

31. Carrasco, C.; Fuentaja, A.; Medel, P.; Carro, M.D. Effect of malate form (acid or disodium/calcium salt) supplementation on performance, ruminal parameters and blood metabolites of feedlot cattle. *Anim. Feed Sci. Technol.* **2012**, *176*, 140–149. [CrossRef]

32. Martinez, M.E.; Ranilla, M.J.; Tejido, M.L.; Ramos, S.; Carro, M.D. Comparison of Fermentation of Diets of Variable Composition in the Rumen of Sheep and Rusitec Fermenters: I. Digestibility, Fermentation Parameters and Efficiency of Microbial Protein Synthesis. *J. Dairy Sci.* **2010**, *93*, 3684–3698. [CrossRef]

33. Romero-Huelva, M.; Ramos-Morales, E.; Molina-Alcaide, E. Nutrient utilization, ruminal fermentation, microbial abundances, and milk yield and composition in dairy goats fed dietsincluding tomato and cucumber waste fruits. *J. Dairy Sci.* **2012**, *95*, 6015–6026. [CrossRef]

34. Romero-Huelva, M.; Ramirez-Fenosa, M.A.; Planelles-Gonzalez, R.; Garcia-Casado, P.; Molina-Alcaide, E. Can by-products replace conventional ingredients in concentrate of dairy goat diet? *J. Dairy Sci.* **2017**, *100*, 4500–4512. [CrossRef] [PubMed]

35. Wallace, R.J.; Cotta, M.A. Metabolism of nitrogen-containing compounds. In *The Rumen Microbial Ecosystem*; Elsevier Applied Science: London, UK, 1988; pp. 217–250.

36. Hume, I.D. Synthesis of microbial protein in the rumen. II. A response to higher volatile fatty acids. *Austr. J. Agric. Res.* **1970**, *21*, 297–304. [CrossRef]

37. Satter, L.D.; Slyter, L.L. Effect of ammonia concentration on rumen microbial protein production in vitro. *Br. J. Nutr.* **1974**, *32*, 199–208. [CrossRef] [PubMed]

38. Carrasco, C.; Carro, M.D.; Fuentaja, A.; Medel, P. Performance, carcass and ruminal fermentation characteristics of heifers fed concentrates differing in energy level and cereal type (corn vs. wheat). *Span. J. Agric. Res.* **2017**, *15*, e0606. [CrossRef]

39. Machado, L.; Magnusson, M.; Paul, N.A.; Kinley, R.; Nys, R.; Tomkins, N. Identification of bioactives from the red algae *Asparagopsis taxiformis* that promote antimethanogenic activity in vitro. *J. Appl. Phycol.* **2016**, *28*, 3117–3126. [CrossRef]

40. Haro, A.N.; Carro, M.D.; De Evan, T.; Gonzalez, J. Protecting protein against rumen degradation could contribute to reduce methane production. *J. Anim. Physiol. Anim. Nutr.* **2018**, *102*, 1482–1487. [CrossRef]

41. Vanegas, J.L.; Gonzalez, J.; Carro, M.D. Influence of protein fermentation and carbohydrate source on in vitro methane production. *J. Anim. Physiol. Anim. Nutr.* **2017**, *101*, e288–e296. [CrossRef]

Article

Rumen In Vitro Fermentation and In Situ Degradation Kinetics of Winter Forage Brassicas Crops

José Daza [1], Daniel Benavides [2], Rubén Pulido [3], Oscar Balocchi [1], Annick Bertrand [4] and Juan Keim [1,*]

[1] Animal Production Institute, Faculty of Agricultural Sciences, Universidad Austral de Chile, Valdivia PO Box 567, Chile; jose_daza11@hotmail.com (J.D.); obalocch@uach.cl (O.B.)
[2] Graduate School, Faculty of Agricultural Sciences, Universidad Austral de Chile, Valdivia PO Box 567, Chile; dabepez89@hotmail.com
[3] Animal Science Institute, Faculty of Veterinary Sciences, Universidad Austral de Chile, Valdivia PO Box 567, Chile; rpulido@uach.cl
[4] Soils and Crops Research and Development Centre, Agriculture and Agri-Food Canada, Québec City, QC G1V 2J3, Canada; Annick.Bertrand@AGR.GC.CA
* Correspondence: juan.keim@uach.cl; Tel.: +56-6-3229-3659

Received: 13 September 2019; Accepted: 25 October 2019; Published: 1 November 2019

Simple Summary: Winter brassica crops such as kales and swedes are used to supply feed in times of seasonal shortage. However, to the best of our knowledge, there is little information about the fermentation characteristics of these forages in the rumen. This study assessed the nutrient concentration, in vitro fermentation and in situ rumen degradation characteristics of *Brassica oleracea* (L.) ssp. *acephala* (kales) and *Brassica napus* (L.) ssp. *napobrassica* (swedes). The kales and swedes both showed different nutrient concentrations and fermented fast and extensively in the rumen. However, in vitro fermentation of swedes resulted in lower acetate and greater proportions of butyrate and propionate. Varieties of swedes showed more differences in terms of degradation and fermentation in the rumen compared to kale varieties.

Abstract: The aim of the present study was to evaluate the nutritional value, the rumen in vitro fermentation, and the in situ degradation of *Brassica oleracea* (L.) ssp. *acephala* (kales) and *Brassica napus* (L.) ssp. *napobrassica* (swedes) for winter use. Five varieties of each brassica were used in three field replicates and were randomized in a complete block nested design. All forage varieties were harvested at 210 days post-sowing to analyze the chemical composition, in vitro gas production, volatile fatty acid (VFA) production and in situ dry matter (DM) and crude protein (CP) degradability. Kales presented higher DM and neutral detergent fiber (NDF) content ($p < 0.01$), whereas swedes showed higher CP, metabolizable energy (ME), glucose, fructose, total sugars, NFC, and nonstructural carbohydrate (NSC) content ($p < 0.01$). The kale and swede varieties differed in their CP and sugar concentrations, whereas the kale varieties differed in their DM and raffinose content. The rates of gas production were higher for swedes than for kales ($p < 0.01$). No differences between the brassica species ($p > 0.05$) were observed in the total VFA production, whereas kales had a higher proportion of acetate and swedes had higher proportions of butyrate ($p < 0.05$). Only the swede varieties showed differences in VFA production ($p < 0.05$). The soluble fraction "a", potential and effective in situ DM degradability were higher in swedes ($p < 0.01$), but kales presented greater DM and CP degradation rates. Differences were observed between brassica species in the chemical composition, degradation kinetics, and ruminal fermentation products, whereas differences among varieties within species were less frequent but need to be considered.

Keywords: kale; swede; volatile fatty acids; degradation rates

---

1. Introduction

Brassicas such as kales (*Brassica oleracea* (L.) ssp. *acephala*) and swedes (*Brassica napus* (L.) ssp. *napobrassica*) are used for ruminant feed during winter [1], which is a season with low pasture growth in humid temperate regions [2]. These forages can offer high dry matter (DM) production and nutritional quality in a short time, which is related to high metabolizable energy (ME), water-soluble carbohydrates (WSC), and low neutral detergent fiber (NDF) content [3,4]. Winter brassicas have been used successfully in sheep [4], dry cows [5], and lactating dairy cows [6]. In addition, forage brassicas have an environmental advantage; they reduce the amount of enteric methane (CH_4) per unit of DM intake compared to ryegrass pasture [4,7]. Although nutrient concentrations in brassicas have been widely described, the nutritive value of these forages depends on the quantity of nutrients available to the animal, which is determined by fermentation processes [8] and the presence of secondary compounds such as glucosinolates and S-methyl-cysteine sulfoxide that are present in brassicas [9], thus meaning animal responses can be affected.

Complementary evaluation methods, such as the ruminal digestibility of nutrients or products of ruminal fermentation and metabolism, have been suggested to determine the real nutritive value of forages [10,11]. The ruminal in situ incubation technique is considered a reference method to estimate degradation parameters, when adjusted to suitable nonlinear models [12]. These parameters are used by feeding evaluation models to estimate nutritive value, nutrient supply, and animal performance [8]. On the other hand, the in vitro gas production technique (IVGPT) allows the determination of fermentation kinetics [13]; estimates of DM, protein, and fiber degradation; ruminal volatile fatty acid (VFA) content; and microbial protein synthesis [14]. The popularity of in vitro gas production (GP) stems mainly from the ability to exercise experimental control, the capacity to nondestructively screen a large number of substrates, the kinetic information obtained, and relatively low costs [10]. Thus, IVGPT offers a unique tool for researchers to address a wide range of nutritional issues in ruminants [15].

Whereas degradation kinetics and ruminal fermentation of summer brassica species (rape and turnip) and varieties [16] have been reported in the literature, few reports exist on the effect of winter brassica species (kales and swedes) and varieties on the in situ degradation kinetics and fermentation end products. For example, Sun et al. [4] have observed that sheep fed swedes showed modified VFA profiles in their rumen fluid and lowered methane yield in contrast with those fed kales or perennial ryegrass. Keogh et al. [1] have reported no effects on the rumen VFA concentration from increases in the dietary proportion of kales in the diets of dry cows. Valderrama and Anrique [17] have reported DM and crude protein (CP) degradation kinetics of kale leaves; however, to the best of our knowledge, such data have not been reported for swedes. Moreover, the nutritive value of brassicas varies among species and varieties within species [3,16], and, therefore, information is still lacking about rumen fermentation and the kinetics of winter brassica species such as kales and swedes.

Hence, the aim of this study was to determine the nutritive value of forage brassica species (kales and swedes) and varieties for winter use, based on their nutrient concentration, in vitro ruminal fermentation, and in situ rumen degradation kinetics.

2. Materials and Methods

All animal procedures were performed in accordance with the UK Animals (Scientific Procedures) Act and associated guidelines, and approved by the Animal Ethics Committee of the Austral University of Chile (approval number 144/2013).

2.1. Site and Experimental Design

This experiment was carried out at the Agricultural Research Station (39°47′ S, 73°13′ W) of the Austral University of Chile on a Typic Hapludand soil with an initial water pH of 5.8, Olsen-P of 19.1 mg/kg, exchangeable potassium of 214 mg/kg, and aluminum saturation of 3.1% (measured for the first 20 cm of the soil profile).

Prior to soil preparation and sowing, the weeds were controlled chemically with glyphosate at a dosage of 2025 g/ha of active ingredient. Two brassica species were evaluated (kales and swede), and five varieties were sown for each species: Caledonian (K1), Elba (K2), Sovereign (K3), Regal (K4), and Coleor (K5) for the kales and Major Plus (S1), Aparima Gold (S2), Highlander (S3), Dominion (S4), and Invitation (S5) for swedes. The plot sizes were 6 m by 4 m, with three replicates for each variety, and plots were arranged in field blocks. The varieties were established in October 2014 at a seed dosage of 4.0 (kales) and 1.5 kg/ha (swedes). Fertilizers were applied at sowing to correct any soil nutrient deficiencies. A fertilizer mixture (7 % N-30 % P_2O_5 -12 % K_2O) at doses of 500 kg/ha (35 kg N/ha, 150 kg P_2O_5/ha, and 60 kg K_2O/ha) and 46 kg/ha boronatrocalcite were applied at sowing. After emergence, weeds were controlled chemically by applying Lontrel 3A (clopyralid 475 g of active ingredient (a.i.)/L) and Tordon 24 k (picloram 240 g a.i./L) at doses of 300 and 200 cc/ha respectively. Five weeks after sowing, when the plants had two or three leaves, 125 kg N/ha (urea) was applied, and in January 2015, applications of 750 cc/ha of Aramo (tepraloxydim 200 g a.i./L) and 200 cc/ha of karate (50 g a.i./L Lambda-cyhalothrin) were made.

During the trial, five cuts were made, with an approximate interval of 30 days between cuts, with the first harvest occurring at 90 days after plant emergence. In each cut, 4 m^2 of each crop was harvested. The kale varieties were cut to 20 cm above the ground level, the swede varieties were collected manually, and soil attached to the roots was removed. Plants were weighed and then separated into the main components (leaf and stem for kales and leaf and bulb for swedes). The samples were then dried in a forced-air oven at 60 °C for 48 h for determination of the dry matter (DM).

Samples for nutrient concentrations and in vitro and in situ incubations were harvested at 210 days post-sowing, and the plants were separated into their morphological components (leaf and stems for kale varieties and leaf and bulbs for swede varieties) before being chopped and then being frozen at −20 °C. Later, they were lyophilized (Virtis 10-45 MR-BA, Gardiner, New York, NY, USA) and then ground (Wiley mill, 158 Arthur H. Thomas, Philadelphia, Pennsylvania, PA, USA) to 5 mm for in situ incubations and to 1 mm for nutrient concentrations analyses and in vitro gas production. For the in vitro and in situ incubations, samples of brassica species were composed of a leaf to stem ratio of 35:65 for kales and a leaf to bulb ratio 30:70 for swedes. The ratios were the average proportions of organ components at harvest obtained in this study.

2.2. In Situ Incubations

Three dry Holstein-Friesian cows (one for each block) fitted with ruminal cannulas (4′ Pliable Rumen Cannula w/Stopper and U Bolt, Ankom Technologies, Macedon, New York, NY, USA) were used. At the time of the experiment, the ruminal pH (6.55 ± 0.32) of each cow was measured. Cows were offered grass silage (7.5 kg DM), summer turnips (4.5 kg DM) and commercial concentrate (2.0 kg DM). Samples of each variety were incubated in duplicate (~4 g DM) in Dacron bags (10 cm by 20 cm; pore size of 40–60 μm) and sealed. Up to 20 bags were deposited inside a lingerie bag (30 cm by 40 cm in size). Brassica samples from each block were incubated in a different cow and a control sample (commercial concentrate) was incubated in each cow to evaluate cow-to-cow variation.

Prior to incubation in the rumen, the bags were soaked in warm water (40 °C) for 20 min. Nine incubation times were considered: 0, 2, 4, 8, 10, 12, 14, 24, and 48 h. The samples corresponding to 0 h were not introduced into the rumen and were used to determine the soluble fraction. After the incubation, the bags containing the residue were removed from the rumen and were washed under running cold water until no further color appeared; then, they were frozen at −20 °C for 24 h to stop fermentative activity. Thereafter, the bags were defrosted, thawed in water at 4 °C, and washed with a commercial washing machine for 30 min at a "normal" wash setting. Finally, residues were oven-dried at 60 °C for 48 h. The residues were weighed, the DM was calculated by placing the samples in an oven at 105 °C for 12 h, and the CP concentration was determined to calculate nutrient loss.

A correction for small particle loss was made as follows. The samples of each brassica variety (1.5 g) were weighed in a beaker. Then, 40 mL of tap water was added and the mix was stored at room temperature (20 °C) for 1 h; afterward, the mix was filtered through a nitrogen-free filter paper and washed eight times with 20 mL of water. The residues were oven-dried and analyzed individually. Degradation parameters (a, b, and c), potential degradation (PD) and effective degradability (ED) were corrected according to Hvelplund and Weisbjerg [18].

2.3. In Vitro Incubations

Duplicates (1 g) of each sample were incubated in 160 mL glass bottles. Each brassica variety, a control (commercial concentrate) sample and two blanks (bottles without substrate) were used. Subsequently, 85 mL of Goering-Van Soest medium and 4 mL of reducing agent (NaOH 2.5 mM and cysteine-HCl 2.5 mM) were added at 39 °C under continuous gasification (CO_2) to maintain anaerobic conditions and the bottles were covered with rubber stoppers and aluminum seal.

The inoculum was extracted from two dry Holstein-Friesian cows with ruminal cannulas with a live weight (PV) of 560 ± 20 kg and a ruminal pH of 6.6 ± 0.53; at the time of the extraction, the animals were offered the same diet as in the in situ trial. Rumen fluid was obtained before the cows were fed in the morning and was stored in a thermos flask to preserve the temperature until being transferred to the laboratory. Once there, the fluid was filtered through four layers of cheesecloth while being maintained at a temperature of 39 °C and under a constant flow of CO_2. Rumen fluids from the two donor cows were mixed in equal proportions and then inoculated (10 mL) into the bottles. After inoculation, the bottles were placed in a water bath at 39 °C under continuous horizontal movement at 50 rpm.

Once the rumen fluid was inoculated, the initial gas was extracted from the bottles. The gas pressure in the headspace of the bottles above atmospheric pressure was measured manually with a pressure transducer (PCE Instruments, Tobarra, Albacete, Spain) at 2, 3, 4, 5, 6, 8, 10, 12, 18, 24, 36, and 48 h, and the volume of gas produced was measured by extraction using syringes connected through a three-way Luer valve from the bottles until the visual display of the transducer read zero, and once the volume of gas produced was recorded, it was eliminated. Fermentations were stopped after 48 h by placing the bottles on ice. Each field block (five varieties of kales and five varieties of swedes) was incubated at different runs. Thus, the first block was incubated at a first run, those corresponding to block number two were incubated at the second run, and samples from block three were incubated at the third run. A control standard (commercial concentrate) was incubated at each incubation run to control the day-to-day variation.

Once the in vitro incubation was finished, the samples were kept on ice to stop fermentative processes and residue duplicates from each sample were collected and then centrifuged at 15,000× *g* and 4 °C. After centrifugation, 0.9 mL of the supernatant was extracted to determine the VFA concentrations with a GG-2010 gas chromatograph (Shimadzu Corporation, Kyoto, Japan).

2.4. Analyses

The dry matter content was measured by weighing the samples before and after drying with a forced-air oven, initially at 60 °C for 48 h and then at 105 °C for 12 h. The CP concentration was determined by combustion (Leco Model FP-428, Nitrogen Determinator, Leco Corporation, St Joseph, Minnesota, MI, USA) based on the DUMAS method (nitrogen × 6.25); digestible organic matter on a dry matter basis (DOMD) was measured according to Tilley and Terry [19]; neutral detergent fiber (aNDF) was measured by using a heat stable amylase [20]; and ash and ether extract (EE) were analyzed according [21] (Methods ID 942.05 and ID 920.39 for ash and EE respectively). Sugars (raffinose, sucrose, glucose, and fructose) were analyzed by Waters ACQUITY ultrahigh-performance liquid chromatography (UPLC, Waters, Milford, Massachusetts, MA, USA), and starch quantification was determined by colorimetric detection of non-soluble residues after enzymatic digestion with amyloglucosidase according to Pelletier et al. [22]. The sum of sugars and starch yielded the content of total nonstructural carbohydrates (NSC). An estimation of the combined organic acids plus neutral detergent soluble fiber (OA + NDSF) was calculated according to Hall et al. [23], where OA + NDSF =

NFC − NSC. Non-fibrous carbohydrates were calculated as follows: NFC = 100 − CP − aNDF − EE − ash + neutral detergent insoluble crude protein (NDICP).

2.5. Calculations

The in situ disappearance of DM and CP was determined using the non-linear model described by Ørskov and McDonald [12] to determine the potential degradation according to the exponential model

$$PD = a + b \times (1 - e^{-kt})$$

where a is the soluble fraction (fraction washed out at $t = 0$; this value resulted from the incubation of 0 h bags corrected for particle loss and fixed into the model), b is the insoluble but potentially degradable fraction, k is the degradation rate (per hour), and t is the time (h).

The effective degradability (ED) was calculated assuming a fractional passage rate (kp) of 2%, 5%, and 8% per hour according to the following equation:

$$ED = a + b \times c/(c + kp)$$

These parameters were corrected for the losses of small particles which are degraded in a similar way to the particles remaining in the bag, as reported by Hvelplund and Weisbjerg [18].

After correcting for gas production of the blanks, the obtained GP data were adjusted to the generalized Michaelis-Menten model without a lag phase [11], as seen in the equation

$$GP = A \times [T^n/(T^n + K^n)]$$

where GP is the gas production at time T, A is the asymptote of GP (mL), n is the determined value of the shape of the curve, and K is the time taken to produce half of A.

The following parameters were calculated according to Groot et al. [24] and France et al. [11], i.e.,

fermentation rate at half-life (C) = $n/(2 \times K)$
maximal fermentation rate (MDR) = $(n - 1)^{((n-1)/n)/k}$
time to ferment x% of the substrate (t_x) = $K \times ((X/(1 - X))^{(1/n)})$

where X = 0.25, 0.75, and 0.90 of A.

2.6. Statistical Analyses

Parameters of in vitro GP, in situ degradation kinetics, VFA content and nutrient concentration were averaged for analytical replicates and analyzed with the MIXED procedure of SAS (SAS Institute, Cary, North Carolina, NC, USA).

Data were analyzed under a nested design, with three replicates organized in complete randomized blocks. Varieties were nested within species and the random effect of the field replicate was included as a block. When significant differences ($p < 0.05$) were found, the Tukey-Kramer multiple-comparison test was used in the LSMEANS procedure statement in SAS.

3. Results

3.1. Nutrient Concentration and Sugar Profile

Kales had greater concentrations of DM, EE (+4 g/kg), and aNDF (+123 g/kg) than swedes ($p < 0.01$; Table 1). Swedes showed greater CP (+25 g/kg; $p < 0.01$) and total sugar concentrations (+75 g/kg; $p < 0.01$) than kales, as well as individual sugars, such as glucose and fructose, NSC, OA + NDSF and DOMD. Raffinose and sucrose concentrations were greater in kales ($p < 0.01$). The ash and starch concentrations did not vary between the species ($p > 0.05$).

The kale varieties differed in their DM, CP, EE, raffinose, glucose, fructose, sugars, starch, and NSC concentrations ($p < 0.01$). Coleor had greater concentrations of CP (+30 g/kg) than Regal; Elba and Sovereign greater EE than Coleor; Coleor greater raffinose than Coledonian and Sovereign; and Regal and Coledonian greater concentrations of glucose and fructose than Sovereign. Finally, Regal showed greater concentrations of total sugars than Elba and Sovereign; Sovereign and Coleor greater concentrations of starch than Coledonian; and Regal the greatest NSC concentration of all varieties (231 g/kg), whereas Elba and Sovereign had the lowest (190 g/kg and 194 g/kg, respectively).

Furthermore, the swede varieties differed in their CP, glucose, fructose, sugars, starch, and NSC concentrations ($p < 0.01$). The concentrations of ash, aNDF, sucrose, OA + NDSF, and DOMD for varieties of both species were not different ($p > 0.05$). Invitation showed greater concentrations of CP (+30 g/kg) than Aparima Gold. Major Plus and Highlander showed the greater concentration of glucose, fructose, and sugars (280 g/kg and 288 g/kg, respectively, compared to 233 g/kg, 241 g/kg, and 234 g/kg in Aparima Gold, Dominion and Invitation). For starch, Aparima Gold and Dominion had greater concentrations (32 g/kg and 26 g/kg, respectively) than Highlander (7 g/kg). Finally, the concentrations of NSC in Major Plus and Highlander (291 g/kg and 295 g/kg, respectively) were greater than that for Invitation.

3.2. In Situ Degradation Parameters

Dry matter degradation parameters differed between brassica species (Table 2) except for the insoluble but potentially degradable fraction "b" ($p > 0.05$). Swedes had a higher soluble fraction "a" compared to kales (591 g/kg and 499 g/kg, respectively; $p < 0.01$) but a lower degradation rate "c" (0.25 h^{-1} and 0.34 h^{-1}, respectively; $p < 0.01$). However, swedes showed greater PD (+90 g/kg) and ED with ruminal passage rates of 2%, 5%, and 8% per hour compared to kales ($p < 0.01$). No significant differences were found in the in situ degradation parameters for the kale varieties ($p > 0.05$). Within the swede varieties, fraction "a" was greater for Major Plus (634 g/kg) compared to Aparima Gold (545 g/kg), Dominion (581 g/kg), and Invitation (586 g/kg; $p < 0.01$).

For the in situ CP degradation parameters, a species effect was observed on the fractional degradation rate "c", with kales having a faster degradation rate compared to swedes (0.48 h^{-1} and 0.36 h^{-1}, respectively; $p < 0.01$), whereas no brassica species effect was observed on the other in situ CP degradation parameters with average values of 559 g/kg, 380 g/kg, and 939 g/kg for "a", "b", and PD, respectively ($p > 0.05$). Fractions "a" and "b" were affected by the kale varieties: Coleor presented a higher "a" than Regal and Caledonian (+159 g/kg and +139 g/kg, respectively; $p < 0.01$). On the other hand, Caledonian and Regal had a greater "b" fraction compared to Coleor ($p < 0.05$). Effective degradability at 8% per hour of Sovereign (904 g/kg) and Coleor (900 g/kg) were higher than that of Regal (839 g/kg; $p < 0.05$). No effects for the swede varieties were found for any of the CP degradation parameters ($p > 0.05$).

Table 1. Nutrient concentration (g/kg dry matter) of five varieties of kale (from K1 to K5) and swedes (from S1 to S5).

Parameter	Kales	Swedes	[1] SEM	Kales					Swedes					[1] SEM	p Value	
				K1	K2	K3	K4	K5	S1	S2	S3	S4	S5		Species	Variety
DM (g/kg sample)	122	74	0.3	108 [b]	123 [ab]	135 [a]	107 [b]	136 [a]	71	81	72	71	77	0.6	<0.01	<0.01
Ash	79	77	0.2	88	83	76	73	74	74	75	73	81	82	0.5	NS	NS
CP	111	136	0.4	114 [ab]	111 [ab]	116 [ab]	88 [b]	128 [ab]	127 [ab]	123 [b]	124 [ab]	151 [ab]	153 [a]	0.9	<0.01	<0.01
EE	13	9	0.04	13 [ab]	14 [a]	14 [a]	13 [ab]	10 [b]	10	9	9	9	9	0.09	<0.01	<0.05
aNDF	300	176	0.7	271	309	328	297	293	169	196	165	169	182	1.6	<0.01	NS
Raffinose	7	2	0.4	6 [b]	7 [ab]	5 [b]	7 [ab]	10 [a]	2	1	2	2	2	0.9	<0.01	<0.01
Sucrose	59	15	4.6	47	52	75	50	70	11	17	12	16	22	10.4	<0.01	NS
Glucose	63	139	3.1	74 [ab]	59 [bc]	44 [c]	83 [a]	56 [bc]	157 [a]	127 [b]	156 [a]	123 [b]	127 [b]	6.8	<0.01	<0.01
Fructose	51	99	3.0	58 [ab]	50 [abc]	35 [c]	69 [a]	42 [bc]	109 [a]	88 [b]	115 [a]	100 [ab]	82 [b]	6.6	<0.01	<0.01
Sugars	180	255	5.3	185 [ab]	168 [b]	160 [b]	209 [a]	179 [ab]	280 [a]	233 [b]	288 [a]	241 [b]	234 [b]	11.9	<0.01	<0.01
Starch	25	19	3.7	14 [b]	21 [ab]	34 [a]	22 [ab]	31 [a]	11 [bc]	32 [a]	7 [c]	26 [ab]	20 [abc]	8.4	NS	<0.05
NSC	205	274	7.1	199 [ab]	190 [b]	194 [b]	231 [a]	210 [ab]	291 [a]	265 [ab]	295 [a]	267 [ab]	253 [b]	15.8	<0.01	<0.05
OA + NDSF	292	327	5.4	315	293	272	298	284	328	331	332	322	321	12.2	<0.01	NS
DOMD	727	831	0.7	740	715	698	732	749	841	823	841	828	824	1.7	<0.01	NS

[a, b, c] within a row, different letters represent the significant differences at p value < 0.05. [1] SEM standard error of the mean; DM, dry matter; CP, crude protein; EE, ether extract; aNDF, neutral detergent fiber with a heat stable amylase; NSC, non-structural carbohydrates, representing the sum of sugars and starch; OA + NDSF, organic acids and neutral detergent soluble fiber; DOMD, digestible organic matter on DM basis. Kale varieties: K1, Caledonian; K2, Elba; K3, Sovereign; K4, Regal; K5, Coleor. Swede varieties: S1, Major Plus; S2, Aparima; S3, Highlander; S4, Dominion; S5, Invitation.

Table 2. In situ ruminal digestion kinetics, potential degradability (PD) and effective degradability (ED, considering a passage rate of 2%, 5%, and 8% per hour) of five varieties of kale (from K1 to K5) and swedes (from S1 to S5).

Parameter	Kales	Swedes	[1] SEM	Kales					Swedes					[1] SEM	*p* Value	
				K1	K2	K3	K4	K5	S1	S2	S3	S4	S5		Species	Variety
							Dry matter degradation parameters									
a	499	591	0.6	519	499	479	504	491	634 [a]	545 [b]	612 [ab]	581 [b]	586 [b]	1.3	<0.01	<0.01
b	370	368	0.6	363	374	374	376	363	349	387	352	367	384	1.4	NS	NS
c	0.34	0.25	0.02	0.35	0.32	0.37	0.30	0.35	0.22	0.29	0.28	0.22	0.22	0.05	<0.01	NS
PD	869	959	0.6	882	873	853	881	854	982	931	964	964	969	1.9	<0.01	NS
ED, 2%/h	846	931	0.8	860	849	831	855	835	952	907	940	919	936	1.7	<0.01	NS
ED, 5%/h	817	895	0.7	831	819	802	823	809	915	875	910	881	896	1.5	<0.01	NS
ED, 8%/h	793	867	0.7	807	794	779	797	786	886	848	885	851	864	1.5	<0.01	NS
							Crude protein degradation parameters									
a	550	568	1.4	494 [bc]	554 [abc]	596 [ab]	474 [c]	633 [a]	580	532	520	594	615	3.2	NS	<0.01
b	383	377	1.4	429 [a]	387 [ab]	354 [ab]	431 [a]	313 [b]	384	398	390	361	353	3.1	NS	<0.05
c	0.48	0.36	0.02	0.53	0.42	0.53	0.43	0.47	0.34	0.40	0.41	0.31	0.32	0.05	<0.01	NS
PD	933	945	1.1	923	941	950	906	946	964	930	911	955	968	2.5	NS	NS
ED, 2%/h	918	927	1.0	909	924	937	888	933	943	912	898	935	947	2.3	NS	NS
ED, 5%/h	897	900	0.9	887	900	920	862	916	915	886	873	906	921	2.1	NS	NS
ED, 8%/h	878	877	0.9	867 [ab]	879 [ab]	904 [a]	839 [b]	900 [a]	890	864	852	882	898	2.0	NS	<0.05

[a,b,c] within a row, different letters represent the significant differences at *p* value < 0.05. [1] SEM standard error of the mean; a, soluble fraction; b, insoluble but potentially degradable fraction; c, fractional disappearance rate constant at which b is degraded. Kale varieties: K1, Caledonian; K2, Elba; K3, Sovereign; K4, Regal; K5, Coleor. Swede varieties: S1, Major Plus; S2, Aparima; S3, Highlander; S4, Dominion; S5, Invitation.

3.3. In Vitro Fermentation Products

The in vitro GP parameters were affected by the brassica species, whereas for varieties within species no effect was observed (Table 3). The GPs at 24 h and 48 h and "A" were higher for swedes compared to kales (256 mL/g, 285 mL/g, and 285 mL/g against 233 mL/g, 260 mL/g, and 261 mL/g DM, respectively; $p < 0.01$). Fermentation rate parameters "C" and "MDR" (h^{-1}) were slightly faster for swedes (0.15 h^{-1}) compared to kales (0.14 h^{-1}; $p < 0.05$). However, no differences were observed for the time to fermentation of 25%, 50%, and 75% of the substrate (h), whereas 90% of the substrate was fermented 1.1 h earlier for swedes ($p < 0.05$).

Table 3. In vitro gas production kinetics of five varieties of kales (from K1 to K5) and swede (from S1 to S5).

Parameter	Kales	Swedes	[1] SEM	*p* Value	
				Species	Variety
24 h GP	232	256	3.0	<0.01	NS
48 h GP	260	285	3.1	<0.01	NS
A	261	285	3.3	<0.01	NS
K	6.1	6.1	0.08	NS	NS
C	0.14	0.15	0.01	<0.05	NS
MDR	0.14	0.15	0.01	<0.05	NS
t25	3.2	3.2	0.04	NS	NS
t75	11.8	11.5	0.2	NS	NS
t90	22.6	21.5	0.5	<0.05	NS

[1] SEM standard error of the mean; 24 h GP (mL/g DM), gas production after 24 h of incubation; 48 h GP (mL/g DM), gas production after 48 h of incubation; A, asymptotic gas production (mL/g DM); K, time to ferment 50% of the substrate (h); C, degradation rate at half-life (per h); MDR, maximal degradation rate (per h); t25, t75, and t90, time to ferment 25%, 75%, and 90% of the substrate, respectively (h).

For the total volatile fatty acids (tVFA) concentrations, the propionate and branched-chain VFA (BCVFA) molar proportions of tVFAs showed no effects for brassica species ($p > 0.05$; Table 4). The acetate molar proportions of tVFAs (+48 mmol/mol) and A:P ratio were greater for kales, whereas the fermentation of swedes increased the molar proportions of butyrate (+33 mmol/mol). Kale varieties showed no differences for tVFA and the relative proportions of the different VFAs, whereas the tVFAs and the relative proportion of acetate and butyrate were affected by swede varieties. Major Plus showed a higher concentration of tVFAs (61.3 mM) compared to Dominion (45.5 mM), and Major Plus, Aparima Gold, and Invitation had higher relative proportions of acetate (555 mmol/mol, 564 mmol/mol, and 572 mmol/mol, respectively) compared to Dominion, which showed a higher relative proportion of butyrate (162 mmol/mol) than Aparima Gold and Invitation (128 mmol/mol and 120 mmol/mol, respectively).

Table 4. In vitro rumen volatile fatty acids (VFA) of five varieties of kale (from K1 to K5) and swede (from S1 to S5).

Parameter	Kales	Swedes	[1] SEM	Kales					Swedes					[1] SEM	p Value	
				K1	K2	K3	K4	K5	S1	S2	S3	S4	S5		Species	Variety
VFA concentration																
Total VFA (mM)	52.5	51.2	1.7	52.6	49.6	49.1	56.9	54.5	61.3 [a]	49.1 [ab]	49.6 [ab]	45.5 [b]	50.4 [ab]	3.8	NS	<0.05
Acetate (mmol/mol)	597	549	4.2	595	591	597	597	602	555 [ab]	564 [ab]	537 [bc]	520 [c]	572 [a]	7.2	<0.05	<0.01
Propionate (mmol/mol)	245	258	1.9	245	249	243	243	244	255	254	266	258	257	4.3	NS	NS
Butyrate (mmol/mol)	106	139	2.3	107	104	106	111	103	141 [ab]	128 [b]	143 [ab]	162 [a]	120 [b]	4.9	<0.01	<0.01
BCVFA (mmol/mol)	52	54	1.5	53	55	54	48	50	48	54	54	60	52	2.7	NS	NS
A:P ratio	2.4	2.1	0.04	2.4	2.4	2.5	2.5	2.5	2.2	2.2	2.0	2.0	2.2	0.09	<0.01	NS

[a,b,c] within a row, different letters represent the significant differences at *p* value < 0.05. [1] SEM standard error of the mean; BCVFA, branched-chain VFA (isobutyrate + isovalerate + valerate); A:P ratio, acetate to propionate ratio. Kale varieties: K1, Caledonian; K2, Elba; K3, Sovereign; K4, Regal; K5, Coleor. Swede varieties: S1, Major Plus; S2, Aparima; S3, Highlander; S4, Dominion; S5, Invitation.

4. Discussion

4.1. Nutrient Concentration

The current study reports the nutritive value of five varieties of two winter forage brassica species that are used to complement the feeding base used in grazing systems when pasture availability is not sufficient to fulfill livestock requirements. Forage brassicas have a higher demonstrated DOMD than perennial pastures; Sun et al. [4], for example, have reported that the DOMD of kales and swedes (883 g/kg and 918 g/kg DM) is 25% higher than that in perennial ryegrass pastures (703 g/kg DM). However, DOMD values observed by the current study are lower than those previously reported for kales and swedes (727 g/kg and 831 g/kg DM).

Sun et al. [4] and Westwood and Mulcock [3] have observed that kales contain higher values of DM, EE, and aNDF than swedes, which is similar to the effect observed in the present study. Regarding CP, swedes showed a higher concentration (136 g/kg) compared to kales (111 g/kg). Westwood and Mulcock [3] have reported similar values for swedes (137 g/kg) but lower values for kales (97 g/kg); however, they have observed that CP is highly variable among kale varieties, ranging from 63 g/kg to 138 g/kg, which is in accordance with our study where the CP concentration of the kale varieties ranged from 88 g/kg to 128 g/kg. These differences are mainly associated with the leaf to stem ratio, as forage brassicas concentrate more CP in their leaves [25]. For example, Valderrama and Anrique [17] have reported CP concentrations of 225 g/kg in kale leaves.

The content of sugars was similar to that reported by Sun et al. [4], with swedes having a higher concentration of sugars compared to kales. Winter brassicas have higher sugars compared to grass-based permanent pastures and perennial ryegrass, where values range from 73 g/kg to 118 g/kg when harvested during winter and early spring, the times when winter brassicas are used [26,27].

The amount of raffinose and sucrose were higher in kales; however, these amounts were not enough to compensate for the higher concentration of fructose and glucose in swedes. Importantly, varieties of both kales and swedes differed in the concentrations of fructose and glucose. In the case of kales, the Regal concentrated almost twofold the amounts of glucose and fructose compared to Sovereign, whereas for swedes, the Major Plus and Highlander concentrated approximately 20 g/kg and 30 g/kg more than the other three varieties. This approach is important, as it considers the type and amount of sugars demonstrated to affect fermentation in the rumen and the end products such as the VFAs [28].

Additionally, the concentration of OA + NDSF was evaluated, and this was higher for swedes (327 g/kg DM) than for kales (291 g/kg DM). This parameter was evaluated in summer brassicas by Keim et al. [16], who found it to be 313 g/kg and 314 g/kg DM for turnip and rape, respectively. Neutral detergent soluble fiber is mainly composed of galactans, β-glucans, soluble hemicelluloses, and pectic substances [23], while organic acids (OA) are mainly carbohydrate derivatives and may contain lactate, citric acid cycle components and secondary plant compounds such as oxalate and shikimate [23]. Previous studies have reported concentrations of pectins between 77 g/kg and 129 g/kg DM for brassicas in general [29]. Likewise, Sun et al. [4] have reported 80 g/kg and 69 g/kg DM for kales and swedes, respectively. In addition to sugars, these components of the NDSF could have an effect on ruminal fermentation and VFA concentrations, and although NDSFs are readily and extensively broken down in the rumen, they do not mimic the pH-lowering effect of starch because they generally produce little or no lactate, and their fermentation ceases at low pH [30]. On the other hand, organic acids do not support microbial growth [23] and therefore have little impact on rumen metabolism.

4.2. In Situ Degradation Parameters

Brassica forages are characterized by their high concentration of readily fermentable carbohydrates and high digestibility [9], which is in accordance with the high soluble fraction and fast degradation rates observed in our study for both species. To the best of our knowledge, limited data exist from reports on degradation kinetics of kales [17] and no published literature has been found regarding the degradation kinetics of swedes. The greater soluble fraction observed in swedes is in accordance with its greater concentration of sugars and nonstructural carbohydrates compared to kales. Both, kales and swedes showed a fast degradation rate and extensive degradation of DM and CP compared with other forages used for livestock feeding during winter such as grass pastures, silages, and hay [31,32], and they showed a similar potential degradability but faster degradation rate compared to other winter fodder crops, such as fodder beets [33], and summer brassicas, such as turnips and forage rape [16]. The faster degradation rates observed in our study compared to those observed by Keim et al. [16] in summer brassicas and the values reported by Valderrama and Anrique [17] for kales might be explained by the composition of the diet offered to the cannulated cows. In this experiment, cows were offered brassicas in their diet, and, therefore, rumen microbes might have been adapted to degrade these kind of ingredients, which would have increased the fermentation rate [34].

Even though kales showed a faster degradation rate of crude protein than swedes, due to the similar soluble (a) and insoluble but potentially degradable (b) fractions, PD and ED were similar among species. The crude protein ED of kales was similar to the values reported by Valderrama and Anrique [17] when calculated at a passage rate of 2% per hour and was slightly greater with passage rates of 5% and 8% per hour because of the fast degradation rate observed in our study (0.48 h^{-1}). Importantly, the CP fractions "a" and "b" differed among the kale varieties and need to be considered for the ration formulation models. This variation in degradation characteristics associated with genetic differences has been reported previously by Sun et al. [35] with perennial ryegrass. The varieties with greater CP concentration (Coleor) also showed the highest soluble fraction of crude protein.

4.3. In Vitro Fermentation

Higher 24 h, 48 h, and asymptotic GP in swedes reflects greater digestibility compared to kales, since total GP is an indicator of forage digestibility [36] and has been related to DM degradability [37], which is in accordance with the greater DM potential and effective degradability observed for swedes compared with kales. This difference may be attributed to the chemical composition, as forages with lower aNDF content and higher NSC, such as swedes, present greater GP [38]. As observed for DM potential and effective in situ degradability, the in vitro GP parameters were not affected by varieties within species, as a good correlation between the GP measurements and in situ degradability has been found [39]. Contrary to what was observed with in situ DM degradation rates (faster for kales than swedes), in vitro gas production rates (C and MDR) were faster for swedes. This is because in vitro gas production rates take into account the gas produced from the soluble fraction, whereas the in situ degradation rate comes only from the insoluble but potentially degradable fraction. Nevertheless, C and MDR values of both swedes and kales (0.14 and 0.15, respectively) demonstrated faster fermentation compared to other typical feedstuffs used for ruminant feeding during winter, such as concentrate (0.11 h^{-1} for both c and MDR), hay (0.03 and 0.05 h^{-1} for c and MDR), silage (0.07 and 0.08 h^{-1} for c and MDR) and pasture (0.09 and 0.10 h^{-1} for c and MDR) [40].

Fast fermentation of both kales and swedes is reflected by parameters t25, k, t75, and t90, where, for example, 90% of both species ferment in less than 23 h, mainly due to the presence of readily fermentable carbohydrates such as sugars and NDSF. Hall et al. [41] have shown that fermentation of NDSF is faster than fermentation from NDF, indicating that forages high in NDSF tend to exhibit a rapid fermentation.

In contrast to the GP, the total VFA production was similar among species. This finding could be because the digested OM is fermented to VFA and GP or converted to microbial protein and therefore total VFA and GP are not always well correlated [42]. Total VFA production was different across swede varieties and related to the in situ DM soluble fraction; that is, the Major Plus variety presented the greater soluble fraction and tVFA production, whereas Dominion showed the lowest values for both tVFA and DM soluble fraction. The acetate and acetate to propionate ratio were greater for kale, which is related to the high aNDF and lower NSC concentrations in kales compared to swedes, as has been reported previously [27,42]. In addition, the highest proportion of butyrate found in swedes coincides with the higher sugar concentration, since the fermentation of sugars generally increases the amount of butyrate in the rumen [28]. The greater acetate and lower butyrate proportions of tVFA for kales compared with swedes are in agreement with in vivo studies with sheep, where acetate was reduced from 515 mmol/mol to 412 mmol/mol, while butyrate was increased from 118 mmol/mol to 176 mmol/mol for sheep fed swedes compared with those fed kales [4].

Kale varieties had no effect on VFA production and the relative proportions of VFAs, whereas differences in the acetate and butyrate relative proportions of tVFA were observed for varieties of swedes. The varieties with greater butyrate relative proportions of tVFA were those with greater sucrose and fructose concentrations as well as a higher DM soluble fraction.

4.4. Implications

Although nutrient concentrations of winter brassicas and variations among varieties have been widely described [3] and their use in sheep [4], dry cows [5] and lactating dairy cows [6] have been reported, to the best of our knowledge few studies have evaluated the rumen fermentation processes of winter brassicas.

The main products of rumen fermentation are VFAs, and among these, propionate is a substrate for gluconeogenesis and is the main source of glucose in the animal, whereas the non-glucogenic acetate and butyrate are sources for long-chain fatty acid synthesis [43]. Glucogenic and lipogenic nutrient supply and VFA profile have been associated with animal energy balance. It has been suggested that in lactating cows increased energy intake is channeled largely through increases in rumen production of butyric and propionic acids and their yield of ATP to the host animal [43,44]. From an environmental point of view, production of acetate and butyrate liberates hydrogen, whereas propionate serves as a net hydrogen sink [45]. Consequently, diets that increase propionate and decrease acetate in the rumen are often associated with a reduction in ruminal CH_4 production [46].

Therefore, the results we obtained for in vitro fermentation end products may lead us to infer some animal responses. However, these extrapolations must be done carefully, because the mechanisms governing microbial efficiency and VFA molar proportions in vitro are not necessarily valid in vivo. For example, in the rumen itself, feed and microbial biomass are subject to passage and VFA subject to passage and absorption [10], processes that do not occur under in vitro conditions.

One of our main findings is that swedes increased the relative molar proportion of butyrate at the expense of acetate compared with kales, however there are differences among varieties of swedes that must be considered and depend basically on their sugar concentration and type of sugars. There is generally a positive association of feeding sugars, such as sucrose, and increased milk and milk fat production, which may be due to the greater molar proportion of butyrate produced from those sugars [44,47]. However, to the best of our knowledge no studies have reported milk production responses of dairy cows fed swedes. In vivo results reported by Sun et al. [4] are similar to our study, with total rumen VFA concentrations being similar for sheep fed swedes or kales and greater butyrate and lower relative molar proportions of acetate found for sheep fed swedes. Complementarily, dry matter intake (DMI) was greater when kales were offered but energy intake was greater when feeding swedes resulting in lower methane emissions; however, no animal responses were reported. Conversely, Keogh et al. [6] have observed no differences in body weight and body condition score change between pre-calving dairy cows supplemented with kales or swedes during the dry period, but rumen fermentation data was not shown in their work.

Finally, the in situ ruminal degradation parameters of kales and swedes generated in this study can be used by researchers and nutritionists in feeding evaluation models to estimate the nutritive value, nutrient supply, and animal performance of livestock fed with winter brassicas.

5. Conclusions

Relative to winter forage brassica crops, our study demonstrated high digestibility for ruminant feeding in times when pasture availability is low. Differences between the two species were observed in terms of chemical composition, gas production, and dry matter degradation parameters, with swedes exhibiting a faster and more extensive degradation due to their greater concentrations of readily fermentable carbohydrates and lower NDF, which resulted in a lower acetate to propionate ratio and greater butyrate concentrations in the rumen. Additionally, differences among varieties within species were observed and must be considered when selecting certain varieties for use. For example, varieties of swedes showed differences regarding their in situ DM soluble fraction and in vitro tVFA and acetate and butyrate relative proportions of tVFA, whereas kale varieties differed in their in situ soluble CP and insoluble but potentially degradable CP fractions. Continuing with studies under in vivo conditions when feeding winter brassicas, especially to lactating dairy cows, is important because less information is available.

Author Contributions: Conceptualization, J.K., O.B., and R.P.; data curation, D.B., J.D., A.B., and J.K.; funding acquisition, J.K., O.B., and R.P.; investigation, A.B., D.B., J.K., and J.D.; methodology, J.K., O.B., and R.P.; project administration, J.K.; resources, J.K., O.B., and R.P.; supervision, J.K.; validation, J.K., O.B., and R.P.; writing—original draft, D.B., J.D., and J.K.; writing—review and editing, A.B., O.B., and R.P.

Funding: This research was funded by Dirección de Investigación y Desarrollo, Universidad Austral de Chile, Valdivia, Chile (DID-S2014-17).

Conflicts of Interest: The authors declare no conflict of interest.

References

1. Keogh, B.; French, P.; Murphy, J.J.; Mee, J.F.; McGrath, T.; Storey, T.; Grant, J.; Mulligan, F.J. A note on the effect of dietary proportions of kale (*Brassica oleracea*) and grass silage on rumen pH and volatile fatty acid concentrations in dry dairy cows. *Livest. Sci.* **2009**, *126*, 302–305. [CrossRef]

2. Keim, J.P.; López, I.F.; Balocchi, O.A. Sward herbage accumulation and nutritive value as affected by pasture renovation strategy. *Grass Forage Sci.* **2015**, *70*, 283–295. [CrossRef]

3. Westwood, C.T.; Mulcock, H. Nutritional evaluation of five species of forage brassica. *Proc. N. Z. Grassl. Assoc.* **2012**, *74*, 31–38.

4. Sun, X.Z.; Waghorn, G.C.; Hoskin, S.O.; Harrison, S.J.; Muetzel, S.; Pacheco, D. Methane emissions from sheep fed fresh brassicas (*Brassica spp.*) compared to perennial ryegrass (*Lolium perenne*). *Anim. Feed Sci. Technol.* **2012**, *176*, 107–116. [CrossRef]

5. Rugoho, I.; Edwards, G.R. Dry matter intake, body condition score, and grazing behavior of nonlactating, pregnant dairy cows fed kale or grass once versus twice daily during winter. *J. Dairy Sci.* **2018**, *101*, 257–267. [CrossRef] [PubMed]

6. Keogh, B.; French, P.; McGrath, T.; Storey, T.; Mulligan, F.J. Comparison of the performance of dairy cows offered kale, swedes and perennial ryegrass herbage in situ and perennial ryegrass silage fed indoors in late pregnancy during winter in Ireland. *Grass Forage Sci.* **2009**, *64*, 49–56. [CrossRef]

7. Sun, X.Z.; Pacheco, D.; Luo, D.W. Forage brassica: A feed to mitigate enteric methane emissions? *Anim. Prod. Sci.* **2016**, *56*, 451–456. [CrossRef]

8. Hackmann, T.J.; Sampson, J.D.; Spain, J.N. Variability in in situ ruminal degradation parameters causes imprecision in estimated ruminal digestibility. *J. Dairy Sci.* **2010**, *93*, 1074–1085. [CrossRef]

9. Barry, T.N. The feeding value of forage brassica plants for grazing ruminant livestock. *Anim. Feed Sci. Technol.* **2013**, *181*, 15–25. [CrossRef]

10. Dijkstra, J.; Kebreab, E.; Bannink, A.; France, J.; López, S. Application of the gas production technique to feed evaluation systems for ruminants. *Anim. Feed Sci. Technol.* **2005**, *123–124*, 561–578. [CrossRef]

11. France, J.; Dijkstra, J.; Dhanoa, M.S.; Lopez, S.; Bannink, A. Estimating the extent of degradation of ruminant feeds from a description of their gas production profiles observed in vitro: Derivation of models and other mathematical considerations. *Br. J. Nutr.* **2000**, *83*, 143–150. [CrossRef] [PubMed]

12. Ørskov, E.R.; McDonald, I. The estimation of protein degradability in the rumen from incubation measurements weighted according to rate of passage. *J. Agric. Sci.* **1979**, *92*, 499–503. [CrossRef]

13. Theodorou, M.K.; Williams, B.A.; Dhanoa, M.S.; McAllan, A.B.; France, J. A simple gas production method using a pressure transducer to determine the fermentation kinetics of ruminant feeds. *Anim. Feed Sci. Technol.* **1994**, *48*, 185–197. [CrossRef]

14. Niderkorn, V.; Baumont, R.; Le Morvan, A.; Macheboeuf, D. Occurrence of associative effects between grasses and legumes in binary mixtures on in vitro rumen fermentation characteristics. *J. Anim. Sci.* **2011**, *89*, 1138–1145. [CrossRef]

15. Getachew, G.; DePeters, E.J.; Robinson, P.H.; Fadel, J.G. Use of an in vitro rumen gas production technique to evaluate microbial fermentation of ruminant feeds and its impact on fermentation products. *Anim. Feed Sci. Technol.* **2005**, *123–124*, 547–559. [CrossRef]

16. Keim, J.P.; Cabanilla, J.; Balocchi, O.A.; Pulido, R.N.G.; Bertrand, A. In vitro fermentation and in situ rumen degradation kinetics of summer forage brassica plants. *Anim. Prod. Sci.* **2019**, *59*, 1271–1280. [CrossRef]

17. Valderrama, X.; Anrique, R. In situ rumen degradation kinetics of high-protein forage crops in temperate climates. *Chil. J. Agric. Res.* **2011**, *71*, 572–577. [CrossRef]

18. Hvelplund, T.; Wesiberg, M. In situ techniques for the estimation of protein degradability and postrumen availability. In *Forage Evaluation in Ruminant Nutrition*; Givens, D.I., Ed.; CABI Publishing: Wallingford, UK, 2000; pp. 233–258.

19. Tilley, J.M.A.; Terry, R.A. A two-stage technique for the in vitro digestion of forage crops. *Grass Forage Sci.* **1963**, *18*, 104–111. [CrossRef]

20. Van Soest, P.J.; Robertson, J.B.; Lewis, B.A. Methods for Dietary Fiber, Neutral Detergent Fiber, and Nonstarch Polysaccharides in Relation to Animal Nutrition. *J. Dairy Sci.* **1991**, *74*, 3583–3597. [CrossRef]

21. AOAC. *Official Methods of Analysis*, 16th ed.; Association of Official Analytical Chemists: Washington, DC, USA, 1996.

22. Pelletier, S.; Tremblay, G.F.; Belanger, G.; Bertrand, A.; Castonguay, Y.; Pageau, D.; Drapeau, R. Forage Nonstructural Carbohydrates and Nutritive Value as Affected by Time of Cutting and Species. *Agron. J.* **2010**, *102*, 1388–1398. [CrossRef]

23. Hall, M.B.; Hoover, W.H.; Jennings, J.P.; Webster, T.K.M. A method for partitioning neutral detergent-soluble carbohydrates. *J. Sci. Food Agric.* **1999**, *79*, 2079–2086. [CrossRef]

24. Groot, J.C.J.; Cone, J.W.; Williams, B.A.; Debersaques, F.M.A.; Lantinga, E.A. Multiphasic analysis of gas production kinetics for in vitro fermentation of ruminant feeds. *Anim. Feed Sci. Technol.* **1996**, *64*, 77–89. [CrossRef]

25. Rowe, B.A.; Neilsen, J.E. Effects of irrigating forage turnips, *Brassica rapa* var. *rapa* cv. Barkant, during different periods of vegetative growth. 2. Nutritive characteristics of leaves and roots. *Crop Pasture Sci.* **2011**, *62*, 571–580. [CrossRef]

26. Loaiza, P.A.; Balocchi, O.; Bertrand, A. Carbohydrate and crude protein fractions in perennial ryegrass as affected by defoliation frequency and nitrogen application rate. *Grass Forage Sci.* **2017**, *72*, 556–567. [CrossRef]

27. Keim, J.P.; López, I.F.; Berthiaume, R. Nutritive value, in vitro fermentation and methane production of perennial pastures as affected by botanical composition over a growing season in the south of Chile. *Anim. Prod. Sci.* **2014**, *54*, 598–607. [CrossRef]

28. Oba, M. Review: Effects of feeding sugars on productivity of lactating dairy cows. *Can. J. Anim. Sci.* **2011**, *91*, 37–46. [CrossRef]

29. Cassida, K.A.; Turner, K.E.; Foster, J.G.; Hesterman, O.B. Comparison of detergent fiber analysis methods for forages high in pectin. *Anim. Feed Sci. Technol.* **2007**, *135*, 283–295. [CrossRef]

30. Zhao, X.H.; Liu, C.J.; Liu, Y.; Li, C.Y.; Yao, J.H. Effects of replacing dietary starch with neutral detergent–soluble fibre on ruminal fermentation, microbial synthesis and populations of ruminal cellulolytic bacteria using the rumen simulation technique (RUSITEC). *J. Anim. Physiol. Anim. Nutr.* **2013**, *97*, 1161–1169. [CrossRef]

31. Fulkerson, W.J.; Horadagoda, A.; Neal, J.S.; Barchia, I.; Nandra, K.S. Nutritive value of forage species grown in the warm temperate climate of Australia for dairy cows: Herbs and grain crops. *Livest. Sci.* **2008**, *114*, 75–83. [CrossRef]

32. Aufrere, J.; Graviou, D.; Demarquilly, C. Ruminal degradation of protein of cocksfoot and perennial ryegrass as affected by various stages of growth and conservation methods. *Anim. Res.* **2003**, *52*, 245–261. [CrossRef]

33. Znidarsic, T.; Verbic, J.; Babnik, D.; Velikonja-Bolta, S. The effect of supplementing highly wilted grass silage with maize silage, fodder beet or molasses on degradation of the diets and the efficiency of microbial protein synthesis in the rumen of sheep. *Ital. J. Anim. Sci.* **2010**, *9*, 449–459. [CrossRef]

34. Broderick, G.A.; Cochran, R.C. In vitro and In situ Methods for Estimating Digestibility with Reference to Protein Degradability. In *Feeding Systems and Feed Evaluation Models*; Theodorou, M.K., France, J., Eds.; CAB International: Wallingford, UK, 2000; pp. 53–85.

35. Sun, X.Z.; Waghorn, G.C.; Hatier, J.H.B.; Easton, H.S. Genotypic variation in in sacco dry matter degradation kinetics in perennial ryegrass (*Lolium perenne* L.). *Anim. Prod. Sci.* **2012**, *52*, 566–571. [CrossRef]

36. Xu, M.; Rinker, M.; McLeod, K.R.; Harmon, D.L. Yucca schidigera extract decreases in vitro methane production in a variety of forages and diets. *Anim. Feed Sci. Technol.* **2010**, *159*, 18–26. [CrossRef]

37. Rymer, C.; Huntington, J.A.; Williams, B.A.; Givens, D.I. In vitro cumulative gas production techniques: History, methodological considerations and challenges. *Anim. Feed Sci. Technol.* **2005**, *123–124*, 9–30. [CrossRef]

38. Van Gelder, A.H.; Hetta, M.; Rodrigues, M.A.M.; De Boever, J.L.; Den Hartigh, H.; Rymer, C.; Van Oostrum, M.; Van Kaathoven, R.; Cone, J.W. Ranking of in vitro fermentability of 20 feedstuffs with an automated gas production technique: Results of a ring test. *Anim. Feed Sci. Technol.* **2005**, *123–124*, 243–253. [CrossRef]

39. Seifried, N.; Steingass, H.; Schipprack, W.; Rodehutscord, M. Variation in ruminal in situ degradation of crude protein and starch from maize grains compared to in vitro gas production kinetics and physical and chemical characteristics. *Arch. Anim. Nutr.* **2016**, *70*, 333–349. [CrossRef]

40. Keim, J.P.; Alvarado-Gilis, C.; Arias, R.A.; Gandarillas, M.; Cabanilla, J. Evaluation of sources of variation on in vitro fermentation kinetics of feedstuffs in a gas production system. *Anim. Sci. J.* **2017**, *88*, 1547–1555. [CrossRef]

41. Hall, M.B.; Pell, A.N.; Chase, L.E. Characteristics of neutral detergent-soluble fiber fermentation by mixed ruminal microbes. *Anim. Feed Sci. Technol.* **1998**, *70*, 23–39. [CrossRef]

42. Getachew, G.; Robinson, P.H.; DePeters, E.J.; Taylor, S.J. Relationships between chemical composition, dry matter degradation and in vitro gas production of several ruminant feeds. *Anim. Feed Sci. Technol.* **2004**, *111*, 57–71. [CrossRef]

43. Morvay, Y.; Bannink, A.; France, J.; Kebreab, E.; Dijkstra, J. Evaluation of models to predict the stoichiometry of volatile fatty acid profiles in rumen fluid of lactating Holstein cows. *J. Dairy Sci.* **2011**, *94*, 3063–3080. [CrossRef]

44. Seymour, W.M.; Campbell, D.R.; Johnson, Z.B. Relationships between rumen volatile fatty acid concentrations and milk production in dairy cows: A literature study. *Anim. Feed Sci. Technol.* **2005**, *119*, 155–169. [CrossRef]

45. Moss, A.; Jouany, J.P.; Newbold, J. Methane production by ruminants: Its contri-bution to global warming. *Ann. Zootech.* **2000**, *49*, 231–253. [CrossRef]

46. Beauchemin, K.; McAllister, T.; McGinn, S.M. Dietary mitigation of enteric methane from cattle. *CAB Rev.* **2009**, *4*. [CrossRef]

47. Hall, M.B.; Mertens, D.R. A 100-Year Review: Carbohydrates-Characterization, digestion, and utilization. *J. Dairy Sci.* **2017**, *100*, 10078–10093. [CrossRef] [PubMed]

 *animals*

Article

Comparative Grain Yield, Straw Yield, Chemical Composition, Carbohydrate and Protein Fractions, In Vitro Digestibility and Rumen Degradability of Four Common Vetch Varieties Grown on the Qinghai-Tibetan Plateau

Yafeng Huang, Fangfang Zhou and Zhibiao Nan *

State Key Laboratory of Grassland Agro-ecosystems, Key Laboratory of Grassland Livestock Industry Innovation of Ministry of Agriculture and Rural Affairs, College of Pastoral Agriculture Science and Technology, Lanzhou University, Lanzhou 730020, China
* Correspondence: zhibiao@lzu.edu.cn; Tel.: +86-931-8912-582

Received: 6 July 2019; Accepted: 23 July 2019; Published: 31 July 2019

Simple Summary: The common vetch (*Vicia sativa* L.) is an important legume crop of mixed crop-livestock systems that provides high-quality grains used as food/feed and straw used as ruminant feed. The objective of this study was to determine the variability in grain yield, straw yield, straw chemical composition, carbohydrate and protein fractions, in vitro gas production, and in situ ruminal degradability of four different varieties of common vetch grown on the Qinghai-Tibetan Plateau. The results showed that grain yield, straw yield, and straw nutrient value varied significantly among the four varieties. Overall, the findings indicated that in terms of straw yield and nutritive quality, variety Lanjian No. 1 has the greatest potential as a crop for supplementing ruminant diets in the smallholder mixed crop–livestock systems on the Qinghai-Tibetan Plateau.

Abstract: Four varieties of common vetch, including three improved varieties (Lanjian No. 1, Lanjian No. 2, and Lanjian No. 3) and one local variety (333A), were evaluated for varietal variations in grain yield, straw yield and straw quality attributes on the Qinghai-Tibetan Plateau. Crops were harvested at pod maturity to determine grain yield, straw yield, harvest index, and potential utility index (PUI). Straw quality was determined by measuring chemical composition, carbohydrate and protein fractions, in vitro gas production and in situ ruminal degradability. Results showed a significant effect ($p < 0.01$) of variety on the grain yield [875.2–1255 kg dry matter (DM)/ha], straw yield (3154–5556 kg DM/ha), harvest index (15.6–28.7%) and PUI (53.3–63.2%). Variety also had a significant effect on chemical composition, carbohydrate and protein fractions ($p < 0.05$) except non-structural carbohydrates and rapidly degradable sugars. Significant differences ($p < 0.05$) were observed among the varieties in potential gas production [188–234 mL/g DM], in vitro organic matter (OM) digestibility (43.7–54.2% of OM), and metabolizable energy (6.40–7.92 MJ/kg DM) of straw. Significant differences ($p < 0.001$) were also observed among the varieties in rapidly degradable DM fraction and effective DM degradability of straw; however, no difference was observed in other DM degradation parameters and neutral detergent fiber degradation parameters. In conclusion, based on straw yield and quality, Lanjian No. 1 has the greatest potential among the tested varieties as a crop for supplementing ruminant diets for smallholder farmers on the Qinghai-Tibetan plateau.

Keywords: common vetch; straw; nutritive value; varietal effect; ruminants

---

1. Introduction

Ruminants provides the majority of food worldwide, contributing approximately 25.9% of global meat production and nearly 100% of global milk production [1,2]. The demand for cattle and sheep meat has been growing at a global growth rate of 2.0% per year from 2016 and is predicted to continue till 2026 [3]. The Qinghai-Tibetan plateau supports approximately 41 million sheep and over 14.3 million cattle, one of the largest livestock systems in Asia [4,5]. The plateau is also the headwater region of most of Asia's five major rivers, including Yangtze, Yellow, Mekong, Ganges, and Indus rivers [4]. Thus, sustainable management of the Plateau is important for the livelihood of over 9.8 million nomadic populations and for protecting these crucial river systems.

Unfortunately, harsh climatic conditions (e.g., low temperatures) and short growing season on the Qinghai-Tibetan plateau limit crop production [4–6]. Combined with grassland degradation, the annual fodder gap in this region is expected to reach 2 million tonnes, which implies that each year 2.7 million sheep units with insufficient feedstuffs will result in 20–30% liveweight loss during winter and early spring [5,7]. An effective remedy to stabilize feed supply is to incorporate nitrogen-fixing annual cool-season feed legumes such as common vetch (*Vicia sativa* L.) into fallow lands that will improve soil properties and lead to increased grain and straw production [8].

The use of common vetch in the mixed crop-livestock systems has expanded greatly in the last decade [4,9–11]. Several studies showed high generation of crop residues after harvesting common vetch that has substantial potential to be used for feeding ruminant livestock [9,12]. Previous studies also reported that the nutritive quality of common vetch straw is relatively high, containing an average of 9.41% crude protein (CP), 55.2% in vitro organic matter digestibility (IVOMD) and 7.3 MJ/kg metabolizable energy (ME) [9,13]. This makes common vetch straw a good feed supplement for ruminants, which are offered low-quality cereal straw-based diets in many small-holder crop-livestock systems [9,14]. In addition, the fatty acid composition of grains has increased the use of common vetch as a feed for ruminants [15]. Although common vetch is important as a feed for ruminants, limited information is available on the varietal variations in straw composition (carbohydrate and protein fractions), in vitro gas production, and in situ ruminal degradability, because most of the earlier studies focused on grain yield, straw yield, and straw quality including only chemical composition, IVOMD and ME [9–13]. Russel et al. [16] suggested that carbohydrate and protein fractions can be used as a reliable indicator to accurately predict biological value and performance of feed in ruminants. Marcos et al. [17], and Blümmel and Ørskov [18] demonstrated that the chemical composition in combination with in vitro digestibility and in situ ruminal degradability can be used as crucial parameters to evaluate the nutritive value of feed.

Accordingly, the objective of this study was to evaluate grain and straw yields, as well as chemical composition, carbohydrate and protein fractions, in vitro gas production and in situ ruminal degradability of straw of four different common vetch varieties grown on the Qinghai-Tibetan Plateau.

2. Materials and Methods

2.1. Location, Experimental Design, and Sampling

A detailed description of the location, experimental design and sampling is in the companion paper on the nutritive value of common vetch grain [4] and are summarized here.

This experiment was conducted during the cropping season of 2015 at the Xiahe Experimental Station of Lanzhou University, Gansu, China (35°45′ N, 102°34′ E; altitude 2880 m). The region under this study is situated at the eastern margin of the Qinghai-Tibetan Plateau. The location received an annual average precipitation of 452 mm (80% in May and September) and recorded a mean annual air temperature of 3.5 °C (1984–2014, 31 years). The previous crop was rape (*Brassica campestris* L.).

Three improved varieties (Lanjian No. 1, Lanjian No. 2, Lanjian No. 3) and one local variety (333A) were utilized in the field study. These improved varieties, developed by the Common Vetch Breeding Program of Lanzhou University, are well-adapted and extensively grown by smallholder farmers in the region surrounding the research site. Agronomic characteristics of these tested varieties are shown in Table 1. All four varieties were planted under the same agronomic conditions following a completely randomized design with four replicates. The individual plot area was 40 m^2 (8 m × 5 m) with a row spacing of 20 cm, and grains were sown by hand at a depth of 3–5 cm at a density of 150 viable grains m-2. Grains were sown (6 May 2015) before inoculation with rhizobium (CCBAU01069, China Agricultural University, Beijing, China), which was recommended based on the symbiont performance of these varieties [19]. Irrigation and fertilizers were not applied after sowing, and weeds in each plot were adequately controlled manually.

Table 1. Agronomic characteristics of the common vetch varieties utilized in this study.

Agronomic Characteristic	333A	Lanjian No.1	Lanjian No.2	Lanjian No.3
Days to mature (day)	134	145	132	124
1000 grains weight (g)	54	79	71	76
Plant height (cm)	92	106	80	69
Altitude (m.a.s.l)	–	<3000	<3500	<4000
Year of release	1987	2014	2015	2011

Source: Ministry of Agriculture, Beijing, China.

Harvesting of the common vetch plants was done at the pod maturity (6, 15, 15, and 26 September for Lanjian No. 3, 333A, Lanjian No. 2 and Lanjian No. 1, respectively). The crops were manually harvested from two representative subplots (1 m × 1 m) of each plot for each variety and threshed to obtain grain and straw samples. Harvest index was calculated as follows: Harvest index (%) = Grain yield × 100 / (Grain yield + Straw yield). All samples were oven-dried at 65 °C for 48 h and ground to pass through a 2-mm sieve for analysis of in situ ruminal incubation and to pass through a 1-mm sieve for chemical analysis and in vitro gas production measurement. All procedures involved animals were approved by the Animal Ethics Committee of Lanzhou University (protocol AEC-LZU-2016-01).

2.2. In Vitro Gas Production

Four adult Dorper rams (approximately 33-month-old; 58.4 ± 1.24 kg body weight) fitted with flexible rumen cannulas were used as donors of ruminal fluid. The rams were kept in individual stalls and had free access to fresh water and mineral/vitamin licks. The rams were daily fed 1.2 kg of 550 g/kg DM sheepgrass [*Leymus chinensis*, (Trin.) Tzvel], 140 g/kg DM soybean meal, 294 g/kg DM maize (*Zea mays* L.) seed, 8.6 g/kg DM calcium hydrophosphate, 5.0 g/kg DM salt and 2.4 g/kg DM mineral-vitamin mix at maintenance energy level in equal portions at 08:00 and 16:30 hours. In vitro gas production (GP) was measured as described by Blümmel and Ørskov [18]. Rumen fluids were obtained before morning feeding and strained through four layers of cheesecloth into a preheated, insulated bottle. Briefly, approximately 200 mg DM of each sample (in duplicate) was weighed into calibrated glass syringes (100 mL). Each syringe was preheated at 39 °C before injecting 30 mL rumen fluid/buffer mixture [20]. Then, the syringes were placed vertically in a water bath at 39 °C with three syringes without a sample used as blank. The volume of GP was manually recorded after 0, 4, 6, 8, 12, 24, 36, 48 and 72 h of incubation and was blank corrected. The incubation run was repeated three times. All operations involving rumen fluid were conducted under a continuous flush of CO_2 to ensure anaerobic conditions.

The GP data were fitted with time using the following equation [17], as follows: $y = b \times [1 - e^{-c(t - lag)}]$, where y = the volume of GP at time t; b = the potential GP (mL/g DM); c (h^{-1}) = the fractional rate of GP (h^{-1}); lag = the initial delay in the onset of gas production (h); t = incubation time (h). Parameters b, c and lag were determined by an iterative least square method using the NLIN procedure of SAS 9.2 (SAS Inst. Inc., Cary, NC). The IVOMD [% organic matter (OM)] and ME (MJ/kg DM) were estimated using the equations of Menke and Steingass [21] as: IVOMD = 14.88 + 0.889 GP + 0.45 CP + 0.0651 ASH; ME = 2.20 + 0.136 GP + 0.057 CP + 0.0029 EE^2, where GP = the net gas production after

24 h of incubation (mL/200mg DM), ASH = the ash content (% DM), CP = the crude protein content (% DM), and EE = the ether extract content (% DM).

2.3. In Situ Ruminal Incubation

In situ ruminal degradability of DM and neutral detergent fiber (NDF) of straw samples was determined using the nylon bag technique described by Nandra et al. [22]. Briefly, approximately 5.0 g of each sample (in duplicate) was weighted into nylon bags (9 × 5 cm; 50-μm pore size) and incubated in the ventral sacs of the rumen of the same four rams used for the production of ruminal fluid in Section 2.2. The bags were inserted into the rumen for 0 (control), 4, 8, 12, 24, 48 and 72 h. Following removal, the bags were briefly washed under cold running water and frozen (–20°C) until further analysis. All bags were defrosted, manually washed in cold tap water until the water was clear, oven-dried at 60°C for 48 h, and weighed. The dried undigested residues of replicates per same time within sheep were pooled to measure DM and NDF. Straw DM and NDF disappearance rates were estimated from the difference in straw weight before and after incubation. The kinetic parameters of DM and NDF degradation were determined using the exponential equation described by Ørskov and McDonald [23]. The effective degradability of DM (EDDM) and NDF (EDNDF) were calculated as ED= A + [(B × C)/(C + k)], where A = the soluble fraction, B = the potentially degradable fraction, C = the rate of degradation of fraction B, k = the rumen outflow rate (0.031 h^{-1}) [4].

2.4. Laboratory Analysis

Determination of DM (ID 930.15), nitrogen (N; ID 988.05), ether extract (EE; ID 920.85), ash (ID 938.08), acid detergent fiber (ADF; ID 973.18) and acid detergent lignin (ADL; ID 973.18) were analyzed following the methods of the Association of Official Agricultural Chemists [24]. The CP content was calculated by multiplying the nitrogen value by 6.25. The NDF content was determined following the method by Van Soest et al. [25] using heat-stable α-amylase and sodium sulfite. Contents of NDF and ADF were expressed inclusive of residual ash. Acid detergent insoluble protein (ADIP), neutral detergent insoluble protein (NDIP) were measured by Kjeldahl analysis of the ADF and NDF bag residues, respectively, using the procedure described by Licitra et al. [26].

Carbohydrate fractions of straw samples from the four common vetch varieties were determined as proposed by the Cornell Net Carbohydrate and Protein system (CNCPS) [27]. The system divides carbohydrate into four fractions in terms of their degradation rate as follow: C_A, rapidly degradable sugars; CB_1, intermediately degradable pectin and starch; CB_2, slowly degradable cell wall; and C_C, undegradable/lignin-bound cell wall. Total carbohydrates (TCHO) content was calculated as TCHO = 100 – (CP + EE + Ash) [28]. Non-structural carbohydrates (NSC) and structural carbohydrates (SC) were estimated using the equations given by Caballero et al. [29] as: NSC = TCHO – SC and SC = NDF – NDIP. Starch content was determined by enzymatic hydrolysis of α-linked glucose polymers [30].

The CP of straw samples was fractionated into five different fractions according to CNCPS as described by Licitra et al. [26] and Sniffen et al. [27]. These fractions include: fraction P_A, non-protein nitrogen (NPN), calculated as the difference between total N and true protein N, analyzed using sulphuric acid (0.5 M) and sodium tungstate (0.30 M); fraction P_{B1}, buffer-soluble protein, estimated by subtracting buffer-insoluble protein precipitated with freshly prepared (1 g/10 mL) sodium azide and borate-phosphate buffer (pH 6.7–6.8) solution from true protein; fraction P_{B2}, neutral detergent- soluble protein, calculated by subtracting NDIP from buffer-insoluble protein; fraction P_{B3}, acid detergent-soluble protein, calculated by subtracting ADIP from NDIP; and fraction P_C, ADIP, is indigestible protein. All measurements were performed in duplicate and appropriate chemical standards were included in each analytical run.

2.5. Calculations and Statistical Analyses

Potential utility index (PUI) was estimated from the amount of utilizable portion of the total biomass yield for grain and straw regardless of the economic value as described by Alkhtib et al. [31]:

$$\text{PUI (\%)} = 100 \times [\text{grain yield} + 0.01 \times \text{IVOMD (\%)} \times \text{straw yield}]/ \text{total biomass yield}$$

Data collected were subjected to one-way ANOVA using SPSS software (Version 21.0. IBM Corporation, Armonk, NY, USA). The fermentation parameters were subjected to separate analysis of variance with varieties as fixed effect and the incubation run as random effect. Differences between means were compared using the Duncan significant difference test at $p < 0.05$.

3. Results

3.1. Grain Yield, Straw Yield, and PUI

Table 2 shows a significant effect of variety on grain yield and straw yield ($p < 0.01$). The grain yield ranged from 875.2 to 1255 kg DM/ha with an average value of 1088 kg DM/ha. The grain yield of 333A and Lanjian No. 1 was significantly less than that of other improved varieties. The straw yield varied from 3154 to 5556 kg DM/ha. The least straw yield was observed for Lanjian No. 3 and the greatest for Lanjian No. 1.

Variety had a significant ($p < 0.001$) effect on harvest index and PUI. Harvest index varied from 15.6 to 28.7%. Harvest index of Lanjian No. 2 was less than that of Lanjian No. 3, but greater than that of the other varieties ($p < 0.01$). The PUI varied from 53.3 to 63.2%. The PUI of the local variety was less than the improved varieties, which had similar PUI (average 62.1%).

Table 2. Influence of variety on grain yield, straw yield, harvest index, and potential utility index (PUI) of four common vetch varieties.

Dependent Variable	333A	Lanjian No. 1	Lanjian No. 2	Lanjian No. 3	SEM [1]	*p*-Value
Grain yield (kg DM/ha)	875.2 [b]	1024 [b]	1196 [a]	1255 [a]	45.1	0.001
Straw yield (kg DM/ha)	4319 [b]	5556 [a]	4158 [bc]	3154 [c]	267.6	0.003
Harvest index (%)	17.0 [c]	15.6 [c]	22.6 [b]	28.7 [a]	1.39	<0.001
PUI (%)	53.3 [b]	61.4 [a]	62.8 [a]	63.2 [a]	0.660	<0.001

[a,b] Within a raw, different letters represent the significant differences at *p*-value < 0.05. [1] DM, dry matter; SEM, standard error of the mean.

3.2. Chemical Composition

Table 3 shows a significant effect of variety on DM, ash, CP and EE contents of the straw samples ($p < 0.01$). The DM content among varieties varied from 90.1 to 90.5% and was the greatest in Lanjian No. 2. The ash content varied from 10.2 to 13.5% DM. The least ash content was for 333A and the greatest for Lanjian No. 1. The CP content varied from 9.76 to 13.8% DM. The local variety had significantly less CP content compared to the improved varieties. There were significant differences in CP content among the improved varieties, which was greater for Lanjian No. 1. The EE content varied from 0.459 to 1.11% DM. The EE content of 333A was similar to that of Lanjian No. 3 but was significantly greater than that of Lanjian No. 1 and Lanjian No. 2.

Significant differences ($p < 0.05$) were observed in the cell wall contents of different varieties. The NDF, ADF, and ADL contents varied from 45.0 to 54.1% DM, 27.4 to 33.2% DM, and 6.08 to 9.56% DM, respectively. The NDF, ADF and ADL contents were the greatest in 333A and the least in Lanjian No. 1. Hemicellulose and cellulose contents varied from 17.6 to 21.4% DM and 21.4 to 23.7% DM, respectively. The hemicellulose and cellulose contents of variety 333A were higher than those of Lanjian No. 1, but similar to those of other varieties. Phosphorus content varied from 0.185 to 0.296% DM. The phosphorus content was considerably greater ($p < 0.05$) in Lanjian No. 3 than in Lanjian No. 2

and 333A, with Lanjian No. 1 being intermediate. Calcium content was also significantly influenced by variety ($p < 0.001$). It varied from 1.00 to 1.54% DM and was the greatest in Lanjian No. 1.

Table 3. Influence of variety on chemical composition (% dry matter unless stated otherwise) of straw in four common vetch varieties.

Dependent Variable	333A	Lanjian No. 1	Lanjian No. 2	Lanjian No. 3	SEM [1]	*p*-Value
Dry matter %	90.2 [bc]	90.4 [ab]	90.5 [a]	90.1 [c]	0.055	0.009
Ash	10.2 [c]	13.5 [a]	11.9 [b]	11.0 [bc]	0.393	0.003
Crude protein	9.76 [c]	13.8 [a]	12.4 [b]	11.5 [b]	0.424	<0.001
Ether extract	1.06 [a]	0.459 [b]	0.515 [b]	1.11 [a]	0.088	<0.001
Neutral detergent fiber	54.1 [a]	45.0 [c]	49.4 [b]	51.0 [b]	0.915	<0.001
Acid detergent fiber	33.2 [a]	27.4 [c]	30.3 [b]	29.6 [b]	0.603	<0.001
Acid detergent lignin	9.56 [a]	6.08 [c]	6.53 [bc]	7.05 [b]	0.369	<0.001
Hemicellulose	20.9 [a]	17.6 [b]	19.2 [ab]	21.4 [a]	0.537	0.025
Cellulose	23.6 [a]	21.4 [b]	23.7 [a]	22.5 [ab]	0.353	0.039
Phosphorus	0.185 [b]	0.249 [ab]	0.206 [b]	0.296 [a]	0.0149	0.019
Calcium	1.13 [bc]	1.54 [a]	1.00 [c]	1.32 [ab]	0.061	<0.001

[a,b] Within a raw, different letters represent the significant differences at *p*-value < 0.05. [1] SEM, standard error of the mean.

3.3. Carbohydrate and Protein Fractions

As shown in Table 4, the TCHO content was significantly different ($p < 0.001$) among varieties and varied from 72.2 to 79.0% DM. The greatest TCHO content was observed in 333A and the least in Lanjian No. 1. The NSC content varied from 29.3 to 31.2% DM with no difference ($p > 0.05$) among varieties. The SC content varied from 41.1 to 49.7% DM. The variation in SC content was the greatest in 333A and the least in Lanjian No. 1. Significant differences ($p < 0.05$) were observed in CHO fractions except C_A fraction (Table 4). The C_{B1} fraction was greatest ($p < 0.001$) in Lanjian No. 1 (26.0% CHO), intermediate in Lanjian No. 2 (22.9% CHO), and least in Lanjian No. 3 (20.1% CHO) and 333A (17.7% CHO). The C_{B2} fraction varied from 33.9 to 39.5% CHO. Variety 333A had significantly less ($p < 0.05$) C_{B2} fraction than Lanjian No. 2 and Lanjian No. 3 but similar to Lanjian No. 1. The Cc fraction varied from 20.2 to 29.0% CHO. The C_C fraction of variety 333A was significantly greater ($p < 0.001$) compared to the improved varieties, which had similar C_C fraction.

Table 4. Influence of variety on carbohydrate and protein fractions of straw in four common vetch varieties.

Dependent Variable	333A	Lanjian No. 1	Lanjian No. 2	Lanjian No. 3	SEM [1]	*p*-Value
Carbohydrates (% dry matter)						
TCHO	79.0 [a]	72.2 [c]	75.2 [b]	76.4 [b]	0.665	<0.001
NSC	29.3	31.2	29.8	29.6	0.474	0.575
SC	49.7 [a]	41.1 [c]	45.4 [b]	46.7 [b]	0.886	<0.001
Carbohydrate fractions (% CHO)						
C_A	19.4	17.1	18.1	18.7	0.645	0.435
C_{B1}	17.7 [c]	26.0 [a]	22.9 [b]	20.1 [c]	0.885	<0.001
C_{B2}	33.9 [b]	36.7 [ab]	39.5 [a]	39.0 [a]	0.835	0.042
C_C	29.0 [a]	20.2 [b]	20.8 [b]	22.1 [b]	0.998	<0.001
Protein fractions (% CP)						
P_A	7.16 [c]	11.5 [a]	9.43 [b]	8.55 [bc]	0.499	0.004
P_{B1}	37.3 [a]	25.8 [c]	30.1 [bc]	31.8 [b]	1.31	0.004
P_{B2}	10.8 [c]	34.0 [a]	27.9 [b]	22.3 [b]	2.36	<0.001
P_{B3}	23.6 [a]	13.3 [c]	14.8 [c]	18.4 [b]	1.10	<0.001
P_C	21.1 [a]	15.4 [c]	17.8 [b]	18.8 [b]	0.628	0.001

[a,b] Within a raw, different letters represent the significant differences at *p*-value < 0.05. [1] C_A, rapidly degradable sugars; C_{B1}, intermediately degradable pectin and starch; C_{B2}, slowly degradable cell wall; C_C, unavailable/lignin bound cell wall; CP, crude protein; TCHO, total carbohydrates; NSC, non-structural carbohydrates; P_A, non-protein nitrogen; P_{B1}, buffer soluble protein; P_{B2}, neutral detergent soluble protein; P_{B3}, acid detergent soluble protein; P_C, indigestible protein; SC, structural carbohydrates; SEM, standard error of the mean.

Variety had a significant influence on the protein fractions of straw ($p < 0.01$; Table 4). The P_A fraction varied from 7.16 to 11.5% CP, and the least value was recorded in 333A and the greatest in Lanjian No. 1. The P_{B1} fraction varied from 25.8 to 37.3% CP; the greatest value was recorded in 333A. The P_{B2} fraction varied from 10.8 to 34.0% CP. The variation in P_{B2} fraction was the least in 333A and the greatest in Lanjian No. 1. The P_{B3} fraction varied from 13.3 to 23.6% CP. The P_{B3} fraction was significantly greater for 333A than for Lanjian No. 3, and significantly greater for Lanjian No. 3 than the other varieties. The P_C fraction varied from 15.4 to 21.1% CP. The P_c fraction was greatest in 333A and least in Lanjian No. 1.

3.4. In Vitro Gas Production

Table 5 shows the gas production parameters, IVOMD and ME of straw in four common vetch varieties. The potential gas production differed considerably ($p < 0.001$) between the varieties and varied from 188 to 234 mL/g DM. The potential gas production of Lanjian No. 3 was greater than that of 333A; however, it was less than that of the other varieties. The fractional rate of GP was not affected by variety ($p > 0.05$), averaging 0.0631 h^{-1}. The *lag* value was similar ($p > 0.05$) for the four varieties (average 0.633 h).

Variety had a significant influence on the IVOMD content of straw ($p < 0.001$; Table 5). Straw IVOMD varied from 43.7 to 54.2% OM, and the varieties were ranked in order Lanjian No. 1 > Lanjian No. 2 > Lanjian No. 3 > 333A. Significant difference in the ME content was also observed among the different varieties ($p < 0.001$), and it varied from 6.40 to 7.92 MJ/kg DM. The least value was recorded for 333A and the greatest for Lanjian No. 1 and Lanjian No. 2.

Table 5. Influence of variety on gas production parameters, in vitro organic matter digestibility (IVOMD) and metabolizable energy (ME) of straw in four common vetch varieties.

Dependent Variable	333A	Lanjian No. 1	Lanjian No. 2	Lanjian No. 3	SEM [1]	*p*-Value
b (mL/g DM)	188 [c]	234 [a]	228 [a]	208 [b]	3.39	<0.001
c (h^{-1})	0.0608	0.0650	0.0649	0.0617	0.000740	0.334
lag (h)	0.608	0.656	0.656	0.612	0.00981	0.131
IVOMD (% OM)	43.7 [d]	54.2 [a]	52.0 [b]	48.4 [c]	0.676	<0.001
ME (MJ/kg DM)	6.40 [c]	7.92 [a]	7.61 [a]	7.08 [b]	0.0995	<0.001

[a,b] Within a raw, different letters represent the significant differences at *p*-value < 0.05. [1] *b*, potential gas production; *c*, fractional rate of gas production; DM, dry matter; *lag*, initial delay in the onset of gas production; OM, organic matter; SEM, standard error of the mean.

3.5. In Situ Ruminal Degradability

The soluble DM fraction was significantly different ($p < 0.001$) among varieties and varied from 22.8 to 28.4% DM (Table 6). The soluble DM fraction of 333A was similar to that of Lanjian No. 3, but less than that of Lanjian No. 1 and Lanjian No. 2. The potentially degradable DM fraction and rate of DM degradation were similar ($p > 0.05$) for the four varieties and recorded an average of 43.2% DM and 0.0442 h^{-1}, respectively. The EDDM value was significantly different ($p < 0.001$) among the four varieties. Average EDDM value was 51.0% DM and ranged from 46.7 to 55.2% DM. Variety 333A had significantly less EDDM value compared to Lanjian No. 1 and Lanjian No. 2, but it was similar to Lanjian No. 3.

The NDF degradation profiles in situ are given in Table 5. Soluble NDF fraction, potentially degradable NDF fraction, and rate of NDF degradation were not influenced ($p > 0.05$) by variety and recorded an average of 11.5% NDF, 46.7% NDF and 0.0374 h^{-1}, respectively. No difference was observed in EDNDF value among varieties, but there was a trend toward greater EDNDF for Lanjian No. 1 ($p = 0.078$).

Table 6. Influence of variety on in situ ruminal degradation kinetics of dry matter (DM) and neutral detergent fiber (NDF) of straw in four common vetch varieties.

Dependent Variable	333A	Lanjian No. 1	Lanjian No. 2	Lanjian No. 3	SEM [1]	*p*-Value
DM						
A (% DM)	22.8 [b]	28.4 [a]	26.8 [a]	24.4 [b]	0.630	<0.001
B (% DM)	41.1	45.1	43.9	42.6	0.581	0.066
C (h^{-1})	0.0430	0.0453	0.0444	0.0442	0.0005	0.377
EDDM (% DM)	46.7 [c]	55.2 [a]	52.7 [ab]	49.5 [bc]	0.953	<0.001
NDF						
A (% NDF)	10.6	12.2	12.0	11.2	0.535	0.735
B (% NDF)	45.3	48.8	46.7	46.0	0.536	0.101
C (h^{-1})	0.0366	0.0388	0.0371	0.0373	0.0004	0.142
EDNDF (% NDF)	35.1	39.3	37.4	36.2	0.618	0.078

[a,b] Within a raw, different letters represent the significant differences at *p*-value < 0.05. [1] *A*, soluble fraction; *B*, potentially degradable fraction; *C*, rate of degradation of fraction B; EDDM, effective dry matter degradability; EDNDF, effective neutral detergent fiber degradability; SEM, standard error of the mean.

4. Discussion

4.1. Grain Yield, Straw Yield, and PUI

In smallholder crop-livestock systems, improvement in crop straw yield implies increase in milk and meat production from ruminants [6,32]. Substantial variability differences among varieties were found for grain and straw yields, partly due to differences in days to pod maturity and harvest index as suggested by Abd El-Moneim [8], Larbi et al. [9], and Kafilzadeh and Maleki [32]. Our findings are consistent with the earlier reports on common vetch [9] as well as faba bean (*Vicia faba* L.) [31], lentil (*Lens culinaris*) [32] and chickpea (*Cicer arietinum*) [33]. The grain yield recorded in this study (875.2–1255 kg DM/ha) is within the reported range (287–1783 kg DM/ha) [9], but less than ranges reported value of 1340–2240 kg DM/ha [12]. Meanwhile, the straw yield (3154–5556 kg DM/ha) is greater than that reported by Larbi et al. [9] (629–2226 kg DM/ha), but slightly less than that reported by Albayrak et al. [12] (4620–7320 kg DM/ha). The yields vary between studies [9,12] as consequence of the differences in the varieties, agronomic practices and growing conditions (e.g., soil type and climate) [8,12,34]. The grain yield of the improved varieties was significantly greater than that of the local variety in this study. Similar results were reported for other leguminous crops such as faba bean [31] and lentil [35]. The local variety demonstrated inferior PUI compared to the improved varieties. This is consistent with the findings in faba bean [31]; however, contrary to the findings of Tolera et al. [34] who observed less PUI for the improved varieties compared with the local varieties.

4.2. Chemical Composition

Higher CP and lower cell wall contents (NDF, ADF, cellulose and ADL) can be used as indicators of good feed quality [28]. Large varietal differences in straw chemical composition observed in our study is in agreement with the earlier findings in common vetch [9,11,13], as well as faba bean [31], lentil [32] and chickpea [33]. The range of CP content observed in this study (9.76–13.8% DM) was more than the threshold content (8.0% DM CP) required for optimum activity of rumen microorganisms in ruminants [36]. The NDF content among the varieties varied from 45.0 to 54.1% DM with a mean value of 49.9% DM. Van Soest [37] indicated that NDF content over 65% leads to effects on voluntary intake and production by ruminants. This makes common vetch straw a good source of CP supplements for ruminants in smallholder crop-livestock/agro-pastoral systems. Common vetch straw NDF and ADF contents of 52.2 and 36.1% DM reported by Makkar et al. [13] are consistent with our results; however, the CP contents (6.2% DM) is less than our results. Phosphorus and calcium contents observed in this study are similar to those reported by Abreu and Bruno-Soares [38]. The differences in chemical composition of straw between studies may be due to varietal variability, differences in growing condition (e.g., soil type and climate), or differences in harvesting and postharvest handling practices.

4.3. Carbohydrate and Protein Fractions

The significant varietal differences in carbohydrate fractions of common vetch straw are in agreement with reported for sorghum (*Sorghum bicolor* [L.] Moench) [28], timothy (*Phleum pratense* L.) [39] and wheat (*Triticum aestivum* L.) [40]. The values for TCHO, NSC and SC obtained in this study are comparable with the previous studies on other legume crops such as *Trifolium alaxendrinum* [41]. Among the CHO fractions, C_{B2} fraction was the highest in the straw of common vetch varieties analyzed. Others have reported similar results in sorghum, berseem (*Trifolium alexandrium*) and cowpea (*Vigna sinensis*) [28,41]. However, there is limited data available on the CHO fractions of common vetch straw. The pattern of CHO fractions observed in our study for common vetch straw is comparable with the earlier reports on other forage crops [41–43].

Wide range of protein fractions observed in the varieties is consistent with the findings in sorghum [28] and wheat [40]. Swarna et al. [44] indicated that P_{B2} and P_{B3} fractions represent a bypass protein of forage, while P_c fraction represents the non-degraded fraction. Compared to other varieties, Lanjian No. 1 had lower P_c fraction and higher P_{B2} + P_{B3} fractions. These observations on protein fractions suggest that straw of variety Lanjian No. 1 could be used as the better nitrogen source for ruminants. There is limited data available on the contents of protein fractions of common vetch. The pattern of protein fractions revealed here is similar to the reports on lucerne (*Medicago sativa* L.) [43] and black gram [*Vigna mungo* (L.) Hepper] [44].

4.4. In Vitro Gas Production

The gas produced by in vitro fermentation reflects the degree of feed fermentation and digestibility [34]. In this study, we observed significant differences among varieties in potential gas production, but not in gas production rate and *lag* time. Our results are partially in line with previous studies on other crops such as spineless cacti (*Opuntia* spp) [20] and chickpea [33]. Studies on gas production from common vetch straws are scarce, with the exception of the Spain studies [45]. López et al. [45] reported greater potential gas production, and less gas production rate and *lag* time in comparison with those observed in the current study. Cone and van Gelder [46] indicated that despite high degradability, feed with high CP typically produce less gas during fermentation as protein fermentation produces ammonia, which affects the carbonate buffer balance by neutralizing H^+ ions from volatile fatty acids without releasing carbon dioxide. The differences in gas production parameters between studies may be attributed to the differences in straw CP content, varieties, season, and location.

The IVOMD and ME contents of straw varied among the varieties, which is partly due to varietal variations in straw ADF and NDF and may be due to the proportions of straw morphological fractions, which were not measured in this study. In this study, varieties Lanjian No. 1 and Lanjian No. 2 recorded straw CP and IVOMD as high as 8.0% DM and 50.0% OM, respectively, which suggests that the straw of these varieties may be effectively used as a CP supplement to ruminants fed low-quality cereal straw-based diets [9]. The straw IVOMD range in this study is comparable to the range reported earlier in other vetch varieties [9], while the ME range is slightly less than those reported for vetches and other legume straws [13]. The straw quality varied between studies possibly due to differences in varieties, cell wall lignification, leaf to stem ratio, and the stage at which the straw was harvested. Larbi et al. [9] and Makkar et al. [13] earlier reported that straw IVOMD and ME contents of common vetch are influenced by varieties, growing season, and stage of straw harvest.

4.5. In Situ Ruminal Degradability

Higher ruminal degradability of high-fiber forages is satisfying because it implies improved the nutrient availability to rumen microbes [47]. The observed differences in DM degradation profiles of the straw varieties may be related to their varietal traits reflected as substantial differences in morphological and chemical composition. Our results are consistent with those reported for other crops such as maize [47]. For DM degradation parameters, the value of *A* fraction observed in this study for common vetch straw is greater and the values of *B* and *C* fractions are similar to those reported by Bruno-Soares et al. [48]. In this study, the value of *A* fraction of Lanjian No. 1 was significantly greater than other varieties, while the values for *B* and *C* fractions were similar for the four varieties analyzed. The differences in EDDM between the varieties can be mainly attributed to the differences in *A* fraction and not to *B* and *C* fractions, which is consistent with the literature [47]. The EDDM reported for common vetch straw in our study are similar to that reported for chickpea straw (51.8% DM) [49], but greater than that reported for fenugreek (*Trigonella foenum-graecum*) straw (32.2% DM) [50]. The degradation profiles of NDF were not influenced by variety in our study, which is in agreement with the earlier reports on maize and chickpea [47,49]. Bruno-Soares et al. [48] reported less rapidly degradable NDF fraction and rate of NDF degradation, and similar potentially degradable NDF fraction in comparison with those observed in this study. The EDNDF of straw in this study was less than reported by Abbeddou et al. [51] for lentil straw (45.9% NDF), but greater than that reported by Mustafa et al. [50] for fenugreek straw (20.3% NDF). Different studies recorded different ruminal degradation kinetics due to differences in varieties/species, straw composition, and animal species [4,49–51].

5. Conclusions

The results of this study showed varietal differences in grain and straw yields, and straw nutrient value in common vetch. Evaluation of common vetch varieties showed that Lanjian No. 1 had straw yield, straw CP, non-protein nitrogen, neutral detergent soluble protein, IVOMD, ME, and EDDM greater than other varieties, despite its less grain yield. Variety Lanjian No. 3 demonstrated early maturity and greater grain yield; however, straw yield and quality were less than Lanjian No. 1. Variety Lanjian No. 2 had greater grain yield and moderate straw CP; however, straw IVOMD, ME, EDDM and EDNDF contents were comparable with Lanjian No. 1. Based on these results, variety Lanjian No. 1 is the best option among varieties examined for smallholder farmers on the Qinghai-Tibetan plateau.

Author Contributions: Y.H. and Z.N. conceived and designed the experiment, Y.H. and F.Z. performed the experiments, Y.H. analyzed the data, and Y.H. and Z.N. wrote the manuscript.

Funding: This research was funded by the National Basic Research Program of China (No. 2014CB138706).

Conflicts of Interest: The authors declare no conflict of interest.

References

1. Food and Agriculture Organization of the United Nations Home Page. Available online: http://www.fao.org/docrep/015/am081m/am081m00.html (accessed on 1 November 2018).
2. Wathes, C.M.; Buller, H.; Maggs, H.; Campbell, M.L. Livestock production in the UK in the 21st century: A perfect storm averted? *Animals* **2013**, *3*, 574–583. [CrossRef] [PubMed]
3. Mottet, A.; Teillard, F.; Boettcher, P.; Dé Besi, G.; Besbes, B. Review: Domestic herbivores and food security: Current contribution, trends and challenges for a sustainable development. *Animal* **2018**, *12*, s188–s198. [CrossRef]
4. Huang, Y.F.; Li, R.; Coulter, J.A.; Zhang, Z.X.; Nan, Z.B. Comparative grain chemical composition, ruminal degradation in vivo, and intestinal digestibility in vitro of *Vicia sativa* L. varieties grown on the Tibetan Plateau. *Animals* **2019**, *9*, 212. [CrossRef] [PubMed]
5. Shang, Z.H.; Gibb, M.J.; Leiber, F.; Ismail, M.; Ding, L.M.; Guo, X.S.; Long, R.J. The sustainable development of grassland-livestock systems on the Tibetan plateau: Problems, strategies and prospects. *Rangel. J.* **2014**, *36*, 267–296. [CrossRef]
6. Nan, Z.B.; Abd El-Moneim, A.M.; Larbi, A.; Nie, B. Productivity of vetches (*Vicia* spp.) under alpine grassland conditions in China. *Trop. Grassl.* **2006**, *40*, 177–182.
7. Zhang, Q.P.; Bell, L.W.; Shen, Y.Y.; Whish, J.P.M. Indices of forage nutritional yield and water use efficiency amongst spring-sown annual forage crops in north-west China. *Eur. J. Agron.* **2018**, *93*, 1–10. [CrossRef]

8.   Abd El Moneim, A.M. Agronomic potential of three vetches (*Vicia* spp.) under rainfed conditions. *J. Agron. Crop Sci.* **1993**, *170*, 113–120. [CrossRef]

9.   Larbi, A.; Hassan, S.; Kattash, G.; El-Moneim, A.M.; Jammal, B.; Nabil, H.; Nakkul, H. Annual feed legume yield and quality in dryland environments in north-west Syria: 2. grain and straw yield and straw quality. *Anim. Feed Sci. Technol.* **2010**, *160*, 90–97. [CrossRef]

10.   Huang, Y.F.; Gao, X.L.; Nan, Z.B.; Zhang, Z.X. Potential value of the common vetch (*Vicia sativa* L.) as an animal feedstuff: A review. *J. Anim. Physiol. Anim. Nutr.* **2017**, *101*, 807–823. [CrossRef] [PubMed]

11.   Fortina, R.; Gasmi-Boubaker, A.; Lussiana, C.; Malfatto, V.; Tassone, S.; Renna, M. Nutritive value and energy content of the straw of selected *Vicia* L. taxa from Tunisia. *Ital. J. Anim. Sci.* **2015**, *14*, 280–284. [CrossRef]

12.   Albayrak, S.; Sevimay, C.S.; Töngel, Ö. Effects of inoculation with rhizobium on seed yield and yield components of common vetch (*Vicia sativa* L.). *Turk. J. Agric. For.* **2006**, *30*, 31–37.

13.   Makkar, H.P.S.; Goodchild, A.V.; Abd El-Moneim, A.M. Cell-constituents, tannin levels, by chemical and biological assays and nutritive value of some legume foliage and straw. *J. Food Agric.* **1996**, *71*, 129–136. [CrossRef]

14.   Haddad, S.G.; Husein, M.Q. Nutritive value of lentil and vetch straws as compared with alfalfa hay and wheat straw for replacement ewe lambs. *Small Rumin. Res.* **2001**, *40*, 255–260. [CrossRef]

15.   Renna, M.; Gasmi-Boubaker, A.; Lussiana, C.; Battaglini, L.M.; Belfayez, K.; Fortina, R. Fatty acid composition of the seed oil of selected *Vicia* L. taxa from Tunisia. *Ital. J. Anim. Sci.* **2014**, *13*, 308–316. [CrossRef]

16.   Russel, J.B.; O'Connor, J.D.; Fox, D.G.; Van Soest, P.J.; Sniffen, C.J. A net carbohydrate and protein system for evaluating cattle diets. I. Ruminal fermentation. *J. Anim. Sci.* **1992**, *70*, 3551–3561. [CrossRef] [PubMed]

17.   Marcos, C.N.; García-Rebollar, P.; de Blas, C.; Carro, M.D. Variability in the chemical composition and in vitro ruminal fermentation of olive cake by-products. *Animals* **2019**, *9*, 109. [CrossRef] [PubMed]

18.   Blümmel, M.; Ørskov, E.R. Comparison of in vitro gas production and nylon bag degradability of roughage in predicting feed intake in cattle. *Anim. Feed Sci. Technol.* **1993**, *40*, 109–119. [CrossRef]

19.   Fu, P. Effects of Inoculation with Rhizobium and Phosphate on Growth of Vetch. Master's Thesis, Lanzhou University, Lanzhou, China, 2015; pp. 1–52.

20.   Batista, A.M.; Mustafa, A.F.; McAllister, T.; Wang, Y.X.; Soita, H.; McKinnon, J.J. Effects of variety on chemical composition, in situ nutrient disappearance and in vitro gas production of spineless cacti. *J. Sci. Food Agric.* **2003**, *83*, 440–445. [CrossRef]

21.   Menke, K.H.; Steingass, H. Estimation of the energetic feed value obtained from chemical analysis and gas production using rumen fluid. *Anim. Res. Dev.* **1988**, *28*, 7–55.

22.   Nandra, K.S.; Dobos, R.C.; Orchard, B.A.; Neutze, S.A.; Oddy, V.H.; Cullis, B.R.; Jones, A.W. The effect of animal species on in sacco degradation of dry matter and protein of feeds in the rumen. *Anim. Feed Sci. Technol.* **2000**, *83*, 273–285. [CrossRef]

23.   Ørskov, E.R.; McDonald, I. The estimation of protein degradability in the rumen from incubation measurements weighted according to rate of passage. *J. Agric. Sci.* **1979**, *92*, 499–503. [CrossRef]

24.   AOAC (Association of Official Analytical Chemists). *Official Methods of Analysis*, 15th ed.; AOAC International: Arlington, VA, USA, 1990; pp. 71–90.

25.   Van Soest, P.J.; Robertson, J.B.; Lewis, B.A. Methods for dietary fibre, neutral detergent fibre, and non-starch polysaccharides in relation to animal nutrition. *J. Dairy Sci.* **1991**, *74*, 3583–3597. [CrossRef]

26.   Licitra, G.; Hernandez, T.M.; Van Soest, P.J. Standardizations of procedures for nitrogen fractionation of ruminant feeds. *Anim. Feed Sci. Technol.* **1996**, *57*, 347–358. [CrossRef]

27.   Sniffen, C.J.; O'Connor, J.D.; Van Soest, P.J.; Fox, D.G.; Russell, J.B. A net carbohydrate and protein system for evaluating cattle diets: II. Carbohydrate and protein availability. *J. Anim. Sci.* **1992**, *70*, 3562–3577. [CrossRef]

28.   Singh, S.; Bhat, B.V.; Shukla, G.P.; Gaharana, D.; Anele, U.Y. Nutritional evaluation of different varieties of sorghum stovers in sheep. *Anim. Feed Sci. Technol.* **2017**, *227*, 42–51. [CrossRef]

29.   Caballero, R.; Alzueta, C.; Ortiz, L.T.; Rodríguez, M.L.; Barro, C.; Rebolé, A. Carbohydrate and protein fractions of fresh and dried common vetch at three maturity stages. *Agron. J.* **2001**, *93*, 1006–1013. [CrossRef]

30.   Rode, L.M.; Yang, W.Z.; Beauchemin, K.A. Fibrolytic enzyme supplements for dairy cows in early lactation. *J. Dairy Sci.* **1999**, *82*, 2121–2126. [CrossRef]

31.   Alkhtib, A.S.; Wamatu, J.A.; Wegi, T.; Rischkowsky, B.A. Variation in the straw traits of morphological fractions of faba bean (*Vicia faba* L.) and implications for selecting for food-feed varieties. *Anim. Feed Sci. Technol.* **2016**, *222*, 122–131. [CrossRef]

32. Wamatu, J.; Mersha, A.; Tolera, A.; Beyan, M.; Alkhtib, A.; Eshete, M.; Ahmed, S.; Rischkowsky, B. Selecting for food-feed traits in early and late maturity lentil genotypes (*Lens culinaris*). *J. Exp. Biol. Agri. Sci.* **2017**, *5*, 697–705.

33. Kafilzadeh, F.; Maleki, E. Chemical composition, in vitro digestibility and gas production of straws from different varieties and accessions of chickpea. *J. Anim. Physiol. Anim. Nutr.* **2012**, *96*, 111–118. [CrossRef]

34. Tolera, A.; Tsegaye, B.; Berg, T. Effects of variety, cropping year, location and fertilizer application on nutritive value of durum wheat straw. *J. Anim. Physiol. Anim. Nutr.* **2008**, *92*, 121–130. [CrossRef]

35. Alkhtib, A.; Wamatu, J.; Ejeta, T.; Rischkowsky, B. Integrating the straw yield and quality into multi-dimensional improvement of lentil (*Lens culinaris*). *J. Sci. Food Agric.* **2017**, *97*, 4135–4141. [CrossRef]

36. Belachew, Z.; Yieshak, K.; Taye, T.; Janssens, G. Chemical composition and in sacco ruminal degradation of tropical trees rich in condensed tannins. *Czech J. Anim. Sci.* **2013**, *4*, 176–192. [CrossRef]

37. Van Soest, P.J. *Nutritional Ecology of the Ruminant*, 2nd ed.; Cornell University Press: Ithaca, NY, USA, 1994.

38. Abreu, J.M.F.; Bruno-Soares, A.M. Characterization and utilization of rice, legume and rape straws. *Toxicon* **1998**, *34*, 292–293.

39. Yu, P.; Christensen, D.A.; McKinnon, J.J.; Markert, J.D. Effect of variety and maturity stage on chemical composition carbohydrate and protein subfractions, in vitro rumen degradability and energy values of timothy and alfalfa. *Can. J. Anim. Sci.* **2003**, *83*, 279–290. [CrossRef]

40. Mustafa, A.F.; Christensen, D.A.; McKinnon, J.J.; Hucl, P.J. Chemical characterization of selected Western Canadian Spring and durum wheat cultivars using the Cornell net carbohydrate and protein analysis. *J. Sci. Food Agric.* **1999**, *79*, 1659–1665. [CrossRef]

41. Gupta, A.; Singh, S.; Kundu, S.S.; Jha, N. Evaluation of tropical feedstuffs for carbohydrate and protein fractions by CNCP system. *Indian J. Anim. Sci.* **2011**, *81*, 1154–1160.

42. Singh, S.; Kushwaha, B.P.; Nag, S.K.; Mishra, A.K.; Singh, A.; Anele, U.Y. In vitro ruminal fermentation, protein and carbohydrate fractionation, methane production and prediction of twelve commonly used Indian green forages. *Anim. Feed Sci. Technol.* **2012**, *178*, 2–11. [CrossRef]

43. Kamble, A.B.; Puniya, M.; Kundu, S.S.; Shelke, S.K.; Mohini, M. Evaluation of forages in terms of carbohydrate, nitrogen fractions and methane production. *Indian J. Anim. Nutr.* **2011**, *28*, 231–238.

44. Swarna, V.; Dhulipalla, S.K.; Elineni, R.R.; Dhulipalla, N.N. Evaluation of certain crop residues for carbohydrate and protein fractions by cornell net carbohydrate and protein system. *J. Adv. Vet. Anim. Res.* **2015**, *2*, 213–216. [CrossRef]

45. López, S.; Davies, D.R.; Giraldez, F.J.; Dhanoa, M.S.; Dijkstra, J.; France, J. Assessment of nutritive value of cereal and legume straws based on chemical composition and in vitro digestibility. *J. Sci. Food Agric.* **2005**, *85*, 1550–1557. [CrossRef]

46. Cone, J.W.; van Gelder, A.H. Influence of protein fermentation on gas production profiles. *Anim. Feed Sci. Technol.* **1999**, *76*, 251–264. [CrossRef]

47. Zeller, F.M.E.; Edmunds, B.L.; Schwarz, F.J. Effect of genotype on chemical composition, ruminal degradability and in vitro fermentation characteristics of maize residual plants. *J. Anim. Physiol. Anim. Nutr.* **2014**, *98*, 982–990. [CrossRef]

48. Bruno-Soares, A.M.; Abreu, J.M.F.; Guedes, A.; Dias-da-Silva, A. Chemical composition, DM and NDF degradation kinetics in rumen of seven legume straws. *Anim. Feed Sci. Technol.* **2000**, *83*, 75–80. [CrossRef]

49. Kılıçalp, N.; Hızlı, H.; Mart, D. Chemical composition and rumen degradation characteristics of different chickpea (*Cicer Arietinum* L.) lines straw. *Turk. J. Agric. Food Sci. Technol.* **2017**, *5*, 459–463. [CrossRef]

50. Mustafa, A.F.; Christensen, D.A.; McKinnon, J.J. In vitro and in situ evaluation of fenugreek (*Trigonel foenum-graecum*) hay and straw. *Can. J. Anim. Sci.* **1996**, *76*, 625–628. [CrossRef]

51. Abbeddou, S.; Rihawi, S.; Hess, H.D.; Iniguez, L.; Mayer, A.C.; Kreuzer, M. Nutritional composition of lentil straw, vetch hay, olive leaves, and saltbush leaves and their digestibility as measured in fat-tailed sheep. *Small Rumin. Res.* **2011**, *96*, 126–135. [CrossRef]

Article

Comparative Grain Chemical Composition, Ruminal Degradation In Vivo, and Intestinal Digestibility In Vitro of *Vicia Sativa* L. Varieties Grown on the Tibetan Plateau

Yafeng Huang [1,2,3], **Rui Li** [1,2,3], **Jeffrey A. Coulter** [4], **Zhixin Zhang** [1,2,3] **and Zhibiao Nan** [1,2,3,*]

[1] State Key Laboratory of Grassland Agro-ecosystems, Lanzhou University, Lanzhou 730020, China; huangyafeng316@163.com (Y.H.); lir15@lzu.edu.cn (R.L.); zhixinzhang.phd@gmail.com (Z.Z.)

[2] Key Laboratory of Grassland Livestock Industry Innovation of Ministry of Agriculture and Rural Affairs, Lanzhou University, Lanzhou 730020, China

[3] College of Pastoral Agriculture Science and Technology, Lanzhou University, Lanzhou 730020, China

[4] Department of Agronomy and Plant Genetics, University of Minnesota, St. Paul, MN 55108, USA; jeffcoulter@umn.edu

[*] Correspondence: zhibiao@lzu.edu.cn; Tel.: +86 931 8912582

Received: 29 March 2019; Accepted: 30 April 2019; Published: 2 May 2019

Simple Summary: Common vetch (*Vicia sativa* L.) grain is an important source of protein in rations for ruminants, but little information is available on the protein value of common vetch grains, both in terms of chemical composition and protein degradability, and regarding variation between intra-species and year. The objective of this study was to evaluate grain chemical composition, ruminal protein degradability in vivo, and intestinal protein digestibility in vitro of four common vetch varieties over two cropping years on the Tibetan Plateau. This study was also conducted to establish correlations of grain chemical composition with ruminal degradability parameters of grain protein and with intestinal digestibility of grain protein. Results of this study demonstrated that grain quality characteristics varied significantly among varieties and years. The relationship between grain chemical composition and intestinally absorbable digestible protein (IADP) was best described by a linear regression equation, and coefficients of determination remained very high ($R^2 = 0.891$). Overall, the results indicated that in terms of effective crude protein degradability and IADP of grain, common vetch varieties Lanjian No. 2 and Lanjian No. 3 have the greatest potential among varieties examined for supplementing ruminant diets when grown on the Tibetan Plateau.

Abstract: Four varieties of common vetch, three improved varieties and one local variety, were evaluated for grain chemical composition, rumen protein degradability, and intestinal protein digestibility over two cropping years on the Tibetan Plateau. This study also examined correlations of grain chemical composition with rumen degradability parameters of grain protein and with intestinal digestibility of grain protein. Results of this study showed that grain quality attributes varied ($p < 0.05$) among varieties and cropping years. Significant intra-species variation was observed for concentrations (g/kg dry matter) of crude protein (CP; range = 347–374), ether extract (range = 15.8–19.6), neutral detergent fiber (aNDF; range = 201–237), acid detergent fiber (range = 58.2–71.6), ash (range = 27.6–31.0), effective CP degradability (EDCP; range = 732–801 g/kg CP), and intestinally absorbable digestible protein (IADP; range = 136–208 g/kg CP). The relationship between grain chemical composition and IADP was best described by the linear regression equation IADP = –0.828CP + 8.80ash + 0.635aNDF + 70.2 ($R^2 = 0.891$), indicating that chemical analysis offers a quick and reliable method for IADP of common vetch grain. In terms of EDCP and IADP of grain, common vetch varieties, Lanjian No.2 and Lanjian No. 3, have the greatest potential among varieties tested for supplementing ruminant diets when grown on the Tibetan Plateau.

Keywords: common vetch; grain; nutritive value; ruminants

1. Introduction

Livestock sector is the largest land-use system worldwide and is an engine of economic growth that has received significant attention within the last decade [1,2]. The Tibetan Plateau covers an area of 256 million ha^2 and is the largest grassland ecosystem in Eurasia [3]. It supports nearly 41 million Tibetan sheep (*Ovies aries*) and 14.3 million yak (*Bos grunniens*), cattle (*Bos taurus*), and cattle-yak hybrids, which are a predominant part of the local economy for the nomadic population of over 9.8 million [3]. The Tibetan Plateau is also the source of Asia's major river systems, including the Mekong, Yangtze, Yellow, Ganges, and Indus. Therefore, sustainable development and management of the Tibetan Plateau is of vital importance, not only for supporting the livelihood of millions of people, but also for protection of these critical river systems.

Legume grains are important sources of protein in ruminant nutrition [4–6]. Demand for plant-based protein has grown steadily, since the influence of the bovine spongiform encephalopathy which resulted in the banning of meat meal for feeding ruminants in the European Union in 1994 [7]. The Tibetan Plateau has an inherently extreme and unstable climate, and a short growing season, which seriously limits crop production [8], especially for leguminous crops [9]. Nan et al. [9] reported that many annual legume crops [e.g., woolly-pod vetch (*Vicia villosa* ssp. *dasycarpa Roth*) have low grain yield potential in this region. Common vetch was introduced to the Tibetan Plateau of China in 1998 and has shown promising potential for grain production [9]. Recently, new varieties of common vetch designated as Lanjian No.1, Lanjian No.2, and Lanjian No.3, which represent diverse levels of maturity with acceptable grain production on the Tibetan Plateau, were developed by the common vetch (*Vicia sativa* L.) breeding program at Lanzhou University [10].

Common vetch is a cool-season annual legume cultivated in many countries that can satisfy grain demand for feed and food with low requirement for nitrogen input due to biological nitrogen fixation [5,11]. The nutritive value of common vetch grain is relatively high, with an average of 314 g/kg crude protein (CP) and 961 g/kg CP total digestible protein [7,12,13]. This makes common vetch grain a potentially important source of protein in rations for ruminants, such as cattle and sheep, and is of great interest in mixed crop-livestock systems [5,7,13–15], particularly due to the banning of soybean (*Glycine max* (L.) Merr.) meal for feeding organic livestock [16]. Koumas and Economides [14] and Gül et al. [15] reported satisfactory ruminant growth with diets containing common vetch grain as a replacement for soybean (*Glycine max* (L.) Merr.) meal.

For the feed application of common vetch grain, previous studies focused on grain chemical composition [4,11,17] or nutritional values of a single variety [7]. The nutrient value of common vetch for ruminants varies widely in the literature and depends on the common vetch variety and animal species, and chemical composition of common vetch grain. These phenomena indicate the need to analyze the grain of common vetch varieties for nutritive value before using in ration formulations. However, little information is available on protein value of common vetch grains, regarding variation between intra-species and the growing season. Rumen degradability of protein is a key parameter for the evaluation of feed protein value in ruminants [7,18], and the benefits of protein resources are dependent on protein digestion in the small intestine [7,19]. Minerals are a minor ingredient of ruminant feeds, but they are essential for growth and development [20]. Existing data on the mineral composition of common vetch grain is limited. Woods et al. [21,22] reported that chemical analysis of concentrate feedstuffs could be used for accurate prediction of ruminal and intestinal protein degradability. These observations could be more cost and time effective than using in situ nylon bags and in vitro techniques with surgically prepared animals, which are relatively expensive, labor-intensive, and subject to animal welfare problems [21,22]. However, no available information has been reported on the relationship between grain chemical composition with grain rumen degradation parameters of CP and grain intestinal protein digestibility for common vetch.

Understanding the varietal differences in nutritive value of common vetch grain grown on the Tibetan Plateau could increase its use as a protein supplement in ruminant diets and lead to greater integration of common vetch within agro-pastoral farming systems in this region, thereby reducing the need for protein feed imports and enhancing the income of farmers. Against this background, the objectives of this study were: (i) to evaluate the potential of common vetch for use as a protein feed source in ruminant diets, by studying grain chemical composition, ruminal degradability kinetics of CP, and intestinal protein digestibility of four common vetch varieties over two cropping years; (ii) to investigate relationships between grain protein, fiber fractions, and ash concentrations, and grain ruminal degradability parameters of CP and intestinally digestible protein of common vetch varieties grown in rainfed conditions on the Tibetan Plateau.

2. Materials and Methods

2.1. Experimental Site

Common vetch was planted in rainfed conditions during the 2015 and 2016 cropping seasons at the Xiahe experimental station of Lanzhou University, China (35°19′ N, 102°58′E; 2880 m above sea level (asl)), which is located on the eastern margin of the Tibetan Plateau. The soil type of the site is classified as chernozem and is slightly acidic (pH 6.8), low in phosphorous (0.73 g/kg) but adequate in potassium (13.5 g/kg). The preceding crop was rape (*Brassica campestris* L.) in both years. The daily meteorological data (air temperature and rainfall) were recorded via an automatic meteorological station (PC200W, Campbell Scientific) installed near the experimental field. During the cropping period (April through September) in 2015 and 2016, mean air temperature was 10.8 and 12.0 °C, respectively, and total precipitation was 314 and 417 mm, respectively. Monthly mean temperature and total precipitation in July and August were 2.1 to 3.4 °C, respectively, and 22 to 34 mm greater in 2016 than 2015 (Table 1).

Table 1. Monthly mean air temperature and total precipitation during the 2015 and 2016 cropping seasons at the Xiahe experimental station, Gansu, China.

Month	Mean Air Temperature (°C)		Total Precipitation (mm)	
	2015	2016	2015	2016
April	4.4	6.2	23	25
May	9.3	9.3	38	72
June	12.4	12.7	49	54
July	13.4	15.5	69	103
August	14.2	17.6	62	84
September	11.2	10.6	73	79
Sum			314	417

2.2. Plant Material, Experimental Design, and Sampling

Three improved varieties of common vetch (Lanjian No. 1, Lanjian No. 2, and Lanjian No. 3) and one local variety were evaluated in a field experiment. The local variety originated in the Gansu province of China and is now widely cultivated in Gansu, Qinghai, and provinces located in the middle and lower reaches of the Yangtze River in China. These varieties were selected because they are grown extensively in the region surrounding the research site [10]. Agronomic characteristics of these varieties are in Table 2 [23]. Plots were planted at 150 viable seeds m^{-2}, with four replicates per variety in a completely randomized design. Each plot was 8 × 5 m and contained 26 rows of plants, spaced 20 cm apart. Common vetch was planted on 6 May 2015 and 28 April 2016 following rhizobial inoculation (CCBAU01069, China Agricultural University, Beijing, China), which was recommended based on the symbiont performance for these varieties [10]. Irrigation and fertilizer were not applied to the experimental plots during either growing season and weeds were controlled by hand.

Table 2. Agronomic characteristics of the common vetch varieties used in this experiment.

Agronomic Characteristic	Local Variety	Lanjian No.1	Lanjian No.2	Lanjian No.3
Days to mature (d)	134	145	132	124
1000 grains weight (g)	54	79	71	76
Plant height (cm)	92	106	80	69
Altitude (m.a.s.l)	–	<3000	<3500	<4000
Year of release	1987	2014	2015	2011

Source: Ministry of Agriculture, Beijing, China [23].

At the pod maturity stage of common vetch, two representative 1 × 1 m sections from each plot were manually harvested and threshed to obtain grain samples. Grain was oven-dried at 60 °C for 48 h and then ground to pass through a 2-mm sieve to obtain samples for analyses of ruminal incubation and intestinal digestibility, and to pass through a 1-mm sieve to obtain samples for chemical analysis.

2.3. Chemical Analysis

Laboratory dry matter (DM) concentration was measured from 2.0 g of ground grain from each sample by drying in a forced-air oven at 135 °C for 2 h (method 930.15; AOAC) [24]. Subsequently, grain ash concentration was measured by incineration in a muffle furnace at 550 °C for 5 h (method 938.08; AOAC). Total N was measured using the Kjeldahl method (method 988.05) [24] and CP concentration was calculated as total N × 6.25. Ether extract (EE) concentration was measured by extraction with petroleum ether (method 920.85; AOAC) [24]. Concentrations of acid detergent fiber (ADF) and neutral detergent fiber (aNDF) were measured by sequential analysis with α-amylase and sodium sulfite, and expressed with residual ash excluded [25]. Mineral concentrations were determined by atomic absorption spectroscopy (method 985.01; AOAC) [26] and phosphorus (P) concentration was determined by colorimetry (method 965.17; AOAC) [24]. All measurements were performed in triplicate and chemical standards were included in each analytical run as appropriate.

2.4. Ruminal Degradability of CP

Four adult fistulated dorper sheep rams 30–31 months of age and 58 ± 1 kg live body weight were used to determine ruminal degradation of CP of common vetch grain using the nylon bag technique described by Nandra et al. [27]. Briefly, 5.0 g of dry weight grain from each sample (in duplicate) was placed in a nylon bag (9 × 5 cm; 50-μm pore size) and incubated for 0 (control), 2, 4, 8, 12, 24, and 48 h in the rumen of four adult fistulated dorper rams. After removal, the bags were washed thoroughly with tap water and frozen (−20 °C) until further analysis. Prior to analysis, all bags were defrosted and manually washed with tap water until the water ran clear, before being oven-dried at 65 °C for 48 h and weighed. The dried undigested residues of replicates per time within rams were pooled prior to analysis. Degradability parameters of CP were determined using the exponential model described by Ørskov and McDonald [28]. The effective degradable fraction of CP was calculated as effective CP degradability (EDCP) = $A + [(B \times C)/(C + k)]$, where A is the soluble fraction of grain CP, B is the potentially degradable fraction of grain CP, C is the rate of degradation of fraction B (h^{-1}), and k is the rumen outflow rate (0.031 h^{-1}) [7]. The animals were housed in individual stalls and daily fed 1200 g of 550 g/kg DM sheepgrass (*Leymus chinensis*, (Trin.) Tzvel), 294 g/kg DM maize (*Zea mays* L.) grain, 140 g/kg DM soybean meal, and 16 g/kg DM of a concentrate mix at maintenance energy level [7] in two equal meals at 08:30 am and 16:30 pm. The rams had free access to water and mineral/vitamin licks. The experimental protocols were approved by the Animal Ethics Committee of Lanzhou University (protocol number: AEC-LZU-2016-01).

2.5. Intestinal Digestibility of CP

A modified 3-step in vitro procedure described by Gargallo et al. [29] was used to determine intestinal digestibility of rumen undegradable protein (IDP). Briefly, dried duplicate undegradable residues from the 16 h in situ ruminal incubation were pooled and ground to pass through a 1-mm sieve. Six sub-samples of dried sample residues (500 mg each) were then weighed into Ankom F57 filter bags and heat-sealed. Twenty-four sample bags were incubated in each incubation jar of a Daisy^{II} incubator (ANKOM Technology, Fairport, NY, USA) containing 2 L of 0.1 M pre-warmed HCl solution (pH = 1.9) and 1 g L^{-1} of pepsin (P-7000, Sigma, St. Louis, MO, USA), with constant rotation at 39 °C for 1 h. After incubation and washing, sample bags were reintroduced into the same incubation jars containing 2 L of pre-warmed pancreatin (0.5 M KH$_2$PO$_4$ buffer, pH = 7.75, 50 mg/kg of thymol, and 3 g/L of pancreatin (P-7545, Sigma)) and incubated with constant rotation at 39 °C for 24 h. After removal, bags were rinsed with tap water until the water ran clear, dried at 55 °C for 48 h, and reweighed. Undegradable residues from ruminal and intestinal incubations were analyzed for nutrient concentration.

2.6. Statistical Analyses

Ruminal undegradable protein (RUP) concentration was calculated as 1000 – EDCP (g/kg CP), and intestinally absorbable digestible protein (IADP) was determined as RUP (g/kg of CP) × IDP (g/kg of RUP), as described by Lawrence and Anderson [30]. Total digestible protein (TDP) was calculated as TDP = EDCP + IADP. Statistical analyses were performed using SPSS software (Version 21.0. IBM Corporation, Armonk, NY, USA). Data were analyzed using analysis of variance to assess the significance of the main effects of common vetch variety and year, and their interaction. When the F-tests were significant, variances among means were compared using the Duncan significant difference test at $p < 0.05$. Pearson's correlation analysis was used to evaluate relationships between chemical composition variables and A, B, C, and IADP of grain samples across varieties and years ($n = 32$). Chemical composition data of grain samples of common vetch varieties across years ($n = 32$) were analyzed to predict the rumen degradability parameters of CP (A, B, and C) and IADP using stepwise multiple linear regression [21,22].

3. Results

3.1. Chemical Composition

The effects of year and variety significantly influenced all proximate composition variables of common vetch grain with the exception of grain ash concentration, which was only affected by variety (Table 3). The interaction between year and variety did not significantly affect proximate composition variables. Averaged across varieties, CP concentration was less and EE, aNDF, and ADF concentrations were greater in 2016 compared to 2015. Averaged over both years, CP concentration among varieties ranged from 353 g/kg DM for Lanjian No. 3 to 370 g/kg DM for the local variety. Crude protein concentration of Lanjian No.1 was less than that of the local variety, but greater than that of the other improved varieties. Inversely, EE ranged from 16.1 g/kg DM for the local variety to 19.0 g/kg DM for Lanjian No. 3. The local variety had significantly less EE concentration compared to Lanjian No. 2 and Lanjian No. 3 but was similar to Lanjian No. 1. Averaged across years, aNDF and ADF among varieties ranged from 207 to 229 g/kg DM and 59.5 to 68.8 g/kg DM, respectively. Neutral detergent fiber and ADF of the local variety were not significantly different than that of Lanjian No. 1 but were less than that of other improved varieties. Ash concentration ranged from 27.8 g/kg DM for the local variety to 31.0 g/kg DM for Lanjian No. 3. The local variety had significantly less ash concentration compared to the improved varieties. There were significant differences in ash concentration among improved varieties, which was greater for Lanjian No. 2 and Lanjian No. 3 than Lanjian No. 1.

Phosphorus (P), calcium (Ca), magnesium (Mg), and iron (Fe) concentrations in common vetch grain varied significantly between years and among varieties, but were not affected by the year × variety interaction (Table 3). Averaged across varieties, grain concentration of all measured macronutrients was greater in 2016 than in 2015. Means over both years, P, Mg, Ca, and Fe concentrations among varieties ranged from 1.25 to 3.42, 3.18 to 3.85, 1.39 to 1.95, and 0.431 to 0.801 g/kg DM, respectively.

Phosphorus and Fe concentrations were lower and Mg and Ca concentrations were greater in grain of the local variety compared with the improved varieties, and there were no significant differences among improved varieties. Micronutrient levels also varied among varieties, but were not significantly affected by year or the interaction between year and variety. Averaged across years, grain concentrations of zinc (Zn), manganese (Mn), and copper (Cu) among varieties ranged from 36.3 to 48.6, 12.2 to 14.3, and 6.41 to 9.59 mg/kg DM, respectively. Micronutrient concentrations of grain of the local variety were greater than those of the improved varieties, and there were no significant differences among improved varieties.

3.2. Ruminal Degradation Kinetics of CP

Soluble fraction of grain CP of common vetch was significantly influenced by year and variety, but not by their interaction (Table 4). Averaged across varieties, A was greater in 2016 than 2015 (346 and 325 g/kg, respectively). Averaged across years, A among varieties ranged from 312 g/kg for Lanjian No. 2 to 359 g/kg for the local variety, and A of the local variety did not differ significantly from that of Lanjian No. 1 and was greater than that of other improved varieties. There were significant differences in the B and C among varieties, although the effect of year and of year × variety interaction were not significant. Averaged across years, B among varieties ranged from 552 g/kg for Lanjian No. 3 to 608 g/kg for the local variety. Potentially degradable protein of the local variety was greater than that of Lanjian No. 2 and Lanjian No. 3, and did not differ significantly from that of Lanjian No.1. In contrast, C ranged from an average of 0.0770 h^{-1} for the local variety and Lanjian No. 1 to an average of 0.0966 h^{-1} for Lanjian No. 2 and Lanjian No. 3. The rate of protein degradation of the local variety was not significantly different from that of Lanjian No. 1, but was less than that of other improved varieties.

Effective CP degradability of common vetch grain differed significantly between years (772 and 755 g/kg CP in 2015 and 2016, respectively) and among varieties, but was not significantly influenced by the year × variety interaction. Averaged across years, EDCP among varieties ranged from 736 g/kg CP for Lanjian No. 3 to 792 g/kg CP for the local variety. The EDCP of the local variety was significantly greater than that of Lanjian No. 2 and Lanjian No. 3, but did not differ significantly from that of Lanjian No. 1.

3.3. Intestinal Digestibility of CP

The IDP and IADP of grain were significantly affected by year and variety of common vetch, but not by the year × variety interaction (Table 4). Averaged across varieties, IDP and IADP were less in 2015 than in 2016. Averaged across years, IDP and IADP among varieties ranged from 692 to 768 g/kg of RUP, and 144 to 203 g/kg CP, respectively. The IDP and IADP of the local variety were less than that of Lanjian No. 2 and Lanjian No. 3, but not significantly different than that of Lanjian No. 1. Total digestible protein of grain was not significantly affected by year, variety of common vetch, or the interaction between year and variety, and averaged 937 g/kg CP across years and varieties.

Table 3. Effects of cropping year, variety, and the year × variety interaction on the chemical composition of four common vetch grain varieties grown on the Tibetan Plateau.

Dependent Variable	Mean across Years					Year										p-Value		
						2015					2016							
	Local Variety	Lanjian No.1	Lanjian No.2	Lanjian No.3	SEM [1]	Local Variety	Lanjian No.1	Lanjian No.2	Lanjian No.3	SEM	Local Variety	Lanjian No.1	Lanjian No.2	Lanjian No.3	SEM	Year	Variety	Year × Variety
Proximate composition (g/kg DM)																		
CP	370 [a]	360 [b]	354 [c]	353 [c]	1.57	374 [a]	365 [b]	358 [b]	358 [b]	1.88	366 [a]	356 [b]	350 [c]	347 [c]	1.96	< 0.001	<0.001	0.827
EE	16.1 [c]	17.5 [bc]	18.5 [ab]	19.0 [a]	0.278	15.8 [b]	17.2 [ab]	17.8 [a]	18.5 [a]	0.372	17.3 [b]	17.8 [ab]	19.3 [a]	19.6 [a]	0.368	0.011	0.002	0.818
aNDF	207 [c]	217 [bc]	225 [ab]	229 [a]	2.46	201 [b]	209 [ab]	220 [a]	220 [a]	2.87	214 [b]	225 [ab]	231 [a]	237 [a]	3.19	< 0.001	<0.001	0.942
ADF	59.5 [b]	61.3 [b]	68.3 [a]	68.8 [a]	1.12	58.2 [b]	59.0 [ab]	65.8 [a]	66.1 [a]	1.38	60.8 [c]	63.5 [bc]	70.8 [ab]	71.6 [a]	1.63	0.014	<0.001	0.929
Ash	27.8 [c]	29.3 [b]	30.7 [a]	31.0 [a]	0.261	27.9 [c]	29.6 [b]	30.7 [ab]	31.0 [a]	0.353	27.6 [c]	29.1 [b]	30.8 [a]	31.0 [a]	0.395	0.548	<0.001	0.845
Macronutrient minerals (g/kg DM)																		
P	1.25 [b]	3.06 [a]	3.42 [a]	3.33 [a]	0.192	0.993 [b]	2.78 [a]	3.24 [a]	3.08 [a]	0.268	1.50 [b]	3.33 [a]	3.61 [a]	3.57 [a]	0.270	0.042	<0.001	0.993
Mg	3.85 [a]	3.20 [b]	3.18 [b]	3.23 [b]	0.0736	3.75 [a]	3.05 [b]	3.09 [b]	3.11 [b]	0.108	3.96 [a]	3.35 [b]	3.28 [b]	3.35 [b]	0.0947	0.047	<0.001	0.989
Ca	1.95 [a]	1.39 [b]	1.54 [b]	1.47 [b]	0.0655	1.90 [a]	1.20 [b]	1.36 [b]	1.29 [b]	0.105	2.00 [a]	1.58 [b]	1.72 [ab]	1.65 [b]	0.0612	0.007	0.003	0.396
Fe	0.431 [b]	0.793 [a]	0.801 [a]	0.750 [a]	0.0479	0.345	0.630	0.635	0.600	0.0467	0.517 [b]	0.955 [a]	0.967 [a]	0.900 [a]	0.0684	< 0.001	0.002	0.820
Micronutrient minerals (mg/kg DM)																		
Zn	48.6 [a]	38.4 [b]	37.7 [b]	36.3 [b]	1.30	50.0 [a]	39.2 [b]	38.7 [b]	40.2 [b]	1.76	47.2 [a]	37.7 [b]	36.7 [b]	32.5 [b]	1.88	0.097	<0.001	0.625
Mn	14.3 [a]	12.2 [b]	13.2 [b]	13.1 [b]	0.220	14.0	12.2	13.3	14.1	0.291	14.6 [a]	12.2 [b]	13.0 [b]	12.0 [b]	0.332	0.201	0.002	0.056
Cu	9.59 [a]	6.81 [b]	6.68 [b]	6.41 [b]	0.357	10.1 [a]	7.79 [ab]	6.30 [b]	6.05 [b]	0.589	9.10 [a]	5.83 [b]	7.05 [b]	6.77 [b]	0.420	0.527	0.002	0.271

[a,b] Within a row, different letters represent the significant differences at p-value < 0.05 by the Duncan significant difference test. [1] SEM, standard error of the mean; aNDF, neutral detergent fiber; ADF, acid detergent fiber; Ca, calcium; CP, crude protein; Cu, copper; DM, dry matter; EE, ether extract; Fe, iron; Mg, magnesium; Mn, manganese; P, phosphorus; Zn, zinc.

Table 4. Effects of cropping year, variety, and the year × variety interaction on the ruminal degradation kinetics of crude protein and intestinal protein digestibility of four common vetch varieties grown on the Tibetan Plateau.

Dependent Variable	Mean across Years					Year										p-Value		
						2015					2016							
	Local Variety	Lanjian No.1	Lanjian No.2	Lanjian No.3	SEM [1]	Local variety	Lanjian No.1	Lanjian No.2	Lanjian No.3	SEM	Local Variety	Lanjian No.1	Lanjian No.2	Lanjian No.3	SEM	Year	Variety	Year × Variety
A, g/kg CP	359 [a]	352 [a]	312 [b]	320 [b]	5.03	367 [a]	360 [a]	327 [b]	331 [b]	6.13	352 [a]	343 [a]	298 [b]	309 [b]	7.22	0.004	<0.001	0.873
B, g/kg CP	608 [a]	604 [a]	570 [b]	552 [b]	5.86	610 [a]	610 [a]	565 [ab]	537 [b]	10.3	606 [a]	597 [ab]	575 [bc]	567 [c]	5.88	0.508	<0.001	0.355
C, h^{-1}	0.0766 [b]	0.0774 [b]	0.0975 [a]	0.0956 [a]	0.00202	0.0727 [b]	0.0759 [b]	0.0981 [a]	0.0985 [a]	0.00321	0.0761 [b]	0.0788 [b]	0.0968 [a]	0.0928 [a]	0.00256	0.560	<0.001	0.587
EDCP, g/kg CP	792 [a]	782 [a]	744 [b]	736 [b]	5.02	801 [a]	793 [a]	755 [b]	739 [b]	7.38	782 [a]	771 [a]	732 [b]	733 [b]	6.30	<0.001	<0.001	0.514
IDP, g/kg of RUP	692 [b]	706 [b]	756 [a]	768 [a]	6.73	683 [b]	705 [b]	746 [a]	754 [a]	8.39	702 [b]	708 [b]	765 [a]	782 [a]	10.3	0.023	<0.001	0.514
IADP, g/kg CP	144 [b]	154 [b]	194 [a]	203 [a]	4.99	136 [b]	146 [b]	183 [a]	197 [a]	7.00	153 [b]	162 [b]	205 [a]	208 [a]	6.70	<0.001	<0.001	0.661
TDP, g/kg CP	936	937	937	939	1.03	937	940	938	936	1.12	935	934	937	942	1.76	0.700	0.861	0.796

[a,b] Within a row, different letters represent the significant differences at p-value < 0.05 by the Duncan significant difference test. [1] SEM, standard error of the mean; A, soluble fraction of grain CP; B, potentially degradable fraction of grain CP; C, rate of degradation of fraction B; CP, crude protein; EDCP, effective CP degradability (k = 0.031 h^{-1}); RUP, rumen undegraded protein; IDP, intestinal digestible protein after 16 h of rumen incubation and 1 h of pepsin and 24 h of pancreatin digestion; IADP, intestinally absorbable digestible protein; TDP, total digestible protein. RUP = 1000 − EDCP (g/kg of CP); IADP = RUP (g/kg of CP) × IDP (g/kg of RUP); TDP = EDCP + IADP.

3.4. Prediction of A, B, C, and IADP

Correlations between grain chemical composition and A, B, C, and IADP were all significant at $p < 0.05$, with absolute values of the correlation coefficient ranging from 0.350 to 0.787 (Table 5). Therefore, stepwise multiple linear regression was used to predict A, B, C, and IADP. Goodness of fit of the regression models was evaluated based on the coefficient of determination (R^2) and root mean squared error (RMSE). For each parameter, the selected regression model produced the highest R^2 and lowest RMSE, with R^2 values of 0.805, 0.712, 0.748, and 0.891 for prediction of A, B, C, and IADP, respectively (Table 6).

Table 5. Pearson's correlations between chemical composition, ruminal degradability parameters of crude protein, and intestinally absorbable digestible protein of common vetch grains.

Dependent Variable [1]	CP	EE	NDF	ADF	Ash
A	0.581 ***	−0.463 **	−0.751 ***	−0.672 ***	−0.597 ***
B	0.605 ***	−0.524 **	−0.350 *	−0.461 **	−0.619 ***
C	−0.611 ***	0.549 **	0.533 **	0.620 ***	0.710 ***
IADP	−0.787 ***	0.663 ***	0.735 ***	0.727 ***	0.781 ***

[1] A, soluble fraction of grain CP; B, potentially degradable fraction of grain CP; C, rate of degradation of fraction B; aNDF, neutral detergent fiber; ADF, acid detergent fiber; CP, crude protein; EE, ether extract; IADP, intestinally absorbable digestible protein. * $p < 0.05$; ** $p < 0.01$; *** $p < 0.001$.

Table 6. Equations to predict the ruminal degradability parameters A, B, (g/kg CP), and C (h^{-1}) of crude protein and intestinally absorbable digestible protein (IADP; g/kg CP) of common vetch grains.

Dependent Variable [1]	Equation	R^2	RMSE
	= −1.53 aNDF + 672	0.751	18.8
A	= −1.23aNDF − 5.73ash + 777	0.795	17.3
	= −1.38aNDF − 7.18ash + 3.27EE + 794	0.805	16.9
	= −13.9ash + 996	0.619	26.1
B	= −8.81ash + 1.33CP + 367	0.676	24.4
	= −8.99ash + 2.01CP + 0.591aNDF − 0.231	0.696	23.8
	= −6.99ash + 2.11CP + 1.22 aNDF − 1.73ADF − 125	0.712	23.3
C	= 0.00549ash − 0.0765	0.710	0.00806
	= 0.00409ash + 0.000536ADF − 0.0692	0.748	0.00759
	= −2.51CP + 1076	0.787	17.4
IADP	= −1.55CP + 9.00ash − 465	0.867	14.1
	= −0.828CP + 8.80ash + 0.635aNDF + 70.2	0.891	12.8

[1] Units: g/kg dry matter for grain chemical composition of CP, EE, aNDF, and ash; A, soluble fraction of grain CP; B, potentially degradable fraction of grain CP; C, rate of degradation of fraction B (h^{-1}); aNDF, neutral detergent fiber; CP, crude protein; EE, ether extract; R^2, coefficients of determination; RMSE, root mean squared error.

4. Discussion

4.1. Chemical Composition

In this study, CP concentration of common vetch grain was greater in 2015 than 2016, while concentrations of EE, aNDF, and ADF were less in 2015 compared to 2016. Dornbos and Mullen [31] reported that drought stress during grain filling of soybean increased grain protein concentration and decreased lipid concentration, and Al-Karaki and Ereifej [32] reported lower protein and greater EE concentration of field pea (*Pisum sativum* L.) with greater precipitation during pod filling. Ítavo et al. [33] reported that EE concentration of legume grain is positively associated with energy release. These observations help to explain differences in the grain chemical composition between years in this study, which may be attributed to greater mean air temperature and total precipitation during July and August of 2016 when common vetch was in flowering to pod-filling stages of phenological development.

In this study, all varieties produced relatively high CP, with values ranging from 353 to 370 g/kg DM. These CP values were greater than those reported from Syria (266–316 g/kg DM) [11] and Turkey (249–279 g/kg DM) [34], but less than that of varieties grown in Jordan (mean = 393 g/kg DM) [12]. Neutral detergent fiber (207–229 g/kg DM) and ADF (59.5–68.8 g/kg DM) concentrations of grain of all varieties fell within the range of values for aNDF and ADF reported by Ramos-Morales et al. [7], Seifdavati and Taghizadeh [13], and Huang et al. [5] (aNDF = 144–423 g/kg DM, ADF = 52–124 g/kg DM). Ether extract of grain of common vetch varieties in this study (16.1–19.0 g/kg DM) was greater than that reported from Spain (14.3 g/kg DM) [7], but within the range reported by Karadag and Yavus [34] in Turkey (11.6–32.3 g/kg DM). Such differences in the proximate composition of grain of common vetch among studies may be attributed to variation in varieties, site, growing environment, and crop management [4,11,34]. The relative high levels of grain protein for all common vetch varieties evaluated in this study demonstrate that they could be a valuable supplement to low-quality ruminant diets, particularly for the local variety [7].

Macronutrient concentrations of common vetch grain in this study were greater in 2016 than 2015, which may be partially due to greater precipitation and warmer air temperature during grain filling in 2016, resulting in accelerated translocation of macronutrients from vegetative tissues and roots to the developing grain. Similarly, Thavarajah et al. [35] reported that warmer air temperature during grain filling was associated with greater Fe concentration of lentil (*Lens culinaris* L.) grain. Uzun et al. [36] reported that 1000-seed weight of common vetch grain was negatively correlated with most mineral elements, such as Ca, Mg, and Mn. Previous studies of these varieties tested showed that seed weight of the local variety was less than that the improved varieties (Table 2) [23]. In this study, large varietal differences in grain concentrations for most minerals may be attributed to a starch dilution effect in lager seeds [36]. Macro and micronutrient concentrations of grain of common vetch varieties in this study are within the ranges reported from studies in Turkey [36] and China [17]. Concentrations of P, Mg, Ca, and Fe in grain of common vetch grown in Cyprus [37] were 5.7, 1.70, 1.40, and 0.14 g/kg DM, respectively, in contrast to those in this study (mean = 2.77, 3.37, 1.59, and 0.69 g/kg DM, respectively). Grain concentrations of Cu and Mn in the present study are similar to values reported for common vetch grown in Jordan [12], although the concentration of Zn was greater in the present study. Grain Zn concentration in this study was similar to that of common vetch grown in Cyprus [37], but the levels of Cu and Mn were comparatively greater in that study. Differences in grain mineral concentrations among studies may be related to differences in varieties planted [36], the stage of grain maturity at harvest [12], and other factors such as site conditions, growing environment, and agronomic practices [17].

Overall, the large varietal differences in chemical composition of common vetch grain indicate that farmers can integrate grain of these varieties into ruminant diets in terms of quality traits to meet the nutrient demands of diverse livestock classes.

4.2. Ruminal Degradability of CP

In this study, the values for *A* and *B* are in accordance with earlier reports on grain of common vetch [7,38], pea, and lupin (*Lupinus albus* var. *multolupa*) [7,39]. The average value for *C* (0.0867 h^{-1}) of grain of the common vetch varieties in this study were in line with the previous reports on common vetch (0.081 h^{-1}) as well as soybean meal (0.082 h^{-1}) [38]. Mean EDCP of common vetch grain in this study (764 g/kg CP) was less than that (857 g/kg CP) reported from Greece studies by Zagorakis et al. [40], but in agreement with earlier observations from Spain (753 g/kg CP; [41]. Such differences in the EDCP of common vetch grain could be ascribed in part to methodological differences such as animal species [42], basal diet [38], rumen outflow rate [39], and protein availability among varieties.

Effective CP degradability of grain was relatively high for all varieties in this study (736–792 g/kg CP). Therefore, these varieties could be considered as a source of degradable protein for ruminal microorganisms and then for microbial protein synthesis [7]. Zagorakis et al. [40] reported an optimum rumen degradable: undegradable protein ratio of 60:40 in feeds for highly productive ruminants. As a consequence, results of the present study suggest that if the tested varieties are used as a protein source in highly productive ruminant diets, they should first be treated to reduce CP degradability, such as with heat treatment [39]. Seifdavati and Taghizadeh [13] reported that autoclaved common vetch grain had 7.07% less EDCP compared to the raw grain.

4.3. Intestinal Digestibility of CP

In this study, large variability in grain IDP among years and varieties (683–782 g/kg RUP) may be partially related to varietal differences in concentrations of ash and ADF. Fu et al. [43] reported that IDP of concentrated feedstuffs is positively associated with ADF concentration, while it was negatively associated with ash concentration. Mean IDP (731 g/kg RUP) of common vetch varieties in this study was greater than that reported for other varieties (600 g/kg RUP by Ramos-Morales et al. [7] and 668 g/kg RUP by Seifdavati and Taghizadeh [13]), but less than that reported for soybean meal (905 g/kg RUP) [30]. These differences among studies of common vetch may be ascribed to differences in common vetch varieties.

In ruminants, the nutritive value of a protein supplementation depends on protein digestion in the small intestine [7,19]. In the present study, IADP of the local variety was less than that of Lanjian No. 2 and Lanjian No. 3, but did not differ significantly from that of Lanjian No. 1. This implies that Lanjian No. 2 and Lanjian No. 3 may be more suitable sources of protein in highly productive ruminant diets than the local variety and Lanjian No. 1. Reports on IADP in grain of common vetch are scarce. However, grain IADP of common vetch varieties in this study was similar to that reported for other annual grains such as camelina (*Camelina sativa* (L.) Crantz; 191 g/kg CP) [30], but lower than for soybean meal (376 g/kg CP) [30].

4.4. Prediction of Degradability Parameters and Intestinal Digestibility of Grain

A significant positive or negative correlation was observed between chemical composition and *A*, *B*, *C*, and IADP of common vetch grain in this study. Limited information has been reported on these relationships for the grain of legumes. To our knowledge, the information available is limited, with the exception of that from studies in Ireland, where concentrations of ash, CP, EE, NDF, ADF, and acid detergent lignin were significantly correlated with degradability parameters of CP in 12 concentrated feedstuffs [21,22].

The use of the chemical composition of concentrated feedstuffs to predict the degradability parameters and intestinal digestibility was previously proposed by Woods et al. [21,22]. Woods et al. [21,22] indicated that R^2 values greater than 0.90 are desirable for equations predicting degradability parameters and intestinal digestibility of concentrated feeds. In the present study, the best fit equations for predicting *A*, *B*, and *C* based on chemical composition had R^2 values of 0.805, 0.712, and 0.748, respectively. Therefore, the equations developed from this study may not be sufficiently precise to accurately predict degradability parameters based on chemical analyses. For equations predicting IADP, the R^2 value increased from 0.787 when only CP was included in the prediction equation to 0.891 when CP, ash, and aNDF were included. Since the R^2 value of this equation is close to 0.9, it may be feasible to use these variables, derived from chemical analysis, to predict grain IADP of common vetch. This result may be incorporated into future breeding programs for improving grain quality of common vetch.

5. Conclusions

All common vetch varieties evaluated in this study produced high grain CP concentration. However, when grain of these varieties is used as a protein source in diets of highly productive ruminants, it should be pre-treated to reduce ruminal degradability. This study also shows that CP, ash, and aNDF can be used to predict IADP of common vetch grain. Additionally, large varietal differences in nutritive value in this and other studies demonstrate that it is necessary to analyze grain of common vetch varieties for nutritive value before using in ration formulations. In terms of grain EDCP and

IADP, the varieties Lanjian No. 2 and Lanjian No. 3 have great potential for supplementing ruminant diets when grown on the Tibetan Plateau.

Author Contributions: Y.H. and Z.N. conceived and designed the experiment; Y.H. and R.L. performed the experiments; Y.H. and Z.Z. analyzed the data; Y.H., J.A.C., and Z.N. wrote the manuscript.

Funding: This research was funded by the National Basic Research Program of China (No. 2014CB138706).

Conflicts of Interest: The authors declare no conflicts of interest.

References

1. Herrero, M.; Havlík, P.; Valin, H.; Notenbaert, A.; Rufino, M.C.; Thornton, P.K.; Blümmel, M.; Weiss, F.; Grace, D.; Obersteiner, M. Biomass use, production, feed efficiencies, and greenhouse gas emissions from global livestock systems. *PNAS* **2013**, *110*, 20888–20893. [CrossRef]

2. Wathes, C.M.; Buller, H.; Maggs, H.; Campbell, M.L. Livestock production in the UK in the 21st century: A perfect storm averted? *Animals* **2013**, *3*, 574–583. [CrossRef]

3. Shang, Z.H.; Gibb, M.J.; Leiber, F.; Ismail, M.; Ding, L.M.; Guo, X.S.; Long, R.J. The sustainable development of grassland-livestock systems on the Tibetan plateau: Problems, strategies and prospects. *Rangeland J.* **2014**, *36*, 267–296. [CrossRef]

4. Larbi, A.; Hassan, S.; Kattash, G.; El-Moneim, A.M.; Jammal, B.; Nabil, H.; Nakkul, H. Annual feed legume yield and quality in dryland environments in north-west Syria: 2. grain and straw yield and straw quality. *Anim. Feed Sci. Technol.* **2010**, *160*, 90–97. [CrossRef]

5. Huang, Y.F.; Gao, X.L.; Nan, Z.B.; Zhang, Z.X. Potential value of the common vetch (*Vicia sativa* L.) as an animal feedstuff: A review. *J. Anim. Physiol. An. N.* **2017**, *101*, 1–17. [CrossRef] [PubMed]

6. Machado Filho, L.C.P.; D'Ávila, L.M.; da Silva Kazama, D.C.; Bento, L.L.; Kuhnen, S. Productive and economic responses in grazing dairy cows to grain supplementation on family farms in the South of Brazil. *Animals* **2014**, *4*, 463–475. [CrossRef]

7. Ramos-Morales, E.; Sanz-Sampelayo, M.; Molina-Alcaide, E. Nutritive evaluation of legume seeds for ruminant feeding. *J. Anim. Physiol. An. N.* **2010**, *94*, 55–64. [CrossRef] [PubMed]

8. Zhang, J.; Guo, G.; Chen, L.; Yuan, X.J.; Yu, C.Q.; Shimojo, M.; Shao, T. Effect of applying lactic acid bacteria and propionic acid on fermentation quality and aerobic stability of oats-common vetch mixed silage on the Tibetan plateau. *Anim. Sci. J.* **2015**, *86*, 595–602. [CrossRef]

9. Nan, Z.B.; Abd El-Moneim, A.M.; Larbi, A.; Nie, B. Productivity of vetches (*Vicia* spp.) under alpine grassland conditions in China. *Trop. Grassl.* **2006**, *40*, 177–182.

10. Fu, P. Effects of inoculation with rhizobium and phosphate on growth of vetch. M.S. Thesis, Lanzhou University, Lanzhou, China, 2015; pp. 1–52.

11. Larbi, A.; Abd El-Moneim, A.M.; Nakkoul, H.; Jammal, B.; Hassan, S. Intra-species variations in yield and quality determinants in Vicia species: 3. Common vetch (*Vicia sativa* ssp. *sativa* L.). *Anim. Feed Sci. Technol.* **2011**, *164*, 241–251. [CrossRef]

12. Samarah, N.H.; Ereifej, K. Chemical composition and mineral content of common vetch seeds during maturation. *J. Plant Nutr.* **2009**, *32*, 177–186. [CrossRef]

13. Seifdavati, J.; Taghizadeh, A. Effects of moist heat treatment on ruminal nutrient degradability of and *in vitro* intestinal digestibility of crude protein from some of legume seeds. *J. Food Agric. Env.* **2012**, *10*, 390–397.

14. Koumas, A.; Economides, S. Replacement of soybean meal by broad bean or common vetch grain in lamb and kid fattening diets. *Cyprus Agric. Res. I. Tech. Bull.* **1987**, *88*, 1–7.

15. Gül, M.; Yörük, M.A.; Macit, M.; Esenbuga, N.; Karaoglu, M.; Aksakal, V.; Irfan Aksu, M. The effects of diets containing different levels of common vetch (*Vicia sativa*) seed on fattening performance, carcass and meat quality characteristics of Awassi male lambs. *J. Sci. Food Agric.* **2005**, *85*, 1439–1443. [CrossRef]

16. Musco, N.; Cutrignelli, M.I.; Calabrò, S.; Tudisco, R.; Infascelli, F.; Grazioli, R.; Lo Presti, V.; Gresta, F.; Chiofalo, B. Comparison of nutritional and antinutritional traits among different species (*Lupinus albus* L. *Lupinus luteus* L. *Lupinus angustifolius* L.) and varieties of lupin seeds. *J. Anim. Physiol. An. N.* **2017**, *101*, 1–15. [CrossRef] [PubMed]

17. Mao, Z.X. Studies on the nutritive quality of common vetch. M.S. Thesis, Lanzhou University, Lanzhou, China, 2012; pp. 1–107.

18. NRC (Nutrient Requirements of Small Ruminants). *Nutrient Requirements of Dairy Cattle*, 7th ed.; National Academy of Sciences: Washington, DC, USA, 2001.

19. McNiven, M.A.; Prestløkken, E.; Mydland, L.T.; Mitchell, A.W. Laboratory procedure to determine protein digestibility of heat-treated feedstuffs for dairy cattle. *Anim. Feed Sci. Technol.* **2002**, *96*, 1–13. [CrossRef]

20. Phipps, D.A. Metal ion physiology: Role, action and mechanism. In *Handbook of Metal–Ligand Interactions in Biological Fluids*, 1st ed.; Berthon, G., Ed.; Marcel Dekker: New York, NY, USA, 1995; Volume 1, pp. 89–106.

21. Woods, V.B.; Moloney, A.P.; Calsamiglia, S.; Mara, F.P.O. The nutritive value of concentrate feedstuffs for ruminant animals Part II: In situ ruminal degradability of crude protein. *Anim. Feed Sci. Technol.* **2003**, *110*, 131–143. [CrossRef]

22. Woods, V.B.; Moloney, A.P.; Calsamiglia, S.; Mara, F.P.O. The nutritive value of concentrate feedstuffs for ruminant animals Part III. Small intestinal digestibility as measured by in vitro or mobile bag techniques. *Anim. Feed Sci. Technol.* **2003**, *110*, 145–157. [CrossRef]

23. Gao, X.L. Responses of four *Vicia sativa* cultivars to phosphorus and potassium fertilizers. M.S. Thesis, Lanzhou University, Lanzhou, China, 2018; pp. 1–97.

24. AOAC (Association of Official Analytical Chemists). *Official Methods of Analysis*, 15th ed.; AOAC International: Arlington, VA, USA, 1990; pp. 71–90.

25. Van Soest, P.J.; Robertson, J.B.; Lewis, B.A. Methods for dietary fiber, neutral detergent fiber, and non-starch polysaccharides in relation to animal nutrition. *J. Dairy Sci.* **1991**, *74*, 3583–3597. [CrossRef]

26. AOAC (Association of Official Analytical Chemists). *Official Methods of Analysis*, 17th ed.; AOAC: Gaithersburg, MD, USA, 2000; pp. 84–90.

27. Nandra, K.S.; Dobos, R.C.; Orchard, B.A.; Neutze, S.A.; Oddy, V.H.; Cullis, B.R.; Jones, A.W. The effect of animal species on *in sacco* degradation of dry matter and protein of feeds in the rumen. *Anim. Feed Sci. Technol.* **2000**, *83*, 273–285. [CrossRef]

28. Ørskov, E.R.; McDonald, I. The estimation of protein degradability in the rumen from incubation measurements weighted according to rate of passage. *J. Agric. Sci.* **1979**, *92*, 499–503. [CrossRef]

29. Gargallo, S.; Calsamiglia, S.; Ferret, A. Technical note: A modified three-step *in vitro* procedure to determine intestinal digestion of protein. *J. Anim. Sci.* **2006**, *84*, 2163–2167. [CrossRef]

30. Lawrence, R.D.; Anderson, J.L. Ruminal degradation and intestinal digestibility of camelina meal and carinata meal compared with other protein sources. *Prof. Anim. Sci.* **2017**, *34*, 10–18. [CrossRef]

31. Dornbos, D.L., Jr.; Mullen, R.E. Soybean seed protein and oil contents and fatty acid composition adjustments by drought and temperature. *J. Am. Oil Chem. Soc.* **1992**, *69*, 228–231. [CrossRef]

32. Al-Karaki, G.N.; Ereifej, K.I. Relationships between seed yield and chemical composition of field peas grown under semi-arid mediterranean conditions. *J. Agron. Crop Sci.* **1999**, *182*, 279–284. [CrossRef]

33. Ítavo, L.C.V.; Soares, C.M.; Ítavo, C.C.B.F.; Dias, A.M.; Petit, H.V.; Leal, E.S.; Souza, A.D.V.V. Calorimetry, chemical composition and *in vitro* digestibility of oilseeds. *Food Chem.* **2015**, *185*, 219–225. [CrossRef]

34. Karadag, Y.; Yavus, M. Seed yields and biochemical compounds of common vetch (*Vicia sativa* L.) lines grown in semi-arid regions of Turkey. *Afr. J. Biotechnol.* **2010**, *9*, 8334–8338.

35. Thavarajah, D.; Thavarajah, P.; See, C.T.; Vandenberg, A. Phytic acid and Fe and Zn concentration in lentil (*Lens culinaris* L.) seeds is influenced by temperature during seed filling period. *Food Chem.* **2010**, *122*, 254–259. [CrossRef]

36. Uzun, A.; Gücer, S.; Acikgoz, E. Common vetch (*Vicia sativa* L.) germplasm: Correlations of crude protein and mineral content to seed traits. *Plant Foods Hum. Nutr.* **2011**, *66*, 254–260. [CrossRef]

37. Hadjipanayiotou, M.; Economides, S.; Koumas, A. Chemical composition, digestibility and energy content of leguminous grains and straws grown in a Mediterranean region. *Ann. Zootech.* **1985**, *34*, 23–30. [CrossRef]

38. Rotger, A.; Ferret, A.; Calsamiglia, S.; Manteca, X. *In situ* degradability of seven plant protein supplements in heifers fed high concentrate diets with different forage to concentrate ratio. *Anim. Feed Sci. Technol.* **2006**, *125*, 73–87. [CrossRef]

39. Aguilera, J.F.; Bustos, M.; Molina, E. The degradability of legume seed meals in the rumen: Effect of heat treatment. *Anim. Feed Sci. Technol.* **1992**, *36*, 101–112. [CrossRef]

40. Zagorakis, K.; Liamadis, D.; Milis, C.; Dotas, V.; Dotas, D. Nutrient digestibility and *in situ* degradability of alternatives to soybean meal protein sources for sheep. *Small Rumin. Res.* **2015**, *124*, 38–44. [CrossRef]

41. González, J.; Andrés, S. Rumen degradability of some feed legume seeds. *Anim. Res.* **2003**, *52*, 17–25. [CrossRef]

42. Molina Alcaide, E.; YáñezRuiz, D.R.; Moumen, A.; Martín García, A.I. Ruminal degradability and *in vitro* intestinal digestibility of sunflower meal and *in vitro* digestibility of olive by-products supplemented with urea or sunflower meal Comparison between goats and sheep. *Anim. Feed Sci. Technol.* **2003**, *110*, 3–15. [CrossRef]

43. Fu, L.X.; Ma, T.; Diao, Q.Y.; Cheng, S.R.; Sun, Z.L.; Li, C. Correlation analysis of ruminal degradation characteristics and *in vitro* small intestinal digestibility of rumen undegraded protein of common concentrates for mutton sheep. *Chinese J. Anim. Nutr.* **2018**, *30*, 2641–2651.

Communication

Ruminal In Vitro Protein Degradation and Apparent Digestibility of Energy and Nutrients in Sheep Fed Native or Ensiled + Toasted Pea (*Pisum sativum*) Grains

Martin Bachmann [1], Christian Kuhnitzsch [1,2], Paul Okon [1], Siriwan D. Martens [2], Jörg M. Greef [3], Olaf Steinhöfel [1,2] and Annette Zeyner [1,*]

[1] Institute of Agricultural and Nutritional Sciences, Martin Luther University Halle-Wittenberg, Theodor-Lieser-Straße 11, 06120 Halle (Saale), Germany
[2] Saxon State Office for Environment, Agriculture and Geology, Am Park 3, 04886 Köllitsch, Germany
[3] Institute for Crop and Soil Science, Federal Research Centre for Cultivated Plants, Julius Kühn Institute (JKI), Bundesallee 58, 38116 Braunschweig, Germany
* Correspondence: annette.zeyner@landw.uni-halle.de

Received: 9 May 2019; Accepted: 26 June 2019; Published: 1 July 2019

Simple Summary: Pea grains may partially replace soybean or rapeseed meals and cereals in ruminant diets, but this is limited by high solubility of pea protein in the rumen. Hydro-thermic treatments such as toasting may stabilize the protein and shift digestion from the rumen to the small intestine. The effect of toasting of ensiled pea grains on rumen-undegraded protein was tested in vitro and on apparent digestibility of organic matter, gross energy, and proximate nutrients in a digestion trial with sheep. Ensiling plus toasting increased rumen-undegraded protein from 20 to 62% of crude protein, but it also increased acid detergent insoluble protein, which is unavailable for digestive enzymes in the small intestine from 0.5 to 2.6% of crude protein. Ensiling plus toasting did not, however, affect total tract apparent digestibility of organic matter, energy, crude protein, or any other nutrient fraction, nor did it alter the concentration of metabolizable energy or net energy lactation in the peas. The technique can be implemented on farms and might have a positive impact on field pea production.

Abstract: Pea grains may partially replace soybean or rapeseed meals and cereals in ruminant diets, but substitution by unprocessed peas is limited by high ruminal protein solubility. The effect of combined ensiling and toasting of peas using a mobile toaster (100 kg/h throughput rate, 180 to 190 °C supplied air temperature) on rumen-undegraded protein (RUP) was tested in vitro using the *Streptomyces griseus* protease test. The effects of ensiling plus toasting on apparent digestibility of organic matter (OM), gross energy (GE), and proximate nutrients were examined in a digestion trial. Concentrations of metabolizable energy (ME) and net energy lactation (NEL) were calculated. Native peas had 38 g RUP/kg dry matter (DM), which was 20% of crude protein (CP). Rumen-undegraded protein increased three-fold after ensiling plus toasting ($p < 0.001$). Acid detergent insoluble protein increased five-fold. Apparent digestibility was 0.94 (OM), 0.90 (CP), and above 0.99 (nitrogen-free extract, starch, and sugars) and was not altered by the treatment. The ME (13.9 MJ/kg DM) or the NEL (8.9 MJ/kg DM) concentration was similar in native and ensiled plus toasted peas. This technique can easily be applied on farms and may increase RUP. However, it needs to be clarified under which conditions pea protein will be damaged.

Keywords: field peas; ensiling; hydro-thermic treatment; nutrient digestibility; rumen-undegraded protein; *Streptomyces griseus* protease test

---

1. Introduction

The production of non-genetically modified foods and animal feeds is increasingly prominent in the public and the political spotlights. In the livestock sector, soybean meal is, apart from rapeseed meal (after oil extraction), undoubtedly the most intensively used protein feed, but it nearly completely originates from genetically modified sources [1]. The European Union (EU-28) currently imports about 95% of the utilized soybeans [2]. National programs now intend to reduce soybean imports and partly replace the classical protein feeds by more cost-effective and sustainable indigenous protein plants such as lupines, faba beans, or field peas. This supports local feed production and nutrient cycles as well. Field pea grains combine high percentages of protein and starch—210 to 261 and 404 to 496 g/kg dry matter (DM), respectively [3–5]. Although post-extraction soybean and rapeseed meals have higher protein contents (508 to 564 and 365 to 411 g/kg DM, respectively [6,7]), they do not provide comparable amounts of starch or other non-structural carbohydrates. The one-to-one substitution of soybean meal and cereals by field pea grains was successful in dairy cows on moderate and high production levels and did not affect feed intake and (fat-corrected) milk yields as long as equal levels of rumen-undegraded protein (RUP) were considered [8,9]. However, continuously high and increasing levels of milk production require a better utilization of feed energy and nutrients. Starch, but especially protein, from field peas is highly degradable in the rumen when the grains are unprocessed, which is limiting small intestinal availability [10]. Increasing RUP might therefore enable the substitution of soybean or rapeseed meals in high-yielding animals [8].

Previous in vitro [11] and in vivo [12] studies demonstrated that the combination of ensiling and subsequent toasting of pea grains is suitable to obtain harvesting and storage stability regardless of the weather conditions. Contrary to our expectations, ensiling of pea grains still led to reduction of protein solubility and an increase of RUP in vitro [11]. Toasting of dry and unprocessed peas after harvesting did not decrease protein solubility, and did not increase RUP or post-ruminal crude protein (CP). Therefore, we further used the combination of ensiling and toasting. We hypothesized that hydro-thermic treatment (toasting) of ensiled field pea grains with 180 to 190 °C supplied air temperature (i.e., 85 to 90 °C grain temperature) and 100 kg/h throughput rate in the toaster would increase RUP without damaging the protein. Apparent digestibility of organic matter (OM), crude nutrient fractions, starch, sugars, and gross energy (GE) was expected to remain unaffected, which would (regarding pea protein) increase the amount that is available for digestion in the small intestine. The effect of ensiling plus toasting on RUP contents of field pea grains was tested in vitro using the *Streptomyces griseus* protease test. Treatment effects on apparent digestibility of energy, OM, and proximate nutrients were proven in a standard digestion trial with sheep. Additionally, the data were used to calculate GE, metabolizable energy (ME), and net energy lactation (NEL) of the pea treatments, applying official equations of the Society of Nutrition Physiology (GfE) [13,14] and the National Research Council (NRC) [15].

2. Materials and Methods

2.1. Pea Treatments

The field pea cultivar Alvesta (KWS SAAT SE, Einbeck, Germany) was used. It was grown and harvested in 2017 in Köllitsch, Saxony. A total quantity of 27.3 t DM of the native grains (i.e., as harvested; not ensiled or toasted) were re-moistened from 779 to 749 g/kg by adding 120 g of a homo-fermentative lactic acid bacteria inoculant including *Lactobacillus plantarum* and *Pediococcus acidilactici* strains (LAB; together 6.8 × 10^6 colony forming units per g fresh matter; Josilac® classic, Josera GmbH & Co. KG, Kleinheubach, Germany) in 200 L water (i.e., 0.6 g LAB/L), crushed using a Murska 2000 S2x2 (Murska, Ylivieska, Finland), and ensiled in a silage plastic bag (BAG Budissa Agroservice GmbH, Malschwitz, Germany) for 9 months. The grain silage had 749 g DM/kg, 34 g crude ash (CA)/kg DM, 189 g CP/kg DM, 20 g acid ether extract (AEE)/kg DM, and 63 g crude fiber (CF)/kg DM. Further ensiling characteristics are specified in Table 1. The aerobic stability of the field pea grain silage was tested following the procedure of H. Honig [16]. Subsequently, the ensiled grains were toasted using a mobile toaster (EcoToast 100, Agrel GmbH, Arnstorf, Germany) at atmospheric pressure with a throughput rate of 100 kg/h, 180 to 190 °C supplied air temperature, and 85 to 90 °C grain temperature.

Table 1. Fermentation characteristics of the field pea grain silage.

pH [1]	6.1
pH [2]	6.3
Lactic acid	2.3
Acetic acid	0.3
Propionic acid	<0.2
iso-Butyric acid	<0.3
n-Butyric acid	<0.1
iso-Valeric acid	<0.1
n-Valeric acid	<0.1
Ethanol	9.4
1,2-Propanediol	<0.3
1-Propanol	<0.4
Aerobic stability	≥7

The fatty acid and alcohol concentrations are given in g/kg dry matter, and the aerobic stability is given in days until the temperature difference between material and ambient exceeds 3 °C. [1] After ensiling. [2] After 7 days of aerobic storage.

2.2. In Vitro Estimation of RUP

The concentrations of RUP in native (i.e., not ensiled or toasted) and ensiled plus toasted pea grains were estimated in vitro using the *Streptomyces griseus* protease test [17] with 10 analytical replicates each. Briefly, 0.5 g of material was weighed into the 136 mL glass bottles, 40 mL of borate phosphate buffer consisting of 12.20 g/L $NaH_2PO_4 \times H_2O$ + 8.91 g/L $Na_2B_4O_7 \times 10H_2O$ (pH 6.7 to 6.8) was added, and the solution was incubated for 1 h at 39 °C in a shaking water bath (80 rpm). A solution of nonspecific type XIV *Streptomyces griseus* protease (Sigma-Aldrich Chemie GmbH, Munich, Germany; ≥3.5 U/mg) with an activity of 0.58 U/mL was made following Licitra et al. [18]. The protease preparation combined aminopeptidase and caseinolytic activities [19]. One unit was defined to hydrolyze casein producing color equivalent (Folin-Ciocalteu reagent) to 1.0 μmol (i.e., 181 μg) of tyrosine per min at pH 7.5 and 37 °C. After 1 h, 3.63 and 3.55 mL of protease solution were added to the bottles (i.e., 24 U/g true protein) based on 179 and 175 g true protein/kg DM in native and ensiled plus toasted peas, respectively (calculated relative to a soybean standard with 493 g true protein/kg DM, which requires 10 mL of the solution) [17]. True protein was calculated from protein fractionation according to Licitra et al. [20] as CP—the non-protein nitrogen (A). The incubation time was set to 24 h considering a lag time of approximately 2 h [18]. After incubation, the bottles' contents were filtered through Whatman #41 filter circles, and each was washed with 200 mL deionized water. The filters were air-dried, and nitrogen was determined in the residues and the blank filters using a FOSS Kjeltec™ 8400 (FOSS GmbH, Hamburg, Germany).

The RUP content of the pea treatments was calculated as follows considering a weighed portion of 0.5 g of each treatment:

$$\text{RUP, g/kg DM} = \left(\frac{((N_{\text{residue}} - N_{\text{blank}}) \times 6.25 \times 10)}{0.5 \times DM_{\text{feed}}} \right) \times 10 \tag{1}$$

where N_{residue} is nitrogen measured in the filter residues (mg); N_{blank} is mean nitrogen measured in the blank filters (mg); and DM_{feed} is the DM content of the native or ensiled plus toasted field pea grains (%).

2.3. In Vivo Determination of OM, GE, and Nutrient Digestibilities

The sheep used in this study were kept and cared for by the Research Centre for Agricultural and Nutritional Sciences, Martin Luther University Halle-Wittenberg, Wettin/Löbejün, Saxony-Anhalt, Germany. The experiment was carried out with approval by the Saxony-Anhalt Federal Administration Authority (approval no. 2-1524 MLU).

Eight adult Pomeranian coarsewool wethers were used as model animals for the digestion trial. All animals were clinically healthy and under regular veterinary supervision.

The composition of the offered diets is given in Table 2. Tap water was offered ad libitum. The feeding level of the sheep was close to the energy maintenance level [13,21]. The recommended feed protein content for digestibility trials with sheep is a minimum of 120 g CP/kg diet DM [22]. This was met during the current experiment. The diets were offered in two equal meals per day. The chemical compositions of the diet components and the mixed diets are given in Table 3.

Table 2. Composition of mixed diets offered during the experiment.

Component (g/Day as Fed)	Control Diet	Test Diet (Native Peas)	Test Diet (Ensiled + Toasted Peas)
Lucerne (chopped)	450	225	225
Barley (crushed, Ø 3.5 mm)	450	225	225
Wheat straw (chopped, Ø 6.0 mm)	100	50	50
Native peas	0	500	0
Ensiled + toasted peas	0	0	500
Mineral feed [1]	10	10	10

[1] basu-kraft® Top-Mineral (BASU Heimtierspezialitäten GmbH, Bad Sulza, Germany).

The digestion trial was run as a difference test in two consecutive periods following the guidelines of GfE [22]. Each period consisted of 14 days adaptation to the diet followed by 6 days of total collection of feces. In total, 4 sheep received the native pea grains (i.e., not ensiled or toasted; 4 analytical replicates), and 5 sheep received the ensiled plus toasted peas (i.e., 5 analytical replicates). During the experiment, the sheep were individually housed in metabolic cages. All feed components were weighed for each mealtime and animal before the experiment started, and composited samples were taken. During total collection, the animals were fitted with harnesses for feces collection, which were emptied each morning prior the first meal. Daily defecation was quantified, and an aliquot of 0.2 was taken. Feed residuals were recorded. The fecal samples were composited individually per sheep per collection period. The feed and the fecal samples were stored dry or frozen at −20 °C, respectively.

The animals' body weights were recorded in each period before adaptation, before feces collection, and at the end of the experiment. The initial body weight was 78 ± 9.7 kg. After a light decline, it was kept constant during the experiment (76 ± 7.5 kg before the first collection, 74 ± 7.8 kg before the second adaptation, 73 ± 11 kg before the second collection, and 73 ± 10 kg at the end of the experiment).

Table 3. Chemical composition of the feeds and the diets used in the experiment.

	Lucerne	Wheat Straw	Barley	Native Peas	Ensiled + Toasted Peas	Mixed Diet (Control)	Mixed Diet (Including Native Peas) [1]	Mixed Diet (Including Ensiled + Toasted Peas) [1]
Dry matter	927	945	888	779	970	912	851	942
Crude ash	71	71	22	31	32	50	41	41
Organic matter	929	929	978	969	968	950	959	959
Crude protein	146	46	115	186	186	122	152	155
Acid ether extract	21	10	25	13	10	22	18	16
Starch	8	14	537	533	496	240	375	372
Sugars	47	12	42	77	38	41	58	40
Crude fiber	363	448	52	62	61	236	157	146
Neutral detergent fiber	524	843	210	128	197	420	285	305
Acid detergent fiber	389	506	64	80	79	259	176	166
Acid detergent lignin	85	59	8	5	15	49	29	31
Cellulose	56	447	304	75	64	210	147	135
Hemicellulose	135	337	146	48	118	161	109	139
Nitrogen-free extract	399	425	786	708	711	570	632	642
ESOM	555	344	880	953	944	675	803	814
Gross energy	19.1	17.9	18.6	18.4	18.3	18.8	18.6	18.5

ESOM = enzyme-soluble organic matter. [1] Pea treatments were included at an amount of 500 g dry matter per day. Dry matter (DM) is given in g/kg, gross energy is given in MJ/kg DM, and all other analytes are given in g/kg DM.

2.4. Analyses

Dry matter, CA, CP, AEE, CF, and Van Soest detergent fiber contents of feeds and feces were analyzed according to the German key book for feed analysis (VDLUFA methods no. 3.1, 4.1.1, 5.1.1 B, 6.1.1, 6.5.1, 6.5.2, 6.5.3, and 8.1) [23]. Neutral detergent fiber (NDF) was determined after 1 h treatment with heat stable amylase. Neutral detergent fiber and acid detergent fiber (ADF) were expressed exclusive of residual ash. Organic matter was calculated as OM = 1000 − CA, and the nitrogen-free extract (NFE) was calculated as NFE = OM − CP − AEE − CF. The amount of cellulose was calculated as cellulose = ADF − acid detergent lignin, and hemicellulose (HEM) was calculated as HEM = NDF − ADF. Neutral detergent insoluble CP (NDICP) was determined in the feeds according to VDLUFA method no. 4.13.1 [24]. Gross energy was determined in the feeds and the feces by bomb calorimetry using a C7000 Oxygen Bomb Calorimeter (IKA® Werke, Staufen, Germany). Starch was determined in the feeds and the feces according to the amyloglucosidase method (VDLUFA method no. 7.2.5) [23]. The sugar content of the feeds and the feces was analyzed using anthrone reagent [25]. Enzyme-soluble organic matter (ESOM) was analyzed according to VDLUFA method no. 6.6.1 [23]. The protein fractions A (i.e., non-protein nitrogen), B1 (i.e., true protein, which is soluble in borate phosphate buffer at pH 6.7 to 6.8 but precipitable), B2 (i.e., true protein, which is insoluble in the borate phosphate buffer minus true protein, which is insoluble in neutral detergent), B3 (i.e., true protein, which is insoluble in neutral detergent but soluble in acid detergent), and C (i.e., true protein, which is insoluble in acid detergent) were determined in native and ensiled plus toasted field peas according to Licitra et al. [21] (method no. LKS FMUAA 1402015-11). For each fraction, residual nitrogen was determined according to the Kjeldahl method. The protein fractions were used to calculate the true protein content (i.e., B1 + B2 + B3 + C) and the soluble protein (i.e., A + B1). The protein insoluble in pepsin (CP_{ip}) was also analyzed using the Kjeldahl method after 48 h of incubation in a pepsin-hydrochloric acid solution (method no. LKS FMUAA 1112014-07) [26]. The organic acids and the alcohols produced during the fermentation of the silage were determined by high performance liquid chromatography and refractive index detection (method no. LKS FMUAA 1662018-05) using a Shimadzu LC-20A Prominence (Shimadzu Corp., Kyoto, Japan) and a Hi-Plex H 8 µm column (300 × 7.7 mm; Agilent Technologies Inc., Santa Clara, CA, USA). All LKS methods were accredited according to DIN EN ISO/IEC 17025:2005.

2.5. Calculations and Statistical Analysis

Apparent digestibility coefficients of OM, CA, CP, AEE, CF, NDF, ADF, NFE, starch, sugars, and GE were calculated as the difference between intake and fecal output divided by the intake on a daily basis for each individual. Feed residuals were consistently lower than 2% of what was offered to the sheep. In accordance with the method prescription [22], feed residuals were thus not considered for digestibility calculation.

The current measured GE data and apparent digestibility coefficients were used to calculate GE, ME, and NEL according to GfE [13] as follows: GE, MJ/kg DM = $0.0239 \times CP + 0.0398 \times AEE + 0.0201 \times CF + 0.0175 \times NFE$; ME, MJ/kg DM = $0.0312 \times$ digestible AEE + $0.0136 \times$ digestible CF + 0.0147 (digestible OM − digestible AEE − digestible CF) + $0.00234 \times CP$; and NEL, MJ/kg DM = 0.6 (1 + 0.004 (q − 57)) × ME, where q = ME/GE × 100. These equations were derived from the large data pool of total metabolic trials provided by the Institute for Animal Nutrition "Oskar Kellner" in Rostock (Germany) [27]. Moreover, GE, ME, and NEL were calculated on the basis of crude nutrient analyses using equations provided by GfE [13,14] and NRC [15].

Statistical analysis was performed using SAS 9.4 (SAS Institute Inc., Cary, NC, USA). The pooled *t*-test was used to compare RUP estimates and digestibility coefficients between the field pea treatments at a significance level of $p < 0.05$. Homogeneity of the sample variances was confirmed using the folded *F*-test, and the studentized residuals were confirmed to have Gaussian distribution using the UNIVARIATE procedure.

3. Results

Native field pea grains had 38 g RUP/kg DM, which was 20% of CP. The concentration of RUP was increased to a three-fold in the ensiled plus toasted grains (115 g/kg DM, i.e., 62% of CP; $p < 0.001$), as shown in Figure 1. The protein fractionation procedure revealed that, after toasting the ensiled grains, protein solubility was five-fold decreased, soluble protein (B1) was nine-fold decreased, and the insoluble fractions B2 and B3 were two-fold and 27-fold increased, respectively (Table 4). The CP_{ip} was slightly increased. However, the C fraction was also increased to a five-fold after ensiling plus toasting (Table 4).

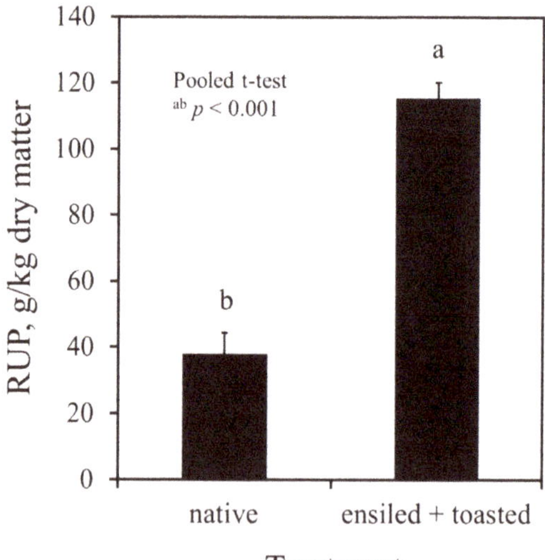

Figure 1. In vitro estimation of rumen-undegraded protein (RUP) in native and ensiled + toasted field pea grains using *Streptomyces griseus* protease (3.63 and 3.55 mL of 0.58 U/mL protease solution, respectively, and 24 h incubation time); native peas had 779 g DM/kg, 186 g CP/kg DM, and 179 g TP/kg DM; ensiled + toasted peas had 970 g DM/kg, 186 g CP/kg DM, and 175 g TP/kg DM; CP = crude protein, DM = dry matter, TP = true protein.

Table 4. Crude protein composition of the field pea treatments.

	Native Peas	Ensiled + Toasted Peas
Crude protein (CP)	186	186
True protein [1]	179	175
Protein solubility [2]	74	16
Protein fraction A	6.5	9.0
Protein fraction B1	67.7	7.2
Protein fraction B2	24.5	56.7
Protein fraction B3	0.9	24.5
Protein fraction C	0.5	2.6
CP_{ip}	5.2	7.2
NDICP	1.1	61.6

A = non-protein nitrogen, B1 = buffer-soluble true protein, B2 = buffer-insoluble true protein—true protein that is insoluble in neutral detergent, B3 = true protein that is insoluble in neutral detergent, but soluble in acid detergent, C = true protein that is insoluble in acid detergent, CP_{ip} = protein insoluble in pepsin, NDICP = neutral detergent-insoluble CP. [1] True protein was calculated as CP − A. [2] Protein solubility was calculated as A + B1. CP, true protein, and NDICP are given in g/kg DM, and protein solubility, the protein fractions (A, B1, B2, B3, and C), and CP_{ip} are given in % of CP.

The apparent digestibility coefficients determined for OM, CA, CP, CF, ADF, NFE, starch, sugars, and GE did not differ between the treatments (Table 5). Native and ensiled plus toasted field peas did differ in AEE digestibility by tendency (0.49 vs. 0.61; p = 0.0548). They significantly differed in apparent digestibility of the NDF fraction (0.69 vs. 0.81; $p < 0.05$; Table 5).

Table 5. Total tract apparent digestibility of energy and proximate nutrients in native and ensiled + toasted field pea grains.

	Native Peas	Ensiled + Toasted Peas
Organic matter	0.94 (0.019)	0.94 (0.026)
Crude ash	0.38 (0.200)	0.39 (0.140)
Crude protein	0.90 (0.033)	0.89 (0.042)
Acid ether extract	0.49 (0.049)	0.61 (0.088)
Crude fiber	0.61 (0.055)	0.65 (0.078)
Neutral detergent fiber	0.69 (0.059) [b]	0.81 (0.055) [a]
Acid detergent fiber	0.65 (0.069)	0.66 (0.074)
Nitrogen-free extract	0.99 (0.008)	0.99 (0.015)
Starch	1.00 (0.0007)	1.00 (0.001)
Sugars	1.00 (0.003)	0.99 (0.006)
Gross energy	0.91 (0.022)	0.91 (0.027)

Standard deviations are given in parentheses. [ab] Different superscripts mark significant differences between the treatments ($p < 0.05$).

Gross energy contents measured by bomb calorimetry and calculated according to GfE [13] were similar in the native and the ensiled plus toasted field peas (18.4 and 18.3 vs. 18.6 MJ/kg DM; Table 6). The native field peas had a digestible energy concentration of 16.7 MJ/kg DM, and the ensiled plus toasted peas had a digestible energy concentration of 16.6 MJ/kg DM, which was calculated based on measured GE and GE digestibility. The ME contents calculated on the basis of measured GE and nutrient digestibilities were 13.8 and 13.9 MJ/kg DM in native and ensiled plus toasted peas, respectively. A similar ME content was calculated in the native peas using GfE and NRC equations. In ensiled plus toasted peas, ME was slightly underestimated by 0.2 MJ/kg DM using GfE and by 0.4 MJ/kg DM using NRC equations (Table 6). This was similar with the NEL contents (Table 6).

Table 6. Gross energy (GE), metabolizable energy (ME), and net energy lactation (NEL) in native and ensiled + toasted field pea grains.

	Native Peas			Ensiled + Toasted Peas		
	Measured/Calculated [1]	GfE [2]	NRC [3]	Measured/Calculated [1]	GfE [2]	NRC [3]
GE	18.4	18.6	n.a.	18.3	18.6	n.a.
ME	13.8 [4]	13.9	13.9	13.9 [4]	13.7	13.5
NEL	8.9 [4]	8.9	9.0	8.9 [4]	8.8	8.7

n.a. = not applicable. [1] Calculated on the basis of measured GE concentrations and apparent digestibility of crude nutrients according to GfE [13]. [2] Calculated according to GfE [13,14]. [3] Calculated according to NRC [15]. [4] Given as the mean of 4 and 5 measurements in native and ensiled + toasted peas, respectively. All items are given in MJ/kg dry matter.

4. Discussion

The social, the political, and the scientific interest in using regionally grown legumes as protein feeds for farm animals is increasing. Faba beans, field peas, lupines, clover, or lucerne can offer an alternative to soybean meal and rapeseed meal in Europe. Using field peas as at least a partial replacement of soybean or rapeseed meals and/or cereals was possible without addressed digestibility reduction, performance depression, or impaired quality of the end products in growing and finishing pigs [28], broiler chickens [29], laying hens [30], growing lambs and beef cattle [31,32], and dairy cows [8,9]. However, a complete replacement was not always successful in either monogastric [33,34] or ruminant livestock [8,32] when the peas remained unprocessed. A main factor that is limiting the substitution efficiency of pea grains is the higher ruminal protein solubility compared to soybean [35] or rapeseed meal. The RUP content of unprocessed peas is around 20% of CP, but in post-extraction soybean and rapeseed meals, 25 to 69% of CP was reported [36–39]. Moreover, ruminal starch degradation of unprocessed peas is approximately 56%, while it is 44% in maize and 83% in barley [40].

Ensiling may generally lead to proteolysis and carbohydrate degradation [41–43] but should not affect ruminal degradation of protein, starches, and other carbohydrates as long as no perishable fermentation (e.g., yeast fermentation) and concomitant heating occurs. Despite a good sensorial quality, the current pea grain silage was not reduced in pH (6.1), and ethanoic acid and ethanol contents were increased compared to previously used silages (by 1.4 and 6.8 g/kg DM) [11]. However, the proportion of true protein was comparable [11,12].

In native pea grains, storage proteins are mainly composed of the globulins 7S vicilin and 11S legumin, which are soluble to more than 70% in the digestive tract (74% in this study) [11,44]. Heat or heat plus pressure treatments such as dry roasting, steam flaking, autoclaving, toasting, extrusion, or expander treatment clearly decrease protein degradability in the rumen [4–6,10,45–48]. Feed protein structures are stabilized by complex Maillard reactions [46,47,49–53]. Next, proteolysis in the rumen can be inhibited or slowed down, but total tract digestibility of OM or protein should not be affected [4,54]. These effects largely depend on the type and the conditions of the treatment (e.g., temperature, throughput rate/duration of heat exposure, feed quantity, and moisture content) [11,53]. The native field pea grains in our study had a RUP content of about 20% of CP, which was similar to what Masoero et al. [5] measured. Then, RUP tripled in the ensiled plus toasted pea grains. Protein solubility decreased to one-fifth, B1 to one-ninth, and especially B3 increased by 27 times. We, however, also found that the C fraction increased five-fold after ensiling plus toasting, which might point to some protein damage.

The determination of total tract apparent digestibility of OM and crude nutrients is the official method to obtain ME and NEL of feedstuffs in Germany [13,21]. In native peas, total tract apparent digestibility coefficients of OM, GE, and proximate nutrients were similar to those reported by the German Agricultural Society (DLG) [39]. In our case, sheep were fed slightly below the maintenance requirement, which might have caused a minor overestimation of digestibility. Total tract apparent digestibility of OM, CA, CP, CF, ADF, NFE, starch, sugars, and GE was not affected by ensiling plus toasting, whereas AEE and NDF digestibility increased by approximately 10%. Total tract apparent digestibility does not allow conclusions on partial (e.g., ruminal) nutrient digestibility. The latter, however, might be altered by ensiling and toasting of peas, even if total tract digestibility remains unaffected. The increase in AEE and NDF digestibility was probably more an analytical issue and not a physiological response. Increased AEE digestibility was probably caused by the low AEE contents in both feed and feces, which needs to be set into relation to the inherent analytical error. Following ensiling plus toasting, NDICP was increased by approximately 30% of NDF. This probably led to an overestimation of the NDF intakes by the sheep and thus to an overestimation of NDF digestibility. When we reduced the NDF contents in the feed by the respective NDICP amounts, apparent digestibilities of NDF were 0.68 ± 0.063 and 0.74 ± 0.063 in native and ensiled plus toasted peas, respectively (p = 0.22). Although the samples were treated with amylase during NDF analysis, it is still possible that parts of starch contributed to an overestimation of the NDF intakes, at least when parts of the starch from the feed were altered by the feed treatment. Fiber, ash, and AEE digestibility coefficients had a large variation among the animals (standard deviation ranged from 0.05 to 0.20). Thus, the apparent increase of AEE and NDF digestibility was due to analytical uncertainties.

No effect of ensiling plus toasting on ME or NEL estimations was found. The available equations for ME and NEL estimation provided by the GfE [14] and the NRC [15] had a high conformity with calculations based on measured GE and ME calculated on the basis of measured GE and nutrient digestibilities.

5. Conclusions

The combination of ensiling and toasting of field pea grains on farms using a mobile toaster with 100 kg/hour throughput rate and 180 to 190 °C temperature of the supplied air (i.e., 85 to 90 °C grain temperature) led to a three-fold increase of RUP. However, boundary conditions for heat damage of proteins have to be clarified. Protein fractions that are fully insoluble to digestive enzymes (e.g., CPip or acid detergent insoluble protein) also increased after ensiling plus toasting, which should be avoided. Total tract apparent digestibility of OM, CA, CP, CF, ADF, NFE, starch, sugars, and GE was not affected by ensiling plus toasting. The apparent increase in AEE digestibility was probably an analytical issue. The NDF apparent digestibility probably increased due to increasing amounts of NDICP after toasting. Contents of GE, ME, and NEL remained unaffected. A high conformity was found among GE, ME, and NEL, which were measured, calculated, or estimated using GfE or NRC equations, respectively.

Author Contributions: Conceptualization, M.B.; data curation, M.B.; formal analysis, M.B., C.K., P.O., and J.M.G.; funding acquisition, O.S. and A.Z.; investigation, M.B., C.K., P.O., and S.D.M.; methodology, M.B., C.K., and J.M.G.; project administration, O.S. and A.Z.; resources, J.M.G.; supervision, O.S. and A.Z.; validation, J.M.G.; visualization, M.B.; writing—original draft, M.B.; writing—review & editing, M.B., C.K., P.O., J.M.G., S.D.M., O.S., and A.Z., M.B. and C.K. contributed equally to the work.

Funding: This work was funded by the German Federal Ministry of Food and Agriculture (BMEL) through the Federal Office for Agriculture and Food (BLE) (grant number 2815EPS058). Moreover, we acknowledge the financial support within the funding program Open Access Publishing by the German Research Foundation (DFG).

Conflicts of Interest: The authors declare that they have no conflict of interest.

References

1. ISAAA. *Global Status of Commercialized Biotech/GM Crops: 2016*; ISAAA Brief No. 52; ISAAA: Ithaca, NY, USA, 2016.
2. Tillie, P.; Rodríguez-Crezo, E. Markets for Non-Genetically Modified, Identity-Preserved Soybean in the EU. 2015, pp. 1–72. Available online: http://publications.jrc.ec.europa.eu/repository/handle/JRC95457 (accessed on 6 March 2019).
3. Abreu, J.M.F.; Bruno-Soares, A.M. Chemical composition, organic matter digestibility and gas production of nine legume grains. *Anim. Feed Sci. Technol.* **1998**, *70*, 49–57. [CrossRef]

4. Goelema, J.O.; Spreeuwenberg, M.A.M.; Hof, G.; van der Poel, A.F.B.; Tamminga, S. Effect of pressure toasting on the rumen degradability and intestinal digestibility of whole and broken peas, lupins and faba beans and a mixture of these feedstuffs. *Anim. Feed Sci. Technol.* **1998**, *76*, 35–50. [CrossRef]

5. Masoero, F.; Pulimeno, A.M.; Rossi, F. Effect of extrusion, expansion and toasting on the nutritional value of peas, faba beans and lupins. *Ital. J. Anim. Sci.* **2005**, *4*, 177–189. [CrossRef]

6. Konishi, C.; Matsui, T.; Park, W.; Yano, H.; Yano, F. Heat treatment of soybean meal and rapeseed meal suppresses rumen degradation of phytate phosphorus in sheep. *Anim. Feed Sci. Technol.* **1999**, *80*, 115–122. [CrossRef]

7. Getachew, G.; Robinson, P.H.; DePeters, E.J.; Taylor, S.J. Relationships between chemical composition, dry matter degradation and in vitro gas production of several ruminant feeds. *Anim. Feed Sci. Technol.* **2004**, *111*, 57–71. [CrossRef]

8. Corbett, R.R.; Okine, E.K.; Goonewardene, L.A. Effects of feeding peas to high-producing dairy cows. *Can. J. Anim. Sci.* **1995**, *75*, 625–629. [CrossRef]

9. Khorasani, G.R.; Okine, E.K.; Corbett, R.R.; Kennelly, J.J. Nutritive value of peas for lactating dairy cattle. *Can. J. Anim. Sci.* **2001**, *81*, 541–551. [CrossRef]

10. Yu, P.; Holmes, J.H.G.; Leury, B.J.; Egan, A.R. Influence of dry roasting on rumen protein degradation characteristics of whole faba bean (*Vicia faba*) in dairy cows. *Asia. Austral. J. Anim. Sci.* **1998**, *11*, 35–42. [CrossRef]

11. Bachmann, M.; Kuhnitzsch, C.; Thierbach, A.; Michel, S.; Bochnia, M.; Greef, J.M.; Martens, S.D.; Steinhöfel, O.; Zeyner, A. Effects of toasting temperature and time on ruminal gas production kinetics and post-ruminal crude protein from field pea (*Pisum sativum*) grain silages measured *in vitro*. In Proceedings of the 73rd Conference of the Society of Nutrition Physiology, Göttingen, Germany, 13–15 March 2019; DLG-Verlag: Frankfurt (Main), Germany, 2019; p. 112.

12. Kuhnitzsch, C.; Martens, S.D.; Bachmann, M.; Zeyner, A.; Hofmann, T.; Steinhöfel, O. Vergleichende Untersuchungen zum Einsatz siliert und getoasteter Erbsen bzw. Erbsenschröpfschnitt-GPS in der Fütterung hochleistender Milchkühe. In Proceedings of the Forum angewandte Forschung in der Rinder- und Schweinefütterung, Fulda, Germany, 2–3 April 2019; Self-Publishing: Bad Sassendorf, Germany, 2019; pp. 61–64.

13. Society of Nutrition Physiology (GfE). *Empfehlungen zur Energie- und Nährstoffversorgung der Milchkühe und Aufzuchtrinder*; DLG-Verlag: Frankfurt (Main), Germany, 2001.

14. Society of Nutrition Physiology (GfE). New equations for predicting metabolisable energy of compound feeds for cattle. In Proceedings of the 63rd Conference of the Society of Nutrition Physiology, Göttingen, Germany, 10–12 March 2009; DLG-Verlag: Frankfurt (Main), Germany, 2009; pp. 143–146.

15. National Research Council (NRC). *Nutrient Requirements of Dairy Cattle*, 7th ed.; National Academy Press: Washington, DC, USA, 2001.

16. Honig, H. Evaluation of aerobic stability. In Proceedings of the EUROBAC Conference, Uppsala, Sweden, 12–16 August 1986; Lindgren, S., Pettersson, K., Eds.; Swedish University of Agricultural Sciences: Uppsala, Sweden, 1990; pp. 72–78.

17. Licitra, G.; Lauria, F.; Carpino, S.; Schadt, I.; Sniffen, C.J.; Van Soest, P.J. Improvement of the *Streptomyces griseus* method for degradable protein in ruminant feeds. *Anim. Feed Sci. Technol.* **1998**, *72*, 1–10. [CrossRef]

18. Licitra, G.; Van Soest, P.J.; Schadt, I.; Carpino, S.; Sniffen, C.J. Influence of the concentration of the protease from *Streptomyces griseus* relative to ruminal protein degradability. *Anim. Feed Sci. Technol.* **1999**, *77*, 99–113. [CrossRef]

19. Jurášek, L.; Johnson, P.; Olafson, R.W.; Smillie, L.B. An improved fractionation system for pronase on CM-Sephadex. *Can. J. Biochem.* **1971**, *49*, 1195–1201. [CrossRef] [PubMed]

20. Licitra, G.; Hernandez, T.M.; Van Soest, P.J. Standardization of procedures for nitrogen fractionation of ruminant feeds. *Anim. Feed Sci. Technol.* **1996**, *57*, 347–358. [CrossRef]

21. Society of Nutrition Physiology (GfE). Formeln zur Schätzung des Gehaltes an Umsetzbarer Energie und Nettoenergie-Laktation in Mischfuttern. In *Proceedings of the 50th Conference of the Society of Nutrition Physiology*; DLG-Verlag: Frankfurt (Main), Germany, 1996; pp. 153–155.

22. Society of Nutrition Physiology (GfE). Leitlinien für die Bestimmung der Verdaulichkeit von Rohnährstoffen an Wiederkäuern. *J. Anim. Physiol. Anim. Nutr.* **1991**, *65*, 229–234. [CrossRef]

23. VDLUFA. *Die Chemische Untersuchung Von Futtermitteln Methodenbuch*, 3rd ed.; VDLUFA-Verlag: Darmstadt, Germany, 2012.

24. VDLUFA. *Methode 4.13.1, Bestimmung des Neutral-Detergenzien-Löslichen Rohproteins (NDLXP)*; VDLUFA-Verlag: Darmstadt, Germany, 2016.

25. Yemm, E.W.; Willis, A.J. The estimation of carbohydrates in plant extracts by anthrone. *Biochem. J.* **1954**, *56*, 508–514. [CrossRef] [PubMed]

26. Weissbach, F.; Prym, R.; Peters, G.; Lengerken, J.V. Pepsin-insoluble crude protein-Criterion of green forage silage quality. *Tierzucht* **1985**, *39*, 346–349.

27. Schiemann, R.; Nehring, K.; Hoffmann, L.; Jentsch, W.; Chudy, A. *Energetische Futterbewertung und Energienormen*; VEB Deutscher Landwirtschaftsverlag: Berlin, Germany, 1972.

28. White, G.A.; Smith, L.A.; Houdijk, J.G.M.; Homer, D.; Kyriazakis, I.; Wiseman, J. Replacement of soya bean meal with peas and faba beans in growing/finishing pig diets: Effect on performance, carcass composition and nutrient excretion. *Anim. Feed Sci. Technol.* **2015**, *209*, 202–210. [CrossRef]

29. Dotas, V.; Bampidis, V.A.; Sinapis, E.; Hatzipanagiotou, A.; Papanikolaou, K. Effect of dietary field pea (*Pisum sativum* L.) supplementation on growth performance, and carcass and meat quality of broiler chickens. *Livest. Sci.* **2014**, *164*, 135–143. [CrossRef]

30. Frau-Nji, F.; Niess, E.; Pfeffer, E. Effect of graded replacement of soybean meal by faba beans (*Vicia faba* L.) or field peas (*Pisum sativum* L.) in rations for laying hens on egg production and quality. *J. Poult. Sci.* **2007**, *44*, 34–41. [CrossRef]

31. Loe, E.R.; Bauer, M.L.; Lardy, G.P.; Caton, J.S.; Berg, P.T. Field pea (*Pisum sativum*) inclusion in corn-based lamb finishing diets. *Small Rum. Res.* **2004**, *53*, 39–45. [CrossRef]

32. Reed, J.J.; Lardy, G.P.; Bauer, M.L.; Gilbery, T.C.; Caton, J.S. Effect of field pea level on intake, digestion, microbial efficiency, ruminal fermentation, and in situ disappearance in beef steers fed growing diets. *J. Anim. Sci.* **2004**, *82*, 2123–2130. [CrossRef]

33. Valencia, D.G.; Serrano, M.P.; Centeno, C.; Lázaro, R.; Mateos, G.G. Pea protein as a substitute of soya bean protein in diets for young pigs: Effects on productivity and digestive traits. *Livest. Sci.* **2008**, *118*, 1–10. [CrossRef]

34. Valencia, D.G.; Serrano, M.P.; Jiménez-Moreno, E.; Lázaro, R.; Mateos, G.G. Ileal digestibility of amino acids of pea protein concentrate and soya protein sources in broiler chicks. *Livest. Sci.* **2009**, *121*, 21–27. [CrossRef]

35. Vander Pol, M.; Hristov, A.N.; Zaman, S.; Delano, N.; Schneider, C. Effect of inclusion of peas in dairy cow diets on ruminal fermentation, digestibility, and nitrogen losses. *Anim. Feed Sci. Technol.* **2009**, *150*, 95–105.

36. Cone, J.W.; Van Gelder, A.H.; Steg, A.; Van Vuuren, A.M. Prediction of in situ rumen escape protein from in vitro incubation with protease from *Streptomyces griseus*. *J. Sci. Food Agric.* **1996**, *72*, 120–126. [CrossRef]

37. Can, A.; Hummel, J.; Denek, N.; Südekum, K.-H. Effects of non-enzymatic browning reaction intensity on in vitro ruminal protein degradation and intestinal protein digestion of soybean and cottonseed meals. *Anim. Feed Sci. Technol.* **2011**, *163*, 255–259. [CrossRef]

38. Böttger, C.; Weber, T.; Südekum, K.-H. Proteinwert von Rapsextraktionsschroten für Wiederkäuer—Schätzung mittels chemischer und in vitro-Verfahren. In Proceedings of the 129th VDLUFA-Kongress, Freising, Germany, 12–15 September 2017; VDLUFA-Verlag: Darmstadt, Germany, 2017; p. 107.

39. Deutsche Landwirtschafts-Gesellschaft (DLG). *DLG-Futterwerttabellen Wiederkäuer*, 7th ed.; DLG-Verlag: Frankfurt (Main), Germany, 1997.

40. Ljøkjel, K.; Skrede, A.; Harstad, O.M. Effects of pelleting and expanding of vegetable feeds on in situ protein and starch digestion in dairy cows. *J. Anim. Feed Sci.* **2003**, *12*, 435–449. [CrossRef]

41. Mustafa, A.F.; Seguin, P. Characteristics and in situ degradability of whole crop faba bean, pea, and soybean silages. *Can. J. Anim. Sci.* **2003**, *83*, 793–799. [CrossRef]

42. Hoedtke, S.; Gabel, M.; Zeyner, A. Protein degradation in feedstuffs during ensilage and changes in the composition of the crude protein fraction. *Übers. Tierernährg.* **2010**, *38*, 157–179.

43. Gefrom, A.; Ott, E.M.; Hoedtke, S.; Zeyner, A. Effect of ensiling moist field bean (*Vicia faba*), pea (*Pisum sativum*) and lupine (*Lupinus* spp.) grains on the contents of alkaloids, oligosaccharides and tannins. *J. Anim. Physiol. Anim. Nutr.* **2013**, *97*, 1152–1160. [CrossRef] [PubMed]

44. Duranti, M. Grain legume proteins and nutraceutical properties. *Fitoterapia* **2006**, *77*, 67–82. [CrossRef] [PubMed]

45. Focant, M.; Van Hoecke, A.; Vanbelle, M. The effect of two heat treatments (steam flaking and extrusion) on the digestion of *Pisum sativum* in the stomachs of heifers. *Anim. Feed Sci. Technol.* **1990**, *28*, 303–313. [CrossRef]

46. Aufrère, J.; Graviou, D.; Melcion, J.P.; Demarquilly, C. Degradation in the rumen of lupin (*Lupinus albus* L.) and pea (*Pisum sativum* L.) seed proteins. Effect of heat treatment. *Anim. Feed Sci. Technol.* **2001**, *92*, 215–236. [CrossRef]

47. Azarfar, A.; Tamminga, S.; Pellikaan, W.F.; van der Poel, A.F.B. In vitro gas production profiles and fermentation end-products in processed peas, lupins and faba beans. *J. Sci. Food Agric.* **2008**, *88*, 1997–2010. [CrossRef]

48. Vaga, M.; Hetta, M.; Huhtanen, P. Effects of heat treatment on protein feeds evaluated in vitro by the method of estimating utilisable crude protein at the duodenum. *J. Anim. Physiol. Anim. Nutr.* **2017**, *101*, 1259–1272. [CrossRef] [PubMed]

49. Hellwig, M.; Henle, T. Baking, ageing, diabetes: A short history of the Maillard reaction. *Angew. Chem. Int. Ed.* **2014**, *53*, 10316–10329. [CrossRef] [PubMed]

50. Holum, J.R. *Fundamentals of General, Organic, and Biological Chemistry*, 2nd ed.; Wiley: New York, NY, USA, 1982.

51. Hurrell, R.E.; Finot, R.A. Effect of food processing on protein digestibility and amino acid availability. In *Digestibility and Amino Acid Availability in Cereals and Oilseeds*; Finley, J.W., Hopkins, D.T., Eds.; American Association of Cereal Chemists: St. Paul, MN, USA, 1985; pp. 516–527.

52. Broderick, G.A.; Wallace, R.J.; Ørskov, E.R. Control of Rate and Extent of Protein Degradation. In *Physiological Aspects of Digestion and Metabolism in Ruminants, Proceedings of the 7th International Symposium on Ruminant Physiology, Clermont—Ferrand, France, 3–7 September 1979*; Tsuda, T., Sasaki, Y., Kawashima, R., Eds.; Academic Press: San Diego, CA, USA, 1991; pp. 541–592.

53. Yu, P.; Goelema, J.O.; Leury, B.J.; Tamminga, S.; Egan, A.R. An analysis of the nutritive value of heat processed legume seeds for animal production using the DVE/OEB model: A review. *Anim. Feed Sci. Technol.* **2002**, *99*, 141–176. [CrossRef]

54. Poncet, C.; Rémond, D. Rumen digestion and intestinal nutrient flows in sheep consuming pea seeds: The effect of extrusion or chestnut tannin addition. *Anim. Res.* **2002**, *51*, 201–216. [CrossRef]

Article

Models Based on the Mitscherlich Equation for Describing Typical and Atypical Gas Production Profiles Obtained from In Vitro Digestibility Studies Using Equine Faecal *Inoculum*

Christopher D. Powell [1,*], Mewa S. Dhanoa [2], Anna Garber [3], Jo-Anne M. D. Murray [3], Secundino López [4,5,*], Jennifer L. Ellis [1] and James France [1]

[1] Department of Animal Biosciences, University of Guelph, Guelph, ON N1G 2W1, Canada; jellis@uoguelph.ca (J.L.E.); jfrance@uoguelph.ca (J.F.)
[2] Institute of Grassland and Environmental Research, Plas Gogerddan, Aberystwyth SY23 3EB, UK; dandhanoa@hotmail.com
[3] College of Medical, Veterinary and Life Sciences, School of Veterinary Medicine, University of Glasgow, Glasgow G61 1QH, UK; Anna.Garber@glasgow.ac.uk (A.G.); Jo-Anne.Murray@glasgow.ac.uk (J.-A.M.D.M.)
[4] Departamento de Producción Animal, Universidad de León, E-24007 León, Spain
[5] Instituto de Ganadería de Montaña, CSIC-Universidad de León, Finca Marzanas s/n, 24346 Grulleros, Spain
* Correspondence: cpowell@uoguelph.ca (C.D.P.); s.lopez@unileon.es (S.L.); Tel.: +1-(519)-824-4120 (ext. 56688) (C.D.P.); +34-987-291-291 (S.L.)

Received: 29 December 2019; Accepted: 10 February 2020; Published: 17 February 2020

Simple Summary: Feedstuff evaluation through animal trials is time consuming and expensive. An alternative, the gas production method, measures the amount of fermentation gas produced from incubating feedstuffs with microbes from ruminal fluid or faecal samples. Models can be applied to gas production profiles to determine extent of feedstuff degradation either in the rumen or in the hindgut. Typical gas production profiles show a monotonically increasing monophasic pattern. However, atypical gas production profiles exist whereby at least two consecutive phases of gas production are present; these profiles are much less well described. Two models are proposed to fit these biphasic profiles, a sum of two Mitscherlich equations, and sum of Mitscherlich + linear equations. Additionally, two models that describe typical monophasic gas production curves, the simple Mitscherlich and the generalised Mitscherlich (root-t) model, were assessed for comparison. Models were fitted to 25 gas production profiles resulting from incubating feedstuffs with faecal *inocula* from equines. Of these 25 profiles, 17 displayed atypical biphasic patterns, and 8 displayed typical monophasic patterns. The two biphasic models were found to describe both the atypical and typical gas production profiles accurately. These models allow for the evaluation of feedstuffs using cost- and time-efficient methods.

Abstract: Two models are proposed to describe atypical biphasic gas production profiles obtained from in vitro digestibility studies. The models are extensions of the standard Mitscherlich equation, comprising either two Mitscherlich terms or one Mitscherlich and one linear term. Two models that describe typical monophasic gas production curves, the standard Mitscherlich and the France model [a generalised Mitscherlich (root-t) equation], were assessed for comparison. Models were fitted to 25 gas production profiles resulting from incubating feedstuffs with faecal *inocula* from equines. Seventeen profiles displayed atypical biphasic patterns while the other eight displayed typical monophasic patterns. Models were evaluated using statistical measures of goodness-of-fit and by analysis of residuals. Good agreement was found between observed atypical profiles values and fitted values obtained with the two biphasic models, and both can revert to a simple Mitscherlich allowing them to describe typical monophasic profiles. The models contain kinetic fermentation parameters that can be used in conjunction with substrate degradability information and digesta passage rate to calculate extent of substrate degradation in the rumen or hindgut. Thus, models link the in vitro gas production technique to nutrient supply in the animal by providing information relating to digestion and nutritive value of feedstuffs.

Keywords: gas production technique; in vitro digestibility; Mitscherlich equation; feedstuff evaluation; fermentation kinetics; substrate degradation

---

1. Introduction

The in vitro gas production technique [1,2] is widely applied in animal nutrition for ranking and evaluating feedstuffs. This technique is based upon the assumption that the gas produced from incubating a feedstuff with a microbial *inoculum* is the consequence of the anaerobic fermentation of that feedstuff [3]. In ruminant nutrition, gas production profiles generated have been used in conjunction with the retention time of digesta (derived from the rate of passage) to determine extent of degradation in the rumen [3–9]. In equine nutrition, the technique has been proposed as an in vitro surrogate for determining the digestibility and nutritive value of feedstuffs using in vivo methods [10–14].

Typical gas production profiles are diminishing returns or sigmoidal in shape (see [5] for illustration), and France et al. [4] derived a purpose-built function in the form of a generalised Mitscherlich equation with an additional root-*t* term to represent a variable fractional rate of degradation for fitting to a wide range of curve shapes. This model is commonly referred to as the "France" model, and this term will be used herein. However, atypical patterns have also been recorded. Groot et al. [15] reported biphasic profiles and selected a function comprising two generalised rectangular hyperbolae to fit them, while other atypical patterns have been observed by research workers though not formally reported in the scientific literature. Interpretation of these atypical patterns include the autonomous fermentation of feed components in incubated feedstuffs, with these feed components representing chemical or nutritional fractions, with total gas produced being a summation of gas produced from each fermented feed component [16].

The Mitscherlich equation has a long history of application in the agricultural sciences and in applied biology generally, both as a response function and as a growth function [17,18]. The Mitscherlich, which is an expression of the principle of the Law of Diminishing Increments as originally applied to the effect of fertilization on crop yields, is a function that reaches an asymptotic maximum and represents diminishing returns behaviour in rising to the asymptote. It is a special case of the function proposed by France et al. [4]. In this paper, we consider four types of gas production profile (diminishing returns, sigmoidal, biphasic and asymptotic, biphasic but non-asymptotic). The profiles considered were obtained from incubating feedstuffs with faecal *inocula* from equines using the gas production method of Theodorou et al. [2]. These data were taken from two experiments with either grazing horses or ponies fed primarily grass hay. The main objective of this paper was to assess the ability of the simple Mitscherlich, and three extensions of this classical function, to describe both typical and atypical gas production profiles. The functions were derived to describe gas production profiles on the basis of substrate degradation, rather than on the basis of gas produced, permitting the estimation of fermentation kinetic parameters. Using relatively simple equations, proposed herein, extent of feedstuff degradation in the hindgut of equines can be calculated using model parameter estimates in conjunction with information regarding substrate degradability and digesta passage rate. Therefore, a secondary objective was to compare how model fits, and by extension model derived parameters, affect extent of feedstuff degradation values when these models are applied to mono- and bi-phasic gas production profiles.

2. Materials and Methods

2.1. Datasets

2.1.1. Experiment 1: Inoculum from Horses

In a study to assess the fermentative capacity of faecal *inocula*, Murray et al. [19] sourced *inoculum* from 14 grass-kept horses (maintained on grass 24 h a day) from the International League for the Protection of Horses in Norfolk, UK. *Inocula* were prepared from these 14 horses—7 of them predisposed to laminitis and the other 7 clinically normal—so that the effect of laminitis on hindgut fermentative activity could be evaluated. Grass hay was the substrate incubated in vitro. Due to the large distance to the laboratory, *inoculum* was stored at −20 °C for transportation on ice. *Inocula* were subsequently thawed and incubated at 38 °C. Gas production was recorded using the method of Theodorou et al. [2] and three replicates per *inoculum* were used. Standard in vitro gas production results were described by Murray et al. [19]. The grass hay data yielded 14 gas production profiles, one for each horse, as the average over the three replicates. Visual inspection of these profiles revealed a predominance of atypical patterns.

2.1.2. Experiment 2: Inoculum from Ponies

The study comprised a total of eleven different *inocula* (see Table 1 for details). Garber et al. [20] sourced eight *inocula* from Welsh Section A geldings arranged in a 4 × 4 Latin square experimental design aiming to investigate the in vitro fermentation of high fibre/high concentrate diets supplemented with yeast (control diets with no yeast). Another 3 faecal *inocula* were obtained in an experiment in which ponies were fed a grass hay only diet (control), or the same grass hay supplemented with increasing concentrations of a fibrolytic enzyme (either 0.75 or 3.75 mL of enzyme solution per kg DM hay). Gas production was recorded using the ANKOM RF gas production system [21] and three replicates per *inoculum* were used. Preliminary results were reported by Garber et al. [20]. The data yielded 11 gas production profiles, one for each treatment, after averaging the three replicates for each *inoculum*. Visual inspection revealed both typical and atypical patterns.

The entirety of the observed gas production values (Dataset 1–25) used in this study can be found in the Supplementary Information section of this paper, Table S1.

Table 1. Details of the eleven treatments used in Experiment 2.

	Control and Enzyme Treatments	
Dataset-Substrate-*Inoculum*	Substrate (Chopped)	*Inoculum* from
15-C-I1	**C** Grass hay untreated (same as ponies were fed).	**I1** Ponies fed control diet consisting of grass hay fed at 100%
16-E1-I2	**E1** Grass hay treated with enzyme 0.75 mL/kg DM hay	**I2** Ponies fed grass hay treated with enzyme 0.75 mL/kg DM hay
17-E2-I3	**E2** Grass hay treated with enzyme 3.75 mL/kg DM hay	**I3** Ponies fed grass hay treated with enzyme 3.75 mL/kg DM hay
	Other treatments	
Dataset-Substrate-*Inoculum*	Substrate (Ground to pass 1 mm screen)	*Inoculum* from
18-A-I4	**A** Grass hay (50%) + alfalfa (50%)	**I4** Ponies fed 25% alfalfa and 75% grass hay (dry matter (DM) basis).
19-A1-I4	**A1** Grass hay (75%) + alfalfa (25%)	Not supplemented with yeast.
20-B-I5	**B** Grass hay (50%) + alfalfa (50%) + Yeast (0.011 g)	**I5** Ponies fed 25% alfalfa and 75% grass hay (DM basis).
21-B1-I5	**B1** Grass hay (75%) + alfalfa (25%) + Yeast (0.011 g)	Supplemented daily with 30 g yeast per pony was mixed with alfalfa.
22-C-I6	**C** Grass hay (50%) + concentrate (50%)	**I6** Ponies fed 25% concentrate and 75% grass hay (DM basis).
23-C1-I6	**C1** Grass hay (75%) + concentrate (25%)	Not supplemented with yeast.
24-D-I7	**D** Grass hay (50%) + concentrate (50%) + Yeast (0.011 g)	**I7** Ponies fed 25% concentrate and 75% grass hay (DM basis).
25-D1-I7	**D1** Grass hay (75%) + concentrate (25%) + Yeast (0.011 g)	Supplemented daily with 30 g yeast per pony mixed with concentrate.

2.2. Models Fitted

The classical Mitscherlich equation used in crop science has the general form:

$$y = A - (A - B)e^{-ct}$$

where the ordinate y is crop yield, the abscissa t is fertilizer rate, and A, B and c are constants. The parameter A represents the asymptotic value of y (i.e., maximum yield) and B the minimum yield (i.e., no fertilizer application).

For application to the gas production technique, the ordinate becomes cumulative gas production (mL) and the abscissa becomes time since inoculation (h). Cumulative gas production at zero time can be considered negligible and a lag $T \geq 0$ (h) may occur before onset of fermentation, so the Mitscherlich equation becomes:

$$y = A\left(1 - e^{-c(t-T)}\right); t \geq T \tag{1}$$

In this equation, A would represent the asymptotic gas production (mL) and c (h^{-1}) the fractional rate of fermentation. In this paper, we explore Equation (1) and three different extensions (Equations (2)–(4) below) of this classical function for use in describing typical and atypical gas production profiles.

Gas production profiles are typically monophasic, asymptotic and often sigmoidal (e.g., [5]). France et al. [4] derived the following equation from rate:state principles to describe such profiles:

$$y = A\left\{1 - \exp\left[-c(t - T) - d\left(\sqrt{t} - \sqrt{T}\right)\right]\right\}; t \geq T \tag{2}$$

Here, A (mL) is the asymptotic value of y, and c (h^{-1}) and d ($h^{-0.5}$) are fractional rate constants. Equation (1) is a special case of Equation (2) (i.e., $d = 0$). This equation is commonly referred to as the France model.

Biphasic, asymptotic gas production profiles have also been observed (e.g., [15]), and these would appear to lend themselves to description by the sum of two Mitscherlich terms:

$$y = A_1(1 - e^{-c_1(t-T_1)}) + A_2(1 - e^{-c_2(t-T_2)}); t \geq T_1, t \geq T_2 \tag{3}$$

The first term in Equation (3) is zero until time T_1 and likewise the second term until time T_2. Equation (1) is also a special case encompassed by Equation (3) (i.e., $A_2 = 0$). This latter equation will be referred to as the double Mitscherlich model.

As mentioned above, instances of profiles that do not exhibit typical asymptotic behaviour have also been observed but not formally reported. Such profile forms suggest a function resulting from the sum of a Mitscherlich term and a linear term might provide an appropriate description:

$$y = A(1 - e^{-c(t-T_1)}) + \beta(t - T_2); t \geq T_1, t \geq T_2 \tag{4}$$

where the parameter β (mL h^{-1}) is the slope of an underlying linear trend. As for Equation (3), the first term in Equation (4) is zero until time T_1 and likewise the second term until time T_2. Putting $\beta = 0$ in Equation (4) yields Equation (1). This latter equation will be referred to as the Mitscherlich + linear model.

2.3. Extent of Degradation

The extent of degradation (E) of substrate in a specific compartment or region of the gastro-intestinal tract may be calculated from the gas production curve, provided the *inoculum* used to generate the profile is representative of that compartment. If the profile is diminishing returns in shape, first-order kinetics with a constant fractional rate of degradation describes substrate degradation and Equation (1) can be fitted to the profile. Extent of degradation is then given by:

$$E = S_0 e^{-kT} c / [(c + k)(S_0 + U_0)] \tag{5}$$

where S_0 (g) is the amount of the incubated substrate that is potentially degradable, U_0 (g) the amount that is undegradable, T (h) the lag before commencement of degradation, c (h^{-1}) the fractional rate of fermentation, and k (h^{-1}) is the fractional rate of passage out of the compartment [4].

If the profile is sigmoidal, first-order kinetics with a variable fractional degradation rate would account for substrate degradation and the France model (Equation (2)) can be fitted. Extent of degradation is then given by:

$$E = S_0 e^{-kT} (1 - kI_1) / (S_0 + U_0) \tag{6}$$

$$= kS_0 I_2 / (S_o + U_0) \tag{7}$$

where

$$I_1 = \int_T^\infty \exp\left\{-\left[(c + k)(t - T) + d\left(\sqrt{t} - \sqrt{T}\right)\right]\right\} dt$$

$$I_2 = \int_T^\infty e^{-kt}\left\{1 - \exp\left[-c(t - T) - d\left(\sqrt{t} - \sqrt{T}\right)\right]\right\} dt$$

and c (h^{-1}) is the constant portion of the fractional degradation rate and d (h$^{-0.5}$) the coefficient of the variable portion. The integrals I_1 and I_2 are non-analytical and therefore have to be evaluated numerically [4].

If the profile is linear with an abrupt cut-off (i.e., a broken stick), zero-order kinetics with constant rate of degradation independent of substrate remaining can be assumed and a piecewise linear model fitted. The extent of degradation is then given by:

$$E = \beta\left[\ln\left(k\beta^{-1}S_0 + e^{kT}\right) - kT\right] / [k(S_0 + U_0)] \tag{8}$$

where β (mL h^{-1}) is the slope of the line fitting the ascending portion of the profile [22].

If the gas production profile is multiphasic, then the extent of degradation for each phase can be calculated by applying the appropriate equation, and the weighted extents summed to estimate overall extent of degradation. For example, if the profile resolves into two diminishing returns components (1 and 2) as in Equation (3), then Equation (5) can be independently applied to each of the two phases and the overall extent calculated as:

$$E = (w_1 E_1 + w_2 E_2) / (w_1 + w_2) \tag{9}$$

where w_1 and w_2 are the relative weights assigned to the respective phases. If the profile resolves into a diminishing returns and a linear (with abrupt cut-off) component as in Equation (4), then Equations (5) and (8) respectively can be applied to the two phases and the overall extent calculated again using Equation (9). As an arbitrary rule of thumb, the asymptotic gas production values for the two phases (abrupt cut-off value if a phase is linear), viz. A_1 and A_2, can be adopted as the weights w_1 and w_2 respectively.

Thus, for the equine data considered herein, the extent of degradation of substrate in the hindgut can be calculated using Equations (5)–(9) if we assume faecal *inoculum* is representative of that region of the gastro-intestinal tract. Herein, when calculating extent of degradation, the amount of the incubated substrate that is potentially degradable S_0 (g), the amount that is undegradable U_0 (g), and the fractional rate of passage out of the compartment k (h^{-1}), were assumed to be 0.538, 0.465 and 0.019, respectively, for all datasets [23].

2.4. Fitting and Evaluation of Models

Each of the four models (Equations (1)–(4)) was fitted by non-linear regression to the 25 gas production profiles using the NLIN procedure in the statistical software SAS [24]. Initial estimates of parameter values were obtained through visual inspection of the data.

Using various statistical tests, the models were evaluated for goodness-of-fit along with analysis of residuals. Mean square prediction error (MSPE) was calculated as the sum of the squared difference between predicted and observed values divided by the number of observations [25]. The accuracy factor (AF) index is a measure of the average deviation of a model's predictions and is used as a simple index of the level of confidence in these predictions [26]. Agreement between model predictions and observations was further determined using the concordance correlation coefficient (CCC), a single statistic ranging between −1 (perfect disagreement) and +1 (perfect agreement) which contains both accuracy and precision indicators [27,28]. The Akaike information criterion (AIC) is a test for model selection which accounts for goodness-of-fit while penalizing for over-fitting, with the model resulting in the smallest AIC being the most appropriate [29].

The ability of each model to predict gas production without systematically over- or under-estimating was examined using the number of runs test and the Durbin–Watson (DW) test. The runs test examines a sequence of residuals for unusual groupings of positive or negative residuals and tests the null hypothesis that the arrangement of signs (+/−) is random, with too few runs indicating the presence of autocorrelation [30]. The DW test examines dependencies in the error terms by testing for correlations between a residual and the residuals immediately before and after it in the sequence. Compared to the runs test, the DW provides greater information regarding analysis of residuals by not only considering the sign of the residual but also its magnitude. The DW statistic (D), and upper (D_u) and lower (D_l) critical values, were calculated according to [30]. When D is less than the lower critical value D_l, evidence of positive autocorrelation occurs, and when D is greater than the upper critical value D_u, evidence of negative autocorrelation occurs.

3. Results

The ability of the Mitscherlich, and the other derived functions, to describe typical and atypical cumulative gas production profiles was assessed by fitting the four equations (Equations (1)–(4)) to 25 datasets. The profiles examined resulted from incubating forage using faecal *inoculum* from equines following the methodology of Theodorou et al. [2]. Using parameter estimates resulting from fitting these models to the gas production profiles, extents of substrate degradation were calculated and compared.

3.1. Fitting Behaviour

Of the 25 gas production profiles considered, 17 displayed atypical patterns, characterized by more than one phase, while the remaining 8 displayed typical monophasic patterns. No convergence issues were encountered when fitting the simple Mitscherlich (Equation (1)), double Mitscherlich (Equation (3)) and the Mitscherlich + linear (Equation (4)) to any of the datasets. The use of an "if than" statement concerning $t \geq T_2$ and its effect on A_2 and β in SAS allowed both Equations (3) and (4) to revert to the simple Mitscherlich if that resulted in a better fit compared to the extended biphasic equations (i.e., when $A_2 = 0$ in Equation (3) and $\beta = 0$ in Equation (4)). The France equation (Equation (2)) also encompasses the ability to revert to a simple Mitscherlich (Equation (1)) when $d = 0$. When fitted to the 25 gas production profiles, the France equation (Equation (2)) reverted to the simple Mitscherlich (Equation (1)) in four cases as the best fit for these gas production profiles was achieved when $d = 0$. Likewise, the double Mitscherlich (Equation (3)) reverted to the simple Mitscherlich (Equation (1)), i.e., $A_2 = 0$, in five cases as a single Mitscherlich term described these profiles better than two Mitscherlich terms.

When fitting the France model (Equation (2)) to the atypical gas production curves, the convergence criteria had to be relaxed in order to reach successful convergence. When enforcing relaxed convergence criteria, Equation (2) was unable to converge for one of the 25 datasets. In order to achieve biologically meaningful parameters, lag time (T) and fractional rate constant (c) were constrained to be non-negative when fitting each model. Furthermore, in fitting Equation (2) a constraint was placed on parameter d, viz. $d \geq -2c \sqrt{T}$ to ensure the fractional rate of degradation remained non-negative [4].

3.2. Parameter Estimates and Fitted Gas Production Curves

Initial parameter estimates of lag time (T), asymptotic value (A) and slope (β) were determined by visual inspection of the gas production curves, while ranges for the fractional rate constants (c and d) were provided. The final parameter estimates resulting from fitting Equations (1)–(4) to Dataset 1 and 8 of Experiment 1 and Dataset 18 and 22 of Experiment 2 are presented in Tables 2 and 3, respectively. The final estimates resulting from fitting Equations (1)–(4) to the remaining 21 datasets are given in the Supplementary Information section of this paper (Tables S2–S5). Using the parameter estimates in Tables 2 and 3, the gas production profiles resulting from applying Equations (1)–(4) to Dataset 1 and 8 are shown in Figure 1. Both Dataset 1 and 8 show clear atypical biphasic gas production curves which are more faithfully represented by the double Mitscherlich (Equation (3)) and Mitscherlich + linear (Equation (4)) equations than the monophasic simple Mitscherlich (Equation (1)) or the France model (Equation (2)). Examining the four lower panels of Figure 1, the extent to which each phase contributes to the overall gas production curve of the double Mitscherlich (Equation (3)) and Mitscherlich + linear (Equation (4)) are clearly distinguishable.

Table 2. Parameter estimates obtained by fitting Equations (1)–(4) to Dataset 1 (laminitis) and 8 (clinically normal) from Experiment 1. An asterisk (*) denotes the equation that resulted in the best fit, based on AIC, to a particular dataset.

	Simple Mitscherlich (Equation (1))		France (Equation (2))		Double Mitscherlich (Equation (3)) [‡]		Mitscherlich + linear (Equation (4)) [§]	
	Dataset 1	Dataset 8	Dataset 1	Dataset 8	Dataset 1	Dataset 8	Dataset 1	Dataset 8
A	163.5	123.7	246.8	159.5	60.3, 84.3 *	60.8, 66.8 *	63.7	64.6
c	0.014	0.025	0.005	0.008	0.093, 0.026 *	0.140, 0.027 *	0.078	0.111
T	0	0	0	0	3.1, 43.4 *	2.8, 34.6 *	2.8, 35.9	2.5, 28.7
d			0.019	0.049				
β							0.936	0.723

[‡] The two scale parameters of this equation are entered under A in the order A_1, A_2 in this table. Likewise, the two rate parameters under c in the order c_1, c_2, and the two lags under T in the order T_1, T_2. [§] The two lag parameters of this equation are entered under T in the order T_1, T_2.

Table 3. Parameter estimates obtained by fitting Equations (1)–(4) to Dataset 18 (50% grass hay + 50% alfalfa) and 22 (50% grass hay + 50% concentrate), from Experiment 2. An asterisk (*) denotes the equation that resulted in the best fit, based on AIC, to a particular dataset.

	Simple Mitscherlich (Equation (1))		France (Equation (2))		Double Mitscherlich (Equation (3)) [‡]		Mitscherlich + linear (Equation (4)) [§]	
	Dataset 18	Dataset 22	Dataset 18	Dataset 22	Dataset 18	Dataset 22	Dataset 18	Dataset 22
A	120.6	148.7 *	122.9	148.7 [†]	109.7, 38.7	148.7 [Ψ]	109.6 *	147.7
c	0.071	0.078 *	0.057	0.078 [†]	0.090, 0.016	0.078 [Ψ]	0.091 *	0.079
T	0	0.5 *	0	0.5 [†]	0.3, 37.2	0.5 [Ψ]	0.3, 34.8 *	0.5, 33.1
d			0.037					
β							0.473 *	0.040

[‡] The two scale parameters of this equation are entered under A in the order A_1, A_2 in this table. Likewise, the two rate parameters under c in the order c_1, c_2, and the two lags under T in the order T_1, T_2. [§] The two lag parameters of this equation are entered under T in the order T_1, T_2. [†] Best fit by France, Equation (2), achieved with $d = 0$, therefore reverting to a simple Mitscherlich, viz. Equation (1). [Ψ] Best fit by double Mitscherlich, Equation (3), achieved with $A_2 = 0$, therefore reverting to a simple Mitscherlich, viz. Equation (1).

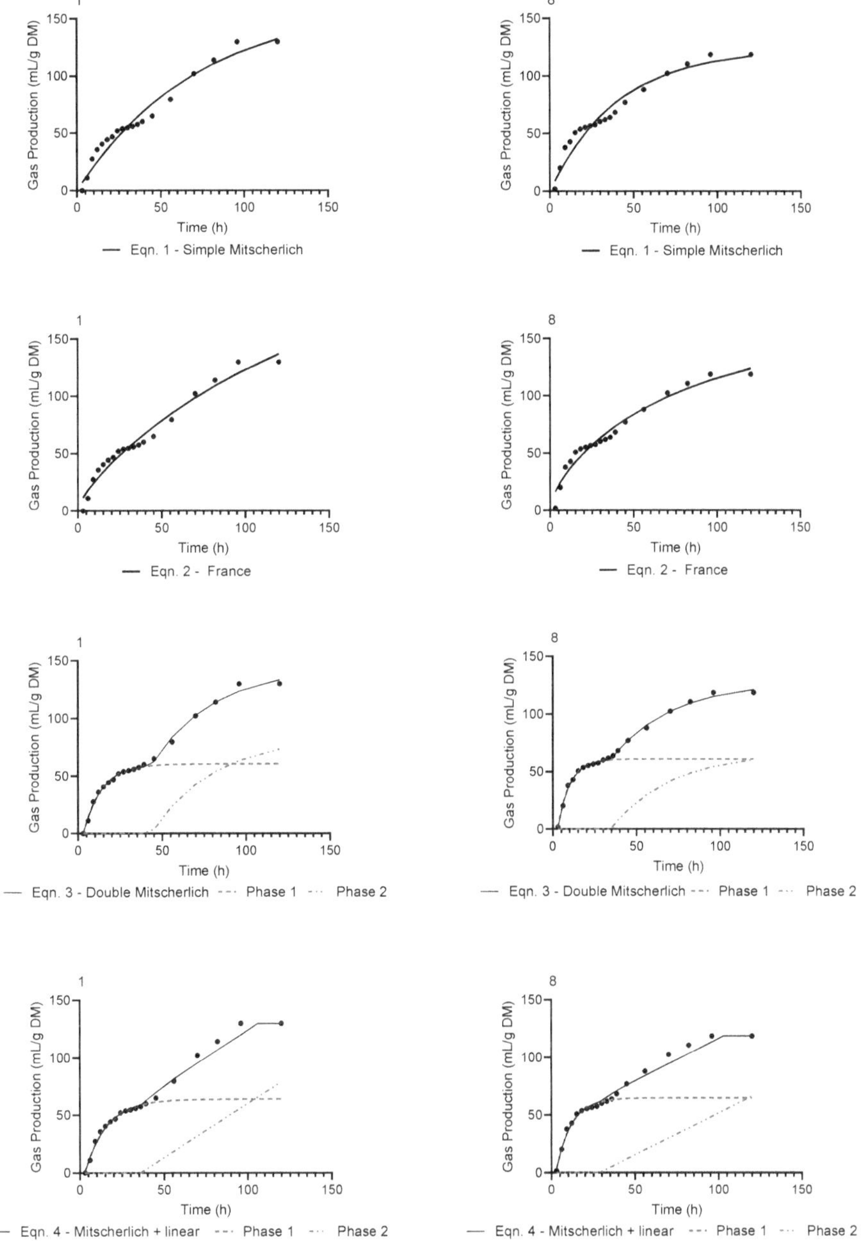

Figure 1. Observed (•) atypical gas production profiles and predicted curves resulting from fitting Equations (1)–(4) to Dataset 1 and 8 of Experiment 1.

In contrast to Dataset 1 and 8, Dataset 18 and 22 of Experiment 2 display more typical monophasic gas production curves as shown in Figure 2. Again, examining the bottom four panels of Figure 2, the second phase of the biphasic models (viz. Equations (3) and (4)) is much less evident, with the second phase being entirely absent when fitting the double Mitscherlich (Equation (3)) to Dataset 22 as

Equation (3) reverts to the simple Mitscherlich with $A_2 = 0$. Additionally, when fitting the equation of France (Equation (2)) to Dataset 22, the best fit was achieved when $d = 0$, and thus Equation (2) reverted to a simple Mitscherlich (Equation (1)) when applied to this dataset. The gas production profiles resulting from fitting Equations (1)–(4) to the remaining datasets are shown in the Supplementary Information (Figures S1–S4).

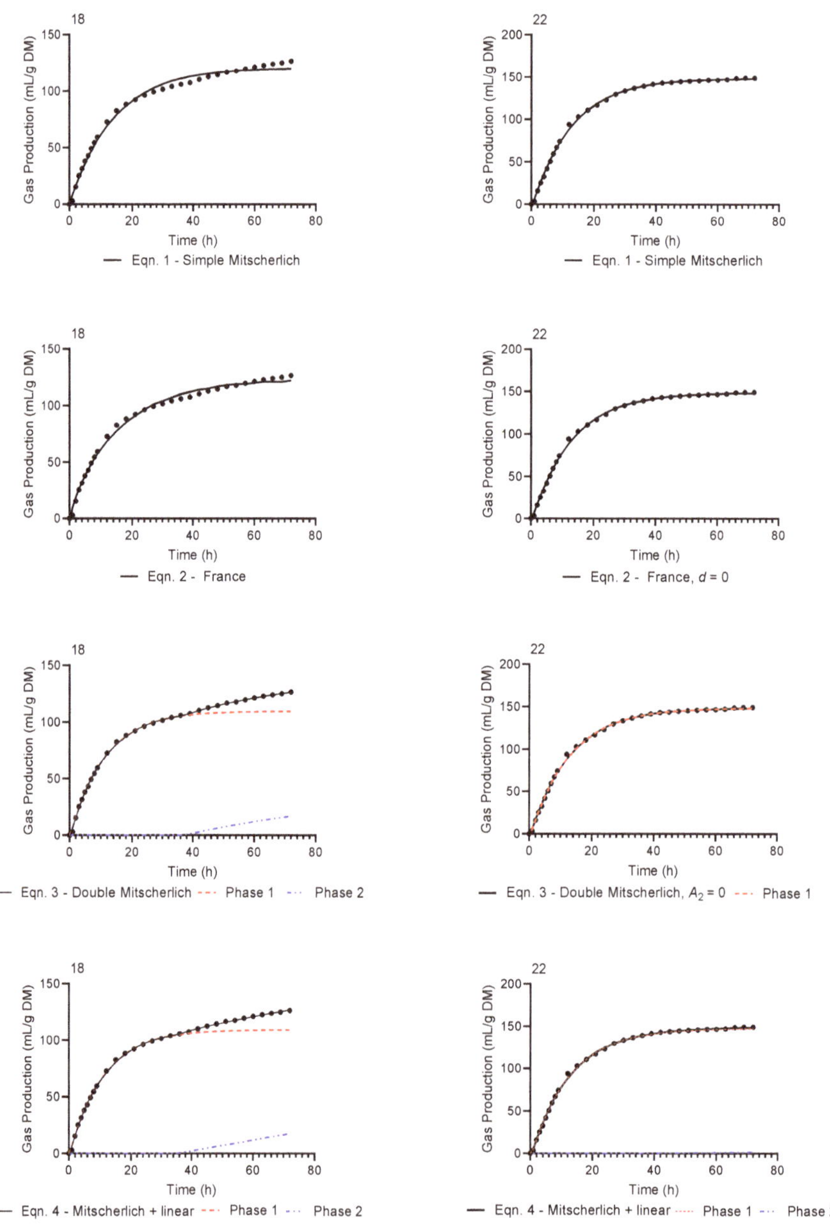

Figure 2. Observed (●) typical gas production profiles and predicted curves resulting from fitting Equations (1)–(4) to Dataset 18 and 22 of Experiment 2.

3.3. Model Evaluation

Goodness-of-fit was assessed using four criteria, namely AIC, MSPE, CCC and AF. The goodness-of-fit values resulting from fitting each of the four models to the 25 datasets were averaged and the models ranked from 1 to 4 based upon their comparative performance with the other models for a given criterion. Individual models averaged goodness-of-fit values, along with their mean rank and the number of times the model ranked first or second under a given criterion, are presented in Table 4.

Table 4. Goodness-of-fit and analysis of residuals from fitting the four equations to the 25 datasets of Experiments 1 and 2.

Criteria	Model			
	Simple Mitscherlich (Equation (1))	France (Equation (2))	Double Mitscherlich (Equation (3))	Mitscherlich + Linear (Equation (4))
Akaike information criterion (AIC)				
Average (±SE)	73.1 (2.5)	66.6 (2.6)	46.7 (2.7)	47.6 (5.2)
Mean rank	3.4	2.9	1.4	1.7
Number of times a model ranked 1st or 2nd	2	5	22	25
Mean square prediction error (MSPE)				
Average (±SE)	31.6 (5.4)	20.9 (3.2)	5.1 (0.5)	10.7 (1.7)
Mean rank	3.4	2.6	1.3	1.6
Number of times a model ranked 1st or 2nd	2	5	22	25
Concordance correlation coefficient (CCC)				
Average (±SE)	0.979 (0.004)	0.985 (0.003)	0.996 (0.001)	0.994 (0.001)
Mean rank	3.5	2.6	1.4	1.7
Number of times a model ranked 1st or 2nd	5	9	22	25
Accuracy factor (AF)				
Average	1.32 (0.04)	1.32 (0.04)	1.18 (0.03)	1.17 (0.03)
Mean rank	2.8	3.2	1.4	1.5
Number of times a model ranked 1st or 2nd	7	3	25	25
Number of runs test				
Too few runs	25	24	7	19
Runs are random	0	0	18	6
Durbin–Watson (DW) test				
Positive correlation	25	23	7	10
No evidence	0	1	2	4
Negative correlation	0	0	16	11

When fitted to the 25 datasets, the double Mitscherlich resulted in the smallest averaged AIC value (46.7 ± 2.7), followed by the Mitscherlich + linear (47.6 ± 5.2) and the France equation (66.6 ± 2.6), with the simple Mitscherlich (73.1 ± 2.5) resulting in the highest average AIC. Based upon AIC, the mean rank of the double Mitscherlich, Mitscherlich + linear, France and the simple Mitscherlich was 1.4, 1.7, 2.9 and 3.4, respectively, with the Mitscherlich + linear being ranked 1st or 2nd 25 times followed by the double Mitscherlich (23), France (5), and the simple Mitscherlich (2). The double Mitscherlich resulted in the lowest average MSPE (5.1 ± 0.5) followed by the Mitscherlich + linear (10.7 ± 1.7), France (20.9 ± 3.2), and the simple Mitscherlich (31.6 ± 5.4). In agreement with AIC, the double Mitscherlich resulted in the highest mean rank (1.3), followed by the Mitscherlich + linear (1.6), France (2.6) and simple Mitscherlich (3.4) with the Mitscherlich + linear ranking 1st or 2nd 25 times compared to the double Mitscherlich (22), France (5), and simple Mitscherlich (2).

Following the same trend as AIC and MSPE, the double Mitscherlich resulted in the highest CCC (0.996 ± 0.001), followed by the Mitscherlich + linear (0.994 ± 0.001), France (0.985 ± 0.003) and simple Mitscherlich (0.979 ± 0.004). Again, the double Mitscherlich yielded the best average rank of 1.4, followed by the Mitscherlich + linear (1.7), France (2.6) and simple Mitscherlich (3.5). The simple Mitscherlich was ranked 1st or 2nd in 5 of the 25 datasets and France in 9 of the 25, whilst the Mitscherlich + linear and double Mitscherlich were ranked 1st or 2nd in 25 and 22 datasets, respectively. On the basis of AF, the Mitscherlich + linear (1.17 ± 0.03) and double Mitscherlich (1.18 ± 0.03) outperformed both the France and simple Mitscherlich (with an averaged AF of 1.32 ± 0.04). The double Mitscherlich was ranked higher than the Mitscherlich + linear, 1.4 vs. 1.5, with both models being ranked 1st or 2nd 25 times. The simple Mitscherlich had a higher rank compared to the France, 2.8 vs. 3.2, with the simple Mitscherlich ranking 1st or 2nd 7 times compared to 3 times for the France equation.

In addition to goodness-of-fit, the runs test and DW test were used for the analysis of residuals. The results of these analyses are presented in Table 4. The runs test determined that too few runs occurred for all datasets when fitting the simple Mitscherlich and France equations. In comparison, runs of residuals were determined to be random in 18 and 6 of the 25 datasets when fitting the double Mitscherlich and Mitscherlich + linear, respectively. Using the DW test there was evidence of positive correlation of the residuals in all of the datasets that the simple Mitscherlich fitted successfully, and in all but one for the France model. In contrast, positive correlation was only found in 7 and 10 of the 25 datasets when fitting the double Mitscherlich and Mitscherlich + linear, with negative correlation being found in 16 and 11 of the 25 datasets, respectively.

3.4. Extent of Degradation

Extent of substrate degradation, calculated from the parameter estimates resulting from fitting the four models to the 25 gas production profiles, is presented in Table 5. When calculating extent of degradation using the biphasic equations, viz. Equations (3) and (4), the relative weights in which each phase contributed to the overall extent of substrate degradation need to be incorporated. Using the double Mitscherlich (Equation (3)) the weights of each phase were simply assumed to be their respective asymptotic gas production values for each phase, A_1 and A_2. For the Mitscherlich + linear, the weight of the first phase was its respective asymptotic value (A_1), while the weight of the second (linear) phase was calculated as the amount of gas produced over the course of this linear segment (i.e., multiplying its slope by the duration of the linear segment). The duration of the linear segment was the difference between time at which the abrupt cut-off value occurred and time at which the linear portion commenced (i.e., T_2, the lag time). The abrupt cut-off was determined in two ways. In Experiment 1, Dataset 1–14, abrupt cut-off values were assumed to occur at the intersection between the linear segment of the equation and the apparent plateau in gas production, which visually occurred between the last two data points. In Experiment 2, Dataset 15–17, a plateau in the observed data was not evident, therefore the abrupt cut-off value was set to the end of incubation. Finally, in Experiment 2, Dataset 18–25, a plateau was eventually reached with the linear segments being horizontal or near horizontal, and abrupt cut-off values were again set to the end of incubation.

Table 5. Calculated extent of degradation (%) from the 14 datasets of Experiment 1 displaying atypical gas production profiles and the 11 datasets of Experiment 2 displaying both typical and atypical profiles. An asterisk (*) denotes the equation that resulted in the best fit, based on AIC, to a particular dataset.

Dataset	Visual Curve Pattern	Simple Mitscherlich (Equation (1))	France (Equation (2))	Double Mitscherlich (Equation (3))	Mitscherlich + Linear (Equation (4))
1	Atypical	22.7	15.5	25.5 *	33.9
2	Atypical	34.3	28.5	35.1 *	40.0
3	Atypical	25.4	12.7	25.1 *	37.5
4	Atypical	26.7	-	25.0 *	35.6
5	Atypical	31.1	21.8	29.1 *	42.0
6	Atypical	30.3	21.7	30.0 *	38.0
7	Atypical	34.9	28.6	32.7 *	39.7
8	Atypical	30.4	24.4	30.0 *	38.0
9	Atypical	23.4	16.0	33.2 *	36.8
10	Atypical	22.4	19.0	24.2 *	34.9
11	Atypical	36.3	32.9	29.1 *	39.6

Table 5. *Cont.*

Dataset	Visual Curve Pattern	Simple Mitscherlich (Equation (1))	France (Equation (2))	Double Mitscherlich (Equation (3))	Mitscherlich + Linear (Equation (4))
12	Atypical	33.4	32.6	30.2 *	38.2
13	Atypical	25.9	22.0	28.1 *	34.5
14	Atypical	33.3	31.3	25.1 *	37.3
15	Atypical	31.6	31.6	31.6 $^\Psi$	38.4 *
16	Atypical	34.2	23.5	34.2 $^\Psi$	38.9 *
17	Atypical	35.5	19.5	35.5 $^\Psi$	42.0 *
18	Typical	42.4	41.6	35.9	41.9 *
19	Typical	42.1	40.5	36.8 *	42.8
20	Typical	43.3 *	43.3 †	43.3 $^\Psi$	43.0
21	Typical	43.2	43.2 †	41.2	42.9 *
22	Typical	42.9 *	42.9 †	42.9 $^\Psi$	42.8
23	Typical	42.5	42.3	41.9	43.0 *
24	Typical	44.2	44.2 †	43.1	43.8 *
25	Typical	43.0	43.1	41.2	42.8 *

† Best fit by France, Equation (2), achieved when $d = 0$, therefore reverting to a simple Mitscherlich, viz. Equation (1).
$^\Psi$ Best fit by double Mitscherlich, Equation (3), achieved with $A_2 = 0$, therefore reverting to a simple Mitscherlich, viz. Equation (1).

Although substrate was primarily grass hay, or a grass hay mix, the range of calculated extent of degradation varied widely, from a minimum of 12.7% to a maximum of 44.2%. The wide range in extent of degradation values can be attributed to the use of parameter estimates from a model that fits a given gas production profile poorly. Therefore, Table 5 includes an indicator of which model fitted the particular dataset values best, based upon AIC. Fitting these four models to 25 datasets that encompass both typical and atypical gas production curves resulted in the double Mitscherlich being the best fitting in 15 of these datasets, the Mitscherlich + linear 8, simple Mitscherlich 2 and France 0. Given that extent of degradation is determined using parameter estimates obtained by fitting these models to a dataset, the importance of model fit in calculating extent of degradation is apparent.

4. Discussion

Gas production profiles generated from incubating a substrate with either ruminal or faecal *inocula* have been widely used to provide information regarding the degradability of forages and supplementary feeds in both ruminants and non-ruminants [3,5,6,10–14]. Typical shapes of these profiles range from diminishing returns to strongly sigmoidal [3]. Various models, e.g., Mitscherlich, Michaelis-Menten, Gompertz and logistic, have been proposed to describe these curves, including generalised models such as Richards and that of France which are able to accommodate both diminishing returns and sigmoidal behaviour [4,31–33]. Deriving these models on the basis of substrate degradation rather than amount of gas produced permits the generation of fermentation kinetic parameters [3]. By fitting these models to gas production profiles, such parameters (e.g., fractional rate of degradation and lag time) can be estimated. These model-derived parameters have been used in conjunction with information regarding substrate degradability and digesta passage rate to calculate extent of substrate degradation in the rumen [3–6,34]. This method has been successfully applied to typical monophasic sigmoidal and diminishing returns gas profiles to evaluate substrates based upon the extent of their degradability [3,5,6,8–14,34].

In addition to the typical sigmoidal and diminishing return patterns displayed by gas production profiles, atypical multiphasic curves have been reported [15,16,19,20,35]. Although multiphasic gas production curves have been described by both Groot et al. [15] and Wang et al. [36], proposed models are based upon the amount of gas produced, rather than the amount of substrate degraded, resulting in the model being unable to link the gas production technique to animal performance [3].

4.1. Profile Shapes and Associated Parameters

The diminishing returns behaviour described by the simple Mitscherlich is a result of the interaction between the constant fractional degradation rate (c) and the amount of degradable substrate (S) available for fermentation. The amount of degradable substrate available is time dependent with the maximal value occurring at time zero (S_0). The instantaneous rate of degradation is calculated by multiplying the constant fractional degradation rate (c) by the amount of degradable substrate (S) at time t [3]. As the fractional degradation rate is constant, and S is maximal at the commencement of the incubation, instantaneous rate of degradation is maximal at the start of incubation, following a lag period if present. As fermentation progresses, the amount of degradable substrate decreases while the fractional rate of degradation remains constant. Therefore, the instantaneous rate of degradation declines continuously, from its maximum at the start of the incubation until it finally reaches zero due to available fermentable material being exhausted. When fermentation ceases, due to a lack of degradable substrate, the instantaneous rate of degradation becomes zero, with no additional gas production occurring, having reached an upper asymptote. Therefore, the characteristic diminishing returns pattern of the simple Mitscherlich describes a scenario whereby rate of fermentation, and thus gas production, is initially at a maximum and continuously decreases, as a function of time, until an asymptote is reached.

Unlike the simple Mitscherlich, the France model is capable of describing both diminishing returns and sigmoidal behaviour. This is achieved by assuming that fractional rate of degradation can vary with time. Depending on the values of the fractional rate constants, viz. c and d in this manuscript, the fractional rate of degradation can remain constant, decrease or increase with time [4]. In the France model, when the fractional rate of degradation is constant diminishing return type behaviour is described (as in the simple Mitscherlich). None of the gas production profiles examined in this study showed clear sigmodal behaviour and therefore the flexibility of the France model in its ability to describe both diminishing returns and sigmoidal shapes was not demonstrated. When describing sigmoidal behaviour, initially the rate of degradation increases resulting in exponential-type behaviour. As it continues to increase, a point of inflexion occurs whereby the rate reaches its maximal value. Following inflexion, the rate of degradation decreases resulting in diminishing returns behaviour with an asymptote being approached.

Of the 25 datasets examined in this study, in all but three the substrate was ground and passed through a 1 mm screen prior to incubation, with the remaining three being chopped (length not reported). A clear increase in gas production, and associated increase in extent of substrate degradation, is observed when comparing the ground (Dataset 18–25) with the chopped substrates (Dataset 15–17) of Experiment 2 (see Supplementary Information Figure S3 vs. Figure S4 and Table 5 for associated gas production profiles and extent of degradation values, respectively). However, 17 of these datasets, including both ground and chopped substrates, are atypical in nature and exhibit a second phase of gas production. In these 17 datasets the first phase is well described by the simple Mitscherlich whereby following a lag, rate of degradation and thus gas production are initially at their maximum and continuously decrease until an asymptote is approached. Following this first phase, a second phase occurs. This second phase can show either diminishing returns or a linear pattern. These phases might be attributed to differences in chemical or nutritional fractions of the feedstuffs [16]. Phase 1 may represent the gas produced from the fermentation of sugars or a soluble readily fermentable fraction, while the second phase consists of gas produced from the fermentation of structural carbohydrates or an insoluble potentially fermentable fraction [35,37]. Alternatively, the occurrence of the second gas production phase can potentially be attributed to chemical or structural barriers implicit in the substrate that must be overcome in order to continue degradation [15]. Furthermore, the possibility of microbial turnover in batch cultures, and the small amount of gas produced from 'self-fermentation' may add to the second phase of gas production [35,38]. Many other factors may influence the profile shape including: inter-animal variability, ration of the donor of the microbial *inoculum*, length of time the donor animal was adapted to the ration, time of day the *inoculum* was collected, ruminal vs. faecal sources of *inoculum* and frozen vs. fresh *inoculum* [5,19,35]. This leads to the conclusion that these factors may have the potential to influence microbial diversity and abundance in the *inoculum*, which in turn influences fermentative ability and by extension influences gas production.

4.2. Extent of Degradation

The ability to describe typical diminishing returns and sigmoidal gas production profiles using a variety of models (e.g., Mitscherlich, Michaelis-Menten, logistic, Gompertz and France) is well established [3,34,36]. However, the description of atypical multiphasic gas production curves is much less established, particularly how to link the in vitro gas production technique to the extent of degradation in the animal. Groot et al. [15] proposed a model that fitted multiphasic profiles using two or more generalised rectangular hyperbolae. Applying this model, some authors have estimated the amount of gas produced and rate of gas production of various feedstuffs [35,39]. Likewise, Wang et al. [36] described single- and multi-phase gas production curves using logistic-exponential equations. However, in these studies differences in feedstuffs were identified on the basis of the amount of gas produced rather than on criteria linked to animal performance.

Two biphasic models are presented in this paper that make use of a simple Mitscherlich term when describing the first phase of gas production and a second phase comprising either an additional Mitscherlich or a simple linear term with abrupt cut-off. Depending on the nature of the profile, both the double Mitscherlich (Equation (3)) and Mitscherlich + linear (Equation (4)) fitted the atypical datasets well, resulting in parameter values that can be used to calculate and compare the extent of degradation of respective substrates and substrate treatments. For example, three of the datasets examined in this study (Dataset 15–17) encompass chopped hay treated with increasing levels of an enzyme (0, 0.75 or 3.7 mL enzyme per kg DM hay, respectively). The Mitscherlich + linear fitted these datasets the best and using the associated parameter values, extent of substrate degradation was demonstrated to increase with increasing levels of enzymatic treatment, viz. 38.4%, 38.9% and 42.0% for Dataset 15–17, respectively. When fitting the Mitscherlich + linear to these datasets, it was assumed that following the linear trend an abrupt cut-off is reached (i.e., an asymptote is reached) and gas production ceases. This can be observed by inspecting Dataset 1 and 8 in Figure 1 whereby gas production ceases to increase between the last two data points. In comparison, examining Dataset 15–17 (see Supplementary Information, Figure S3), visually there appears to be potential for further gas production as an asymptote, in the form of an abrupt cut-off following the linear segment, has yet to be reached. If the linear trend continues after the 76 h incubation period, there is potential for continued substrate degradation and associated gas production. Therefore, extent of substrate degradation calculated using the Mitscherlich + linear would be an underestimate if fermentation continued beyond the 76 h incubation period used to generate these gas production profiles.

As previously mentioned, the extent of degradation calculated using the fermentation kinetic parameters generated from applying the four proposed models to 25 datasets ranged widely from a minimum of 12.7% to a maximum of 44.2%. This wide range in values can partially be attributed to the use of parameter values from a model that fits a particular gas production profile poorly. Even in a given dataset, large variation in calculated extent of degradation values existed. For example, in Dataset 3, extent of degradation using the France model was 12.7% compared to 37.5% with the Mitscherlich + linear. These findings are in contrast with those of Dhanoa et al. [34] whereby the model being applied (generalised Mitscherlich, simple Mitscherlich, generalised Michaelis-Menten, simple Michaelis-Menten, Gompertz and logistic) had very little effect on extent of degradation values. However, the gas production profiles of Dhanoa et al. [34] using mixed rumen microorganisms as the *inoculum* were all monophasic in nature and therefore reasonably well described by the aforementioned models. Indeed, when examining the eight typical gas production profiles of this manuscript, Dataset 18–25, there was very little difference in extent of degradation in a given dataset regardless of model applied. When comparing the standard deviation of extent of degradation determined by the four models when applied to the same typical gas production dataset, Dataset 22 had the lowest value of

0.1% while Dataset 18 had the highest deviation at 3.1%. Examining Dataset 22, the simple Mitscherlich fitted the dataset best with an associated extent of degradation of 42.9%. Both the double Mitscherlich and France models reverted to the simple Mitscherlich, as the simple Mitscherlich fitted this dataset better than their generalized forms, and therefore were in agreement with an extent of degradation of 42.9%. The Mitscherlich + linear was also in agreement with this value, 42.8%. In contrast, examining the 17 atypical biphasic gas production profiles, the model applied had considerable ramifications on calculated extent of degradation. In these datasets, the lowest standard deviation of extent of degradation between the four models when applied to a single dataset occurred in Dataset 12 at 3.4% while the largest deviation occurred in Dataset 3 at 10.1%. In the atypical profile of Dataset 12, the double Mitscherlich fitted the gas production profile the best with a calculated extent of degradation of 30.2%, the simple Mitscherlich, France and Mitscherlich + linear overestimated the extent of degradation, viz. 33.4%, 32.6% and 38.2%, respectively. Overall, this discrepancy in extent of degradation values for a given dataset can be attributed to fitting a monophasic equation (viz. Equations (1) and (2)) to a distinctly biphasic profile, or fitting a linear term to a non-linear segment, resulting in poor kinetic parameter estimates and by extension extent of degradation values.

It is important to note that when calculating extent of degradation, the value of S_0, the amount of incubated substrate that is potentially degradable, was taken from the literature. This value was set to 0.538 regardless of dataset and the associated substrate represented by that dataset. The value of 0.538 is the apparent in vivo dry matter digestibility, using ponies, of ground and pelleted hay consisting of a 50:50 mix of Lucerne hay and Cocksfoot hay [23]. When performing the gas production technique of Theodorou et al. [2], the potentially undegradable fraction of the substrate (U_0) can be obtained by weighing the residual matter after gas production has ceased. Likewise, the potentially degradable value (S_0) can be calculated by subtracting U_0 from the quantity of substrate initially incubated. However, these values were not available at the time of this current study and a constant value was assumed. Therefore, greater differences in calculated values of extent of degradation should be expected as S_0 and U_0 will vary between substrates, substrate composition and the treatment received.

5. Conclusions

Two models, a double Mitscherlich and Mitscherlich + linear with abrupt cut-off, were proposed and derived to describe atypical gas production patterns characterized by two distinct phases of gas production. The models fitted these atypical curves well and due to their hybrid nature are also able to describe typical monophasic gas production profiles through their ability to revert to a simple Mitscherlich. These models contain kinetic parameters that can be used to calculate extent of substrate degradation using relatively simple equations. Given that extent of degradation is linked to nutrient supply, these models provide useful information regarding the evaluation of feedstuffs using in vitro methods [34,40].

Supplementary Materials: The following are available online at http://www.mdpi.com/2076-2615/10/2/308/s1, **Table S1**: Observed Gas Production Values of Datasets 1–25 from Experiment 1 and 2; **Table S2**: Final parameter estimates from fitting the simple Mitscherlich (Eqn. 1) to Dataset 1–25; **Table S3**: Final parameter estimates from fitting the France model (Eqn. 2) to Dataset 1–25; **Table S4**: Final parameter estimates from fitting the double Mitscherlich model (Eqn. 3) to Dataset 1–25; **Table S5**: Final parameter estimates from fitting the Mitscherlich + linear (Eqn. 4) to Dataset 1–25. **Figure S1**: Observed (•) and predicted gas production profiles resulting from fitting Equations (1)–(4) to Dataset 1–7, horses displaying clinical signs of laminitis, from Experiment 1; **Figure S2**: Observed (•) and predicted gas production profiles resulting from fitting Equations (1)–(4) to Datasets 8–14, clinically normal horses, from Experiment 1; **Figure S3**: Observed (•) and predicted gas production profiles resulting from fitting Equations (1)–(4) to Dataset 15–17, datasets exhibiting atypical dual-phase gas production curves, from Experiment 2. **Figure S4**: Observed (•) and predicted gas production profiles resulting from fitting Equations (1)–(4) to Dataset 18–25, datasets exhibiting typical single-phase gas production curves, from Experiment 2.

Author Contributions: Conceptualization, J.F. and M.S.D.; methodology, J.F., M.S.D. and S.L.; software, C.D.P.; validation, C.D.P.; formal analysis, C.D.P.; investigation, C.D.P.; resources, A.G. and J.-A.M.D.M.; data curation, C.D.P.; writing—original draft preparation, C.D.P.; writing—review and editing, C.D.P., M.S.D., A.G., J.-A.M.D.M.,

S.L., J.L.E. and J.F.; visualization, C.D.P.; supervision, J.F. and S.L.; project administration, J.F.; funding acquisition, J.F. All authors have read and agreed to the published version of the manuscript.

Funding: This research in addition to the APC were funded by The Canada Research Chairs program, grant number 045867 (Natural Sciences and Engineering Research Council of Canada, Ottawa).

Conflicts of Interest: The authors declare no conflict of interest. The funders had no role in the design of the study; in the collection, analyses, or interpretation of data; in the writing of the manuscript, or in the decision to publish the results.

References

1. Menke, K.H.; Steingass, H. Estimation of the energetic feed value obtained from chemical analysis and in vitro gas production using rumen fluid. *Anim. Res. Dev.* **1988**, *28*, 7–55.

2. Theodorou, M.K.; Williams, B.A.; Dhanoa, M.S.; McAllan, A.B.; France, J. A simple gas production method using a pressure transducer to determine the fermentation kinetics of ruminant feeds. *Anim. Feed Sci. Technol.* **1994**, *48*, 185–197. [CrossRef]

3. France, J.; Dijkstra, J.; Dhanoa, M.S.; López, S.; Bannink, A. Estimating the extent of degradation of ruminant feeds from a description of their gas production profiles observed *in vitro*: Derivation of models and other mathematical considerations. *Br. J. Nutr.* **2000**, *83*, 143–150. [CrossRef] [PubMed]

4. France, J.; Dhanoa, M.S.; Theodorou, M.K.; Lister, S.J.; Davies, D.R.; Isac, D. A model to interpret gas accumulation profiles associated with in vitro degradation of ruminant feeds. *J. Theor. Biol.* **1993**, *163*, 99–111. [CrossRef]

5. Mauricio, R.M.; Owen, E.; Mould, F.L.; Givens, I.; Theodorou, M.K.; France, J.; Davies, D.R.; Dhanoa, M.S. Comparison of bovine rumen liquor and bovine faeces as inoculum for an in vitro gas production technique for evaluating forages. *Anim. Feed Sci. Technol.* **2001**, *89*, 33–48. [CrossRef]

6. Dhanoa, M.S.; France, J.; Crompton, L.A.; Mauricio, R.M.; Kebreab, E.; Mills, J.A.N.; Sanderson, R.; Dijkstra, J.; López, S. Technical note: A proposed method to determine the extent of degradation of a feed in the rumen from the degradation profile obtained with the in vitro gas production technique using feces as the inoculum. *J. Anim. Sci.* **2004**, *82*, 733–746. [CrossRef]

7. López, S. In vitro and in situ techniques for estimating digestibility. In *Quantitative Aspects of Ruminant Digestion and Metabolism*; Dijkstra, J., Forbes, J.M., France, J., Eds.; CAB International: Wallingford, UK, 2005; pp. 87–121.

8. Calabrò, S.; Cutrignelli, M.I.; Piccolo, G.; Bovera, F.; Zicarelli, F.; Gazaneo, M.P.; Infascelli, F. In vitro fermentation kinetics of fresh and dried silage. *Anim. Feed Sci. Technol.* **2005**, *123–124*, 129–137. [CrossRef]

9. Yao, K.Y.; Gu, F.F.; Liu, J.X. In vitro rumen fermentation characteristics of substrate mixtures with soybean meal partially replaced by microbially fermented yellow wine lees. *Ital. J. Anim. Sci.* **2020**, *19*, 18–24. [CrossRef]

10. Lowman, R.S.; Theodorou, M.K.; Hyslop, J.J.; Dhanoa, M.S.; Cuddeford, D. Evaluation of an in vitro batch culture technique for estimating the in vivo digestibility and digestible energy content of equine feeds using equine faeces as the source of microbial inoculum. *Anim. Feed Sci. Technol.* **1999**, *80*, 11–27. [CrossRef]

11. Murray, J.A.M.D.; Longland, A.; Moore-Colyer, M. In vitro fermentation of different ratios of high-temperature dried lucerne and sugar beet pulp incubated with an equine faecal inoculum. *Anim. Feed Sci. Technol.* **2006**, *129*, 89–98. [CrossRef]

12. Elghandour, M.M.Y.; Vázquez Chagoyán, J.C.; Salem, A.Z.M.; Kholif, A.E.; Martínez Castañeda, J.S.; Camacho, L.M.; Buendía, G. In vitro fermentative capacity of equine fecal inocula of 9 fibrous forages in the presence of different doses of Saccharomyces cerevisiae. *J. Equine Vet. Sci.* **2014**, *34*, 619–625. [CrossRef]

13. Murray, J.M.D.; McMullin, P.; Handel, I.; Hastie, P.M. Comparison of intestinal contents from different regions of the equine gastrointestinal tract as inocula for use in an in vitro gas production technique. *Anim. Feed Sci. Technol.* **2014**, *187*, 98–103. [CrossRef]

14. Garber, A.; Hastie, P.M.; Handel, I.; Murray, J.M.D. In vitro fermentation of different ratios of alfalfa and starch or inulin incubated with an equine faecal inoculum. *Livest. Sci.* **2018**, *215*, 7–15. [CrossRef]

15. Groot, J.C.J.; Cone, J.W.; Williams, B.A.; Debersaques, F.M.A.; Lantinga, E.A. Multiphasic analysis of gas production kinetics for in vitro fermentation of ruminant feeds. *Anim. Feed Sci. Technol.* **1996**, *64*, 77–89. [CrossRef]

16. López, S.; Dijkstra, J.; Dhanoa, M.S.; Bannink, A.; Kebreab, E.; France, J. A generic multi-stage compartmental model for interpreting gas production profiles. In *Modelling Nutrient Digestion and Utilization in Farm Animals*; Sauvant, D., Van Milgen, J., Faverdin, P., Friggens, N., Eds.; Wageningen Academic Publishers: Wageningen, The Netherlands, 2010; pp. 139–147.

17. Mitscherlich, E. Das gesetz des minimums und das gesetz des abnehmenden bodenertrages. *Landwirstsch Jahrb* **1909**, *38*, 537–552.

18. Thornley, J.; France, J. Growth functions. In *Mathematical Models in Agriculture: Quantitative Methods for the Plant, Animal and Ecological Sciences*, 2nd ed.; CAB International: Wallingford, UK, 2007; pp. 136–169.

19. Murray, J.A.M.D.; Scott, B.; Hastie, P.M. Fermentative capacity of equine faecal inocula obtained from clinically normal horses and those predisposed to laminitis. *Anim. Feed Sci. Technol.* **2009**, *151*, 306–311. [CrossRef]

20. Garber, A.; Hastie, P.M.; Farci, V.; Murray, J.M.D. Yeast increases in vitro gas production of high fibre substrate when incubated with equine faecal inoculum. In Proceedings of the 8th European Equine Health & Nutrition Congress, Antwerp, Belgium, 23–24 March 2017; p. 119.

21. ANKOM^{RF}. *Gas Production System–Operator's Manual (Rev F 1/26/15)*; Ankom Technology: Macedon, NY, USA, 2015.

22. Dhanoa, M.S.; López, S.; Sanderson, R.; France, J. Simplified estimation of forage degradability in the rumen assuming zero-order degradation kinetics. *J. Agric. Sci.* **2009**, *147*, 225–240. [CrossRef]

23. Drogoul, C.; Poncet, C.; Tisserand, J.L. Feeding ground and pelleted hay rather than chopped hay to ponies 1. Consequences for in vivo digestibility and rate of passage of digesta. *Anim. Feed Sci. Technol.* **2000**, *87*, 117–130. [CrossRef]

24. SAS Institute Inc. *SAS/STAT® 9.2 User's Guide, Second Edition*; SAS Institute Inc.: Cary, NC, USA, 2009.

25. Bibby, J.; Toutenburg, H. *Prediction and Improved Estimation in Linear Models*; John Wiley and Sons: London, UK, 1977.

26. Ross, T. Indices for performance evaluation of predictive models in food microbiology. *J. Appl. Bacteriol.* **1996**, *81*, 501–508. [CrossRef]

27. Lin, L.I.-K. A concordance correlation coefficient to evaluate reproducibility. *Biometrics* **1989**, *45*, 255–268. [CrossRef]

28. Lin, L.; Torbeck, L.D. Coefficient of accuracy and concordance correlation coefficient: New statistics for methods comparison. *PDA J. Pharm. Sci. Technol.* **1998**, *52*, 55–59. [PubMed]

29. Akaike, H. A new look at the statistical model identification. *IEEE Trans. Automat. Contr.* **1974**, *19*, 716–723. [CrossRef]

30. Draper, N.R.; Smith, H. Fitting a straight line by least squares. In *Applied Regression Analysis*; John Wiley & Sons, Inc.: Hoboken, NJ, USA, 1998; pp. 15–96.

31. Krishnamoorthy, U.; Soller, H.; Steingass, H.; Menke, K.H. A comparative study on rumen fermentation of energy supplements in vitro. *J. Anim. Physiol. Anim. Nutr.* **1991**, *65*, 28–35. [CrossRef]

32. Blümmel, M.; Ørskov, E.R. Comparison of in vitro gas production and nylon bag degradability of roughages in predicting feed intake in cattle. *Anim. Feed Sci. Technol.* **1993**, *40*, 109–119. [CrossRef]

33. Schofield, P.; Pitt, R.E.; Pell, A.N. Kinetics of fiber digestion from in vitro gas production. *J. Anim. Sci.* **1994**, *72*, 2980–2991. [CrossRef]

34. Dhanoa, M.S.; López, S.; Dijkstra, J.; Davies, D.R.; Sanderson, R.; Williams, B.A.; Sileshi, Z.; France, J. Estimating the extent of degradation of ruminant feeds from a description of their gas production profiles observed *in vitro*: Comparison of models. *Br. J. Nutr.* **2000**, *83*, 131–142. [CrossRef]

35. Cone, J.W.; Van Gelder, A.H.; Visscher, G.J.W.; Oudshoorn, L. Influence of rumen fluid and substrate concentration on fermentation kinetics measured with a fully automated time related gas production apparatus. *Anim. Feed Sci. Technol.* **1996**, *61*, 113–128. [CrossRef]

36. Wang, M.; Tang, S.X.; Tan, Z.L. Modeling in vitro gas production kinetics: Derivation of Logistic-Exponential (LE) equations and comparison of models. *Anim. Feed Sci. Technol.* **2011**, *165*, 137–150. [CrossRef]

37. Stefanon, B.; Pell, A.N.; Schofield, P. Effect of maturity on digestion kinetics of water-soluble and water-insoluble fractions of alfalfa and brome hay. *J. Anim. Sci.* **1996**, *74*, 1104–1115. [CrossRef]

38. Theodorou, M.K.; Davies, D.R.; Nielsen, B.B.; Lawrence, M.I.G.; Trinci, A.P.J. Determination of growth of anaerobic fungi on soluble and cellulosic substrates using a pressure transducer. *Microbiology* **1995**, *141*, 671–678. [CrossRef]

39. Van Gelder, A.H.; Hetta, M.; Rodrigues, M.A.M.; De Boever, J.L.; Den Hartigh, H.; Rymer, C.; Van Oostrum, M.; Van Kaathoven, R.; Cone, J.W. Ranking of in vitro fermentability of 20 feedstuffs with an automated gas production technique: Results of a ring test. *Anim. Feed Sci. Technol.* **2005**, *123–124*, 243–253. [CrossRef]
40. Tamminga, S.; Van Straalen, W.M.; Subnel, A.P.J.; Meijer, R.G.M.; Steg, A.; Wever, C.J.G.; Blok, M.C. The Dutch protein evaluation system: The DVE/OEB-system. *Livest. Prod. Sci.* **1994**, *40*, 139–155. [CrossRef]

Article

Effects of Gas Production Recording System and Pig Fecal Inoculum Volume on Kinetics and Variation of In Vitro Fermentation using Corn Distiller's Dried Grains with Solubles and Soybean Hulls

Jae-Cheol Jang [1], Zhikai Zeng [1], Gerald C. Shurson [1] and Pedro E. Urriola [1,2,*

[1]   Department of Animal Science, University of Minnesota, St. Paul, MN 55108, USA;
    jang0046@umn.edu (J.-C.J.); zzeng@umn.edu (Z.Z.); shurs001@umn.edu (G.C.S.)
[2]   Department of Veterinary Population Medicine, University of Minnesota, St. Paul, MN 55108, USA
*   Correspondence: urrio001@umn.edu

Received: 18 September 2019; Accepted: 6 October 2019; Published: 9 October 2019

Simple Summary: Various in vitro methodologies have been developed and used to estimate the digestibility of feed ingredients, such as corn distillers dried grains with solubles (cDDGS) and soybean hulls (SBH) which contain high concentrations of dietary fiber. This study evaluated two in vitro gas production recording systems (manual vs. automated) and two initial fecal inoculum volumes (30 vs. 75 mL) on the parameters of in vitro fermentation of cDDGS and SBH. The results showed that the use of 75-mL inoculum volume with 0.5 g substrate tended to reduce the variation of measurements compared to the 30-mL inoculum volume with 0.2 g substrate regardless of the gas production recording system. These findings suggest that using larger inoculum volume with more substrate increases the precision of measurements. Furthermore, the automated system decreases labor for conducting the assay.

Abstract: An experiment was conducted to investigate the effect of inoculum volume (IV), substrate quantity, and the use of a manual or automated gas production (GP) recording system for in vitro determinations of fermentation of corn distillers dried grains with solubles (cDDGS) and soybean hulls (SBH). A $2 \times 2 \times 2$ factorial arrangement of treatments was used and included the factors of (1) ingredients (cDDGS or SBH), (2) inoculum volume and substrate quantity (IV30 = 0.2 g substrate + 30 mL inoculum or IV75 = 0.5 g substrate + 75 mL inoculum), and (3) GP recording system (MRS = manual recording system or ARS = automated recording system). Feed ingredient samples were pre-treated with pepsin and pancreatin, and the hydrolyzed residues were subsequently incubated with fresh pig feces in a buffered mineral solution. The GP recording was monitored for 72 h, and the kinetics were estimated by fitting data using an exponential model. Compared with SBH, cDDGS yielded less ($p < 0.01$) maximal gas production (G_f), required more time ($p < 0.02$) to achieve half gas accumulation ($T/2$), and had less ($p < 0.01$) fractional rate of degradation (μ) and in vitro fermentability of dry matter (IVDMF). Using the ARS resulted in less IVDMF ($p < 0.01$) compared with MRS (79.0% vs. 81.2%, respectively). Interactions were observed between GP recording system and inoculum volume and substrate quantity for Gf ($p < 0.04$), μ ($p < 0.01$), and $T/2$ ($p < 0.04$) which implies that increasing inoculum volume and substrate quantity resulted in decreased Gf (332 mL/g from IV30 vs. 256 mL/g from IV75), μ (0.05 from IV30 vs. 0.04 from IV75), and $T/2$ (34 h for IV30 vs. 25 h for IV75) when recorded with ARS but not MRS. However, the recorded cumulative GP at 72 h was not influenced by the inoculum volume nor recording system. The precision of G_f (as measured by the coefficient of variation of G_f) tended to increase for IV30 compared with IV75 ($p < 0.10$), indicating that using larger inoculum volume and substrate quantity (IV75) reduced within batch variation in GP kinetics. Consequently, both systems showed comparable results in GP kinetics, but considering convenience and achievement of consistency, 75 mL of inoculum volume with 0.5 g substrate is recommended for ARS.

Keywords: corn distillers dried grains with solubles; gas collection technique; in vitro; pig fecal inoculum; soybean hulls

---

1. Introduction

The use of increasing amounts of dietary fiber (DF) in swine feeding programs contributes to various environmental [1], animal well-being [2], and sustainability [3] impacts. About 46.3 million metric tonnes of feed was fed to pigs in the United States in 2016, consisting of 16% corn distillers dried grains with solubles (cDDGS), and 15% total soybean products [4]. However, these ingredients contain higher amounts of DF and less starch compared with corn, resulting in a greater production of short-chain fatty acids (SCFA) by gut microbiota when pigs are fed diets containing cDDGS or soybean hulls (SBH) than corn. The SCFA affect the intestinal epithelial cells and affect the intestinal integrity by regulating ion absorption and gut motility [5].

Various in vitro methodologies have been developed and used to estimate the digestibility of various feed ingredients, including ingredients that contain high concentrations of DF. The most widely used procedure is a three-step in vitro assay that combines replicated enzymatic hydrolysis from the stomach through small intestine [6] with representative large intestine fermentation using swine feces as a living bacterial inoculum [7]. This procedure has been well accepted to estimate in vitro dry matter digestibility (IVDMD) in the large intestine and total gas production of various feed ingredients for swine [5,8–11]. An automated recording system (ARS) for gas production (GP) was introduced in the early 1990s to reduce the amount of labor, compared with the manual recording system (MRS) when evaluating diets and feed ingredients for ruminants [12]. The ARS technique measures the kinetics of microbial fermentation in an automated fashion by monitoring the gas pressure and ventilation process [12]. Several in vitro studies have investigated the advantages and disadvantages of using ARS to measure the gas production profile and fermentation kinetics in ruminant-based in vitro systems [13,14]. However, the type of feed ingredient, amount of fecal inoculum, quantity of substrate, and the type of recording system may affect the accuracy and precision of the parameters estimated.

Our current study was conducted to determine the effects of inoculum volume and recording system on in vitro gas production and the concentration of SCFA produced from the fermentation of cDDGS and SBH. This investigation was based on the hypothesis that the type of ingredient, volume of fecal inoculum, and amount of substrate in a bottle would affect the accuracy and precision of gas production parameter measurements, including the concentration of SCFA when using the ARS in a pig-based in vitro digestibility system.

2. Materials and Methods

2.1. Experimental Design, Feed Samples, and Enzymatic Hydrolysis

This experiment was conducted using a $2 \times 2 \times 2$ factorial arrangement of treatments to examine the effects of feed ingredients (cDDGS or SBH), fecal inoculum volume (IV30 = 200 mg substrate + 30 mL inoculum or IV75 = 500 mg substrate + 75 mL inoculum), and GP recording system (MRS or ARS) on IVDMF and the production of SCFA. Hydrolyzed corn DDGS and SBH residues were obtained from the two-step procedure involving pepsin and pancreatin hydrolysis in our previous studies [8,15] that was developed by Boisen and Fernandez [6]. Briefly, 2 g of each cDDGS and SBH sample was weighed into a 500-mL Pyrex Erlenmeyer flask and incubated at 39 °C in a water bath. Then, 100 mL of phosphate buffer solution (0.1 M 7:1 KH_2PO_4:Na_2HPO_4, pH 6.0) and 40 mL 0.2 M HCl solution (pH 2.0) were added. The pH was adjusted to 2.0 by adding 1 M HCl or 1 M NaOH. The addition of 2 mL of 5 mg/mL chloramphenicol (C0378; Sigma-Aldrich Corp., St. Louis, MO, USA) solution (dissolved in ethanol) was added to prevent bacterial growth during hydrolysis. A volume of 4 mL of 100 mg/mL fresh porcine pepsin (P7000, 421 pepsin units / mg solids; Sigma-Aldrich Corp.)

solution (dissolved in 0.2 M HCl) was added to each bottle and incubated in a water bath at 39 °C for 2 h. All the flasks were shaken gently by hand for 5 s every 15 min. Subsequently, 40 mL of 0.2 M phosphate buffer (7:1 KH_2PO_4:Na_2HPO_4, pH 6.8) and 20 mL of 0.6 M NaOH were added to each flask. Finally, 4 mL of 100 mg/mL fresh porcine pancreatin (P1750, 4 times the specifications of the United States Pharmacopeia; Sigma-Aldrich Corp.) solution (dissolved in 0.2 M phosphate buffer) was added. The hydrolysis continued for 4 h under the same conditions as used for pepsin hydrolysis. Subsequent in vitro fermentation analysis was performed using these residues according to the procedure developed by Jha et al. [10,11].

2.2. Experimental Design and In Vitro Fermentation Procedures

Before fermentation, samples of cDDGS and SBH were hydrolyzed by enzymatic digestion with pepsin and pancreatin. The residues from enzymatic digestion were then subsequently pooled within each ingredient source for in vitro fermentation. Blank inocula without substrates were used as controls. The experimental scheme was as follows: 8 treatments × 3 replications + 4 blanks repeated over three batches. Briefly, either 0.2 g or 0.5 g of pooled hydrolyzed cDDGS and SBH samples (ground to 1 mm in particle size) was weighed and incubated in a buffer solution containing macro-and micro-minerals [16]. Feces were collected by rectal stimulation from one finishing pig per batch. Pigs were fed a conventional corn-soybean meal-based diet without antibiotics (Innovation Campus, Cargill Animal Nutrition, Elk River, MN, USA). Collected fecal samples were immediately placed in air-tight plastic syringes and kept in a water bath at 39 °C until incubation. The time from fecal collection until incubation was less than 1 h. In the laboratory, the inoculum was formulated by diluting blended feces in an inoculation solution composed of distilled water (474 mL/L), trace mineral solution (0.12 mL/L containing 132 g/L of $CaCl_2$, 100 g/L of $MnCl_3 \cdot 4H_2O$, 10 g/L of $CoCl_2 \cdot 6H_2O$, and 80 g/L of $FeCl_3 \cdot 6H_2O$), in vitro buffer solution (237 mL/L containing 4.0 g/L of NH_4HCO_3 and 35 g/L of $NaHCO_3$), macromineral solution (237 mL/L composed of 5.7 g/L of Na_2HPO_4, 6.2 g/L of KH_2PO_4, 0.583 g/L of $MgSO_4 \cdot 7H_2O$, and 2.22 g/L of NaCl), and resazurin (blue dye, 0.1% wt/vol solution; 1.22 mL/L) and filtered through four layers of cheesecloth. The final inoculum concentration was 0.05 g feces/mL of buffer. Either 30 mL or 75 mL of inoculum aliquots were respectively transferred into bottles containing 200 mg or 500 mg of the hydrolyzed sample substrates to provide an equal inoculum to substrate ratio (6.67 mL/mg) between the two systems. Carbon dioxide (CO_2) was provided to maintain an anaerobic environment during the entire inoculum preparation process.

The headspace gas pressure in the bottles was recorded using either MRS or ARS. The gas was measured manually at 11 time points post-inoculation using an inverted 25-mL burette with its stopcock end attached to the vacuum, and its open end submerged into a water bath (39 °C) in MRS. The ARS was designed to measure the kinetics of microbial fermentation by monitoring the gas pressure automatically every 5 min and recording remotely using a commercial apparatus (Ankom[RF] Gas Production System, Ankom Technology, Macedon NY, USA) equipped with real-time sensors. The headspace volume was 57.5 mL in MRS and 257.5 mL in ARS. For the ARS system, accumulated gas in the headspace was automatically released when the pressure exceeded 35 psi. Recording of headspace pressure was terminated at 72 h post-incubation. At the end of the 72 h, the supernatant from each bottle was collected for SCFA analysis.

2.3. Chemical Analyses

Before liquid chromatography–mass spectrometry (LC-MS) analysis, samples of fermentation supernatants were derivatized with hydroquinone (HQ) for the determination of SCFA concentrations [17]. Briefly, two microliters of the extracted supernatant were mixed with 70 μL of acetonitrile (ACN) containing 7.5 μM acetic acid-d4, 10 μL dipridyl disulfide (DPDS), 10 μL triphenylphosphine (TPP), and 10 μL HQ. The mixture was incubated at 60 °C for 30 min, chilled on ice, and mixed with 100 μL H_2O. The vials were then centrifuged at 21,000 × g at 4 °C for 10 min. The processed HQ-reaction mixture from chemical derivatization of samples was injected into ultra-performance liquid chromatography (UPLC) system (Xevo-G2-S; Waters, Milford, MA, USA). The concentration of individual compounds was determined by calculating the ratio between the peak area of compounds and the peak area of internal standards. Acetic acid-d4 was used as an internal standard calibration curve for precise SCFA quantification. The acquired data were processed by software (QuanLynx, Waters, Milford, MA, USA).

2.4. Calculations

The in vitro fermentability of dry matter (IVDMF) during fecal inoculum fermentation was calculated as follows:

$$\text{IVDMF, \%} = [(\text{dry weight of the hydrolyzed residue} - \text{dry weight of the residue after fermentation})/\text{dry weight of the hydrolyzed residue}] \times 100$$

After correction for the blank units, the recorded cumulative gas pressure (psi) was converted into mL of gas produced per g DM using Avogadro's law as follows:

$$\text{Gas volume, mL} = \text{gas pressure} \times [V/RT] \times 22.4 \text{ L/mol} \times 1000 \text{ mL/L},$$

where V denotes head space volume in the bottle (L), R was the gas constant 8.314472 L k Pa/K/mol, and T represents the temperature in Kelvin (273 °K + Celsius temperature in the bottle).

Gas accumulation curves (mL/g DM) recorded during the 72 h of fermentation were fitted by the following model developed by France et al. [18]:

$$G \text{ (mL/g DM)} = 0, \text{ if } 0 < t < L$$

$$G = G_f \left(1 - \exp\left(-[b\,(t-L) + c\,(\sqrt{t} - \sqrt{L})]\right)\right), \text{ if } t \geq L,$$

where G denotes the gas accumulation at a specific time (t), G_f (mL/g DM) was the maximum gas volume for $t = \infty$, and L (h) represents the lag time before the fermentation began and is determined by the initial delay until the onset of gas production occurs. In the present study, gas accumulation of the cDDGS treatment rapidly reached one-fourth of the maximum accumulation in 2 h, and the parameter L (h) was very close to 0, which resulted in the model failing to converge. Therefore, L (h) data were removed from the final model. The constants b (h^{-1}) and c ($h^{-1/2}$) determine the fractional rate of degradation of the substrate μ (h − 1), which is postulated to vary with time as follows:

$$\mu = b + c/(2\sqrt{t}), \text{ if } t \geq L$$

Kinetics parameters of gas production (G_f, T/2, G72, and μ at T/2) were compared in the statistical analysis, with T/2 representing the time to half asymptote when $G = G_f/2$.

2.5. Statistical Analyses

The kinetics of gas production parameters were fitted based on the individual time series data and were analyzed using PROC NLIN of SAS version 9.4 (SAS Inst. Inc., Cary, NC, USA). The IVDMF, fitted gas production kinetic parameters, and the concentration of SCFA were analyzed using the GLIMMIX procedure of SAS version 9.4 (SAS Inst., Inc., Cary, NC, USA), with individual bottles considered as the experimental unit. The model included substrates (cDDGS and SBH), inoculum volume (30 mL and 75 mL), GP recording system (MRS and ARS), and their interactions (Substrate × Volume, Substrate × System, Volume × System, and Substrate × Volume × System) as the fixed factors and batches of samples as random factors. The average coefficient of variance (CV) was calculated based on the average values of kinetic parameters within each treatment using PROC GLM of SAS version 9.4 (SAS Inst., Inc., Cary, NC, USA). The least square means of individual treatments were separated by the Tukey method. Results were considered significant at $p \leq 0.05$ and trends at $0.05 < p \leq 0.10$.

3. Results and Discussion

3.1. Fermentation Kinetics and Metabolites

Soybean hulls yielded greater ($p < 0.01$) maximal gas production (G_f), required less time ($p < 0.02$) to achieve half gas accumulation ($T/2$), and had greater ($p < 0.01$) fractional rate of degradation (μ) and IVDMF compared to cDDGS (Table 1). Each of ingredients showed similar gas production curves regardless of gas recording system and inoculum volume (Figure 1). Results for IVDMF of cDDGS (69.2%) obtained in the current study were greater than that reported in previous studies (59.6% by Jha et al. [9]; 55.7% by Huang et al. [8]), but maximum gas volume (G_f) of cDDGS (200 mL/g DM) was comparable to those reported by Jha et al. [9] (200 mL/g DM) and Huang et al. [8] (208 mL/g DM). Different kinetics of GP between these two ingredients can be explained by their fiber composition. Soybean hulls contain about 5.5 times more soluble dietary fiber (SDF) than insoluble dietary fiber (IDF), whereas cDDGS contains 1.6 times more SDF than IDF [19]. It has been suggested that apparent ileal digestibility (AID) and apparent total tract digestibility (ATTD) of SDF are a result of greater fermentation compared with IDF in growing-finishing pigs [20]. Moreover, while SDF is mainly fermented in the proximal colon, IDF is fermented primarily in the distal colon [21], which is likely due to the hydrophobic and the crystalline characteristics of these types of DF [22]. Consequently, a greater SDF/IDF ratio in SBH may have resulted in a sharp increase in the fractional rate of degradation during earlier fermentation stage (<8 h) compared with cDDGS in the current study.

Gas production (GP) kinetics parameters were not different between the GP recording systems, whereas IVDMF was less ($p < 0.01$) when recorded in the ARS system compared with the MRS system (79.0% vs. 81.2%, respectively). Moreover, interactions were observed between GP recording system and inoculum volume and substrate quantity for G_f ($p < 0.04$) and μ ($p < 0.01$), and the time to half asymptote ($T/2$, $p < 0.04$). According to the meta-analysis on methodological factors influencing GP during in vitro rumen fermentation, the GP recording apparatus with venting system (i.e., ARS) resulted in greater gas production estimates compared to the MRS GP recording apparatus operating without venting system [23]. Furthermore, the absence of automatic ventilation system in MRS increased headspace pressure, so that it may have caused a partial dissolution of carbon dioxide (CO_2) in the inoculum, and subsequently resulted in the underestimation of GP as well as restricting microbial respiration. Results from the current experiment showed no difference in parameters of GP kinetics between the two systems. One possible explanation for the lack of differences may be due to the differences between the headspace volumes to fermentation inoculum ratio between the systems, which was 4.9 for ARS compared to 2.8 for MRS in the current experiment. This ARS ratio is greater than the ratio used in a previous in vitro study conducted using swine fecal inoculum with ARS (ratio: 3.2, Pastorelli et al. [24]). However, the optimal ratio between headspace and fermentation inoculum has not yet been established. The smaller ratio may result in greater underestimation of GP because of higher pressure [25], whereas the larger ratio may result in lower pressure and cause inhibition of microbial activity [13]. Therefore, based on the current results, it can be expected that relatively larger ratio between headspace volume to inoculum in ARS may interfere with the microbial fermentation in the bottle, resulting in decreased IVDMF, as well as increased within batch variation. However, further investigations are required to determine the optimal ratio between headspace and inoculum volume when using swine fecal inoculum in ARS.

Table 1. Parameters of the fitted kinetics and concentration of short-chain fatty acids (SCFA) after in vitro fermentation of corn distillers dried grains with solubles (cDDGS) and soybean hulls (SBH) using an automatic recording system (ARS) and manual recording system (MRS) using different fecal inoculum volumes (30 and 75 mL) obtained from pigs [1].

Item	ARS		MRS		SEM [2]	Substrates		SEM	*p*-Values [3]			
	30 mL	75 mL	30 mL	75 mL		cDDGS	SBH		Sub	Sys	Vol	Sys × Vol
						Fermentation kinetics						
N [4]	8	8	8	8		20	20					
Gf [5]	332 [a]	256 [b]	281 [ab]	286 [ab]	37.4	200 [B]	362 [A]	34.3	0.011	0.971	0.334	0.047
μ [6]	0.04 [b]	0.05 [a]	0.04 [ab]	0.04 [ab]	0.015	0.02 [B]	0.05 [A]	0.014	0.012	0.446	0.544	0.012
T/2 [7]	34.34 [a]	25.07 [b]	27.16 [ab]	28.16 [ab]	7.832	32.02 [A]	26.10 [B]	7.794	0.028	0.984	0.977	0.041
G72 [8]	211	210	224	240	12.2	159 [B]	285 [A]	9.8	0.016	0.987	0.826	0.110
IVDMF [9]	79.2	79.0	80.8	81.0	1.10	69.2 [B]	90.9 [A]	0.93	0.012	0.012	0.731	0.463
						Fermentation metabolites, mmol/g						
Acetic acid	4.78	4.19	4.32	4.53	0.504	3.71 [B]	5.20 [A]	0.504	0.009	0.906	0.346	0.429
Propionic acid	1.14	1.17	1.10	1.12	0.063	1.08 [B]	1.18 [A]	0.031	0.047	0.284	0.433	0.917
Butyric acid	0.58	0.71	0.54	0.58	0.055	0.64	0.57	0.054	0.175	0.130	0.271	0.414
Valeric acid	0.19	0.16	0.15	0.22	0.034	0.19	0.16	0.024	0.306	0.688	0.570	0.108
Total SCFA	6.69	6.23	6.10	6.46	0.584	5.64 [B]	7.11 [A]	0.413	0.001	0.757	0.428	0.485

[A,B] Means within a row with different uppercase superscript letters differ significantly (main effect, $p < 0.05$). [a,b] Means within a row with different lowercase superscript letters differ significantly (Sys × Vol, $p < 0.05$). [1] The least squares mean value presented based on the three replications per treatment. [2] Standard error of the means. [3] Sub = feed ingredients as substrates; Sys = gas production recording system; Vol = inoculum volume. [4] Number of observations in fermentation. [5] Maximum gas volume (mL/g DM incubated). [6] Fractional rate of degradation (h − 1) at t = *T*/2. [7] *T*/2, half-time to asymptote (h). [8] Volume of gas production at 72 h. [9] In vitro dry matter fermentability.

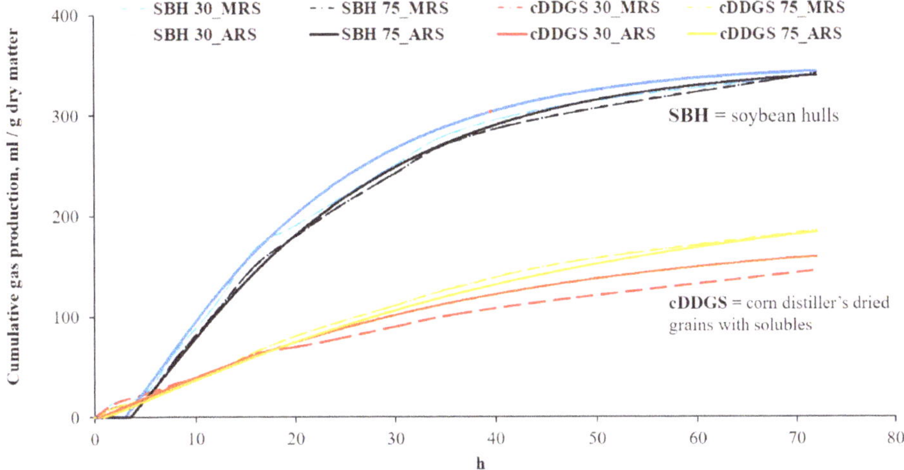

Figure 1. Gas accumulation curves of two ingredients (soybean hulls = SBH; and corn dried distiller's grains with solubles = DDGS) and inoculum volume (30 and 75 mL) incubated either in automatic gas production recording system (ARS) or manual gas production recording system (MRS) during 72 h.

Regardless of substrates, acetic acid was the most abundant SCFA produced during in vitro fermentation. The samples of SBH produced more acetic acid ($p < 0.01$), propionic acid ($p < 0.05$), and total SCFA ($p < 0.01$) compared with cDDGS (Table 1). These results are in agreement with those from a previous in vitro study by Jha and Leterme [26], indicating that both in vitro GP recording systems yielded accurate estimates of microbial fermentation. The greater SCFA production observed during fermentation of SBH compared with cDDGS can be attributed to the solubility of DF in the ingredient. Ferulic acid consists of cross-linked cell wall polysaccharides and other cell wall components such that it might be associated with fiber matrix rigidity [27]. Insoluble DF has been linked to decreased SCFA production resulting from slower fermentation rates compared to soluble DF, and insoluble DF contains 100 times greater ferulic acid content than soluble DF [28]. Thus, there was a greater amount of soluble dietary fiber available for microbiota fermentation in SBH, resulting in increased production of SCFA compared to cDDGS (Figure 1).

3.2. The Average Coefficient of Variance

The hypothesis of this study was that error frequency and severity would be relatively greater in MRS compared to ARS because of the intensive labor involved during the first phase of microbial fermentation. Although we observed no differences between the two GP recording systems for the coefficient of variation (CV) of GP kinetic parameters and IVDMF, the CV tended to be less ($p < 0.10$) when using the greater inoculum volume (IV75) compared to using the smaller inoculum volume (IV30, Table 2). Also, the CV tended to be less in cDDGS on time to half asymptote ($T/2$, $p < 0.07$) and IVDMF ($p < 0.09$) compared to SBH.

Comparison of results from our variability analysis to results from other studies is difficult because each study analyzed results using different mathematical methods. However, results from a previous study evaluating the repeatability and reproducibility of an ARS using rumen fluid from four laboratories indicated that fermentable organic matter had the greatest repeatability and reproducibility (0.2 to 1.9%, and 0.3 to 4.5%, respectively), followed by kinetic parameters (Gf = 1.1 to 2.5% in repeatability and 1.7 to 3.8% in reproducibility; $T/2$ = 4.3 to 13.2% in repeatability and 4.7 to 13.2% in reproducibility; μ = 8.2 to 12.8% in repeatability and 18.6 to 27.5% in reproducibility) [29]. This pattern was similar to the results obtained in the current study, indicating that the CV for IVDMF had the least variation, and kinetic parameters showed comparatively greater variation. Also, a similar CV pattern of kinetics was observed in a recent study using the ring test of in vitro GP recording systems conducted in four different laboratories in Europe (Denmark, United Kingdom, Spain, Italy), using the same wireless apparatus that we used in the current experiment [30]. These researchers also indicated that the least variation among parameters of GP kinetics observed were as follows: GP at 48 h (CV = 4.8%), *Gf* (CV = 6.4%), μ (CV = 11.4%), and *T*/2 (CV = 14.1%).

Table 2. Average coefficient of variation of the in vitro fermentation kinetic parameters for corn distillers dried grains with solubles (cDDGS) and soybean hulls (SBH) using an automatic recording system (ARS) and manual recording system (MRS) with two different pig fecal inoculum volumes (30 and 75 mL) [1].

Item	ARS		MRS		SEM [2]	Substrates		SEM	p-Values [3]			
	30 mL	75 mL	30 mL	75 mL		cDDGS	SBH		Sub	Sys	Vol	Sys × Vol
Gf [4]	29.2	23.7	31.5	13.9	12.53	22.7	26.4	8.16	0.757	0.758	0.100	0.762
μ [5]	43.7	45.7	50.6	38.5	14.49	31.2	58.0	6.06	0.068	0.985	0.688	0.941
T/2 [6]	32.2	48.9	55.6	27.0	16.06	29.1	52.8	9.20	0.122	0.958	0.716	0.590
G72 [7]	33.6	20.9	10.2	9.8	5.53	19.9	17.2	5.92	0.750	0.136	0.447	0.203
IVDMF [8]	4.35	3.15	3.85	4.85	0.03	5.02	3.08	0.81	0.121	0.588	0.929	0.764

[1] The least squares mean value presented based on the three replications per treatment. [2] Standard error of the means. [3] Sub = feed ingredients as substrates; Sys = gas production recording system; Vol = inoculum volume. [4] Maximum gas volume (mL/g DM incubated). [5] Fractional rate of degradation (h − 1) at t = *T*/2. [6] *T*/2, half-time to asymptote (h). [7] Termination gas volume (at 72 h). [8] In vitro dry matter fermentability.

Rymer et al. [31] indicated that the largest source of variation in the GP technique could be attributed to the source of inoculum and its microbial activity. In our study, we assumed that the fecal samples may vary among fecal donor age and may significantly increase the CV in the current experiment. Fecal sampling procedures were irregularly managed because of the bio-security of the company-owned research farm. There is a relatively large variation in age and body weight (60 to 100 kg) of fecal donors between batches. Kim et al. [32] indicated that pig microbial ecosystems in the GIT continue to change as pigs grow, and is influenced by various factors, including genetics, diet, and antibiotics. Therefore, our results reflect the fact that using fecal inocula from pigs of different ages leads to differences in microbial fermentability derived from different microbial communities within the batch of samples, resulting in increased variability of GP kinetic curves. The use of inoculum from one fecal donor per batch can be another factor. In our previous in vitro study, fecal samples were randomly collected from three out of five growing pigs for each batch of feed ingredient samples analyzed [8], resulting in CV's of kinetic parameters (*Gf*, and *T*/2) of 5.2 and 4.5% in SBH, respectively, and 9.8 and 18.5% in cDDGS, respectively, which were 5.18 and 2.81 times less than the CV's obtained from the current experiment. Rymer et al. [31] emphasized that fecal samples should be collected from several animals for in vitro fermentation analysis because each pig has different fecal microflora composition even though they are from the same genetic line and consume the same diet. Evidence from human studies has shown that using inoculum from at least three donors may enhance the predictive value of in vitro colonic fermentation [33]. Based on the results from the current experiment, we suggest collecting fecal samples from more than three pigs is necessary for improving the accuracy of pig in vitro fermentation assays of high fiber ingredients.

4. Conclusions

The results of this experiment demonstrate that both the GP recording systems (manual and automatic) were accurate at recording the gas production during in vitro fermentation similar to results reported in the literature for cDDGS and SBH. These results also suggest that there is an improvement in precision when larger volumes of fecal inocula are used if the ratio of substrate and headspace are kept in proportion between the GP recording systems.

Author Contributions: Z.Z. and P.E.U. designed the experiment. J.-C.J. and Z.Z. conducted the experimental work. All authors contributed to analyzing the data and writing of the manuscript.

Funding: This research was funded by the National Pork Board project #14-045.

Acknowledgments: The authors wish to thank Jinlong Zhu, Edward Zhaohui Yang, and Yuan-Tai Hung for laboratory work.

Conflicts of Interest: The authors declare no conflict of interest.

References

1. Aarnink, A.J.A.; Verstegen, M.W.A. Nutrition, key factor to reduce environmental load from pig production. *Livest. Sci.* **2007**, *109*, 194–203. [CrossRef]

2. De Leeuw, J.A.; Bolhuis, J.E.; Bosch, G.; Gerrits, W.J.J. Effects of dietary fibre on behaviour and satiety in pigs. *Proc. Nutr. Soc.* **2008**, *67*, 334–342. [CrossRef] [PubMed]

3. Basset-Mens, C.; van der Werf, H.M.G. Scenario-based environmental assessment of farming systems: The case of pig production in France. *Agric. Ecosyst. Environ.* **2005**, *105*, 127–144. [CrossRef]

4. United States Department of Agriculture (USDA). *Agricultural Statistics 2017*; Department of Agriculture, National Agricultural Statistics Service: Washington, DC, USA, 2017.

5. Kim, C.H.; Park, J.; Kim, M. Gut microbiota-derived short-chain fatty acids, T cells, and inflammation. *Immune Netw.* **2014**, *14*, 277–288. [CrossRef] [PubMed]

6. Boisen, S.; Fernández, J.A. Prediction of the total tract digestibility of energy in feedstuffs and pig diets by in vitro analyses. *Anim. Feed Sci. Technol.* **1997**, *68*, 277–286. [CrossRef]

7. Bindelle, J.; Buldgen, A.; Lambotte, D.; Wavreille, J.; Leterme, P. Effect of pig faecal donor and of pig diet composition on in vitro fermentation of sugar beet pulp. *Anim. Feed Sci. Technol.* **2007**, *132*, 212–226. [CrossRef]

8. Huang, Z.; Urriola, P.E.; Salfer, I.J.; Stern, M.D.; Shurson, G.C. Differences in in vitro hydrolysis and fermentation among and within high-fiber ingredients using a modified three-step procedure in growing pigs. *J. Anim. Sci.* **2017**, *95*, 5497–5506. [CrossRef]

9. Jha, R.; Woyengo, T.A.; Li, J.; Bedford, M.R.; Vasanthan, T.; Zijlstra, R.T. Enzymes enhance degradation of the fiber-starch-protein matrix of distillers dried grains with solubles as revealed by a porcine in vitro fermentation model and microscopy. *J. Anim. Sci.* **2015**, *93*, 1039–1051. [CrossRef]

10. Jha, R.; Bindelle, J.; Rossnagel, B.; Van Kessel, A.; Leterme, P. In vitro evaluation of the fermentation characteristics of the carbohydrate fractions of hulless barley and other cereals in the gastrointestinal tract of pigs. *Anim. Feed Sci. Technol.* **2011**, *163*, 185–193. [CrossRef]

11. Jha, R.; Bindelle, J.; Van Kessel, A.; Leterme, P. In vitro fibre fermentation of feed ingredients with varying fermentable carbohydrate and protein levels and protein synthesis by colonic bacteria isolated from pigs. *Anim. Feed Sci. Technol.* **2011**, *165*, 191–200. [CrossRef]

12. Davies, Z.S.; Mason, D.; Brooks, A.E.; Griffith, G.W.; Merry, R.J.; Theodorou, M.K. An automated system for measuring gas production from forages inoculated with rumen fluid and its use in determining the effect of enzymes on grass silage. *Anim. Feed Sci. Technol.* **2000**, *83*, 205–221. [CrossRef]

13. Tagliapietra, F.; Cattani, M.; Bailoni, L.; Schiavon, S. In vitro rumen fermentation: Effect of headspace pressure on the gas production kinetics of corn meal and meadow hay. *Anim. Feed Sci. Technol.* **2010**, *158*, 197–201. [CrossRef]

14. He, Z.X.; Zhao, Y.L.; McAllistera, T.A.; Yang, W.Z. Effect of in vitro techniques and exogenous feed enzymes on feed digestion. *Anim. Feed Sci. Technol.* **2016**, *213*, 148–152. [CrossRef]

15. Kerr, B.J.; Gabler, N.K.; Shurson, G.C. Formulating diets containing corn distillers dried grains with solubles on a net energy basis: Effects on pig performance and on energy and nutrient digestibility. *Prof. Anim. Sci.* **2015**, *31*, 497–503. [CrossRef]

16. Menke, K.H.; Steingass, H. Estimation of the energetic feed value obtained from chemical analysis and in vitro gas production using rumen fluid. *Anim. Res. Dev.* **1988**, *28*, 7–55.

17. Lu, Y.; Yao, D.; Chen, C. 2-hydrazinoquinoline as a derivatization agent for LC-MS-based metabolomic investigation of diabetic ketoacidosis. *Metabolites* **2013**, *3*, 993–1010. [CrossRef] [PubMed]

18. France, J.; Dhanoa, M.S.; Theodorou, M.K.; Lister, S.J.; Davies, D.R.; Isac, D. A model to interpret gas accumulation profiles associated with in vitro degradation of ruminant feeds. *J. Theor. Biol.* **1993**, *163*, 99–111. [CrossRef]

19. Jaworski, N.W.; Stein, H.H. Disappearance of nutrients and energy in the stomach and small intestine, cecum, and colon of pigs fed corn-soybean meal diets containing distillers dried grains with solubles, wheat middlings, or soybean hulls. *J. Anim. Sci.* **2017**, *95*, 727–739. [CrossRef]

20. Urriola, P.E.; Shurson, G.C.; Stein, H.H. Digestibility of dietary fiber in distillers coproducts fed to growing pigs. *J. Anim. Sci.* **2010**, *88*, 2373–2381. [CrossRef]

21. Jha, R.; Berrocoso, J.F.D. Dietary fiber and protein fermentation in the intestine of swine and their interactive effects on gut health and on the environment: A review. *Anim. Feed Sci. Technol.* **2016**, *212*, 18–26. [CrossRef]

22. Bach Knudsen, K.E.; Hansen, I. Gastrointestinal implications in pigs of wheat and oat fractions 1. Digestibility and bulking properties of polysaccharides and other major constituents. *Br. J. Nutr.* **1991**, *65*, 217–232. [CrossRef] [PubMed]

23. Maccarana, L.; Cattani, M.; Tagliapietra, F.; Schiavon, S.; Bailoni, L.; Mantovani, R. Methodological factors affecting gas and methane production during in vitro rumen fermentation evaluated by meta-analysis approach. *J. Anim. Sci. Biotechnol.* **2016**, *7*, 1–12. [CrossRef] [PubMed]

24. Pastorelli, G.; Faustini, M.; Attard, E. In vitro fermentation of feed ingredients by fresh or frozen pig fecal inocula. *Anim. Sci. J.* **2014**, *85*, 690–697. [CrossRef] [PubMed]

25. Cattani, M.; Tagliapietra, F.; Maccarana, L.; Hansen, H.H.; Bailoni, L.; Schiavon, S. Technical note: In vitro total gas and methane production measurements from closed or vented rumen batch culture systems. *J. Dairy Sci.* **2014**, *97*, 1736–1741. [CrossRef] [PubMed]

26. Jha, R.; Leterme, P. Feed ingredients differing in fermentable fibre and indigestible protein content affect fermentation metabolites and faecal nitrogen excretion in growing pigs. *Animal* **2012**, *6*, 603–611. [CrossRef] [PubMed]

27. Bunzel, M.; Ralph, J.; Marita, J.M.; Hatfield, R.D.; Steinhart, H. Diferulates as structural components in soluble and insoluble cereal dietary fibre. *J. Sci. Food Agric.* **2001**, *81*, 653–660. [CrossRef]

28. Pedersen, M.B.; Bunzel, M.; Schäfer, J.; Knudsen, K.E.B.; Sørensen, J.F.; Yu, S.; Lærke, H.N. Ferulic acid dehydrodimer and dehydrotrimer profiles of distiller's dried grains with solubles from different cereal species. *J. Agric. Food Chem.* **2015**, *63*, 2006–2012. [CrossRef]

29. Van Laar, H.; Van Straalen, W.M.; Van Gelder, A.H.; De Boever, J.L.; D'heer, B.; Vedder, H.; Kroes, R.; de Bot, P.; Van Hees, J.; Cone, J.W. Repeatability and reproducibility of an automated gas production technique. *Anim. Feed. Sci. Technol.* **2006**, *127*, 133–150. [CrossRef]

30. Cornou, C.; Storm, I.M.L.D.; Hindrichsen, I.K.; Worgan, H.; Bakewell, E.; Yáñez-Ruiz, D.R.; Abecia, L.; Tagliapietra, F.; Cattani, M.; Ritz, C.; et al. A ring test of a wireless in vitro gas production system. *Anim. Prod. Sci.* **2013**, *53*, 585–592. [CrossRef]

31. Rymer, C.; Huntington, J.A.; Williams, B.A.; Givens, D.I. In vitro cumulative gas production techniques: History, methodological considerations and challenges. *Anim. Feed. Sci. Technol.* **2005**, 9–30. [CrossRef]

32. Kim, H.B.; Borewicz, K.; White, B.A.; Singer, R.S.; Sreevatsan, S.; Tu, Z.J.; Isaacson, R.E. Longitudinal investigation of the age-related bacterial diversity in the feces of commercial pigs. *Vet. Microbiol.* **2011**, *153*, 124–133. [CrossRef] [PubMed]

33. McBurney, M.I.; Thompson, L.U. Effect of human faecal donor on in vitro fermentation variables. *Scand. J. Gastroenterol.* **1989**, *24*, 359–367. [CrossRef] [PubMed]

 *animals*

Review

In Vitro Methods of Assessing Protein Quality for Poultry

Dervan D.S.L. Bryan * and Henry L. Classen *

Department of Animal and Poultry Science, University of Saskatchewan, Saskatoon, SK S7N5A8, Canada
* Correspondence: dervan.bryan@usask.ca (D.D.S.L.B.); hank.classen@usask.ca (H.L.C.);
  Tel.: +1-3069-664-122 (D.D.S.L.B)

Received: 12 February 2020; Accepted: 17 March 2020; Published: 25 March 2020

Simple Summary: Over the years, broiler chickens have been selected for rapid growth which makes them very efficient at depositing body protein in a short period of time. This is important since the broiler sector is expected to contribute to the growing global demand for poultry meat. In light of this, the quality of proteins fed to poultry is becoming more important. The concept of protein nutrition is based on the sequential process through which proteins are digested, and the amino acids are absorbed and become available for metabolic processes. The nutritional quality of protein ingredients for poultry is based on their amino acid bioavailability. Animal and plant ingredients are the main sources of protein used in poultry diets and they vary in digestibility and amino acid composition. Although in vivo digestibility assays for poultry are available, they are expensive and time consuming to conduct. In vivo digestibility assays are the optimum tools for characterizing protein sources to be used in commercial production, but it is not always practical to conduct these assays in commercial settings. Commercial production, therefore, relies on the use of other assays such as in vitro assays to evaluate the quality of protein sources.

Abstract: Protein quality assessment of feed ingredients for poultry is often achieved using in vitro or in vivo testing. In vivo methods can be expensive and time consuming. Protein quality can also be evaluated using less expensive and time consuming chemical methods, termed in vitro. These techniques are used to improve the user's efficiency when dealing with large sample numbers, and some mimic the physiological and chemical characteristics of the animal digestive system to which the ingredient will be fed. The pepsin digestibility test is the in vitro method of choice for quick evaluation of protein sample during quality control and in most research settings. Even though the pepsin digestibility test uses enzymes to liberate the amino acids from the protein, it does not mimic normal in vivo digestive conditions. The results obtained with this method may be misleading if the samples tested contain fats or carbohydrates which they often do. Multi-enzyme tests have been proposed to overcome the problem encountered when using the pepsin digestibility test. These tests use a combination of enzymes in one or multiple steps customized to simulate the digestive process of the animal. Multi enzyme assays can predict animal digestibility, but any inherent biological properties of the ingredients on the animal digestive tract will be lost.

Keywords: dietary protein; poultry; digestibility assay; in vitro; pH stat method; pepsin digestibility assay

---

1. Introduction

Over the years, broiler chickens have been selected for rapid growth which makes them very efficient at depositing body protein in a short period of time. This is important since the broiler sector is expected to contribute to the growing global demand for poultry meat. In light of this, the quality of proteins fed to poultry is becoming more important. Animal and plant ingredients are the main sources of protein used in poultry diets and they vary in digestibility and amino acid composition [1–3].

The concept of protein nutrition is based on the sequential process through which proteins are digested, and the amino acids are absorbed and become available for metabolic processes. The nutritional quality of protein ingredients for poultry is based on their amino acid bioavailability. Animal proteins are composed of twenty-two amino acids [4]. Ten of the twenty-two amino acids in poultry meat proteins cannot be synthesized in large enough quantity and, therefore, must be provided in the diet for proper growth and metabolic function [5].

Digestibility is used in practice as an estimator of the amino acid bioavailability in poultry diets [6]. Digestible protein is the proportion of protein that is digested and absorbed in the form of amino acids [6]. On the other hand, amino acid bioavailability is the proportion of an amino acid in a form that is suitable for protein synthesis after the protein has been digested and amino acids absorbed [7]. Since the 1990s, most poultry nutrition research used digestibility assays when evaluating protein feed ingredients instead of bioavailability [5], because they do not require the free form of the amino acid during the evaluation [7]. The digestibility coefficient obtained can be used directly by nutritionist during ration formulation [5].

Although in vivo digestibility assays for assessing protein quality for poultry are available, they are expensive and time consuming to conduct. In vivo digestibility assays are the optimum tool for characterizing protein sources to be used in commercial production, but it is not practical to conduct these assays in a commercial setting. Commercial production therefore, relies on the use of other assays such as in vitro assays to evaluate the quality of protein sources. The pros and cons of in vitro and in vivo assays are covered in the subsequent review. It was clear that there is a need for a poultry specific in vitro protein digestibility assay for assessing protein sources commonly fed to poultry. This review presents a critical overview of current in vitro protein digestibility assays relevant to poultry and the application of their methodology in assessing protein quality of ingredients for poultry. The objectives of this review paper were: (1) To provide a comprehensive review of the in vitro methods currently available which has the potential or has been applied in the assessment of protein quality for poultry, and (2) to explore potential methodological factors which might be important in the assessment protein digestion.

2. Methods of Assessing Protein Quality

Traditionally, protein quality is assessed by evaluating the extent to which amino acids are digested and absorbed from the ingredient. Estimation of protein digestibility is normally achieved by feeding the feed ingredient to the intended animal and assessing protein or amino acid digestibility. This technique is termed in vivo. Protein quality can also be evaluated using less expensive and time consuming in vitro chemical methods. These techniques are used to improve the level of precision while mimicking the physiological and chemical characteristics of the digestive system of the animal to which the ingredient will be feed.

To obtain useful information on the digestibility of nutrients without the use of in vivo assays, researchers often employ the use of in vitro assays. In theory, in vitro digestibility assays should closely simulate the digestive process of the intended animal [8]. Depending on the nature of the research, it is expected that an intended in vitro assay should be reproducible, cheaper than available in vivo assays and simple to perform while giving fast results [9]. Methods for evaluating nutrient digestibility in vitro for simple stomach animals have been reviewed by others [8,10]. Only those methods applicable to protein digestion in poultry will be discussed.

2.1. Chemical In Vitro Methods

Evaluating protein quality using chemical method provides less precision than in vivo techniques but can be used as a routine quality control measure. In the chemical engineering literature, it was known as early as the 1930s that an alkali solution could extract up to 95% of the protein from plant meal sources [11]. In the late 1960s, Rinehart was one of the first to employ the protein solubility technique as a measure of protein quality of soybean meal in the poultry industry [12]. While working at Purina Mills Inc., Rinehart evaluated the suitability of protein from soybean meal derived from different processing systems using potassium hydroxide (KOH).

The ability to predict animal performance is one of the most important criteria of any chemical assay [13]. It was not until the 1950s that Lyman et al. [14] established a relationship between bird performance and the solubility of protein feed ingredients used in poultry diets. The study evaluated the correlation between a chick growth assay and the use of a protein solubility technique using sodium hydroxide as the alkali solution. In the solubility technique, one gram of cottonseed meal with four glass beads was placed in an Erlenmeyer flask with 100 mL of 0.02 N sodium hydroxide solution. The flask was agitated continuously at 37 °C for an h, and then the mixture centrifuged for 5 min at $3000 \times g$. After centrifuging, the solution was filtered and aliquots evaluated for protein concentration [14].

The solubility index method was not adopted as a routine measure of protein quality in the poultry feed industry until the test was validated. A study was reported in which the protein solubility technique was used to evaluate soybean quality in poultry feed [12]. This study provided the foundation for the evaluation of protein quality using the solubility technique. The researcher [13] revived the technique when they proposed the use of sodium tetraborate at 40 °C as a more sensitive test for detecting changes in protein quality due to overcooking of meals. By the end of the late 1990s, protein solubility using KOH became a routine technique in research evaluating dietary protein [13,15–17]. Researchers used the protein solubility index to evaluate canola meal quality and found that the 0.5% sodium hydroxide assay did not accurately predict canola meal lysine digestibility in broiler chickens [18]. This suggests that the relationship between protein solubility and amino acid digestibility is ingredient specific.

Protein dispersibility index (PDI) is another method used to evaluate the quality of protein ingredients. This technique involves high speed mixing of a protein sample in water, followed by the assessment of solubility [17]. In the literature, PDI may be referred to as water dispersible protein or water-soluble protein [19]. In 1970, the PDI technique was published as two official and tentative methods of the American Oil Chemists Society [19]. Veltmann and coworkers [20] evaluated the quality of soybean meal used in poultry diets employing the PDI method. The PDI method was able to distinguish between normal processed meals and meal heat-treated to escape rumen degradation. In that same study, a chick growth assay showed that there was no difference between the bioavailability of the protein from the two meals [20]. This suggested that the PDI method did not correlate well to the bioavailability of protein from the ingredient tested.

In 1978, the American Oil Chemists Society published a revised PDI method, which was corrected in 1979 as method Ba 10-65. In brief, 20 g of protein is mixed for 10 min at $7800 \times g$ with 300 mL of water. A portion of the mixture is centrifuged and the nitrogen content of the solid fraction and the original protein sample measured [21]. The percent dispersed protein is calculated as the protein loss from the original sample to the water. Batal and coworkers [17] compared the revived PDI method against the urease index and KOH solubility test. Of the three tests, the PDI method was more effective and more sensitive in detecting the minimum adequate heat processing conditions required for soybean meal fed to chickens.

Since the 1980s, the PDI method has become a routine technique used worldwide by researchers [17,20,22–24] to assess the quality of protein sources used in monogastric animal feeds. While chemical methods provide an overview of the protein quality of feed ingredients, they do not give a good indication of how much of the nutrient will be absorbed by the animal. Protein solubility index and PDI methods are used as measures of ingredient quality in most poultry nutritional research evaluating high protein ingredients. The information gained from the PDI method and protein solubility index does not provide useful information for diet formulation in a commercial setting, but they are often used in quality control programs.

2.2. pH-Stat/Drop Method

As protein samples are hydrolyzed by digestive enzymes, they release protons from the cleaved peptide bonds, which changes the pH of the reaction media [25]. In the early 1970s, Maga, Lorenz, and Onayemi evaluated the extent to which dietary protein undergoes proteolysis. They realized that there was a close relationship with the initial rate of hydrolysis of the proteins from 0 to 10 min and the digestibility of the protein samples. The rates of hydrolysis of the protein samples were evaluated as an indirect measure of the pH of the reaction mixture over time. In their system, the protein samples were incubated with trypsin at 37 °C in a water bath for 10 min while evaluating the pH change. However, this method lacked precision in predicting the bioavailability of protein [26,27].

To improve precision in predicting bioavailability with the Maga et al. [25] method, Vavak modified the above procedure in a master's thesis while working with distiller's grain protein concentrate [26]. During the modification of the procedure, various enzyme combinations were tested in an effort to gain improvement in the correlation coefficients between pH drop and protein digestibility in rats. The trypsin-chymotrypsin combination gave superior correlation coefficients compared to the initial single trypsin proposed by Maga et al. [25]. Hsu and coworkers [27] suggested that the methods presented by Maga et al. [25] and Vavak [26] were too time consuming and complicated for routine quality control.

A faster method was developed, which could be completed in 1 h [27]. In this method, the trypsin-chymotrypsin enzyme combination was replaced with a multi-enzyme mixture composed of trypsin, chymotrypsin and peptidase. The correlation coefficient with the apparent digestibility of protein for rats was 0.9 using this new multi-enzyme system after evaluating 23 food protein sources. The method was also able to detect the effects of trypsin inhibitor, chlorogenic acid and heat processing on the digestibility of the protein tested. The pH drop method was susceptible to the buffering capacity of the protein source since high ash content affected the digestibility results [27]. A researcher used the pH drop method proposed by Hsu et al. [27] modified by Satterlee et al. [28] to evaluate various high protein feed ingredients while correlating the results to true digestibility in caecectomized cockerels [10]. There was a good correlation with lysine digestibility in caecectomized cockerels and the pH drop test across the ingredients tested. The test, however, showed no relationship to lysine digestibility and protein efficiency ratio in various qualities of feather meal and meat meal samples.

To overcome the susceptibility of the pH drop test to the buffering capacity of protein samples, Pederson and Eggum revised the pH drop method proposed by Hsu et al. [27]. During revision, the consumption of alkali was used as an indirect measure of true protein digestibility values in rats. The pH of the reaction was held constant at 8 during titration with alkali over a 10 min period [29]. The correlation coefficient was improved from 0.9 [27] to 0.96 [29] with a residual error of 1.29 after evaluating 30 protein samples. The authors [29] suggested that the effects of ash content on the test results were due to differences in mineral content, which was mostly due to the influence of calcium. The authors proposed the use of two different regression equations to accurately predict the digestibility of protein samples from plant and animal origins. Using literature derived prediction equation for a specific kind of protein source was unreliable when using the pH-stat method [30]. To measure the degree of hydrolysis of protein, the method requires knowledge of the average dissociation of the α-amino groups of the protein sample and the number of peptide bonds present in the territory structure of the main proteins present in the ingredient [29].

Due to the limitations mentioned, the pH-stat test has been used mostly in food science research to predict the digestibility of highly digestible pure protein sources [29–31]. Such pure protein sources typically have data about the average dissociation of the α–amino groups and the number of peptide bonds present. Since the early 1990s, the pH-stat method has been used to evaluate only aquatic animal feed ingredients [32,33]. To address the limitations of the method, casein average dissociation constant and the number of peptide bonds were used as the standards when calculating the degree of hydrolysis [33]. So far, the data generated with the pH-stat method has been consistent with *in vivo* digestibility assays, especially with the use of purified enzymes extracted from the species to which the ingredient has been fed [32,33]. The pH-stat method has become a valuable tool for aquatic nutritional research, but not for terrestrial animals. The good digestibility correlations seen with aquatic species are probably due to the simple nature of their digestive tract and the use of highly digestible protein sources such as fish meal.

2.3. Closed Enzymatic Methods

These systems are used to evaluate the digestibility of nutrients with multiple or single enzymes while simulating part or all of the in vivo digestive process [8]. The system is flexible, so the procedure and enzymes used may vary to meet the specific needs of the research objectives. Only those procedures used specifically to evaluate the digestibility of protein samples will be reviewed. The digestibility of protein is tied to the amino acid content and to the specificity of the digestive enzyme used to free them from complex peptides [34].

2.3.1. Pepsin Assay

The pepsin digestibility assay is one of the most widely used assays to evaluate the quality of feed and protein ingredients. Gehrt and coworkers and Sheffner and coworkers were the first group of researchers to employ a single enzymatic method to evaluate the digestibility of protein using pepsin [35,36]. In their procedure, 1 g of protein was incubated with 25 mg of pepsin in 30 mL of 0.1 N sulfuric acid at 37 °C for 24 h, during this time, the samples were stirred intermittently [35]. After incubation, the samples were placed in a boiling water bath for 10 min. Samples were cooled and the pH adjusted to 2 followed by the addition of one volume each of 10% sodium tungstate and 2/3 N sulfuric acid. The mixtures were filtered after standing for 10 min, and then the filtrate adjusted to pH 6.8 and analyzed for amino acids. When the digestibility data were regressed against the biological value of the protein samples for rats, there was a 0.998 correlation [36].

The pepsin digestibility assay was not accepted as a routine protein quality evaluation until 1959. The Association of Official Analytical Chemists (AOAC) adopted a revived version of the method proposed by Gehrt et al. and Sheffner et al. [35,36]. Hydrochloric acid was used instead of sulfuric acid, and all the fat was extracted from the samples using ether before digestion. The sodium tungstate and pH steps were eliminated. In 1972, the procedure was revised to improve the filtration step and the pepsin concentration was defined as 0.2%.

Since the 1959 AOAC publication of the pepsin digestibility method, it has been used extensively to evaluate high protein feed ingredient quality of both plant and animal origin [15,37]. Johnston and coworkers were one of the first group of researchers to use this method to evaluate poultry feed ingredients of animal origin [37]. After evaluating 20 commercial animal by-product meals, they were able to get a 0.91 correlation with the net protein utilization and the protein efficiency ratio for chickens. The pepsin digestibility procedure proposed by Johnson et al. [37], adjusted the pepsin concentration to 0.002% while eliminating the preliminary grinding and defatting steps.

In another study, the same group of researchers evaluated various levels of pepsin in order to find a suitable level for use in the assay during routine evaluation of meat and bone meal samples fed to poultry [38]. Lower levels of pepsin (0.002%) were able to detect differences between the quality of the meat and bone meal samples, which was contrary to that of the AOAC 0.2% pepsin. Parsons and coworkers did a comparative study on the ability of 0.2%, 0.002%, and 0.0002% pepsin to detect differences in quality among 14 meat and bone meal samples [1]. They confirmed the findings of Johnson et al. [38] that the 0.002% pepsin level gave the best correlation with lysine digestibility in chickens.

2.3.2. Pancreatin

Some testing systems involve the use of pancreatin as the only enzyme source to digest protein samples. Riesen and coworkers described a single enzymatic method that used pancreatin to evaluate the quality of soybean meal in poultry [39]. The samples were ground in a power-driven mortar,

100 mg or 300 mg of pancreatin was added to 2 g of the ground samples in 50 mL of 0.2 M disodium phosphate buffer at pH 8.3. One mL of toluene was added to the solution, and the mixture incubated for 5 d or 12 h at 37 °C. At the end of each digestion period, the samples were heated with steam for 15 min to facilitate enzyme deactivation. The pH of the mixture was adjusted with glacial acetic acid to precipitate the undigested proteins. This method was able to detect the difference between overheated and the normal heated meals, but not the difference between the normal and under heated soybean meals.

Ingram and coworkers modified the procedure by adding 1.2 g of pancreatin to 12 g of sample in 300 mL of buffer for 6 h [40]. The pattern of amino acid released from the samples correlated with the growth of chickens fed the same samples of soybean meal [40]. In another study by Anwar [41], the pancreatin in vitro test was used to evaluate the quality of cottonseed meal, groundnut meal, meat meal and fish meal [41]. The method was not reliable for fish meal and groundnut meal but gave fair results for meat meals. The one-step pancreatin method has been used routinely by many food scientists to evaluate the digestibility of various protein foods, but not by poultry nutritionists [42].

Pancreatic digestion is controlled by substrate concentration in vivo. An increase in protein concentration will promote an increase in proteolytic enzyme secretion [8]. In vitro digestibility methods using pancreatin as the only enzyme, source keeps the enzyme concentration constant when evaluating a range of protein sources [41]. However, the method lacks precession when evaluating a variety of protein sources [41]. Other researchers have found no difference between in vivo chicken ileal digestibility and the pepsin or pancreatin assay when ranking feather meal digestibility [43].

2.3.3. Multi-Enzymatic Assays

A multi-enzyme method may use two or more enzymes while simulating one, two or all stages of the digestive process [8]. Multi-enzyme methods are more comparable to in vivo conditions since many enzymes are involved in the digestion of proteins. The digestion of proteins starts in the stomach under the action of pepsin and hydrochloric acid. The partially digested protein enters the small intestine where they are hydrolyzed by trypsin, chymotrypsin, elastase and carboxypeptidases [8].

In 1964, Akeson and Stahmann described a method using pepsin and pancreatin as enzyme sources. The method was developed to evaluate large numbers of food protein samples while reducing the labour load associated with the pepsin digestibility assay [44]. The method involved incubating 100 mg of protein sample with 1.5 mg pepsin in 15 mL 0.1 N hydrochloric acid for 3 h at 37 °C [44]. The reaction was neutralized with 7.5 mL of 0.2 N sodium hydroxide solution and then 4 mg pancreatin dissolved in 7.5 mL phosphate buffer with pH 8 was added. Fifty parts per million merthiolate were added to the mixture, which was incubated at 37 °C for 24 h. Samples of the digestion mixture were precipitated with acid and centrifuged at 1000× g for 30 min after which the supernatant was analyzed for amino acids.

In 1973, Saunders and coworkers described a two enzyme system using pepsin and trypsin. The test occurred in a closed system using centrifuge tubes containing 1 g of protein sample suspended in 20 mL of 0.1 N hydrochloric acid and then mixed with 50 mg pepsin dissolved in 1 mL 0.01 N hydrochloric acid [45]. The mixture was incubated at 37 °C while gently shaken for 48 h, centrifuged at 20,000× g for 5 min, and the supernatant removed. The solid was suspended in 10 mL water and 10 mL of 0.1 M phosphate buffer with pH 8 and 5 mg of dissolved trypsin. The mixture was incubated at 23 °C for another 12 h then centrifuged and the solids washed with 30 mL of water five times, with centrifuging and removal of the supernatant each time. The solid was filtered through a 1.2 µm Millipore filter, air-dried, and analyzed for amino acids.

Both the pepsin-pancreatin and pepsin-trypsin methods were able to give good correlation between the in vivo digestibility values for various food proteins using rats [44,45]. The pepsin-pancreatin assay is known to give good correlation with amino acid digestibility of 0.84 in cereal gains with true amino acid availability in chickens but was less reliable for soybean meal and corn gluten meal [46]. However, the pepsin-pancreatin test gave an excellent correlation of 0.91 between the in vivo ileal digestibility of protein of 15 feedstuffs in pigs [47]. The test proposed by Saunders et al. [45] has been used to some extent to evaluate protein digestibility in poultry [48–50].

Dialysis cell method is a non-static system in which products of digestion are removed from the substrate as they become available. When simulating in vivo protein digestion with in vitro techniques, the rate of hydrolysis may be compromised by the accumulation of end products in the system [46,51]. The rate of hydrolysis can be improved if the digestion products are removed from the system as digestion occurs [46]. To prevent the inhibition of proteolysis by the end products, dialysis has been proposed to remove digestion products [52,53]. They conducted their experiments in dialysis bags to facilitate the removal of the end products during incubation of the protein source with the enzymes.

In 1982, Gauthier and coworkers et al. adopted the dialysis principle of [52,53] and presented a method in which the dialysis solution was continually replaced as the incubation proceeded [34]. The content of the dialysis bag was stirred constantly during the digestion process. In brief, 400 mg nitrogen (6.25 × %N) of protein was suspended in a beaker with 100 mL of 0.1 N hydrochloric acid. The beaker was shaken and placed in a water bath at 37 °C for 30 min. The pH of the solution was adjusted to 1.9 then 20 mL of solution containing 5 mg pepsin per mL of 0.1N hydrochloric acid added. The mixture was incubated for 30 min, the pH was adjusted to 8 and transferred to a dialysis bag with a 1000 Da molecular weight cut off. The bag was placed in a U-shaped container with inlets from a peristaltic pump and outlets to a beaker. Twenty mL of a solution containing 5 mg pancreatin per mL sodium phosphate buffer was added to the dialysis bag, which was continuously washed with 37 °C sodium phosphate buffer at a flow rate of 212 mL/h. Samples of dialysate were collected at different time intervals and analyzed. The method was able to detect the effects of heat and alkali treatment on protein digestibility in foods [34].

The digestion unit size plus the use of handmade apparatus were limitations for its use in routine protein evaluation [54]. Savoie and Gauthier modified the design presented by Gauthier et al. [34]. The improvements included the use of a magnetic stir bar and the construction of a cell with an inner compartment fitted with a dialysis membrane. The cell was 100 mm long in comparison to the 298 mm original unit. There was free access to the reaction chamber without disruption of the reaction. Each cell was designed to work as a single unit or part of the multi-unit system. The system developed was very flexible and could be used to measure the release of any product from enzymatic hydrolysis [54].

The dialysis cell method has been applied to study protein digestibility across a number of disciplines [55–57]. This method was able to identify differences in the rate of release of amino acids from different sea bream feed samples [57]. The system was flexible to accommodate the use of crude enzyme extract from sea bream as the digestive enzyme. A comparison between the pH-stat and the dialysis cell method showed that the dialysis cell method was able to identify which products were released from the protein as well as the digestion kinetics of the protein samples [55]. The effects of different processing methods on the digestibility of legume proteins were identified with the dialysis cell method [56]. A detailed description of the availability of different amino acids and the rate at which they were released during digestion was obtained from different protein sources [56,58]. The main disadvantages of the dialysis cell method are the complexity and the number of samples which can be digested in a given run. This method uses custom made dialysis cells, peristalsis pumps, and fraction collectors which can be expensive. Savoie and Gauthier recommend that no more than 6 cells should be used simultaneously due to the manual inputs needed. From a practical point of view, an in vitro method must be simple and easy to implement for it to be adopted by poultry nutritionist [54].

3. Factors Influencing Protein Digestion

The digestibility data obtained by in vitro methods vary even within the same method for the same ingredient. This variation may be due to a number of issues associated with in vitro digestibility systems. Enzymes and their concentration seemed to be one of the most important factors influencing in vitro digestion [1,26,59]. The specificity of enzymes and their ratio to the substrate will determine the level of hydrolysis achieved [8,59].

3.1. Enzyme Specificity

Table 1 shows a list of enzymes involved in protein digestion. The first enzyme responsible for the initiation of protein digestion in poultry is pepsin [8]. This enzyme will only cleave the N-terminal of aromatic amino acids like tyrosine, tryptophan and phenylalanine [4] at low pH. Hydrolysis by pepsin results in smaller peptides that enter the duodenum for further hydrolysis by pancreatic protease [8]. As suggested by Assoumani and Nguyen [10], trypsin will only break a lysyl or arginyl peptide bonds to expose lysine or arginine terminal residues at basic pH. Trypsin binds only to the positive side group of arginine and lysine, where the peptide is cleaved at those amino acids [4].

Table 1. Enzyme and their specific bond cleavage preferences.

Enzymes	Bond Cleave	Reference
Pepsin	N-terminal of aromatic amino acids phenylalanine, tryptophan and tyrosine	[4,8]
Trypsin	Lysyl or arginyl peptide bond to expose lysine or arginine	[10]
Chymotrypsin	Aromatic or large hydrophobic amino acid residues such as tyrosine, phenylalanine, tryptophan, leucyl, methionyl, asparaginyl, and glutamyl	[4,8,60]
Elastase	Glycine and alanine of elastin	[4,8]
Carboxypeptidase A	Peptide bond adjacent to the C-terminal end of a polypeptide chain,	[4,8]
Carboxypeptidase B	Basic amino acids from the C-terminal end of polypeptide chains	[8]
Collagenase	Alpha peptides and hydrogen bonds in the superhelix of tropocollagen and collagen	[61]

The ability of enzymes to hydrolyze substrate may depend on the presence of other enzymes. The activation of chymotrypsin is dependent on the presence of trypsin [4]. Chymotrypsin will act on proteins and peptides, but will also hydrolyze esters and amides [60]. Chymotrypsin cleaves peptides over a wider range of sites than trypsin, both aromatic and hydrophobic side chains of amino acids residues [4]. Peptide bonds involving tyrosine, tryptophan, phenylalanine and glutamyl, leucyl, asparaginyl residues are cleaved by chymotrypsin [4,8].

Lysine or arginine are released from small peptides by carboxypeptidase-B, which is specific for C-terminal basic groups [8]. Animal protein meals may contain high levels of collagen due to the nature of the type of rendering material. Digestion of this meal in vitro may need additional collagenase enzymes during the pancreatic digestion stage [61]. Bonds hydrolyzed in protein feed samples are enzyme-specific, so in vitro digestion models should take this into account by using multiple enzymes [8].

3.2. Protein Structure and Forms

The structure of the protein samples and the food matrix in which the samples are presented will influence protein in vitro digestibility [10]. Protein feed ingredients may contain free amino acids, peptides of various lengths, secondary structure proteins (α-helix, β-pleated sheets, β-turns and superhelix), tertiary structure proteins and quaternary structure proteins [4]. Secondary structure proteins such as scleroproteins, which include collagen, elastin, and keratin, are poorly digested in simple stomach animals [62]. Protein sources containing high levels of these proteins will have limited bioavailability. Higher protein structural configuration requires more time and higher enzyme concentration to achieve greater hydrolysis [4,62].

Secondary structure proteins resist digestion due to the nature of their individual structures. Feather meal, for example, contains high levels of keratin [43], which has highly cross-linked disulphide bonds along the pleated sheet configuration [62]. This makes the protein almost insoluble in water and thereby reduces the action of pepsin and subsequent pancreatic actions [43]. Samples of meat and bone meal may contain elastin and collagen after being produced from tendons, ligaments and bone scraps of animals. Elastin and collagen also contain cross-linking in their helix structures which may influence digestion [62].

The matrix in which the protein is presented in the protein source may limit the access of proteolytic enzymes. Plant proteins are often presented in a matrix with cell walls, lipids, and complex sugars, and may also be organized into specialized storage vacuoles [10,62]. The ability of proteolytic enzymes to access those proteins may depend on the ability of other enzymes to free protein from the matrices [8]. The digestion of protein from plant sources in monogastric animals is closely linked to the protein associated with plant cell wall components [63]. Non-starch polysaccharides are known to protect proteins from enzymatic digestion in a variety of plant feed ingredients in poultry [64]. Solubilization of the cell wall components of plant-sourced protein meals with various carbohydrase enzymes were able to improve the availability of the protein to chickens [63,64].

3.3. Enzyme Activity

In vitro digestion may be influenced by the activity of the enzymes used while enzyme activity is affected by factors such as pH, temperature, ratio of enzyme to substrate, and incubation time [8]. As proteins are hydrolyzed by enzyme in vitro, the pH of the mixture will be reduced by the release of protons from the cleaved peptide bonds [25]. If the original pH of the reaction moisture is further away from the optimum pH of the enzyme, the rate of hydrolysis will be reduced drastically in a short period of time. In the pH-stat method, pH is held constant in the optimal range for the enzyme via automated alkali titration [29]. To achieve optimal reaction conditions, most in vitro assays select appropriate starting pH for the enzyme used [10]. The pepsin digestibility assay requires an acidic condition [36], while the pancreatin assay requires a basic environment [39].

The temperature may play a regulatory role as it relates to enzyme activity. Like all chemical reactions, temperature increases the amount of kinetic energy and increases the velocity at which molecules collide in an enzymatic reaction [4]. In vitro digestibility assays using protease keep the temperature of their reaction between 37 and 45 °C [8]. Enzymes are proteins and all proteins can be denatured at high temperatures; therefore the optimal temperature for a given enzyme is always close to the body temperature of the organism from which the enzyme was derived [4]. In vitro assays should reflect in vivo conditions so the temperature at which the reaction takes place is often that of the animal's internal temperature [33,47].

The ratio of enzyme to substrate and the incubation time varies across individual in vitro assays [1,8,51,59]. Generally, the incubation time can range from 0.5 to 45 h depending on the kind of in vitro assay [10]. The enzyme to substrate ratio is often a function of the specific activity of the enzyme. The specific activity of an enzyme is often defined as the amount of product produced from a specific substrate over time while maintaining the reaction at a fixed pH and temperature range [4]. Enzymes from different preparations with different specific activities are often used for the same in vitro assay [54,58]. The ratio of pepsin used with 4 mg nitrogen of sample in the dialysis method ranged from 5 to 7 mg/mL pepsin [34,54]. Pepsin concentration used in the pepsin digestibility test ranged from 0.02 to 2.5 g/L and the sample size of the protein may be expressed as g of nitrogen per sample [8]. To avoid confusion in the literature, an in vitro method should define the enzyme to substrate ratio and the specific activity of each enzyme in the assay [65].

3.4. Anti Nutritive Agents of Test Samples

Anti-nutritional compounds are often secondary metabolites and structural components of plants that interfere with the metabolic activities of animals when present in feed ingredients [66]. These compounds provide structural support and some metabolites have evolved into defence chemicals to protect plants from insect damage [67]. Some anti-nutritional compounds represent important storage minerals and intermediate molecules used in various pathways by the plant [66]. The main action of these compounds tends to disrupt the digestive process via multiple modes of action.

3.4.1. Sinapine and Tannins

A phenolic compound found in many plant feed ingredients is sinapine, which is a choline ester derived from 3, 5-dimethoxy-4-hydroxyinnamic acid or tannins [68]. Growing plants use sinapine as their main source of sinapic acid and choline [69]. High levels of sinapic acid can react with other compounds to create a colour change and produce a bitter taste in plant feed ingredients [70]. During oxidation, phenolic acids may react with proteins to form indigestible complexes like quinines which bind to the functional group of lysine and methionine [68].

Tannins are another set of water soluble polyphenolic compounds that may be found in protein meals of plant origin [71]. They are normally present in legume seeds, cereal grains, and oilseeds [68,72]. Tannins are generally grouped into hydrolyzable and condensed tannins. Hydrolyzable tannins may have esters of gallic, m-digallic, or hexahydroxydiphenic acids, which are easily hydrolyzed [71]. Condensed tannins resist hydrolysis and are polymers of flavan-2, 4-diol and flavan-3-ol or a mixture of both [72]. Tannins precipitate protein out of solution through the formation of soluble and insoluble complexes [68], and are known to reduce the digestibility of amino acids in poultry [73]. Tannins inhibit the absorption of protein from the digestive tract [72,73]. Low molecular weight tannins may be absorbed from the intestine and cause toxicity through the inhibition of key metabolic pathways [72,73].

3.4.2. Protease Inhibitors

Almost all plant protein sources available for use in animal production contain some type of protease inhibitor [74]. Even commonly consumed foods such as legumes, cereal grains, and tomatoes contain protease inhibitors [72]. Protease inhibitors block the activity of trypsin, chymotrypsin [62], elastase, and carboxypeptidase [75]. Trypsin inhibitor can be found in field pea, peanut, wheat, soybean, rapeseed, lupin, and sunflower seeds [62,75].

Of the plant protein sources used in poultry production, soybean is generally considered to have the highest trypsin inhibitor activity [72]. The inhibitors bind to the active site of the enzyme, thereby reducing their ability to lower the kinetic energy needed during proteolytic cleavage [4]. The two main inhibitors found in soybean are from the Kunitz and Bowman-Birk inhibitor families [62]. Kunitz is about 21.4 kDa with high affinity for trypsin, while Bowman-Birk is about 8 kDa and has high affinity for both trypsin and chymotrypsin [72].

When birds were fed diets containing raw soybean, the granules of the pancreatic acini were totally depleted in 2 h after feeding and the size of the pancreas increased after 8 d [76]. The pancreatic activity of the birds at 16 d was twice the activity before they were given the diet and the birds growth was reduced drastically. Protease inhibitor activities can be reduced through various heat processes, but complete elimination is often not possible in commercial soybean products [15,74].

3.4.3. Phytate

Feed ingredients derived from plants contain some level of phosphorus stored as phytic acid or phytate which are also known as myo-inositol hexaphosphoric acid and myo-inositol hexaphosphate respectively [77]. Phytate is predominantly found in the seeds of plants, which makes animal feed derived from oilseeds and cereal grains a source of phytate [78]. During germination, the inorganic phytate is hydrolyzed by enzymes to produce phosphate which the plant uses for its growth [79]. Phytic acid has strong mineral binding capacity through its six phosphate groups, which actively bind zinc, iron, calcium, and magnesium [79]. Phytate's chelating ability results in complexes with nutrients such as proteins and minerals [80].

The anti-nutritional effects of phytic acid on protein digestion can occur via direct or indirect modes of action. During protein digestion, phytate may bind to metal cofactors needed for the activity of aminopeptidases and carboxypeptidases [4,72]. Phytate may also bind with protein to form complexes in acidic and neutral pH conditions [80], which may inhibit the activities of digestive enzymes [81]. Intestinal phytase activity observed in poultry [82] may depend on magnesium as a cofactor. In such a case intestinal phytase may not be able to hydrolyze a substantial amount of the dietary phytate if sufficient magnesium is not present. However, in practical feeding situation the poultry industry has incorporated exogenous phytase in poultry diets [83]. The exogenous phytase hydrolyzes the ester bond between the inositol ring and phosphate group, thereby releasing phosphorus and reducing the anti-nutritive effects on protein digestibility. This elicits a question of whether exogenous phytase enzymes should be part of an in vitro protein digestibility assay for poultry.

3.4.4. Effects of Ingredient Processing

Proteins used in animal production are often by-products of other processing industries. The nutritional quality of these proteins is a function of the processes used in meal production. Plant-based protein sources generally will contain some form of anti-nutrient and thus require processing to reduce their effects when fed to animals. Protein meals of animal origin are waste products from food processing facilities. As such, the raw materials may contain higher levels of microbial contamination and require additional processing before it is fed to animals.

The major anti-nutritional compounds found in plant-based protein sources can be reduced through some form of heat treatment. Unfortunately, amino acid digestibility in chickens may be compromised if the heat treatment used is excessive [12] or not enough [22]. Autoclaving flaxseed at 120 °C for 20, 40, and 60 min resulted in changes in the α-helix to β-sheet ratio of the protein fraction [84]. Rumen degradable protein is reduced with increased autoclaving time which suggested that the protein resisted digestion as a result of the change in α-helix to β-sheet ratio. This would be true if that same protein was fed to non-ruminants and the effects would be more severe.

During the commercial production of canola meal using the prepress-solvent extraction system, the meal is subjected to toasting during hexane removal [18]. Amino acid digestibility and the content of the meal are reduced after toasting. The elimination of the spurge steam during toasting could alleviate the loss of amino acids [18]. Soybean meal production involves solvent extraction as well. Ideally, the soybean is exposed to 105 °C for half h [85], but if the meal is heated to 121 °C, the concentration and digestibility of amino acids, especially lysine, are reduced [86]. The loss of amino acids during the production of meals from the solvent extraction process may result in poor growth in chickens fed meals processed under such conditions [13,87].

Amino acid loss during heating processing of protein meal may involve Maillard reactions, were a sugar-amine complex is formed from the reaction of sugars and ketones with amino acids, proteins, and peptides in food [88]. Mauron suggested that Maillard reactions involve early, advanced, and final stage reactions. Early Maillard reaction involves a reversible condensation of the carbonyl group of the sugar with the amino group of the amino acid, peptide, or protein to form a hydrolyzable N-substituted glycosylamine and then 1-amino-1-deoxy-2-ketose. At the early stage, food does not have any browning or flavour, but its nutritive value is reduced. During the advanced stage of the reaction, amines are released and are used as catalysts in reactions to form intermediate flavour products such as acetaldehyde and pyruvaldehyde [89]. The final reaction produces a dark brown nitrogen-containing pigment composed of decomposed amino acids, heterocyclic amines, melanoidin polymers and aldol condensation products [88].

The stages of the Maillard reaction requires specific reaction conditions to be successful [88]. Temperature and moisture are the two important parameters which govern each stage of the Maillard reaction [88]. Experimental simulations of Maillard reaction generally take place in solutions and the formation of melanoidin polymers is an exponential function of heating [90]. Reactions of D-xylose and glycine in aqueous solution at 22, 68, and 100 °C produce a temperature-dependent increase in aromaticity or high molecular weight melanoidin polymers [90]. The rate of the Maillard reaction is defined as the function Q_{10} which is the increase in rate for every 10 °C. As the temperature increases from 22 to 100 °C the quantity of high molecular weight melanoidin increases and the low soluble intermediate products of the Maillard reaction decrease [90].

Protein meals of animal origin do not contain the high levels of sugars found in meals of plant origin, so are less likely to undergo Maillard reaction when exposed to heat treatment. The natural soluble carbohydrate concentration of dried animal protein meals range from 0.3 to 1.3% [91], which is far less than what would normally be present in plant-based meals [13,87]. The meals are prone to Maillard reaction if they are exposed to soluble carbohydrate during autoclaving which has been shown to reduce meal digestibility [91].

Large amounts of meat and bone meal are produced by the rendering industry, but the quality of those meals can vary [37]. The variability in the quality of meat and bone meal can limit its use in poultry production [1]. Oxidation and enzymatic denaturing may occur depending on location and source of the raw material used in the rendering process. Polyunsaturated fats are known to react with atmospheric oxygen which results in the production of peroxides and other auto-oxidation products [92]. If the meal is kept in warm conditions, this could increase the formation of peroxides and secondary oxidation products. The application of heat in the presence of oxygen and polyunsaturated fats is known to increase the production of peroxides and secondary oxidation products [92]. This could be a factor during rendering if parameters such a temperature, time, and raw material polyunsaturated fat content are not controlled during meal production.

4. In Vitro Digestibility Systems Validation

One major challenge often encountered when developing in vitro models to evaluate protein digestion is the ability of a single model to effectively assay multiple kinds of feed ingredients. Due to this challenge, multiple quality control assays such as those based on the physicochemical properties of ingredients have been developed to help the feed industry. The in vitro model developed by Bryan et al. [59] evaluated nine different protein sources which are known to have variable digestibility and physicochemical properties. Correlation analysis between the PDI and KOH solubility of the ingredients and in vitro extent of crude protein (CP) digestion were all significant, with correlation coefficients (r) of 0.64 and 0.84, respectively. There was no correlation between the in vitro CP digestibility and the reactive lysine assay. This might be an indication that the reactive lysine assay was not a useful physiochemical candidate assay to compare with the developed in vitro assay. In Figure 1, such correlation with in vivo data is generally used as the last step for validating in vitro systems [93]. It is a common practice to conduct correction analysis in such circumstances but the validation of in vitro digestibility systems requires more analysis.

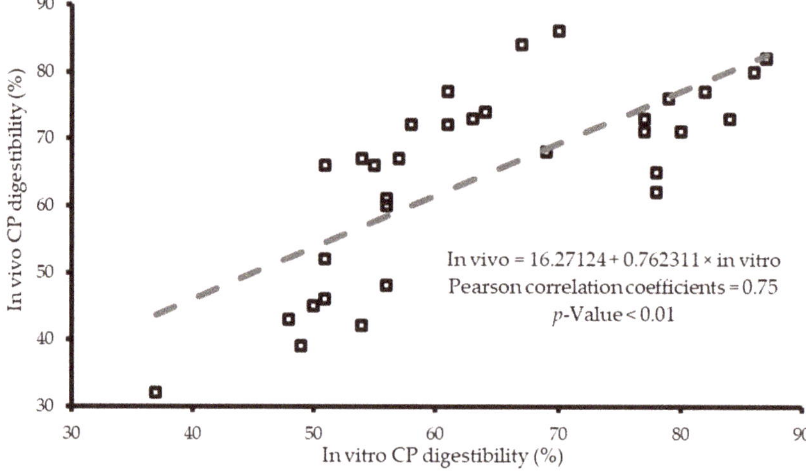

Figure 1. Plot of correlation between in vivo and in vitro crude protein (CP) digestible of nine high protein poultry feed ingredients [93].

In order to know if in vitro CP digestibility data are representative of the in vivo amino acid digestibility, regression and correlation analysis together (Table 2) were performed between the two sets of data generated by Bryan et al. [94,95]. The in vitro CP digestibility was positively correlated with all amino acids except for cysteine (CYS), which had a regression estimate P-value of 0.1. The correlation coefficients ranged from 0.43 to 0.71, except for CYS which was 0.30. The data presented in Table 2 shows the complexity of factors that might affect the interpatient of correlation validation of in vitro systems to in vivo. The in vitro model was developed using soybean meal (SBM) as the model protein source which has both pros and cons. Using SBM might have put the other ingredients at a slight disadvantage since the method optimized SBM digestibility for each stage of digestion and not the other meals. This could have accounted for some of the variation seen in the correlation coefficients of the amino acid with the in vitro CP digestibility. Based on the data presented in Table 2, the in vitro CP digestibility can be used as a predictor of in vivo amino acid digestibility; however, the correlation coefficients varied among amino acids so more samples need to be tested to form stronger prediction equations.

Table 2. Simple linear regression and Pearson correlation of in vitro digestible crude protein (CP) and in vivo standardized ileal amino acids digestibility of the nine meal samples [93].

Item	Regression Coefficients		ANOVA		In Vitro Digestible CP	
	Intercept	In Vitro Digestible CP	R^2	MSE	Correlation Coefficients	*P*-Value
Aspartic acid	3.27	0.83	0.35	252.55	0.59	<0.01
Estimate SE	13.68	0.21	-	-	-	-
Estimate *p*-Value	0.81	<0.01	-	-	-	-
Threonine	27.35	0.55	0.35	115.73	0.59	<0.01
Estimate SE	9.25	0.14	-	-	-	-
Estimate *p*-Value	<0.01	<0.01	-	-	-	-
Serine	38.29	0.43	0.18	168.08	0.43	0.02
Estimate SE	11.15	0.17	-	-	-	-
Estimate *p*-Value	<0.01	0.02	-	-	-	-
Glutamic acid	22.19	0.75	0.50	112.11	0.71	<0.01
Estimate SE	9.11	0.14	-	-	-	-
Estimate *p*-Value	0.02	<0.01	-	-	-	-
Proline	13.72	0.75	0.35	206.60	0.59	<0.01
Estimate SE	12.36	0.19	-	-	-	-
Estimate *p*-Value	0.28	<0.01	-	-	-	-
Glycine	41.58	0.40	0.25	94.69	0.50	<0.01
Estimate SE	8.37	0.13	-	-	-	-
Estimate *p*-Value	<0.01	<0.01	-	-	-	-
Alanine	35.01	0.56	0.39	95.38	0.63	<0.01
Estimate SE	8.40	0.13	-	-	-	-
Estimate *p*-Value	<0.01	0.56	-	-	-	-
Cysteine	26.18	0.41	0.09	326.99	0.30	0.09
Estimate SE	15.55	0.23	-	-	-	-
Estimate *p*-Value	0.10	0.09	-	-	-	-
Valine	42.74	0.40	0.21	113.32	0.46	<0.01
Estimate SE	9.16	0.14	-	-	-	-
Estimate *p*-Value	<0.01	<0.01	-	-	-	-
Methionine	32.256	0.63	0.45	97.07	0.67	<0.01
Estimate SE	8.47	0.13	-	-	-	-
Estimate *p*-Value	<0.01	<0.01	-	-	-	-
Isoleucine	43.74	0.44	0.26	110.68	0.51	<0.01
Estimate SE	9.05	0.14	-	-	-	-
Estimate *p*-Value	<0.01	<0.01	-	-	-	-

Table 2. *Cont.*

Item	Regression Coefficients		ANOVA		In Vitro Digestible CP	
	Intercept	In Vitro Digestible CP	R^2	MSE	Correlation Coefficients	*P*-Value
Leucine	35.38	0.56	0.35	113.76	0.59	<0.01
Estimate SE	9.17	0.14	-	-	-	-
Estimate *p*-Value	<0.01	<0.01	-	-	-	-
Tyrosine	28.97	0.63	0.39	121.17	0.62	<0.01
Estimate SE	9.47	0.14	-	-	-	-
Estimate P-Value	<0.01	<0.01	-	-	-	-
Phenylalanine	39.97	0.5	0.29	120.73	0.54	<0.01
Estimate SE	9.45	0.14	-	-	-	-
Estimate P-Value	<0.01	<0.01	-	-	-	-
Lysine	34.44	0.57	0.50	62.50	0.71	<0.01
Estimate SE	6.80	0.10	-	-	-	-
Estimate *p*-Value	<0.01	<0.01	-	-	-	-
Histidine	12.17	0.84	0.48	150.04	0.70	<0.01
Estimate SE	10.54	0.16	-	-	-	-
Estimate *p*-Value	0.26	<0.01	-	-	-	-
Arginine	33.27	0.63	0.40	119.31	0.63	<0.01
Estimate SE	9.39	0.14	-	-	-	-
Estimate *p*-Value	<0.01	<0.01	-	-	-	-

R^2: R-squared (variance for a dependent variable explained by variables in the regression model); MSE: Means square error; SE: Standard error.

Another approach in the validation step is to add more analysis. A one sample T-Test was performed comparing the difference between the in vitro and in vivo CP digestibility data of Figure 1 to a mean of 0 to see if there were differences between the two methods of assessing CP digestibility. This comparison suggests that there is no difference between in vivo and in vitro CP extent of digestion for the meals evaluated. The Bland Altman plot of the data presented in Figure 2 shows that there was no proportional bias between in vitro and in vivo CP digestibility data for any of the nine meals evaluated and all the data points collected during the assay fell in the 95% confidence limit.

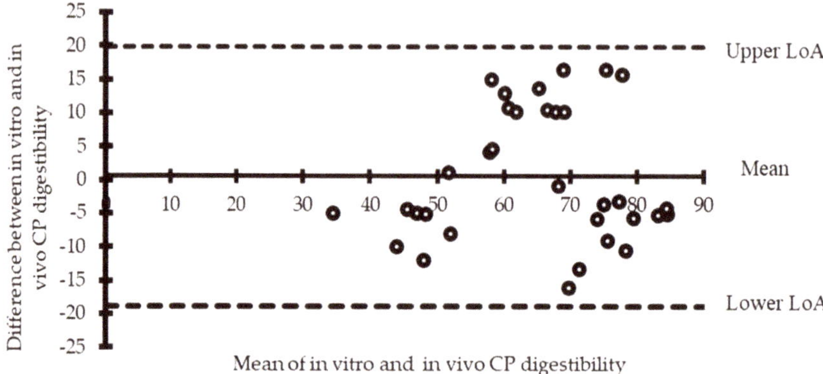

Figure 2. Bland Altman Plot of the difference between in vivo and in vitro crude protein (CP) digestible of nine high protein feed ingredients [93]; LoA: limits of agreement

This indicates that the in vivo and the in vitro CP digestible data were in agreement for the digestibility of nine meals. Based on the correlation, the T-Test, and the Bland Altman plot results, the in vitro assay was able to predict the in vivo CP digestibility of the ingredients. The in vitro assay could, therefore, serve as a tool for assaying CP digestible of meals for broiler chickens.

5. Conclusions

Protein quality assessment of feed ingredients for poultry is often achieved using in vitro or in vivo testing. The disadvantages associated with in vivo methods lead to the commercial acceptance of in vitro methods as the gold standard for assessing protein quality. These techniques are used to improve the user's efficiency when dealing with large numbers of sample and some mimic the physiological and chemical characteristics of the animal digestive system to which the ingredient will be fed. Despite all of the advantages of these in vitro methods, they do not give a true replication of normal in vivo digestive conditions. This is because of the inability of those methods to mimic numerous biological factors involved in in vivo digestion and the complex interaction which exists with various ingredients. Multi enzyme assays can predict animal digestibility of proteins if they are designed properly. However, any inherent biological properties of the ingredients which might impact the animal digestive tract will be lost. Users of in vitro digestibility data should be aware of these disadvantages and take the necessary steps to validate in vitro methods and their data. In any case, in vitro digestibility methods are just estimates of in vivo digestion, which serve as a substitute in situations where in vivo digestion is not possible.

Author Contributions: Conceptualization, D.D.S.L.B. and H.L.C.; writing—original draft preparation, D.D.S.L.B; writing—review and editing, H.L.C.; supervision, H.L.C.; funding acquisition, H.L.C. All authors have read and agreed to the published version of the manuscript.

Funding: "This research was funded by Government of Canada Natural Sciences and Engineering Research Council of Canada Industrial Research Chair Program, Grant number IRCSA 452664-12. Funding for this Chair Program was derived from Aviagen, Canadian Poultry Research Council, Chicken Farmers of Saskatchewan, NSERC, Ontario Poultry Industry Council, Prairie Pride Natural Foods Ltd., Saskatchewan Egg Producers, Saskatchewan Hatching Egg Producers, Saskatchewan Turkey Producers, Sofina Foods Inc. and the University of Saskatchewan" and "The APC was funded by Government of Canada Natural Sciences and Engineering Research Council of Canada Industrial Research Chair Program".

Conflicts of Interest: The authors declare no conflict of interest. The funders had no role in the design of the study; in the collection, analyses, or interpretation of data; in the writing of the manuscript, or in the decision to publish the results.

References

1. Parsons, C.; Castanon, F.; Han, Y. Protein and amino acid quality of meat and bone meal. *Poult. Sci.* **1997**, *76*, 361–368. [CrossRef] [PubMed]
2. Adedokun, S.A.; Adeola, O.; Parsons, C.M.; Lilburn, M.S.; Applegate, T.J. Standardized ileal amino acid digestibility of plant feedstuffs in broiler chickens and turkey poults using a nitrogen-free or casein diet. *Poult.Sci.* **2008**, *87*, 2535–2548. [CrossRef] [PubMed]
3. Kim, E.J.; Utterback, P.L.; Applegate, T.J.; Parsons, C.M. Comparison of amino acid digestibility of feedstuffs determined with the precision-fed cecectomized rooster assay and the standardized ileal amino acid digestibility assay. *Poult. Sci.* **2011**, *90*, 2511–2519. [CrossRef] [PubMed]
4. Bhagavan, N.V.; Bhagavan, N.V. *Medical Biochemistry*; Jones and Bartlett Publishers: Boston, UK, 1992.
5. Ravindran, V.; Bryden, W.L. Amino acid availability in poultry—In vitro and in vivo measurements. *Aust. J. Agric. Res.* **1999**, *50*, 889–908. [CrossRef]
6. Lemme, A.; Ravindran, V.; Bryden, W.L. Ileal digestibility of amino acids in feed ingredients for broilers. *World's Poult. Sci. J.* **2004**, *60*, 423–438. [CrossRef]
7. Batterham, E.S. Availability and utilization of amino acids for growing pigs. *Nutr. Res. Rev.* **1992**, *5*, 1–18. [CrossRef]
8. Boisen, S.; Eggum, B.O. Critical evaluation of in vitro methods for estimating digestibility in simple-stomach animals. *Nutr. Res. Revi.* **1991**, *4*, 141–162. [CrossRef]
9. Clunies, M.; Leeson, S. In vitro estimation of dry matter and crude protein digestibility. *Poult. Sci.* **1984**, *63*, 89–96. [CrossRef]
10. Fuller, M.F. *Vitro Digestion for Pigs and Poultry*; C.A.B. International: Wallingford, UK, 1991; ISBN 0-85198-719-2.

11. Smith, A.K.; Circle, S.J.; Brother, G.H. Peptization of soybean proteins. The effect of neutral salts on the quantity of nitrogenous constituents extracted from oil-free meal. *J. Am. Chem. Soc.* **1938**, *60*, 1316–1320. [CrossRef]
12. Araba, M.; Dale, N.M. Evaluation of protein solubility as an indicator of over processing soybean meal. *Poult. Sci.* **1990**, *69*, 76–83. [CrossRef]
13. Lee, H.; Garlich, J.D. Effect of overcooked soybean meal on chicken performance and amino acid availability. *Poult. Sci.* **1992**, *71*, 499–508. [CrossRef] [PubMed]
14. Lyman, C.M.; Chang, W.Y.; Couch, J.R. Evaluation of protein quality in cottonseed meals by chick growth and by a chemical index method. *J. Nutr.* **1953**, *49*, 679–690. [CrossRef] [PubMed]
15. Parsons, C.M.; Hashimoto, K.; Wedekind, K.J.; Baker, D.H. Soybean protein solubility in potassium hydroxide: An in vitro test of in vivo protein quality. *J. Anim. Sci.* **1991**, *69*, 2918–2924. [CrossRef] [PubMed]
16. Fernandez, S.R.; Zhang, Y.; Parsons, C.M. Determination of protein solubility in oilseed meals using coomassie blue dye binding. *Poult. Sci.* **1993**, *72*, 1925–1930. [CrossRef]
17. Batal, A.; Douglas, M.; Engram, A.; Parsons, C. Protein dispersibility index as an indicator of adequately processed soybean meal. *Poult. Sci.* **2000**, *79*, 1592–1596. [CrossRef]
18. Newkirk, R.W.; Classen, H.L.; Scott, T.A.; Edney, M.J. The digestibility and content of amino acids in toasted and non-toasted canola meals. *Can. J. Anim. Sci.* **2003**, *83*, 131–139. [CrossRef]
19. Johnson, D.W. Functional properties of oilseed proteins. *J. Am. Oil Chem. Soc.* **1970**, *47*, 402–407. [CrossRef]
20. Veltmann, J.R.; Hansen, B.C.; Tanksley, T.D.; Knabe, D.; Linton, A.S. Comparison of the nutritive value of different heat-treated commercial soybean meals: Utilization by chicks in practical type rations. *Poult. Sci.* **1986**, *65*, 1561–1570. [CrossRef]
21. Clarke, E.; Wiseman, J. Effects of variability in trypsin inhibitor content of soya bean meals on true and apparent ileal digestibility of amino acids and pancreas size in broiler chicks. *Anim. Feed Sci. Technol.* **2005**, *121*, 125–138. [CrossRef]
22. De Coca-Sinova, A.; Valencia, D.G.; Jiménez-Moreno, E.; Lázaro, R.; Mateos, G.G. Apparent ileal digestibility of energy, nitrogen, and amino acids of soybean meals of different origin in broilers. *Poult. Sci.* **2008**, *87*, 2613–2623. [CrossRef]
23. Serrano, M.P.; Valencia, D.G.; Méndez, J.; Mateos, G.G. Influence of feed form and source of soybean meal of the diet on growth performance of broilers from 1 to 42 days of age. 1. Floor pen study. *Poult. Sci.* **2012**, *91*, 2838–2844. [CrossRef] [PubMed]
24. Pérez-Calvo, E.; Castrillo, C.; Baucells, M.D.; Guada, J.A. Effect of rendering on protein and fat quality of animal by-products. *J. Anim. Physiol. Anim. Nutr.* **2010**, *94*, e154–e163. [CrossRef] [PubMed]
25. Maga, J.A.; Lorenz, K.; I, O.O. Digestive acceptability of proteins as measured by the initial rate of in vitro proteolysis. *J. Food Sci.* **1973**, *38*, 173–174. [CrossRef]
26. Vavak, D.L.R. A Nutritional Characterization of the Distiller's Grain Protein Concentrates. Ph.D. Thesis, University of Nebraska—Lincoln, Lincoln, NE, USA, 1975.
27. Hsu, H.W.; Vavak, D.L.; Satterlee, L.D.; Miller, G.A. A multienzyme technique for estimating protein digestibility. *J. Food Sci.* **1977**, *42*, 1269–1273. [CrossRef]
28. Satterlee, L.D.; Kendrick, J.G.; Marshall, H.F.; Jewell, D.K.; Ali, R.A.; Heckman, M.M.; Steinke, H.F.; Larson, P.; Phillips, R.D.; Sarwar, G. In vitro assay for predicting protein efficiency ratio as measured by rat bioassay: Collaborative study Milk, chicken, soy protein, cereals, wheat flour, nutritional quality. *J. AOAC Int.* **1982**, *65*, 798–809. [CrossRef]
29. Pedersen, B.; Eggum, B.O. Prediction of protein digestibility by an in vitro enzymatic pH-stat procedure. *Z. Tierphysiol. Tierernähr. Futtermittelkd.* **1983**, *49*, 265–277. [CrossRef] [PubMed]
30. Linder, M.; Rozan, P.; EL Kossori, R.L.; Fanni, J.; Villaume, C.; Mejean, L.; Parmentier, M. Nutritional value of veal bone hydrolysate. *J. Food Sci.* **1997**, *62*, 183–189. [CrossRef]
31. Wang, H.; Faris, R.J.; Wang, T.; Spurlock, M.E.; Gabler, N. Increased in vitro and in vivo digestibility of soy proteins by chemical modification of disulfide bonds. *J. Am. Oil Chem. Soc.* **2009**, *86*, 1093–1099. [CrossRef]
32. Dimes, L.E.; Haard, N.F. Estimation of protein digestibility I. Development of an in vitro method for estimating protein digestibility in salmonids (Salmo gairdneri). *Comp. Biochem. Physiol. A Physiol.* **1994**, *108*, 349–362. [CrossRef]

33. Tibbetts, S.M.; Milley, J.E.; Ross, N.W.; Verreth, J.A.J.; Lall, S.P. In vitro pH-Stat protein hydrolysis of feed ingredients for Atlantic cod, Gadus morhua. 1. Development of the method. *Aquaculture* **2011**, *319*, 398–406. [CrossRef]

34. Gauthier, S.F.; Vachon, C.; Jones, J.D.; Savoie, L. Assessment of protein digestibility by in vitro enzymatic hydrolysis with simultaneous dialysis. *J. Nutr.* **1982**, *112*, 1718–1725. [CrossRef] [PubMed]

35. Gehrt, A.J.; Caldwell, M.J.; Elmslie, W.P. Feed digestibility, chemical method for measuring relative digestibility of animal protein feedstuffs. *J. Agric. Food Chem.* **1955**, *3*, 159–162. [CrossRef]

36. Sheffner, A.L.; Eckfeldt, G.A.; Spector, H. The pepsin-digest-residue (PDR) amino acid index of net protein utilization. *J. Nutr.* **1956**, *60*, 105–120. [CrossRef] [PubMed]

37. Johnson, J.; Coon, C.N. A comparison of six protein quality assays using commercially available protein meals. *Poult. Sci.* **1979**, *58*, 919–927. [CrossRef]

38. Johnson, J.; Coon, C.N. The use of varying levels of pepsin for pepsin digestion studies with animal proteins. *Poult. Sci.* **1979**, *58*, 1271–1273. [CrossRef]

39. Riesen, W.H.; Clandinin, D.R.; Elvehjem, C.A.; Cravens, W.W. Liberation of essential amino acids from raw, properly heated, and overheated soy bean oil meal. *J. Biol. Chem.* **1947**, *167*, 143–150. [PubMed]

40. Ingram, G.R.; Riesen, W.W.; Cravens, W.W.; Elvehjem, C.A. Evaluating soybean oil meal protein for chick growth by enzymatic release of amino acids. *Poult. Sci.* **1949**, *28*, 898–902. [CrossRef]

41. Anwar, A. Evaluation of proteins by in vitro pancreatin digestion. *Poult. Sci.* **1962**, *41*, 1120–1123. [CrossRef]

42. Altangerel, B.; Sengee, Z.; Kramarova, D.; Rop, O.; Hoza, I. The determination of water-soluble vitamins and in vitro digestibility of selected Czech cheeses. *Int. J. Food Sci. Technol.* **2011**, *46*, 1225–1230. [CrossRef]

43. Bielorai, R.; Harduf, Z.; Iosif, B.; Alumot, E. Apparent amino acid absorption from feather meal by chicks. *Brit. J. Nutr.* **1983**, *49*, 395–399. [CrossRef]

44. Akeson, W.R.; Stahmann, M.A. A pepsin pancreatin digest index of protein quality evaluation. *J. Nutr.* **1964**, *83*, 257–261. [CrossRef] [PubMed]

45. Saunders, R.M.; Connor, M.A.; Booth, A.N.; Bickoff, E.M.; Kohler, G.O. Measurement of digestibility of alfalfa protein concentrates by in vivo and in vitro methods. *J. Nutr.* **1973**, *103*, 530–535. [CrossRef] [PubMed]

46. Cave, N.A. Bioavailability of amino acids in plant feedstuffs determined by in vitro digestion, chick growth assay, and true amino acid availability methods. *Poult. Sci.* **1988**, *67*, 78–87. [CrossRef] [PubMed]

47. Boisen, S.; Fernández, J.A. Prediction of the apparent ileal digestibility of protein and amino acids in feedstuffs and feed mixtures for pigs by in vitro analyses. *Anim. Feed Sci. Technol.* **1995**, *51*, 29–43. [CrossRef]

48. Saleh, F.; Ohtsuka, A.; Tanaka, T.; Hayashi, K. Effect of enzymes of microbial origin on in vitro digestibilities of dry matter and crude protein in maize. *J. Poult. Sci.* **2003**, *40*, 274–281. [CrossRef]

49. Saleh, F.; Ohtsuka, A.; Tanaka, T.; Hayashi, K. Carbohydrases are digested by proteases present in enzyme preparations during in vitro digestion. *J. Poult. Sci.* **2004**, *41*, 229–235. [CrossRef]

50. Tahir, M.; Saleh, F.; Ohtsuka, A.; Hayashi, K. An effective combination of carbohydrases that enables reduction of dietary protein in broilers: Importance of hemicellulase. *Poult. Sci.* **2008**, *87*, 713–718. [CrossRef]

51. Robbins, R.C. Effect of ratio of enzymes to substrate on amino acid patterns released from proteins in vitro. *Int. J. Vitam. Nutr. Res.* **1978**, *48*, 44–53.

52. Mauron, J.; Mottu, F.; Bujard, E.; Egli, R.H. The availability of lysine, methionine and tryptophan in condensed milk and milk powder. In vitro digestion studies. *Arch. Biochem. Biophys.* **1955**, *59*, 433–451. [CrossRef]

53. Steinhart, H.; Kirchgessner, M. In vitro digestion apparatus for the enzymatic hydrolysis of proteins. *Arch. Tierernahr.* **1973**, *23*, 449–459. [CrossRef]

54. Savoie, L.; Gauthier, S.F. Dialysis cell for the in vitro measurement of protein digestibility. *J. Food Sci.* **1986**, *51*, 494–498. [CrossRef]

55. Moyano, F.J.; Savoie, L. Comparison of in vitro systems of protein digestion using either mammal or fish proteolytic enzymes. *Comp. Biochem. Physiol. A Mol. Integr. Physiol.* **2001**, *128*, 359–368. [CrossRef]

56. Siddhuraju, P.; Becker, K. Nutritional and antinutritional composition, in vitro amino acid availability, starch digestibility and predicted glycemic index of differentially processed mucuna beans (*Mucuna pruriens* var. utilis): An under-utilised legume. *Food Chem.* **2005**, *91*, 275–286. [CrossRef]

57. Sáenz de Rodrigáñez, M.A.; Gander, B.; Alaiz, M.; Moyano, F.J. Physico-chemical characterization and in vitro digestibility of commercial feeds used in weaning of marine fish. *Aquacult. Nutr.* **2011**, *17*, 429–440. [CrossRef]

58. Savoie, L.; Galibois, I.; Parent, G.; Charbonneau, R. Sequential release of amino acids and peptides during in vitro digestion of casein and rapeseed proteins. *Nutr. Res.* **1988**, *8*, 1319–1326. [CrossRef]

59. Bryan, D.D.S.L.; Abbott, D.A.; Classen, H.L. Development of an in vitro protein digestibility assay mimicking the chicken digestive tract. *Anim. Nutr.* **2018**, *4*, 401–409. [CrossRef]

60. Appel, W. Chymotrypsin: Molecular and catalytic properties. *Clin. Biochem.* **1986**, *19*, 317–322. [CrossRef]

61. Straumfjord, J.V.; Hummel, J.P. Collagen digestion by dog pancreatic juice. *Exp. Biol. Med.* **1957**, *95*, 141–144. [CrossRef]

62. Becker, P.M.; Yu, P. What makes protein indigestible from tissue-related, cellular, and molecular aspects? *Mol. Nutr. Food Res.* **2013**, *57*, 1695–1707. [CrossRef]

63. Theander, O.; Westerlund, E.; Åman, P.; Graham, H. Plant cell walls and monogastric diets. *Anim. Feed Sci. Technol.* **1989**, *23*, 205–225. [CrossRef]

64. Meng, X.; Slominski, B.; Nyachoti, C.; Campbell, L.; Guenter, W. Degradation of cell wall polysaccharides by combinations of carbohydrase enzymes and their effect on nutrient utilization and broiler chicken performance. *Poult. Sci.* **2005**, *84*, 37–47. [CrossRef] [PubMed]

65. Gauthier, S.F.; Vachon, C.; Savoie, L. Enzymatic conditions of an in vitro method to study protein digestion. *J. Food Sci.* **1986**, *51*, 960–964. [CrossRef]

66. Bones, A.M.; Rossiter, J.T. The myrosinase-glucosinolate system, its organisation and biochemistry. *Physiol. Plant.* **1996**, *97*, 194–208. [CrossRef]

67. Chen, S.; Andreasson, E. Update on glucosinolate metabolism and transport. *Plant Physiol. Biochem.* **2001**, *39*, 743–758. [CrossRef]

68. Shahidi, F.; Naczk, M. An overview of the phenolics of canola and rapeseed: Chemical, sensory and nutritional significance. *J. Am. Oil Chem. Soc.* **1992**, *69*, 917–924. [CrossRef]

69. Campbell, L.D.; Smith, T.K. Responses of growing chickens to high dietary contents of rapeseed meal. *Brit. Poult. Sci.* **1979**, *20*, 231–237. [CrossRef]

70. Kozlowska, H.; Naczk, M.; Shahidi, F.; Zadernowski, R. Phenolic acids and tannins in rapeseed and canola. In *Canola and Rapeseed*; Shahidi, F., Ed.; Springer: Boston, MA, USA, 1990; pp. 193–210, ISBN 978-1-4613-6744-4.

71. Mangan, J.L. Nutritional effects of tannins in animal feeds. *Nutr. Res. Rev.* **1988**, *1*, 209–231. [CrossRef]

72. Sarwar Gilani, G.; Wu Xiao, C.; Cockell, K.A. Impact of antinutritional factors in food proteins on the digestibility of protein and the bioavailability of amino acids and on protein quality. *Brit. J. Nutr.* **2012**, *108*, S315–S332. [CrossRef]

73. Elkin, R.G.; Freed, M.B.; Hamaker, B.R.; Zhang, Y.; Parsons, C.M. Condensed tannins are only partially responsible for variations in nutrient digestibilities of sorghum grain cultivars. *J. Agric. Food Chem.* **1996**, *44*, 848–853. [CrossRef]

74. Francis, G.; Makkar, H.P.S.; Becker, K. Antinutritional factors present in plant-derived alternate fish feed ingredients and their effects in fish. *Aquaculture* **2001**, *199*, 197–227. [CrossRef]

75. Friedman, M.; Brandon, D.L. Nutritional and health benefits of soy proteins. *J. Agric. Food Chem.* **2001**, *49*, 1069–1086. [CrossRef] [PubMed]

76. Applegarth, A.; Furuta, F.; Lepkovsky, S. Response of the Chicken Pancreas to Raw Soybeans: Morphologic Responses, Gross and Microscopic, of the Pancreases of Chickens on Raw and Heated Soybean Diets. *Poult. Sci.* **1964**, *43*, 733–739. [CrossRef]

77. Nelson, T.S.; Ferrara, L.W.; Storer, N.L. Phytate phosphorus content of feed ingredients derived from plants. *Poult. Sci.* **1968**, *47*, 1372–1374. [CrossRef] [PubMed]

78. O'Dell, B.L.; De Boland, A.R.; Koirtyohann, S.R. Distribution of phytate and nutritionally important elements among the morphological components of cereal grains. *J. Agric. Food Chem.* **1972**, *20*, 718–723. [CrossRef]

79. Urbano, G.; López-Jurado, M.; Aranda, P.; Vidal-Valverde, C.; Tenorio, E.; Porres, J. The role of phytic acid in legumes: Antinutrient or beneficial function? *J. Physiol. Biochem.* **2000**, *56*, 283–294. [CrossRef]

80. Selle, P.H.; Ravindran, V.; Caldwell, A.; Bryden, W.L. Phytate and phytase: Consequences for protein utilisation. *Nutr. Res. Rev.* **2000**, *13*, 255–278. [CrossRef]

81. Li, Z.; Alli, I.; Kermasha, S. In-vitro α-amylase inhibitor activity-phytate relationships in proteins from Phaseolus beans. *Food Res. Int.* **1993**, *26*, 195–201. [CrossRef]

82. Maenz, D.; Classen, H. Phytase activity in the small intestinal brush border membrane of the chicken. *Poult. Sci.* **1998**, *77*, 557–563. [CrossRef]

83. Adeola, O.; Cowieson, A.J. Board-invited review: Opportunities and challenges in using exogenous enzymes to improve nonruminant animal production. *J. Anim. Sci.* **2011**, *89*, 3189–3218. [CrossRef]

84. Doiron, K.; Yu, P.; McKinnon, J.J.; Christensen, D.A. Heat-induced protein structure and subfractions in relation to protein degradation kinetics and intestinal availability in dairy cattle. *J. Dairy Sci.* **2009**, *92*, 3319–3330. [CrossRef]

85. Ljøkjel, K.; Harstad, O.M.; Skrede, A. Effect of heat treatment of soybean meal and fish meal on amino acid digestibility in mink and dairy cows. *Anim. Feed Sci. Technol.* **2000**, *84*, 83–95. [CrossRef]

86. Parsons, C.M.; Hashimoto, K.; Wedekind, K.J.; Han, Y.; Baker, D.H. Effect of overprocessing on availability of amino acids and energy in soybean meal. *Poult. Sci.* **1992**, *71*, 133–140. [CrossRef]

87. Newkirk, R.; Classen, H. The effects of toasting canola meal on body weight, feed conversion efficiency, and mortality in broiler chickens. *Poult. Sci.* **2002**, *81*, 815–825. [CrossRef] [PubMed]

88. Mauron, J. The Maillard reaction in food; a critical review from the nutritional standpoint. *Prog. Food Nutr. Sci.* **1981**, *5*, 5–35. [PubMed]

89. Hodge, J.E. Dehydrated foods, chemistry of browning reactions in model systems. *J. Agri. Food Chem.* **1953**, *1*, 928–943. [CrossRef]

90. Benzing-Purdie, L.M.; Ripmeester, J.A.; Ratcliffe, C.I. Effects of temperature on maillard reaction products. *J. Agric. Food Chem.* **1985**, *33*, 31–33. [CrossRef]

91. Schroeder, L.J.; Iacobellis, M.; Smith, A.H. Influence of heat on the digestibility of meat proteins. *J. Nutr.* **1961**, *73*, 143–150. [CrossRef]

92. Labuza, T.P.; Ragnarsson, J.O. Kinetic History Effect on Lipid Oxidation of Methyl Linoleate in a Model System. *J. Food Sci.* **1985**, *50*, 145–147. [CrossRef]

93. Bryan, D.D.S.L. Characterization of Protein Sources and Their Effects on Broiler Performance, Digestive Tract Morphology and Caecal Fermentation Metabolites. Ph.D. Thesis, University of Saskatchewan, Saskatoon, SK, Canada, 2018.

94. Bryan, D.D.S.L.; Abbott, D.A.; Van Kessel, A.G.; Classen, H.L. In vivo digestion characteristics of protein sources fed to broilers. *Poult. Sci.* **2019**, *98*, 3313–3325. [CrossRef]

95. Bryan, D.D.S.L.; Abbott, D.A.; Classen, H.L. Digestion kinetics of protein sources determined using an in vitro chicken model. *Anim. Feed Sci. Technol.* **2019**, *248*, 106–113. [CrossRef]

Review

In Vitro Techniques Using the Daisy[II] Incubator for the Assessment of Digestibility: A Review

Sonia Tassone [1,*], Riccardo Fortina [1] and Pier Giorgio Peiretti [2]

[1] Department of Agriculture, Forestry, and Food Sciences, University of Turin, 10095 Grugliasco, Italy; riccardo.fortina@unito.it

[2] Institute of Sciences of Food Production, National Research Council, 10095 Grugliasco, Italy; piergiorgio.peiretti@ispa.cnr.it

* Correspondence: sonia.tassone@unito.it; Tel.: +39-11-6708904

Received: 13 April 2020; Accepted: 22 April 2020; Published: 28 April 2020

Simple Summary: The Ankom Daisy[II] incubator (AD[II]; Ankom Technology Corporation Fairport, NY, USA) has gained acceptance as an alternative to traditional in vitro procedures. It reduces the labour requirement and increases the number of determinations that can be completed by a single operator. The apparatus allows for the simultaneous incubation of several feedstuffs in sealed polyester bags in the same incubation vessel, which is rotated continuously at 39.5 °C. With this method, the material that disappears from the bag during incubation is considered digestible. The method, which was first developed to predict the digestibility of feedstuffs for ruminants, has been modified and adapted to improve its accuracy and prediction capacity. Modifications used by various researchers include the use of different inocula, buffer solutions, and sample weights. Recently, attempts have been made to adapt the method to determine nutrient digestibility of feedstuff in non-ruminant animals, including pets.

Abstract: This review summarises the use of the Ankom Daisy[II] incubator (AD[II]; Ankom Technology Corporation Fairport, NY, USA), as presented in studies on digestibility, and its extension to other species apart from ruminants, from its introduction until today. This technique has been modified and adapted to allow for different types of investigations to be conducted. Researchers have studied and tested different procedures, and the main sources of variation have been found to be: the inoculum source, sample size, sample preparation, and bag type. In vitro digestibility methods, applied to the AD[II] incubator, have been reviewed, the precision and accuracy of the method using the AD[II] incubator have been dealt with, and comparisons with other methods have been made. Moreover, some hypotheses on the possible evolutions of this technology in non-ruminants, including pets, have been described. To date, there are no standardised protocols for the collection, storage, and transportation of rumen fluid or faeces. There is also still a need to standardise the procedures for washing the bags after digestion. Moreover, some performance metrics of the instrument (such as the reliability of the rotation mechanism of the jars) still require improvement.

Keywords: in vitro digestibility; inoculum; rumen fluid; faeces; enzyme; Ankom Daisy[II] incubator

---

1. Introduction

The in vitro digestion method was first developed as an alternative to the costly, labour-intensive, time consuming, and ethically difficult in vivo method to predict nutrient digestibility in ruminants.

The first method, described by Tilley and Terry [1] as a two-stage rumen fluid–pepsin technique (TT), provided satisfactory estimates of in vivo apparent digestibility [2], although some authors found that the TT was just accurate for fresh grasses and not for silages or straw [3–5]. Van Soest et al. [6] (VS) and Goering and Van Soest [7] (GVS) modified the TT by replacing the acid–pepsin step with a neutral detergent digestion step; this version of the method is faster and more accurate than the original TT, and it is able to estimate the in vitro true digestibility of feedstuffs on the basis of the undigested cell-wall constituents.

In an attempt to overcome problems related to the variability of the rumen fluid [8], Czerkawski and Breckenridge [9] developed a continuous-culture system using an apparatus described by Gray et al. [10] and by Aafjes and Nijhof [11] as a starting point: the "RUmen SImulation TEChnique" (Rusitec), which is still successfully used to generate inocula for in vitro studies [12–14].

Other in vitro methods have been developed to estimate the digestibility of feedstuff. Menke and Steingass [15] proposed to measure the gas produced during fermentation and feed composition data to estimate the energy content of feeds. Theodorou et al. [16], considering previous studies [17,18], developed an in vitro method to measure the accumulation of head-space gas; this method was then revised by other authors, who used computerised pressure sensors to monitor the gaseous products of the microbial metabolism and found a clear linear relationship between the disappearance of neutral detergent fibre (NDF) and the production of gas [19,20].

The need for a piece of apparatus that would be capable of automating traditional in vitro digestibility analysis and resolving some analytical errors such as those pertaining to sample handling and manual filtration steps led to the development of the Ankom Daisy^{II} incubator (AD^{II}; Ankom Technology Corporation Fairport, NY, USA).

This review summarises the use of the AD^{II} incubator—from its introduction until today—in digestibility studies on ruminants, compares and correlates it with other digestibility procedures, and discusses the sources of variability of the results and the extension of this technology to other non-ruminant species. Finally, some hypotheses on the future evolution and development of this technology and on the standardisation of the procedure are presented

2. The Ankom Daisy^{II} Incubator

The AD^{II} incubator started out as a project for a Canadian customer and was introduced to the public in 1994 as a wooden and somewhat fragile cabinet [21]. In 1997, a new model was made with a more resistant metal cabinet, exactly as in the currently marketed form (Figure 1). AD^{II} is essentially based on the in vivo simulation of digestion. With this device, it is possible to simultaneously analyse up to 92 samples in a thermostatically controlled chamber that contains four rotating digestion jars. The temperature inside the chamber is maintained at 39 ± 0.5 °C by a heat controller; a timer allows each incubation period to be set. Samples are weighed in F57 filter bags (25 μm pore size) (Ankom Technology Corporation Fairport, NY, USA) and put into the jars (up to 23/jar) together with the inoculum (rumen fluid, faeces, or enzymes) and a buffer solution. Each of the four glass jars, placed on the rotation racks inside the incubator, contains a perforated agitator baffle that divides the internal volume into two parts and allows for the free movement of the digestion medium. The bags are weighed before and after a specific period of incubation, and the material that has disappeared is considered digestible dry matter. The AD^{II} incubator offers advantages in time, efficiency and labour requirements over conventional methods, such as the Tilley and Terry method and the Van Soest method. Because of its design, the AD^{II} design is capable of testing a large number of samples [22–24]. It has been identified as an easy, inexpensive, and efficient instrument for the prediction of the digestibility of several feedstuffs and diets [8,22]. However, compared to other techniques (such as the batch culture technique, the use of the Ankom Gas Production System or the Rumen Simulation Technique), the AD^{II} incubator has been demonstrated to give higher values at different incubation times [25].

One application of the AD^{II} incubator is the estimation of neutral detergent fibre digestibility (NDFD) at single time points (such as 30 or 48 h) [26].

Attempts to address the variability of results have involved the assessment of the vessel type and the sealing, venting, and gassing procedures [27]; the comparisons of different types of fibre-bag and the use of sodium sulphite for long incubation periods [28]; the development of specific in vitro methods to determine indigested NDF and to estimate the individual pool sizes and rates of digestion for application for diet formulation purposes [29]; the evaluation of the storage times and temperatures of rumen fluid before its transfer to the incubation flask [30]; the effects of the priming techniques of rumen fluid [31,32]; comparisons with in situ and various in vitro methods [33,34]; and the quantification of two pools of digestible NDF (fast and slowly digested) with a minimal number of fermentation time points [35].

Figure 1. The Daisy^{II} incubator (Ankom Technology Corporation Fairport, New York, NY, USA).

Recently, the AD^{II} incubator has been used for the in vitro long-term ruminal digestion (240 h) of undigested NDF (uNDF) [28]. To estimate the kinetics of NDF degradation, longer time intervals are essential, especially when using complex models. Complex models may require inputs of fast, slow, and indigestible NDF pools [35,36], which can be determined with ease when using the AD^{II} incubator.

A list of practical recommendations on the use of AD^{II} incubator and a list of the main problems concerning the use of the instrument that require further study are reported in Table S1.

3. Inoculum Applied to the Use of the Ankom Daisy^{II} Incubator

Inoculum is very important for in vitro fermentation studies, but it also represents the greatest source of uncontrolled variation in fermentation systems. The inoculum has to create a similar environment to that of the digestive tract [37], but its digestive capacity may be influenced by the animal species, breed, and individual and within animal variations from time to time [38]. The characteristics and quality of the inoculum is not a specific problem of the AD^{II} incubator. As there is a lack of specific information on the AD^{II} incubator and some authors have studied inoculum for in vitro analysis, we reported their experience with other digestibility systems, because this information may also be useful for Ankom Daisy^{II}.

3.1. Rumen Fluid

As for other systems, the most frequently used inoculum source in the AD^{II} incubator is rumen fluid (RF). The necessity of fistulated and cannulated animals to provide this inoculum raises a number of practical problems, e.g., the need for surgical facilities, constant care to avoid infections, and the costs associated with the long-term maintenance of these animals. Moreover, the use of cannulated animals for this purpose has been criticised on ethical grounds.

Different solutions, in which there is no need to use cannulated animals, have been studied to resolve cost issues and ethical concerns about the well-being of animals. RF can be obtained via the oesophagus, thereby avoiding the need for cannulation, but such samples are often contaminated with saliva, and their collection causes considerable stress to the host animal. Moreover, as a result of the placement of the sampling device, the samples may not be representative of the entire rumen contents [37]. A very different approach with more details on this matter can be found in a paper by Ramos-Morales et al. [39]. These authors assessed in vivo trials conducted with ruminally cannulated sheep and goats to validate the use of stomach probing as an alternative to rumen cannulation in small ruminants with the aim of detecting any differences in ruminal fermentation and in the microbial community between species, diets, and sampling times.

A more ethically acceptable approach that reduces stress and alleviates the suffering of animals by avoiding an invasive procedure is the collection of RF at slaughtering [40]. Alba et al. [41] verified, through the use of an AD^{II} incubator, that the rumen inoculum obtained from slaughtered cattle can be used to replace the use of cannulated animals and that this approach is a viable alternative to digestibility analysis.

This method is accepted by the Rumen Microbial Genomics Network [42] for microbiota studies and has been mentioned as an alternative to sampling via cannula [43].

A supplemental video of the sampling procedures of RF at slaughtering is available online [44]. These procedures involve the collection of the rumen content into plastic bags a few minutes after slaughtering; the rumen content is squeezed, and the RF is filtered and collected into pre-heated plastic bottles. The presence of oxygen is avoided by squeezing the bottles while closing them; the rumen fluid is transported to the lab (max. 1.5 h time) at a temperature of 39–42 °C.

The effects of the source of inoculum with various combinations of donor cow diets generally vary to a great extent [45]. The results of a trial conducted by Holden et al. [22] showed that the source of inoculum affected in vitro dry matter digestibility (DMD). A grass hay donor cow diet resulted in lower digestibility values than a corn silage-based, total mixed ration donor cow diet for alfalfa hay, grass hay, steam flaked corn, and dry ground corn. No influence of the donor diet was found for mixed haylage, corn silage, grain mixture, or high moisture shelled corn. King and Plaizier [46] found that the source of inoculum (steers or cows) did not affect apparent or true DMD to any great extent. They also found that forage digestibility was similar when using the RF from sheep and from cattle [23]. Ammar et al. [47], using an AD^{II} device, found that the RF of sheep and goats was similar under the conditions of the experiment when all the donor animals were fed the same diet and were maintained under the same conditions.

Robinson et al. [30] examined the influence of storage time and temperature on the ability of rumen microorganism to degrade NDF. They reported that within-day delays of up to 6.5 h between the time of collection of rumen inoculum and the time of the initiation of the in vitro incubation had no impact on the measured 48 h digestion of NDF if the RF was maintained at 39 °C under anaerobic conditions during the delay. Similarly, the RF of sheep, preserved for up to 6 h in crushed ice, had no effect on any fermentation parameters [48]. Another possible RF storage system for in vitro incubation is short-term refrigeration [41]. Chaudhry and Mohamed [49] tested thawed RF from frozen rumen contents (stored at −20 °C for 4 w) against fresh RF from the same slaughtered cattle. Though the thawed RF had a lower degradation than the fresh one, it could be used to predict in vitro digestibility, as the values were closely correlated ($R^2 = 0.95$). However, it was still necessary to test its suitability for routine use. Hervas et al. [48] instead found a reduction in fermentative activity as a result of freezing (24 h). Spanghero et al. [14] recently compared inoculum collected at slaughtering with RF samples obtained from a continuous fermenter that were fresh, refrigerated at 4 °C, chilled at −80 °C, and freeze-dried. They evaluated the fermentability by measuring the NDF, crude protein degradability, and gas production. They confirmed that short-term refrigeration is a valuable technique to manage RF, whereas methods based on low temperatures significantly reduce the *Fibrobacter succinogenes*, which are very important for fibre degradation. Denek et al. [50] studied the preservation of microorganisms with a cryoprotectant under different deep-frozen conditions. They showed that RF treated with 5% dimethyl sulphoxide and frozen in liquid nitrogen gave similar results to fresh RF, but they also showed that the incubation time needed to be increased to 72 h to measure the digestibility of roughages. Belanche et al. [51] assessed the relevance of different factors (the diet of the donor animal, the fermentation substrate, microbial fraction, and the inoculum preservation method) to maximize the rumen inoculum activity, and they found that the highest microbial numbers and in vitro fermentation rates were recorded for fresh RF sampled after 3 h from donor animals fed a high concentrate diet.

As far as the microbial population that develops in an AD^{II} incubator is concerned, Soto et al. [52] showed the variations such a population underwent during the incubation process, and they compared the results with those of a Wheaton bottle and a single-flow continuous-culture fermenters using the same goat RF. In an AD^{II} incubator, they monitored the different microbial groups (bacteria, archaea, fungi, and protozoa) for 48 h by means of real time-PCR and terminal-restriction fragment length polymorphism. They observed a general decrease in the microbial population and important changes in microbiota profile, as the methanogens population increased. A similar trend was observed for the Wheaton bottle at 72 h, but there was also a growth of fibrolytic bacteria. However, the continuous-culture fermenters kept the rumen microbiota similar to that sampled from the rumen.

Spanghero et al. [14] found that the fermentation liquid from rumen continuous-fermenters can be used to generate inoculum for in vitro purposes.

Problems can arise for microorganisms, regarding the preparation of inoculum [37], connected with feed particles, the use of multiple layers of cheese cloth, and/or the use of some physical methods (e.g., the Stomacher method or the maceration of the rumen content in a food processor), which may destroy cell integrity.

3.2. Faecal Inocula

Fresh faeces (FF) have been used as an alternative source of ruminal inoculum in many experiments [41]. All these studies have demonstrated that bovine faeces may be used as microbial inocula for in vitro digestion and gas production, but this use has some limitations, such as a lower enzymatic activity than RF [53–55]. According to Akhter et al. [56], cattle faeces could also be used as an alternative to sheep RF.

Tufarelli et al. [57] tested faecal samples of yaks (*Bos grunniens*) as an alternative microbial inoculum source and compared them with RF, which was used as a control. They found that a faecal extract could be utilised instead of RF to estimate in vitro digestibility and that an AD^{II} incubator, with faecal liquor, is able to simply assess the adaptation capability of ruminant species to a pasture. These results were confirmed using camel faeces as a source of inoculum for AD^{II} [58].

Bovine FF may be used to replace bovine RF for incubation times no lower than 48 h [59]. Chiaravalli et al. [60] utilised an AD^{II} incubator to estimate the undigestible NDF of seven substrates using three different inocula (one rumen and two faeces) and considering two incubation times (240 and 360 h). The undigestible NDF results showed that faecal inoculum could be used to replace RF for long incubation times and that faeces can be used as an inoculum for end-point measurements.

The diet of an animal can change its microbial population. Guzmán and Sager [61] compared the microbial inoculum collected from a rumen-fistulated Aberdeen Angus steer fed with alfalfa hay and then with low quality digit hay (*Digitaria eriantha*), as well as the faeces collected from the same animal to evaluate the substrate, inoculum, and digestibility interaction. Using both inoculum sources, the true DMD was found to be affected by the diet of the donor animal, and the RF values ranked higher in the runs. Moreover, Kim et al. [62] suggested considering the diet, because it has an important effect on faecal microbiota, in particular when a forage-based diet is compared with a concentrate.

Faeces have also been extensively used as inoculum for in vitro incubation trials on monogastrics. Lowman et al. [63] were the first to demonstrate that equine faeces can be used as a source of microbial inoculum and that the faecal microflora of equines can remain viable for several hours after excretion. Other authors have confirmed these results. Earing et al. [64] demonstrated that the in vitro methodologies developed for the AD^{II} incubator could produce accurate estimates of in vivo equine apparent DMD and NDFD when equine faeces were used as the inoculum source. They evaluated three incubation periods in their study: 30, 48, and 72 h. Though the 30 and 48 h in vitro estimates were consistently less accurate than the in vivo estimates, they ranked diets in the same order as the in vivo method, and the 72-h period provided the most similar digestibility estimates to the in vivo data. Tassone et al. [65] evaluated the use of the AD^{II} incubator for the apparent and true DMD and NDFD measurements of feedstuffs considering four incubation times (30, 48, 60, and 72 h) using donkey faeces as a source of microbial inoculum. All the digestibility parameters increased significantly after 30–72 h of incubation, with average coefficients of variation for repeatability and reproducibility of 3.4% and 7.3% for apparent DMD; 1.7% and 4.3% for true DMD; and 6.6% and 14.6% for NDFD, respectively.

Table 1 summarises the references pertaining to rumen fluid and fresh faeces inocula applied to the AD^{II} incubator.

Table 1. Rumen fluid (RF) and fresh faeces (FF) inocula applied to the Ankom Daisy^{II} incubator (AD^{II}).

Inoculum	Species	Sample Type	Notes	Ref.
RF	Dairy cattle	10 feeds	Variability of the dry matter digestibility for different donor cow diets as sources of inoculum	[22]
RF	Dairy cattle	By-products	AD^{II} vs. gas production, RF from slaughtered or cannulated cows	[41]
RF	Steers and Dairy cattle	Grains, total mixed ration, silages	Effect of the RF on the apparent and true dry matter digestibility DMD	[46]
RF	Sheep and Goats	Leaves, flowers and fruits of 5 browse plant species	Comparison of the true DMD and gas production kinetics with RF from animals fed the same diet	[47]
FF	Yaks	Forage produced at high altitude	Faeces vs. RF for a comparative digestibility trial	[57]
FF	Sheep and Camels	Fodder species from an arid environment	RF from sheep, and faeces from camels: comparative digestibility trial	[58]
FF-RF	Cattle	Feeds with different neutral detergent fibre (NDF) contents	NDF digestibility and undigested NDF measured with RF and 2 FF from cows fed different diets	[60]
FF-RF	Steers	35-day regrowth alfalfa hay	Comparative evaluation of the true dry matter digestibility; steers fed alfalfa or digit grass	[61]
FF	Horses	4 dietary treatments (hays or hay + oat)	Comparative evaluation of in vivo vs. in vitro DM and NDF digestibility	[64]
FF	Donkeys	7 common feeds for donkeys	Evaluation of the apparent and true DMD and neutral detergent fibre digestibility (NDFD) at 4 incubation times (30, 48, 60, and 72 h)	[65]

3.3. Enzymatic Inoculum

Enzymatic methodologies, in which microbial inoculum is eliminated, were developed to avoid problems associated with variations in rumen fluid over time [37]. This approach can be recommended because it offers an improved standardisation of the methodology, a reduction in the variations that may be attributed to the inoculum source and preparation, and a reduced dependence on surgically modified animals as rumen fluid donors [66]. However, the attempt to use enzymes instead of rumen fluid or other inocula have resulted in problems of variability in their preparation [67], and very little work has been done to optimise enzyme activities or incubation conditions. Though there are no available studies on ruminant digestibility in which enzymes were used in an AD^{II} incubator, many authors have already used enzymes in digestibility studies on pigs [68], rabbits [69,70], and dogs [71].

4. Sample Size, Sample Weight and Bag Type

The sample bags in the AD^{II} incubator constantly rotate in jars (0.95 rpm), and the internal septum leads to the complete immersion of the bags at every spin of the jar; in this way, gases do not accumulate inside the bag, and samples are prevented from floating freely in the flask. The continuous shaking of samples produced significantly higher digestibility results than when shaking occurred only twice daily [72]. As reported by Alende et al. [25], the use of filter bags may be advantageous, because filtration and recovery have been mentioned as sources of variability of the digestibility coefficients. Additionally, jars positioned horizontally render a higher digestibility than vertically placed ones. Holden et al. [22] found no significant differences when grains and forages were incubated in the same digestion vessel.

The first and most extensively used AD^{II} incubator bag is the F57 bag. The F57 bag is made up of an extruded polyethylene fibre with a three-dimensional filtration matrix that facilitates the maximum flow of a solution, thereby obtaining the best substrate interaction and minimum particle loss. The F57 filter bag has an approximately 25 µm pore size, is 50 mm long and 50 mm wide at the open top, and tapers to a bottom width of 30 mm. Sample processing, particularly concerning the grind size, interacts with the pore size of the bag and affects the extent of feed disappearance [73]. The ratio of the sample size to the bag surface area, suggested by Vanzant et al. [74] to increase the accuracy of degradability predictions relative to in vivo ruminal disappearance, is 10 mg/cm^2.

In previous studies, sample sizes of both 0.25 g [28,30] and 0.5 g [33,34] were used in conjunction with Ankom procedures [75]. Coblentz and Akins [76] compared the NDF digestibility values of triticale forages determined with the AD^{II} device, and they considered two sample sizes (0.25 and 0.50 g) and incubation periods of 12, 24, 30, 48, 144, and 240 h. The results were compared with those obtained from a commercial laboratory that used a traditional methodology. With the 0.25 g sample size, the linear equations between the Ankom and the traditional methods did not show differences both 30 and 48 h. There was less agreement, particularly for the 30 h incubation, when a sample of 0.50 g sample was used. The NDF digestibility values were generally greater for the 0.25 g sample size when using the Ankom methods, especially for incubation times of 24, 30, and 48 h.

Cattani et al. [77] evaluated what sample size (0.25 or 0.50 g/bag) allowed for a better correlation to be achieved between the NDFD and true DMD values obtained with the AD^{II} and a conventional batch culture technique. The regressions between the mean values, provided for the various feeds by the two methods, for the NDF and true DMD, had R^2 values of 0.75 and of 0.92,and an RSD (relative standard deviation) of 10.9% and of 4.8%, respectively, for the 0.50 g/bag size. The corresponding regressions for NDFD and true DMD showed R^2 values of 0.94 and of 0.98 and an RSD of 3.0% and of 1.3%, respectively, for the 0.25 g/bag size. This screening analysis therefore indicated that the reduction of the sample size from 0.50 to 0.25 g of feed sample/bag (corresponding to 12 and 6 mg/cm^2 of bag surface), when using an AD^{II} device, allowed for more closely correlated and less variable estimates of NDFD and true DMD to be obtained than those provided by the batch culture technique.

A recent work that evaluated the rate kinetics of triticale forages considered 0.3 g samples sealed within fibre bags as a procedural compromise between the 0.25 g sample size recommended for short incubation times and the necessity of ensuring that an adequate amount of residue remained after a long digestion time (144 and 240 h) [78].

The critics of the Ankom bag method have indicated the potential loss of small indigestible particles through its pores and that any method should decrease the loss of small particles without restricting access to the protozoa and bacterial populations. Ankom recommends F58 for crude fibre, neutral, and acid detergent fibre analyses. A pore sizes of <10 µm can restrict the number of protozoa and bacteria that enter digestion bags, so a smaller bag pore size than that of F58 is not advisable. Wilman and Adesogan [23] verified that soluble matter from samples high in soluble substances is able to escape from F57, thereby influencing the microbial population and increasing cell wall degradation in any samples low in soluble substances that are in the same jar. Valentine et al. [28] compared Ankom F57 bags (25 µm) with F58 bags (8–10 µm pore size) to measure undigested NDF after 240 h of incubation and found that both had significant effects on lowering undegraded NDFom values. In conventional procedures, smaller pore size filters generally tend to have greater average undegraded NDFom values than methods with larger pore size filters. They expected a similar finding, because potentially undigested NDF may be retained by finer filters, whereas potentially indigestible and digestible NDF may inadvertently escape from a coarser filter. They found when using the same technique for in vitro analysis, that Ankom F57 and F58 gave similar digestion rate results.

Adesogan [79] tested alternative bags to Ankom F57. He determined the in vitro apparent dry matter digestibility of the feed samples in an AD^{II} incubator using Ankom F57 bags and dacron bags with pore sizes of 30 and 50 µm, with or without a 5 g glass ball placed in the bags to ensure submersion in the media. He obtained different digestibility estimates when the alternative bags were used instead of the F57 bags, but the Ankom bags gave a more precise prediction of conventionally measured digestibility estimates than the alternative bags. Using Ankom bags ensures more standardised and repeatable results. The characteristics of alternative bags should be disclosed whenever they are used, instead of F57 bags, to estimate digestibility. Anassori et al. [80] also used dacron bags (pore size of 50 µm) in an AD^{II} to measure the organic matter digestibility (OMD) of forage-based sheep diets supplemented with raw garlic, garlic oil, and monensin. They compared AD^{II} with the TT and gas production. The values obtained with the AD^{II} method were always higher than those obtained with the TT and (for diets containing garlic oil) with in vitro gas production methods. According to the authors, in the AD^{II} procedure, a proportion of non-digestible fine particles may have been removed during incubation, boiling, and rinsing, thus reducing the weight of the residue and increasing the estimate of digestibility compared to that obtained with other methods.

Table 2 summarises the references pertaining to the sample size and bag type applied to the AD^{II} incubator.

Table 2. Sample size and bag type applied to the Ankom Daisy^{II} incubator (AD^{II}).

Sample Size (g)	Bag Type	Sample Type	Notes	Ref.
0.25	F57	Forages and plant parts	Particle breakdown: 0.5, 1.0, and 1.5 mm	[23]
0.25	F57 and F58	Temperate and tropical grasses and legumes	uNDF after 240; effect of Na$_2$SO$_3$	[28]
0.25	Polyethylene polyester polymer bags	Low- and high-quality forages and grains	Different time delays and storage time between the collection of RF and the analysis	[30]
0.25	F57 and dacron bags (pore size: 0.30 and 0.50 µm)	Dried samples + 5 g glass balls	DMD of feed samples with alternatives to F57 and weighted to ensure submersion in the media	[79]
0.25	Dacron bags (pore size: 0.50 µm)	5 feeds + garlic or garlic oil vs. Monensin	Sheep RF, the effect of inclusion on organic matter digestibility (OMD)	[80]
0.30	F57	Triticale	Short and long (240 h) incubation times	[78]
0.50	5 × 3 cm pore size 0.45 µm	Pastures, forages and by-products	Comparison of in situ DM and NDF degradation kinetics	[33]
0.50	F0285 (pore size 0.25 µm)	Corn silage	Comparison of in vitro and in situ estimates of indigestible NDF at 2 fermentation end points (120 and 288 h)	[34]
0.25 0.50	F57	Triticale	Comparison of NDFD with 2 sample sizes	[76]
0.25 0.50	F57	7 feeds	Correlation with a conventional batch culture	[77]

5. Buffer Solutions and in Vitro Digestibility Methods Applied to the Ankom Daisy[II] Incubator

Many methods and buffer solutions that are used to study in vitro digestibility, first for ruminants and then for monogastrics, have also been applied to the AD[II] incubator.

A buffer solution (either phosphate, carbonate, or both) is used during incubation to control the pH and to supply nutrients for the inoculum microorganisms. Without a buffer, the short chain of fatty acids would lower the pH [81]. As authors have reported, only phosphate buffers do not require preparation under CO_2. The references of the different buffer solutions used for in vitro digestibility analysis are briefly reported in Table 3. However, a comparison of buffer solutions is still lacking. In 2000, Figueiredo et al. [72] compared buffers that had been described by Marten and Barnes [82] with those that had been described by Minson and McLeod [83], and the authors verified that the solutions could replace each other.

Table 3. Different buffer solutions used in in vitro digestibility trials with the Ankom Daisy[II] incubator (AD[II]) in different animal species.

Buffer Solution References	AD[II] References	Animal Species
[82]	[24]	Ruminants
	[61]	Ruminants
	[64]	Horses
	[65]	Donkeys
[83]	[72]	Ruminants
[84]	[85]	Ruminants
[86]	[25]	Ruminants

6. Precision and Accuracy of the Method Using the Ankom Daisy[II] Incubator

The utilisation and the diffusion of AD[II] to study in vitro digestibility is a result of the reliability and accuracy of the method.

Damiran et al. [87] found a coefficient of variation (CV) of 4.7% for DMD measured with AD[II] and a CV of 12.2% for NDFD. A CV of <1% was observed between sample replicates in other laboratories for the in vitro true digestibility values, but this coefficient normally ranged between 1–3% [21]. However, it is a little higher for NDFD analysis and typically ranges from 2.0–4.5%, depending on the type. Corn silage samples are always a little more variable. If any sample has a CV of over 5%, it should be re-analysed. Figueiredo et al. [72] verified a good reproducibility when measuring digestibility with AD[II]. They reported a low coefficient of variation (CV = 2.65%) between jars and within jars, with values of 3.92, 2.13., 6.12, and 1.94 for jar numbers of 1, 2, 3, and 4, respectively. Tagliapietra et al. [88], in situ and in vitro, studied the rumen fluid of 11 feeds collected by means of oro-ruminal suction from intact donor cows. The reproducibility coefficient of the DMD for AD[II] was 96.0%. The DMD values were underestimated when filter bags were considered, compared to in situ-nylon bags and in vitro conventional bottles. Nevertheless, it was possible to overcome the lower repeatability provided by the filter bags by increasing the number of replicates: three filter bags led to approximately the same standard error as the mean of 2.5 nylon bags and the mean of 2 conventional bottle measurements. The results showed a direct proportionality between the DMD values obtained in situ and in vitro with different techniques (in situ nylon vs. in vitro conventional bottles and in situ synthetic filter bags vs. AD[II]).

Spanghero et al. [89] studied the NDF degradability of 18 hays considering different incubation times (2, 4, 8, 16, 24, 48, and 72 h) and found that the variability (CV) of the AD[II] incubator (including jar repeatability) was 2.8%—that is, a similar value to that generally found for some chemical analyses of feedstuffs [90] and one that is lower than that obtained for in situ measurements (including low repeatability, CV: 3.7%).

Spanghero et al. [91] also evaluated the precision of the AD^{II} device in measuring the in vitro NDF degradability of 162 hay samples from permanent Austrian grasslands. The obtained results showed a within forage standard error of 2.8%. This limited repeatability of the measurement was attributed to various sources of variability (bag porosity, dimensions, amount of substrate, etc.), but not to the different jar positions in the fermenter, because the average values obtained after five incubations for the different jars were not statistically different.

Spanghero et al. [92] also investigated the precision and accuracy of the AD^{II} incubator for NDFD analysis and the accuracy and reproducibility of the associated calculated net energy of lactation. Five laboratories analysed 10 fibrous feed samples each; the fermentation times in the AD^{II} incubator were 30 and 48 h. The precision was measured as the standard deviation (SD) of the reproducibility (SR) and repeatability (Sr) of the between and within laboratory variability. Extending the fermentation time from 30 to 48 h increased the NDFD values (from 42% to 54%) and improved the NDFD precision, in terms of both Sr (12% and 7% for 30 and 48 h, respectively) and SR (17% and 10% for 30 and 48 h, respectively). The 48-h period of incubation improved the accuracy and reproducibility of the calculated net energy of lactation.

The accuracy and precision of NDFD, determined after short or long-time intervals, has recently been of considerable research and industry interest, as the relative consistency of the results.

Cişmileanu and Toma [93] studied the repeatability, reproducibility, and accuracy of an AD^{II} incubator using a new version of the TT. The stages of the method were similar to those of the traditional version: one stage with buffered rumen liquid and one stage with pepsin–HCl. An alfalfa hay sample was tested to establish the OMD by means of the in vivo method, and it was then considered as an internal control feed with a known digestibility. The authors observed that the coefficient of variability was 1.11% for repeatability and 1.85% for reproducibility. The accuracy was the same as that obtained with the conventional method.

Moreover, even if the AD^{II} incubator is fully functional, sometimes the jars do not rotate correctly and suffer from slowdowns, stops, and starts [94]. Some structural adjustments are therefore necessary to better exploit the potential of the AD^{II} incubator and to implement its diffusion and use.

Table 4 summarises the references pertaining to the precision and accuracy of the method using the Ankom Daisy^{II} incubator.

Table 4. Precision and accuracy of the method using the Ankom Daisy^{II} incubator (AD^{II}).

Parameters	Notes	Ref.
DMD and NDFD by means of the two-stage rumen fluid–pepsin technique (TT), the AD^{II} incubator and in situ; 0.25 and 0.50 sample size; 1 and 2 mm grinding size	The digestibility values estimated means of the by AD^{II} incubator and in situ techniques were correlated (R^2 = 0.58–0.88) with values estimated by means of conventional in vitro and in vivo techniques. In most cases, the AD^{II} incubator and in situ techniques overestimated DMD and NDFD	[87]
In vitro DMD vs. Minson and McLeod technique [83]	Good reproducibility between and within the jars in the AD^{II} incubator	[72]
In situ (2 different filter bags) and TDMD (traditional bottles or the AD^{II} incubator)	The AD^{II} incubator underestimated the TDMD values but there was direct proportionality between the in situ and in vitro DMD values	[88]
NDFD of 18 hays	The variability was similar to that of some chemical analysis and lower than the in situ measurements	[89]
NDFD of 162 hays	Similar average values	[91]
NDFD and the associated calculated net energy lactation (NEl) of 10 fibrous feeds; 5 laboratories	Improved NDFD precision and improved accuracy and reproducibility of the calculated NEl for an extended fermentation time (48 h)	[92]
Validation of a modified TT by achieved by testing the repeatability and reproducibility of the new TT as well as the correlation with a previous version of the method	Good repeatability and reproducibility achieved when using the new version of the TT with the AD^{II} incubator; the same accuracy was achieved as that of the conventional method	[93]

7. Comparison with other Methods

Many methods are available to measure in vitro digestibility, but only a few articles have compared the results obtained using an AD[II] incubator with the results of other procedures [25].

The first results on digestibility in ruminants obtained using an AD[II] incubator were presented by Komarek et al. [95] in 1994 at the National Conference on forage quality in Lincoln (USA) [96]. The following year, Ayangbile et al. [97] showed that there were no differences between DMD data obtained from an AD[II] incubator and data obtained by means of the conventional Tilley and Terry methods [1,7]. Traxler et al. [98] determined the true DMD on four forages for different incubation times (48, 72, and 144 h), and even though the conventional Van Soest method [6] was found to be more efficient, the results basically confirmed the conclusions of Ayangbile et al. [97].

Cohen et al. [99] incubated corn silage samples in tubes according to the GVS method [7] and in an AD[II] incubator at different times using unwashed F57 bags or F57 bags washed in acetone before being filled. The NDFD measured with the AD[II] incubator was lower than that in the tubes, probably because of the retention of gas and acid end products within the bags, and the values of the washed filter bags were similar to those obtained by shaking the tubes. Traxler [100] instead noted very few differences between the AD[II] incubator and the GVS method [7].

Over time, other studies have confirmed that the AD[II] incubator can be used to predict the DMD digestibility of forages, grains, and mixed rations for ruminants [7,22–24,26,73,87,101].

Ammar et al. [102] compared the TT and VS methods [6] using an AD[II] incubator for leguminous shrub species. The medium was prepared according to the VS method. After incubation in a buffered rumen fluid, samples were either subjected to a 48 h pepsin–HCl digestion (TT) or gently rinsed and extracted with a neutral detergent solution at 100 °C, as described in the VS method. The apparent digestibility was generally lower than the true digestibility, and the differences were always significant, particularly in leaves.

The same author [103] used the VS method applied to the Ankom technique [104] to obtain the in vitro digestibility of the stems and leaves of grasses and legumes taken from the first and subsequent cuts of a permanent meadow. In this experiment, rumen fluid was withdrawn from adult sheep.

Gargallo et al. [85] verified the use of an AD[II] incubator to determine the intestinal digestion of crude protein using Calsamiglia and Stern's three-step procedure (TSP) [84]. Four tests were conducted to study the effect of the type of pepsin, the type of bags, the amount of sample, and the number of bags per jar on the estimated intestinal digestion using the AD[II] incubator and the TSP techniques on soybean meal samples, heated at different temperatures, and with 12 protein supplements. The results showed that the intestinal digestion of soybean meal and the 12 protein supplements from the TSP and the AD[II] incubator (with R510) were closely correlated. The amount of sample per bag and the number of bags per jar did not affect the estimates, and up to 30 bags (Ankom R510) with 5 g of sample could be used in each jar of an AD[II] incubator to estimate the intestinal digestion of the proteins in ruminants.

In 2017, Cişmileanu and Toma [93] successfully validated a new version of the TT applied to AD[II], in which the stages of the traditional procedure were maintained. Two stages, the first one with buffered rumen liquid and the second with the pepsin–HCl solution, were considered.

Holden et al. [22] compared a modification of the TT and the AD[II] incubator techniques to determine DMD, considering sources of inoculum from two different donor cow diets, as well as all the forage and total mixed rations. Their results showed that the AD[II] incubator did not affect the digestibility values of the forages or grains to any great extent, as well as that the source of inoculum could affect DMD.

Wilman and Adesogan [23] compared the TT and an the AD[II] incubator to estimate apparent and true DMD, apparent and true OMD, and NDFD. The analysed forage samples comprised 72 combinations of two forage species (*Lolium multiflorum* and *Medicago sativa*), three plant parts, three degrees of particle breakdown, two field replicates with rumen fluid from sheep, and two field replicates with rumen fluid from cattle. It was found that the sieve size used when milling did not influence the true OMD. However, small differences were observed between the two forage species: the standard errors and coefficients of variation were higher for the AD^{II} incubator (mean: 4.0%) than for the TT (mean: 2.7%). When they used the TT, they found it was possible to more precisely predict the true digestibility than the apparent digestibility from the AD^{II} incubator results; the difference between apparent and true digestibility, when estimated using the AD^{II} incubator, appeared unrealistically low. The estimated digestibility was similar when rumen fluid from sheep and from cattle was used. In conclusion, the TT gives more precise results than the AD^{II} incubator, albeit at the cost of requiring more labour. Mabjeesh et al. [73] performed the same comparison (AD^{II} vs. TT) on 17 concentrates and protein supplements, and they obtained a satisfactory relationship (R^2 = 0.81), even though the AD^{II} incubator gave higher values for some energy concentrate and protein supplements.

Ricci et al. [105] compared the precision and accuracy of in vitro ruminal DM degradability using the TT, an AD^{II} incubator, and the gas-production technique to estimate the in vivo DM digestibility of tall wheatgrass, hay, and haylage. The goodness-of-fit of all the techniques with the in vivo DM digestibility and the relationships between them were evaluated by means of a simple linear regression analysis. The Pearson correlation coefficient (ρ) was used to evaluate the strength of the association between the observed and in vitro estimated data. The concordance correlation coefficient (ρc) was used as a single indicator to integrate both precision and accuracy (Cb). This indicator (scaled between 0 and 1) is a reproducibility index that evaluates the agreement between two sets of data by measuring the shift in location from the concordance line (the 45° line through the origin) in the observed versus predicted plot. Cb is a bias correction factor that indicates how far the best fit line deviates from the concordance line. Linear relationships were observed between the in vivo and the TT, AD^{II}, and gas production values. The TT had the highest correlation (0.98), and this was followed by the gas-production technique (0.97) and then by AD^{II} (0.96). However, the TT exhibited the lowest accuracy (ρc = 0.341), and AD^{II} exhibited the highest (ρc = 0.850). The regression analysis showed an overestimation of the in vivo dry matter digestibility above 48.8% for AD^{II} and an underestimation below this value. AD^{II} is faster and more accurate than the other techniques, and it therefore appears to be the most suitable for in vitro digestion trials. Figueiredo et al. [72] compared the AD^{II} technique with Minson and McLeod's technique [83], (they modified the TT in 1972) and found higher values when they used the AD^{II} procedure.

Some authors have conducted comparison between an AD^{II} incubator and in situ system. Robinson et al. [30] reported higher NDFD values at 48 h with an AD^{II} incubator. Spanghero et al. [92] showed that the results of an AD^{II} incubator were closely correlated with the results of an in situ method (R^2 = 0.98). Spanghero et al. [89] compared the NDF degradability of 18 hays, measured by means of an in situ method (nylon bag technique) and the AD^{II} incubator. The incubation times were 2, 4, 8, 16, 24, 48, and 72 h. The NDFD values obtained in situ and in vitro with the AD^{II} incubator after 48 h of incubation were closely correlated (R^2 = 0.94). In another study [91], they verified that the NDF degradability of 162 hay samples measured in an AD^{II} incubator was 25–30% higher than the effective in situ values. The regression analysis between the in vitro and in situ NDFD values showed a medium degree of correlation and a low level of accuracy.

Tagliapietra et al. [88] compared four in situ methods with nylon bags and filter bags, as well as in vitro with conventional individual bottles or AD^{II}, to measure the DMD of 11 feeds. The reproducibility coefficients of the dry matter digestibility were 97.9%, 95.1%, 98.8%, and 96.0% for the in situ-nylon, filter bags, conventional bottles, and AD^{II}, respectively. The in situ and in vitro filter bags underestimated the dry matter digestibility values compared to the in situ-nylon bags and conventional bottles. They concluded that in vitro estimates of dry matter digestibility at 48 h with AD^{II}, using rumen fluid collected from intact cows, can produce similar values to those obtained in situ. The filter bags underestimated the dry matter digestibility values compared to the in situ-nylon bags and conventional bottles. However, it was possible to overcome the lower repeatability provided by the filter bags by increasing the number of replicates: three filter bags gave approximately the same standard error as the mean of 2.5 nylon bags and the mean of two CB measurements. The results showed a direct proportionality between the dry matter digestibility values obtained in situ and in vitro with different techniques (in situ-nylon vs. conventional bottles and in situ-filter vs. AD^{II}).

Alende et al. [25] compared three different DMD methods (AD^{II} incubator, batch culture, and Ankom gas production) considering four incubation times (12, 24, 36, and 48 h); the results obtained at 24 h were compared with those obtained from dual-flow, continuous-culture fermenters. The results showed that different methods yield different DMD values. When the incubation time was longer than 12 h, the predicted DMD from the AD^{II} incubator was greater than when the gas production and the batch culture methods were used. The apparent DM digestibility, estimated using the continuous culture fermenter, was similar to that obtained from the batch culture and gas production, but it was lower than that of the AD^{II} incubator. Damiran et al. [87] concluded that the AD^{II} technique is able to accurately predict in vivo and the in situ DMD. Table 5 summarises the references pertaining to comparisons with other methods.

Table 5. Comparison of the Ankom Daisy^{II} incubator (AD^{II}) with other digestibility methods.

Methods	Results (Referred to AD^{II} Technique)	Ref.
TT	True DMD: no differences	[97]
VS	True DMD considering 3 incubation times: the AD^{II} technique was less efficient but there were no significant differences	[98]
GVS	NDFD at different times: AD^{II} always lower than GVS; better results with F57 washed in acetone	[99]
GVS	NDFD: very few differences	[100]
TT, VS	Apparent and true DMD: significant differences	[102,103]
TSP	Intestinal digestibility of crude protein (R510 filter bags, up to 5 g sample): results closely results	[85]
TT	Validation of a modified TT with the AD^{II} technique	[93]
TT	Similar digestion values; the source of inoculum may affect DMD	[22]
TT	Apparent and true DMD, apparent and true OMD, NDFD. The TT gives more precise results but requires more labour	[23]
TT	Good agreement, but the AD^{II} technique gave higher values for some feeds	[73]
TT, gas production, in vivo	The results of 3 in vitro techniques (AD^{II}, TT and gas production) were highly correlated with in vivo; AD^{II} technique is faster and more accurate	[105]
Minson and McLeod [83]	Higher digestibility values were obtained with AD^{II}	[72]
In situ	NDFD was closely correlated	[92]
In situ	NDFD was 25–30% higher than in situ; a medium degree of correlation and low accuracy were achieved	[91]
In situ	Incubation at different times. The digestible NDF values were closely correlated at 48 h incubation, but the AD^{II} values of the NDFD were higher than the in situ values.	[89]
Different in situ and in vitro techniques	Lower reproducibility coefficients for AD^{II} than the other techniques; direct proportionality was observed between the in situ and in vitro DMD for different techniques	[88]
Batch culture, gas production	The AD^{II} dry matter digestibility values were higher than the gas production and batch culture values for longer incubation times than 12 h	[25]
In vivo, in situ, TT	The AD^{II} technique accurately predicted the in vivo DMD but overestimated in situ DMD; AD^{II} less accurately correlated with the TT	[87]

TT = Tilley and Terry; VS = Van Soest; GVS = Goering Van Soest, TSP = three-step procedure.

8. Use of Daisy^{II} Incubator for Non-Ruminants

8.1. Horses

The in vivo standard and the inert marker methods are optimal for the determination and assessment of the digestibility of horse feeds, but they are time consuming. The use of in vitro fermentation procedures, such as enzyme-based essays, for the prediction of pre-caecal starch digestibility [106], and the gas production technique, developed for ruminants [15] using either caecal fluid [107] or faeces as inocula [108] to study diet digestion and fermentative end products has become increasingly more popular in equine nutrition. Abdouli and Attia [109] developed a simple in vitro method that is suitable for both concentrates and forages and that combines both the pre-caecal and hind gut digestion processes. These authors focused on the duration needed to establish feed pre-digestion by pepsin–amylase and its subsequent effect on gas production and organic matter digestibility using horse faeces as a source of microbial inoculum, and they compared the results with those from low-to-high-starch and protein feeds. They concluded that this procedure should be extended and validated with a large array of feeds with known digestibility values, because the enzymatic pre-digestion treatment effects varied between samples (non-pre-digested hay, barley grain, and soybean meal). Equine faeces is a suitable source of microbial inoculum for in vitro gas production studies, and the evaluated in vitro batch culture technique showed a considerable potential for the routine prediction of the nutritive value of a wide range of equine feedstuffs [79].

Lattimer et al. [110,111] studied the effects of *Saccharomyces cerevisiae* on the in vitro fermentation of a high concentrate or high-fibre diet for horses using equine faeces as an in vitro inoculum source in an AD^{II} incubator. These authors demonstrated that the use of 0.25-g samples may yield more accurate and less varied estimates of DM digestibility. Furthermore, the DM digestibility values for the in vivo and in vitro were similar, and they concluded that the AD^{II} incubator could be used to predict the DM digestibility of diets. Earing et al. [64], evaluating the in vitro digestion of four different diets using the AD^{II} incubator, recently confirmed that equine faeces are a suitable source of microbial inoculum for in vitro digestibility studies on horses. They found comparable DM digestibility for diets consisting of timothy hay, timothy hay with oats, and alfalfa hay with oats between in vitro and in vivo methods, while different digestibility values were observed between the two methods for an alfalfa hay diet. These authors stated that further research is needed, using a wider range of forages and methods, to determine whether in vitro and in vivo digestibility methods produce similar results for horses and to establish in vitro digestibility as a viable technique for estimating digestibility in horses.

Blažková et al. [112] compared the in vivo DM digestibility of corn silage for horses with that obtained using equine faeces in an AD^{II} incubator. These authors concluded that DM digestibility is only comparable with data on ruminants, and they showed that horses have a lower DM digestibility of corn silage than ruminants. Moreover, they demonstrated that equine faeces are a suitable source of microbial inoculum for in vitro digestibility.

8.2. Donkeys

Despite the increasing interest in donkeys, studies on this species are very limited. Tassone et al. [65] demonstrated that donkey digestibility can be predicted, with a high repeatability and reproducibility, using an AD^{II} incubator, a closed-system fermentation apparatus, and donkey faeces as a source of microbial inoculum. Moreover, these authors observed that the digestibility of different feeds for donkeys needs different incubation times.

8.3. Camelids

In vitro TTs that use camel rumen liquor as an inoculum require fistulated animals to provide this inoculum [113,114]. Rumen fluid can also be obtained, for the same purpose, from slaughtered dromedaries. Lifa et al. [115] therefore investigated the suitability of this rumen fluid with the aim of evaluating the in vitro degradation characteristics of highly fermentable industrial by-products (citrus, tomato, and apple), fibrous forages, and their mixtures. They concluded that rumen fluid extracted from slaughtered dromedaries is a valuable tool for determining the in vitro degradation of camel feeds. None of these experiments on camelids were conducted using an AD[II] incubator.

The successful use of a liquid suspension of camel faeces, as an alternative inoculum for an in vitro AD[II] incubator, yielded valid in vitro estimates of the DM, NDF, and ADF (acid detergent fibre) digestibility of forages and grains and could make it unnecessary to resort to fistulated animals (particularly in tropical countries) to obtain inoculum; this could solve some practical problems, such as the constant care needed to avoid infections and the costs associated with the long-term maintenance of donor animals, as well as ethical considerations and the necessity of surgical facilities [58].

Laudadio et al. [58] evaluated the in vitro digestibility of the fodder species browsed by camels in pastures in an arid region of Southern Tunisia using an AD[II] incubator. They used different sources of faecal liquor, collected from camels, healthy mature sheep, and goats, as alternative microbial inoculum sources to test the nutrient digestibility of these forages, as well as rumen liquor, collected from sheep, as a control for the in vitro AD[II] incubator. These authors stated that the similarity of the different repetitions for all the fodders in the estimation of nutrient digestibility in the AD[II] incubator reflects its accuracy, making it comparable with traditional methods in regard to digestibility. They concluded that the AD[II] incubator is appropriate for the determination of the in vitro digestibility of nutrients when using camel faecal liquor, which could be used instead of rumen fluid to estimate the in vitro digestibility of forages.

8.4. Rabbits

An AD[II] incubator was also used in rabbit studies to determine the in vitro insoluble fibre [116] and in vitro digestibility of rabbit feedstuffs [69,70,117–120]. Abad et al. [69] adapted the in vitro digestion procedure proposed by Carabaño et al. [121] and compared the quantifications of soluble fibre in rabbit feedstuffs using different chemical and in vitro approaches. The method was modified using Ankom filter bags, which were placed in an AD[II] incubator jar rather than in crucibles (reference method) to facilitate sample filtering. No difference was observed when crucibles and Ankom bags were used (both in single or collective digestion) for two-step pepsin/pancreatin in vitro DM digestibility, corrected for ash and protein. The correlations obtained for in vitro DM digestibility were higher (0.99) than those reported by Vogel et al. [24], who studied the in vitro DM digestibility of forages for ruminants (0.92). The latter authors reported higher in vitro digestibility when using Ankom bags than when using crucibles (0.602 vs. 0.563, respectively), whereas Abad et al. [69] found much less of a difference.

Ferreira et al. [70], in order to evaluate the potential use of dried or autoclaved sugarcane bagasse and enriched or non-enriched with vinasse in the diets of growing rabbits and to determine their in vitro dry matter digestibility, modified the last step of the Abad et al. method [69] using a caecal contents diluted at a ratio of 1:1 (w/v) with a buffered mineral solution [122] as inoculum. Ferreira et al. used the same method to determine the in vitro dry matter digestibility of rabbit diets supplemented with macaúba seed cake meal [117] or with tropical ingredients, co-products, and by-products [118].

The Ramos et al. method [123], which is based on that of Boisen et al. [124], in which Ankom bags are used, and which, in turn, was modified by Abad et al. [69], was used to determine the in vitro dry matter digestibility of rabbit diets supplemented with co-products derived from olive cake [119] or with citrus co-products [120].

8.5. Guinea Pigs

López et al. [125] used an AD[II] incubator to compare two types of "in vitro" digestibility assays, using commercial enzymes and guinea pig caecal liquor with the in vivo assay to identify the assay that resembled the in vivo response the most, and they found that the optimal in vitro method to use for comparisons with the in vivo test is the caecal liquor technique because it presents a smaller difference in results.

8.6. Pigs

Several in vitro feed digestibility estimation methods have been developed and can be divided into three groups, that is single-, two-, or three-step models that simulate gastric digestion, gastric/small intestinal digestion, and gastric/small intestinal/large intestinal digestion, respectively [126]. The Boisen and Fernandez [127] in vitro gastric-ileal digestion procedure was been adapted for use in an AD^{II} incubator and it allows for the simultaneous incubation of different pig feedstuffs in sealed polyester bags (5 × 10 cm bags; R510, Ankom Technology, Macedon, NY) in the same incubation vessel [68].

Fushai [128] determined, with an AD^{II} incubator, the in vitro digestibility of growing pig diets supplemented with exogenous enzymes. Each feed was digested in pepsin, followed by pancreatin, with the recovery of the fibrous residues. The pepsin–pancreatin fibre extracts were digested, by means of Viscozyme and Roxazyme, in a third step to complete the simulated pig gastro-intestinal digestion process.

Torres-Pitarch et al. [129] determined the in vitro ileal digestibility of pig diets by means of a two-step in vitro incubation procedure, adapted from that of Akinsola [68] using an AD^{II} incubator at 39 °C with samples incubated inside Ankom F57 bags. The first step, which simulated the digestion in the stomach, was that of enzymatic hydrolysis with a pepsin solution at pH 2.0 and 39 °C for 5 h, and the second step involved hydrolysis with a multi-enzyme pancreatin at pH 6.8 and 39 °C for 17 h.

Pahm [130] compared the use of an AD^{II} incubator with three Huang et al. [131] in vitro procedures using cellulase in the third step to that of Boisen and Fernandez [127] using Viscozyme or faecal inoculum in the third step. When using the AD^{II} incubator, these authors concluded that, of the three evaluated in vitro procedures, that of increasing the incubation length of the Boisen and Fernandez [127] using Viscozyme in the third step was the one that improved the sensitivity of the assay the most, and it provided a better R^2 between the dry matter digestibility and apparent total tract digestibility of the gross energy, and between the dry matter digestibility and digestible energy, than the procedures that used cellulase or faecal inoculum.

Youssef and Kamphues [132] analysed a commercial swine diet, with lignocellulose A and B, by means an AD^{II} incubator, to determine its in vitro dry matter digestibility, using the fresh faeces of pigs as the inoculum source. The fermentation rates of the tested ingredients were evaluated using the caecum contents of swine as inoculum precursors, and these were then compared with that obtained with faeces inocula. The in vitro results were confirmed in vivo by testing the digestibility rate of the most digestible product of the lignocellulose ingredients. These authors found that the use of faeces/excreta liquor provided a valid estimate of the fermentation or digestibility of feeds, and they concluded that this procedure could be an effective way of approximating the digestibility of pig diets.

8.7. Dogs

Candellone et al. [71] recently performed in vitro analyses of dog pet food using the methods proposed by Hervera et al. [133] and Biagi et al. [134] utilizing Ankom bags and an AD^{II} incubator. They concluded that the two in vitro methods slightly overestimated the digestibility coefficients of the considered dog diets, when compared with the in vivo digestibility values. The in vitro method proposed by Hervera et al. [133] and utilized in this study yielded values closer to the in vivo results, in line with Hervera et al. [135], who showed a higher accuracy approach of in vivo crude protein apparent digestibility ($R^2 = 0.81$) and in vivo digestible energy ($R^2 = 0.94$), respectively.

9. Conclusions

This review summarised the use of the AD^{II} incubator in studies on digestibility in ruminants, as well as its extension to non-ruminants. From its introduction until today, the AD^{II} incubator has proved to be able to allow for the analysis of multiple feedstuffs, to improve the precision and reproducibility of an assay, and to reduce the time and costs of analysis. DMD values from AD^{II} and in situ techniques may be higher than those obtained in vivo [104], but both systems allow for the true digestibility of feedstuffs to be estimated, while the in vivo values only refer to the apparent digestibility.

Even though the use of the AD^{II} incubator is by now standardised, there is still a need for further research, as reported in Table S1, to summarise some practical recommendations concerning the correct use of the AD^{II} incubator. To date, there are no standardised protocols for the collection, storage, and transportation of the rumen fluid or faeces. There is also a need to standardise the procedures for washing the bags after digestion. A major problem is the type of inoculum, which is the main source of variability of the system. Some performance metrics of the instrument (such as the reliability of the rotation mechanism of the jars) also require improvement.

The authors verified the need for caution when comparing data obtained from different methods, because they can yield different results [25]. Table S2 reports the variability of the AD^{II} instrument for 48 h of incubation, as well as the coefficient of variability (CV, %) within and between laboratory, runs, jars, and samples. Table S3 shows the correlation between AD^{II} and in vivo, in situ, and Tilley and Terry digestibility, as well as the respective linear equations.

The authors also verified that there is a lack of a standard terminology in studies and, as such, propose the use of the acronyms reported in Table S4 to make the language homogeneous.

Some potential developments and evolutions in the use of the AD^{II} incubator were also described. Created and developed for digestibility studies on ruminants, before being extended to monogastric and other non-ruminant species, this technology, in the future, could in fact be used for human digestibility studies or to obtain more detailed knowledge on the nutraceutical function of some feeds.

Supplementary Materials: The following are available online at http://www.mdpi.com/2076-2615/10/5/775/s1, Table S1: Practical recommendations on the use of the Ankom Daisy^{II} incubator (AD^{II}), Table S2: Variability (CV, %) of the Ankom Daisy^{II} incubator (AD^{II}) after 48 h of incubation, Table S3: Linear equation between the Ankom Daisy^{II} incubator (AD^{II}) at 48 h and other digestibility systems, Table S4: Acronyms for digestibility trials.

Author Contributions: Conceptualization, P.G.P. and S.T.; writing—original draft preparation, P.G.P., R.F., and S.T.; writing—review and editing, P.G.P., R.F., and S.T. All authors have read and agreed to the published version of the manuscript.

Funding: This research received no external funding.

Acknowledgments: The authors would like to thank M. Jones for the linguistic revision of the manuscript.

Conflicts of Interest: The authors declare no conflict of interest.

References

1. Tilley, J.M.A.; Terry, R.A A two-stage technique for the In vitro digestion of forage crops. *Grass Forage Sci.* **1963**, *18*, 104–111. [CrossRef]
2. Van Soest, P.J. *Nutritional Ecology of the Ruminant*, 2nd ed.; Cornell University Press: Ithaca, NY, USA, 1994.
3. Klopfenstein, T.J.; Krause, V.E.; Jones, M.J.; Woods, W. Chemical treatment of low quality forages. *J. Anim. Sci.* **1972**, *35*, 418–422. [CrossRef]
4. Givens, D.I.; Cottyn, B.G.; Dewey, P.J.S.; Steg, A. A comparison of the neutral detergent-cellulase method with other laboratory methods for predicting the digestibility in vivo of maize silages from three European countries. *Anim. Feed Sci. Technol.* **1995**, *54*, 55–64. [CrossRef]
5. Adesogan, A.T.; Givens, D.I.; Owen, E. Prediction of the in vivo digestibility of whole crop wheat from in vitro digestibility, chemical composition, in situ rumen degradability, in vitro gas production and near infrared reflectance spectroscopy. *Anim. Feed Sci. Technol.* **1998**, *74*, 259–272. [CrossRef]
6. Van Soest, P.J.; Wine, R.H.; Moore, L.A. Estimation of the true digestibility of forages by the in vitro digestion of cell walls. In Proceedings of the 10th International Grassland Congress, Finnish Grassland Association, Helsinki, Finland, 7–8 July 1966; pp. 438–441.
7. Goering, M.K.; Van Soest, P.J. *Forage Fiber Analysis (Apparatus, Reagents, Procedures and some Applications)*; Agricultural Handbook No. 379; USDA: Washington, DC, USA, 1970.

8. Adesogan, A.T. What are feeds worth? A critical evaluation of selected nutritive value methods. In Proceedings of the 13th Annual Florida Ruminant Nutrition Symposium, Gainesville, FL, USA, 11–12 January 2002; pp. 33–47.

9. Czerkawski, J.W.; Breckenridge, G. Design and development of a long-term rumen simulation technique (Rusitec). *Br. J. Nutr.* **1977**, *38*, 371–384. [CrossRef]

10. Gray, F.V.; Weller, A.F.; Pilgrim, A.F.; Jones, G.E. A stringent test for the artificial rumen. *Aust. J. Agric. Res.* **1962**, *13*, 343–349. [CrossRef]

11. Aafjes, J.H.; Nijhof, J.K. A simple artificial rumen giving good production of volatile fatty acids. *Br. Vet. J.* **1967**, *123*, 436–446. [CrossRef]

12. Carro, M.D.; Ranilla, M.J.; Martin-García, A.I.; Molina-Alcaide, E. Comparison of microbial fermentation of high- and low-forage diets in Rusitec, single-flow continuous-culture fermenters and sheep rumen. *Animal* **2009**, *3*, 527–534. [CrossRef] [PubMed]

13. Martínez, M.E.; Ranilla, M.J.; Tejido, M.L.; Saro, C.; Carro, M.D. Comparison of fermentation of diets of variable composition and microbial populations in the rumen of sheep and Rusitec fermenters. II. Protozoa population and diversity of bacterial communities. *J. Dairy Sci.* **2010**, *93*, 3699–3712. [CrossRef]

14. Spanghero, M.; Chiaravalli, M.; Colombini, S.; Fabro, C.; Froldi, F.; Mason, F.; Moschini, M.; Sarnataro, C.; Schiavon, S.; Tagliapietra, F. Rumen inoculum collected from cows at slaughter or from a continuous fermenter and preserved in warm, refrigerated, chilled or freeze-dried environments for in vitro tests. *Animals* **2019**, *9*, 815. [CrossRef]

15. Menke, K.H.; Steingass, H. Estimation of the energetic feed value obtained from chemical analysis and in vitro gas production using rumen fluid. *Anim. Res. Dev.* **1988**, *28*, 7–55.

16. Theodorou, M.K.; Williams, B.A.; Dhanoa, M.S.; McAllan, A.B. A new laboratory procedure for estimating kinetic parameters associated with the digestibility of forages. In Proceedings of the International Symposium on Forage Cell Wall Structure and Digestibility, USD-ARS, Madison, WI, USA, 7–10 October 1991.

17. Hungate, R.E. *The Rumen and Its Microbes*; Academic Press: New York, NY, USA, 1966.

18. Menke, K.H.; Raab, A.; Salewski, H.; Steingass, D.; Fritz, D.; Scneider, W. The estimation of digestibility and metabolizable energy content of ruminant feedstuffs from the gas production when they are incubated with rumen liquor in vitro. *J. Agric. Sci.* **1979**, *193*, 217–225. [CrossRef]

19. Pell, A.N.; Schofield, P. Computerized monitoring of gas production to measure forage digestion in vitro. *J. Dairy Sci.* **1993**, *76*, 1063–1073. [CrossRef]

20. Schofield, P.; Pell, A.N. Measurement and kinetic analysis of the neutral detergent-soluble carbohydrate fraction of legumes and grasses. *J. Anim. Sci.* **1995**, *73*, 3455–3463. [CrossRef]

21. Layton, B.; Ankom Technology Corporation Fairport, NY, USA. Personal communication, 2019.

22. Holden, L.A. Comparison of methods of in vitro dry matter digestibility for ten feeds. *J. Dairy Sci.* **1999**, *82*, 1791–1794. [CrossRef]

23. Wilman, D.; Adesogan, A. A comparison of filter bag methods with conventional tube methods of determining the in vitro digestibility of forages. *Anim. Feed Sci. Technol.* **2000**, *84*, 33–47. [CrossRef]

24. Vogel, K.P.; Pedersen, J.F.; Masterson, S.D.; Toy, J.J. Evaluation of a filter bag system for NDF, ADF, and IVDMD forage analysis. *Crop. Sci.* **1999**, *39*, 276–279. [CrossRef]

25. Alende, M.; Lascano, G.J.; Jenkins, T.C.; Koch Pas, L.E.; Andrae, J.G. Comparison of 4 methods for determining in vitro ruminal digestibility of annual ryegrass. *Prof. Anim. Sci.* **2018**, *34*, 306–309. [CrossRef]

26. Coblentz, W.K.; Akins, M.S.; Ogden, R.K.; Bauman, L.M.; Stammer, A.J. Effects of sample size on neutral detergent fiber digestibility of triticale forages using the Ankom Daisy^{II} Incubator system. *J. Dairy Sci.* **2019**, *102*, 6987–6999. [CrossRef]

27. Hall, M.B.; Mertens, D.R. In vitro fermentation vessel type and method alter fiber digestibility estimates. *J. Dairy Sci.* **2008**, *91*, 301–307. [CrossRef]

28. Valentine, M.E.; Karayilandli, E.; Cherney, J.H.; Cherney, D.J. Comparison of in vitro long digestion methods and digestion rates for diverse forages. *Crop. Sci.* **2019**, *59*, 422–435. [CrossRef]

29. Raffrenato, E.; Ross, D.A.; Van Amburgh, M.E. Development of an in vitro method to determine rumen undigested aNDFom for use in feed evaluation. *J. Dairy Sci.* **2018**, *101*, 9888–9900. [CrossRef]

30. Robinson, P.H.; Campbell Matthews, M.; Fadel, J.G. Influence of storage time and temperature on in vitro digestion of neutral detergent fibre at 48 h, and comparison to 48 h in sacco neutral detergent fibre digestion. *Anim. Feed Sci. Technol.* **1999**, *80*, 257–266. [CrossRef]

31. Goeser, J.P.; Combs, D.K. An alternative method to assess 24-h ruminal in vitro neutral detergent fiber digestibility. *J. Dairy Sci.* **2009**, *92*, 3833–3841. [CrossRef]

32. Goeser, J.P.; Hoffman, P.C.; Combs, D.K. Modification of a rumen fluid priming technique for measuring in vitro neutral detergent fiber digestibility. *J. Dairy Sci.* **2009**, *92*, 3842–3848. [CrossRef]

33. Trujillo, A.I.; Marichal, M.D.J.; Carriquiry, M. Comparison of dry matter and neutral detergent fibre degradation of fibrous feedstuffs as determined with in situ and in vitro gravimetric procedures. *Anim. Feed Sci. Technol.* **2010**, *161*, 49–57. [CrossRef]

34. Bender, R.W.; Cook, D.E.; Combs, D.K. Comparison of in situ versus in vitro methods of fiber digestion at 120 and 288 h to quantify the indigestible neutral detergent fiber fraction of corn silage samples. *J. Dairy Sci.* **2016**, *99*, 5394–5400. [CrossRef]

35. Raffrenato, E.; Nicholson, C.F.; Van Amburgh, M.E. Development of a mathematical model to predict pool sizes and rates of digestion of 2 pools of digestible neutral detergent fiber and an undigested neutral detergent fiber fraction within various forages. *J. Dairy Sci.* **2019**, *102*, 351–364. [CrossRef]

36. Mertens, D.R. Using uNDF to predict dairy cow performance and design rations. In Proceedings of the Four-State Dairy Nutrition and Management Conference, Dubuque, IA, USA, 12–13 June 2016; pp. 12–19.

37. Mould, F.L.; Kliem, K.E.; Morgan, R.; Mauricio, R.M. In vitro microbial inoculum: A review of its function and properties. *Anim. Feed Sci. Technol.* **2005**, *123–124*, 31–50. [CrossRef]

38. López, S. In vitro and in situ techniques for estimating digestibility. In *Quantitative Aspects of Ruminant Digestion and Metabolism*, 2nd ed.; CAB International: Wallingford, UK, 2005; pp. 87–121.

39. Ramos-Morales, E.; Arco-Pérez, A.; Martín-García, A.I.; Yáñez-Ruiz, D.R.; Frutos, P.; Hervás, G. Use of stomach tubing as an alternative to rumen cannulation to study ruminal fermentation and microbiota in sheep and goats. *Anim. Feed Sci. Technol.* **2014**, *198*, 57–66. [CrossRef]

40. Beyihayo, G.A.; Omaria, R.; Namazzi, C.; Atuhaire, A. Comparison of in vitro digestibility using slaughtered and fistulated cattle as sources of inoculum. *Uganda J. Agric. Sci.* **2015**, *16*, 93–98. [CrossRef]

41. Alba, H.D.R.; Oliveira, R.L.; de Carvalho, S.T.; Ítavo, L.C.V.; Ribeiro, O.L.; do Nascimento Júnior, N.G.; Freitas, M.D.; Bezerra, L.R. Can ruminal inoculum from slaughtered cattle replace inoculum from cannulated cattle for feed evaluation research? *Semin. Ciênc. Agrár.* **2018**, *39*, 2133–2144. [CrossRef]

42. RMG Network. A Report in Support of the Rumen Microbial Genomics (RMG) Network Describing Standard Guidelines and Protocols for Data Acquisition, Analysis and Storage. Available online: http://www.rmgnetwork:user/file/37/pdf (accessed on 25 February 2020).

43. Yáñez Ruiz, D.R.; Bannink, A.; Dijkstra, J.; Kebreab, E.; Morgavi, D.P.; O'Kiely, P.; Reynolds, C.K.; Schwarm, A.; Shingfield, K.J.; Yu, Z.; et al. Design, implementation and interpretation of in vitro batch culture experiments to assess enteric methane mitigation in ruminants—A review. *Anim. Feed Sci. Technol.* **2016**, *216*, 1–18. [CrossRef]

44. Bioscreen Technologies Research Laboratories—Artificial Rumen by-Pass. Available online: https://www.youtube.com/watch?v=AAT63svtl0w (accessed on 25 February 2020).

45. Marinucci, M.T.; Dehority, B.A.; Loerch, S.C. In vitro and in vivo studies of factors affecting digestion of feeds in synthetic fiber bags. *J. Anim. Sci.* **1992**, *70*, 296–307. [CrossRef] [PubMed]

46. King, J.; Plaizier, J.C. Effects of source of rumen fluid on in vitro dry matter digestibility of feeds determined using the DAISY^{II} Incubator. *Can. J. Anim. Sci.* **2006**, *86*, 439–441. [CrossRef]

47. Ammar, H.; Lopez, S.; Andres, S.; Ranilla, M.J.; Boda, R.; Gonzalez, J.S. In vitro digestibility and fermentation kinetics of some browse plants using sheep or goat ruminal fluid as the source of inoculum. *Anim. Feed Sci. Technol.* **2008**, *147*, 90–104. [CrossRef]

48. Hervas, G.; Frutos, P.; Giraldez, F.J.; Mora, M.J.; Fernandez, B.; Mantecon, A.R. Effect of preservation on fermentative activity of rumen fluid inoculum for in vitro gas production techniques. *Anim. Feed Sci. Technol.* **2005**, *123–124*, 107–118. [CrossRef]

49. Chaudhry, A.S.; Mohamed, R.A.I. Fresh of frozen rumen contents from slaughtered cattle to estimate in vitro degradation of two contrasting feeds. *Czech. J. Anim. Sci.* **2012**, *6*, 265–273. [CrossRef]

50. Denek, N.; Can, A.; Avci, M. Frozen rumen fluid as microbial inoculum in the two-stage in vitro digestibility assay of ruminant feeds. *S. Afr. J. Anim. Sci.* **2010**, *40*, 251–256. [CrossRef]

51. Belanche, A.; Palma-Hidalgo, J.M.; Nejjam, I.; Serrano, R.; Jiménez, E.; Martín-García, I.; Yáñez-Ruiz, D.R. In vitro assessment of the factors that determine the activity of the rumen microbiota for further applications as inoculum. *J. Sci. Food Agric.* **2019**, *99*, 163–172. [CrossRef]

52. Soto, E.C.; Molina-Alcaide, E.; Khelil, H.; Yáñez-Ruiz, D.R. Ruminal microbiota developing in different in vitro simulation systems inoculated with goats rumen liquor. *Anim. Feed Sci. Technol.* **2013**, *185*, 9–18. [CrossRef]

53. Mauricio, R.M.; Owen, E.; Mould, F.L.; Givens, I.; Theodorou, M.K.; France, J.; Davi, D.R.; Dhanoa, M.S. Comparison of bovine rumen liquor and bovine faeces as inoculum for an in vitro gas production technique for evaluating forages. *Anim. Feed Sci. Technol.* **2001**, *89*, 33–48. [CrossRef]

54. Hughes, M.M.; Mlambo, V.; Lallo, C.H.O.; Jennings, P.G.A. Potency of microbial inocula from bovine faeces and rumen fluid for in vitro digestion of different tropical forage substrates. *Grass Forage Sci.* **2012**, *67*, 263–273. [CrossRef]

55. Ramin, M.; Lerose, D.; Tagliapietra, F.; Huhtanen, P. Comparison of rumen fluid inoculum vs. faecal inoculum on predicted methane production using a fully automated in vitro *gas* production system. *Livest. Sci.* **2015**, *181*, 65–71. [CrossRef]

56. Akhter, S.; Owen, E.; Theodorou, M.K.; Butler, E.A.; Minson, D.J. Bovine faeces as a source of micro-organisms for the in vitro digestibility assay of forages. *Grass Forage Sci.* **1999**, *54*, 219–226. [CrossRef]

57. Tufarelli, V.; Cazzato, E.; Ficco, A.; Laudadio, V. Assessing nutritional value and in vitro digestibility of Mediterranean pasture species using yak (*Bos grunniens*) faeces as alternative microbial inoculum in a Daisy[II] incubator. *J. Food Agric. Environ.* **2010**, *8*, 477–481. [CrossRef]

58. Laudadio, V.; Lacalandra, G.M.; Monaco, D.; Khorchani, T.; Hammadi, M.; Tufarelli, V. Faecal liquor as alternative microbial inoculum source for in vitro (Daisy[II]) technique to estimate the digestibility of feeds for camels. *J. Camelid Sci.* **2009**, *2*, 1–7.

59. Cone, J.W.; Van Gelder, A.H.; Bachmann, H. Influence of inoculum source on gas production profiles. *Anim. Feed Sci. Technol.* **2002**, *99*, 221–231. [CrossRef]

60. Chiaravalli, M.; Rapetti, L.; Rota Graziosi, A.; Galassi, G.; Crovetto, G.M.; Colombini, S. Comparison of faecal versus rumen inocula for the estimation of NDF digestibility. *Animals* **2019**, *9*, 928. [CrossRef]

61. Guzmán, M.L.; Sager, R.L. Ruminant Fecal Inolucum for In Vitro Feed Digestibility Analysis. 2016. Available online: https://unsl.academia.edu/RicardoSager (accessed on 25 February 2020).

62. Kim, M.; Kim, J.; Kuehn, L.A.; Bono, J.L.; Berry, E.D.; Kalchayanand, N.; Freetly, H.C.; Benson, A.K.; Wells, J.E. Investigation of bacterial diversity in the feces of cattle fed different diets. *J. Anim. Sci.* **2014**, *92*, 683–694. [CrossRef]

63. Lowman, R.S.; Theodorou, M.K.; Hyslop, J.J.; Dhanoa, M.S.; Cuddeford, D. Evaluation of an in vitro batch culture technique for estimating the in vivo digestibility and digestible energy content of equine feeds using equine faeces as the source of microbial inoculum. *Anim. Feed Sci. Technol.* **1999**, *80*, 11–27. [CrossRef]

64. Earing, J.E.; Cassill, B.D.; Hayes, S.H.; Vanzant, E.S.; Lawrence, L.M. Comparison of in vitro digestibility estimates using the Daisy[II] incubator with in vivo digestibility estimates in horses. *J. Anim. Sci.* **2010**, *88*, 3954–3963. [CrossRef]

65. Tassone, S.; Renna, M.; Barbera, S.; Valle, E.; Fortina, R. In vitro digestibility measurement of feedstuffs in donkeys using the Daisy[II] incubator. *J. Equine Vet. Sci.* **2019**, *75*, 122–126. [CrossRef]

66. Varadyova, Z.; Baran, M.; Zelenak, I. Comparison of two in vitro fermentation gas production methods using both rumen fluid and faecal inoculum from sheep. *Anim. Feed Sci. Technol.* **2005**, *123–124*, 81–94. [CrossRef]

67. De Boever, J.L.; Cottyn, B.G.; Andries, J.I.; Buysse, F.X.; Vanacker, J.M. The use of a cellulase technique to predict digestibility, metabolizable and net energy of forages. *Anim. Feed Sci. Technol.* **1988**, *19*, 247–260. [CrossRef]

68. Akinsola, M.P. Development of an In Vitro Technique to Determine Digestibility of High Fiber Pig Feed. Ph.D. Thesis, Tshwane University of Technology, Pretoria, South Africa, 2013.

69. Abad, R.; Ibáñez, M.A.; Carabaño, R.; García, J. Quantification of soluble fibre in feedstuffs for rabbits and evaluation of the interference between the determinations of soluble fibre and intestinal mucin. *Anim. Feed Sci. Technol.* **2013**, *182*, 61–70. [CrossRef]

70. Ferreira, F.N.A.; Ferreira, W.M.; Silva Neta, C.S.; Inácio, D.F.S.; Mota, K.C.N.; Costa Júnior, M.B.; Rocha, L.F.; Lara, L.B.; Fontes, D.O. Effect of dietary inclusion of dried or autoclaved sugarcane bagasse and vinasse on live performance and in vitro evaluations on growing rabbits. *Anim. Feed Sci. Technol.* **2017**, *230*, 87–95. [CrossRef]

71. Candellone, A.; Prola, L.; Peiretti, P.G.; Tassone, S.; Longato, E.; Pattono, D.; Russo, N.; Meineri, G. In vivo and in vitro digestibility, palatability and nutritive quality of extruded dog food based on mechanically separated chicken meat or meat by-product. *Ital. J. Anim. Sci.* **2019**, *18* (Suppl. 1), 109.

72. Figueiredo, M.; Mbhele, A.; Zondi, J. An evaluation of the Daisy II-220 technique for determining in vitro digestibility of animal feeds in comparison with the Minson & McLeod technique. *S. Afr. J. Anim. Sci.* **2000**, *30*, 45–46.

73. Mabjeesh, S.J.; Cohen, M.; Arieli, A. In vitro methods for measuring the dry matter digestibility of ruminant feedstuffs: Comparison of methods and inoculum source. *J. Dairy Sci.* **2000**, *83*, 2289–2294. [CrossRef]

74. Vanzant, E.S.; Cochran, R.C.; Titgemeyer, E.C. Standardization of in situ techniques for ruminant feedstuff evaluation. *J. Anim. Sci.* **1998**, *76*, 2717–2729. [CrossRef]

75. In Vitro True Digestibility Using the DaisyII Incubator. Available online: https://www.ankom.com/sites/default/files/document-files/Method_3_Invitro_D200_D200I.pdf (accessed on 25 February 2020).

76. Coblentz, W.; Akins, M. Comparisons of fiber digestibility for triticale forages at two different sample sizes using the Ankom Daisy Incubator II System. In Proceedings of the ADSA Annual Meeting, Cincinnati, OH, USA, 23–26 June 2019. Abstract T69.

77. Cattani, M.; Tagliapietra, F.; Bailoni, L.; Schiavon, S. In vitro rumen feed degradability assessed with Daisy[II] and batch culture: Effect of sample size. *Ital. J. Anim. Sci.* **2009**, *8*, 169–171. [CrossRef]

78. Coblentz, W.K.; Akins, M.S.; Kalscheur, K.F.; Brink, G.E.; Cavadini, J.S. Effects of growth stage and growing degree day accumulations on triticale forages: 1) Dry matter yield, nutritive value, and in vitro dry matter disappearance. *J. Dairy Sci.* **2018**, *101*, 8965–8985. [CrossRef]

79. Adesogan, A.T. Effect of bag type on the apparent digestibility of feeds in Ankom Daisy[II] incubators. *Anim. Feed Sci. Technol.* **2005**, *119*, 333–344. [CrossRef]

80. Anassori, E.; Dalir-Naghadeh, B.; Pirmohammadi, R.; Taghizadeh, A.; Asri-Rezaei, S.; Farahmand-Azar, S.; Besharati, M.; Tahmoozi, M. In vitro assessment of the digestibility of forage based sheep diet, supplemented with raw garlic, garlic oil and monensin. *Vet. Res. Forum* **2012**, *3*, 5–11.

81. Coles, L.T.; Moughan, P.J.; Darragh, A.J. In vitro digestion and fermentation methods incuding gas production techniques, as applied to nutritive evaluation of foods in the hindgut of humans and other simple-stomached animals. *Anim. Feed Sci. Technol.* **2005**, *123–124*, 421–444. [CrossRef]

82. Marten, G.C.; Barnes, R.F. Prediction of energy digestibility of forages in vitro rumen fermentation and fungal enzyme systems. In Proceedings of the International Workshop on Standardization of Analytical Methodology for Feeds, Ottawa, ON, Canada, 12–14 March 1979; Pigden, W.J., Balch, C.C., Graham, M., Eds.; Unipub: New York, NY, USA, 1979; pp. 61–71.

83. Minson, D.J.; Mcleod, M.N. *Division of Tropical Pastures Technical*; Paper No. 8; Commonwealth Scientific and Industrial Research Organization: Melbourne, Australia, 1972.

84. Calsamiglia, S.; Stern, M.D. A three-step in vitro procedure for estimating intestinal digestion of protein in ruminants. *J. Anim. Sci.* **1995**, *73*, 1459–1465. [CrossRef]

85. Gargallo, S.; Calsamiglia, S.; Ferret, A. A modified three-step in vitro procedure to determine intestinal digestion of proteins. *J. Anim. Sci.* **2006**, *84*, 2163–2167. [CrossRef]

86. Slyter, L.L.; Bryant, M.P.; Wolin, M.J. Effect of pH on population and fermentation in a continuously cultured rumen ecosystem. *Appl. Microbiol.* **1966**, *14*, 573–578. [CrossRef]

87. Damiran, D.; DelCurto, T.; Bohnert, D.W.; Findholt, S.L. Comparison of techniques and grinding size to estimate digestibility of forage based ruminant diets. *Anim. Feed Sci. Technol.* **2008**, *141*, 15–35. [CrossRef]

88. Tagliapietra, F.; Cattani, M.; Hindrichsen, I.K.; Hansen, H.H.; Colombini, S.; Bailoni, L.; Schiavon, S. True dry matter digestibility of feeds evaluated in situ with different bags and in vitro using rumen fluid collected from intact donor cows. *Anim. Prod. Sci.* **2012**, *52*, 338–346. [CrossRef]

89. Spanghero, M.; Boccalon, S.; Gracco, L.; Gruber, L. NDF degradability of hays measured in situ and in vitro. *Anim. Feed Sci. Technol.* **2003**, *104*, 201–208. [CrossRef]

90. Bovera, F.; Spanghero, M.; Galassi, G.; Masoero, F.; Buccioni, A. Repeatability and reproducibility of the Cornell Net Carbohydrate and Protein System analytical determinations. *Ital. J. Anim. Sci.* **2003**, *2*, 41–50. [CrossRef]

91. Spanghero, M.; Gruber, L.; Zanfi, C. Precision and accuracy of the NDF rumen degradability of hays measured by the Daisy fermenter. *Ital. J. Anim. Sci.* **2007**, *6*, 363–365. [CrossRef]

92. Spanghero, M.; Berzaghi, P.; Fortina, R.; Masoero, F.; Rapetti, L.; Zanfi, C.; Tassone, S.; Gallo, A.; Colombini, S.; Ferlito, J.C. Precision and accuracy of in vitro digestion of neutral detergent fiber and predicted net energy of lactation content of fibrous feeds. *J. Dairy Sci.* **2010**, *93*, 4855–4859. [CrossRef]

93. Cişmileanu, A.E.; Toma, S. Validation of the in vitro ruminant digestibility method applied on Daisy incubator. *Anim. Sci. Biotech.* **2017**, *50*, 8–10.

94. Ankom. Technical FAQs. Available online: https://www.ankom.com/technical-support/daisy-incubator (accessed on 15 February 2020).

95. Komarek, A.R.; Robertson, J.B.; Van Soest, P.J. Comparison of filter bag technique to conventional filtration in the Van Soest analysis of 21 feeds. In Proceedings of the National Conference on Forage Quality, Evaluation and Utilization, Lincoln, NE, USA, 13–15 April 1994.

96. Cherney, D.J.R.; Traxler, M.J.; Robertson, J.B. Use of Ankom fiber determination systems to determine digestibility. In Proceedings of the NIRS Forage and feed Testing Consortium Annual Conference, Madison, WI, USA, 19–20 February 1997.

97. Ayangbile, O.A.; Meier, J.C.; Vogel, M.K.; Robertson, J.; McElroy, A.R.; Komarek, A.R. Cryogenically protected and fresh rumen inoculum for digestibility study. In Proceedings of the 23rd Biennial Conference on Rumen Function, Chicago, IL, USA, 14–16 November 1995.

98. Traxler, M.J.; Robertson, J.B.; Van Soest, P.J.; Fox, D.G.; Pell, A.N. A comparison of methods for determining IVDMD at three time periods using the filter bag technique versus conventional methods. *J. Dairy Sci.* **1995**, *78*, 274.

99. Cohen, M.A.; Maslanka, H.E.; Kung, L. An evaluation of automated and manual in vitro methods for estimation of NDF digestion. In Proceedings of the Conference on Rumen Physiology, Chicago, IL, USA, 11–13 November 1997.

100. Traxler, M.J. Predicting the Effect of Lignin on the Extent of Digestion and the Evaluation of Alternative Intake Models for Lactating Dairy Cows Consuming High NDF Forages. Ph.D. Thesis, Cornell University, Ithaca, NY, USA, 1997.

101. Brons, E.; Plaizier, J.C. Comparisons of methods for in vitro dry matter digestibility of ruminant feeds. *Can. J. Anim. Sci.* **2005**, *85*, 243–245. [CrossRef]

102. Ammar, H.; López, S.; Gonzales, J.S.; Ranilla, M.J. Seasonal variations in the chemical composition and in vitro digestibility of some Spanish leguminous shrub species. *Anim. Feed Sci. Technol.* **2004**, *115*, 327–340. [CrossRef]

103. Ammar, H.; López, S.; Bochi-Brum, O.; Garcia, R.; Ranilla, M.J. Composition and in vitro digestibility of leaves and stems of grasses and legumes harvested from permanent mountain meadows at different stages of maturity. *J. Anim. Feed Sci.* **1999**, *8*, 599–610. [CrossRef]

104. Ankom Procedures for Fiber and In Vitro Analysis. Available online: http://www.ankom.com (accessed on 25 February 2020).

105. Ricci, P.; Romera, A.J.; Burges, J.C.; Fernández, H.H.; Cangiano, C.A. Case study: Precision and accuracy of methodologies for estimating in vitro digestibility of *Thinopyrum ponticum* (Tall Wheatgrass) hay and haylage fed to beef cattle. *Prof. Anim. Sci.* **2009**, *25*, 625–632. [CrossRef]

106. Rowe, J.; Bird, S.; Brown, W. *Safe and Effective Grain Feeding for Horses: A Report for the Rural Industries Research and Development Corporation*; RIRDC Project No. UNE-62A; RIRDC: Barton, Australia, 2001.

107. Macheboeuf, D.; Jestin, M.; Martin-Rosset, W. Utilization of the gas test method and horse faeces as a source of inoculum. *BSAP Occas. Publ.* **1998**, *22*, 187–189. [CrossRef]

108. Macheboeuf, D.; Jestin, M. Utilization of the gas test method using horse faeces as a source of inoculums. In Proceedings of the Symposium on In Vitro Techniques for Measuring Nutrient Supply to Ruminants, Penicuik, UK, 8–10 July 1997; p. 36.

109. Abdouli, H.; Attia, S.B. Evaluation of a two-stage in vitro technique for estimating digestibility of equine feeds using horse faeces as the source of microbial inoculum. *Anim. Feed Sci. Technol.* **2007**, *132*, 155–162. [CrossRef]

110. Lattimer, J.M.; Cooper, S.R.; Freeman, D.W.; Lalman, D.A. Effects of *Saccharomyces cerevisiae* on in vitro fermentation of a high concentrate or high fiber diet in horses. In Proceedings of the 19th Symposium of the Equine Science Society, Tucson, AZ, USA, 31 May–3 June 2005; pp. 168–173.

111. Lattimer, J.M.; Cooper, S.R.; Freeman, D.W.; Lalman, D.L. Effect of yeast culture on in vitro fermentation of a high-concentrate or high-fiber diet using equine fecal inoculum in a Daisy^{II} incubator. *J. Anim. Sci.* **2007**, *85*, 2484–2491. [CrossRef]

112. Blažková, K.; Homolka, P.; Maršálek, M. The corn silage digestibility by horses. *MendelNet* **2009**, *9*, 171–176.

113. Towhidi, A.; Zhandi, M. Chemical composition, in vitro digestibility and palatability of nine plant species for dromedary camels in the province of Semnan, Iran. *Egypt. J. Biol.* **2007**, *9*, 47–52.

114. Mohammadabadi, T.; Kakar, A.R. Comparison of in vitro digestibility of diets containing Subabul plant as fodder in dromedary camel and cow. *Explor. Anim. Med. Res.* **2019**, *9*, 61–66.

115. Lifa, M.; Haddi, M.L.; Tagliapietra, F.; Cattani, M.; Guadagnin, M.; Sulas, L.; Muresu, R.; Schiavon, S.; Bailoni, L.; Squartini, A. Chemical and fermentative characteristics of agricultural byproducts and their mixtures with roughages incubated with rumen fluid from slaughtered dromedaries. *Turk. J. Vet. Anim. Sci.* **2018**, *42*, 590–599. [CrossRef]

116. Abad-Guamán, R.; Carabaño, R.; Gómez-Conde, M.S.; García, J. Effect of type of fiber, site of fermentation, and method of analysis on digestibility of soluble and insoluble fiber in rabbits. *J. Anim. Sci.* **2015**, *93*, 2860–2871. [CrossRef]

117. Ferreira, W.M.; Ferreira, F.N.A.; Inácio, D.F.D.S.; Mota, K.C.D.N.; Costa Júnior, M.B.D.; Silva Neta, C.S.; Rocha, L.F.D.; Miranda, E.R.D. Effects of dietary inclusion of macaúba seed cake meal on performance, caecotrophy traits and in vitro evaluations for growing rabbits. *Arch. Anim. Nutr.* **2018**, *72*, 138–152. [CrossRef]

118. Ferreira, F.N.A.; Ferreira, W.M.; Inácio, D.F.D.S.; Neta, C.S.S.; Mota, K.C.D.N.; Costa Júnior, M.B.D.; Rocha, L.F.D.; Caicedo, W.O. In vitro digestion and fermentation characteristics of tropical ingredients, co-products and by-products with potential use in diets for rabbits. *Anim. Feed Sci. Technol.* **2019**, *252*, 1–10. [CrossRef]

119. De Blas, J.C.; Rodriguez, C.A.; Bacha, F.; Fernandez, R.; Abad-Guamán, R. Nutritive value of co-products derived from olive cake in rabbit feeding. *World Rabbit Sci.* **2015**, *23*, 255–262. [CrossRef]

120. De Blas, J.C.; Ferrer, P.; Rodríguez, C.A.; Cerisuelo, A.; García-Rebollar, P.; Calvet, S.; Farias, C. Nutritive value of citrus co-products in rabbit feeding. *World Rabbit Sci.* **2018**, *26*, 7–14. [CrossRef]

121. Carabaño, R.; Nicodemus, N.; García, J.; Xiccato, G.; Trocino, A.; Amorós, P.; Maertens, L. In vitro analysis, an accurate tool to estimate dry matter digestibility in rabbits. Intra- and inter-laboratory variability. *World Rabbit Sci.* **2008**, *16*, 195–203. [CrossRef]

122. Theodorou, M.K.; Williams, B.A.; Dhanoa, M.S.; McAllan, A.B.; France, J. A simple gas production method using a pressure transducer to determine the fermentation kinetics of ruminant feeds. *Anim. Feed Sci. Technol.* **1994**, *48*, 185–197. [CrossRef]

123. Ramos, M.A.; Carabaño, R.; Boisen, S. An in vitro method for estimating digestibility in rabbits. *J. Appl. Rabbit Res.* **1992**, *15*, 938–946.

124. Boisen, S. A model for feed evaluation based on in vitro digestibility dry matter and protein. In *In Vitro Digestion for Pigs and Poultry*; Fuller, M.F., Ed.; CAB International: Wallingford, UK, 1991; pp. 135–145.

125. López, S.; Guevara, H.; Duchi, N.; Moreno, G. Evaluation of two "in vitro" digestibility tests with the "in vivo" test of alfalfa (*Medicago sativa*) in guinea pig (*Cavia porcellus*) feeding. *Eur. Sci. J.* **2018**, *14*, 399–404. [CrossRef]

126. Chiang, C.-C.; Croom, J.; Chuang, S.-T.; Chiou, P.W.S.; Yu, B. Development of a dynamic system simulating pig gastric digestion. *Asian Aust. J. Anim. Sci.* **2008**, *21*, 1522–1528. [CrossRef]

127. Boisen, S.; Fernández, J.A. Prediction of the total tract digestibility of energy in feedstuffs and pig diets by in vitro analyses. *Anim. Feed Sci. Technol.* **1997**, *68*, 277–286. [CrossRef]

128. Fushai, F. Fermentability of Dietary Fibre and Metabolic Impacts of Including High Levels of fibrous Feed Ingredients in Maize-Soybean Growing Pig Diets Supplemented with Exogenous Enzymes. Ph.D. Thesis, Tshwane University of Technology, Pretoria, South Africa, 2014.

129. Torres-Pitarch, A.; McCormack, U.M.; Beattie, V.E.; Magowan, E.; Gardiner, G.E.; Pérez-Vendrell, A.M.; Torrallardonaf, D.; O'Dohertye, J.V.; Lawlor, P.G. Effect of phytase, carbohydrase, and protease addition to a wheat distillers dried grains with solubles and rapeseed based diet on in vitro ileal digestibility, growth, and bone mineral density of grower-finisher pigs. *Livest. Sci.* **2018**, *216*, 94–99. [CrossRef]

130. Pahm, S.F.C. In Vivo and In Vitro Disappearance of Energy and Nutrients in Novel Carbohydrates and Cereal Grains by Pigs. Ph.D. Thesis, University of Illinois, Urbana-Champaign, IL, USA, 2011.

131. Huang, G.; Sauer, W.C.; He, J.; Hwangbo, J.; Wang, X. The nutritive value of hulled and hulless barley for growing pigs. 1. Determination of energy and protein digestibility with in vivo and in vitro methods. *J. Anim. Feed Sci.* **2003**, *12*, 759–769. [CrossRef]

132. Youssef, I.M.; Kamphues, J. Fermentation of lignocellulose ingredients in vivo and in vitro via using fecal and caecal inoculums of monogastric animals (swine/turkeys). *Beni Suef Univ. J. Basic Appl. Sci.* **2018**, *7*, 407–413. [CrossRef]

133. Hervera, M.; Baucells, M.D.; Blanch, F.; Castrillo, C. Prediction of digestible energy content of extruded dog food by in vitro analyses. *J. Anim. Physiol. Anim. Nutr.* **2007**, *91*, 205–209. [CrossRef]

134. Biagi, G.; Cipollini, I.; Grandi, M.; Pinna, C.; Vecchiato, C.G.; Zaghini, G. A new in vitro method to evaluate digestibility of commercial diets for dogs. *Ital. J. Anim. Sci.* **2016**, *15*, 617–625. [CrossRef]

135. Hervera, M.; Baucells, M.D.; González, G.; Pérez, E.; Castrillo, C. Prediction of digestible protein content of dry extruded dog foods: Comparison of methods. *J. Anim. Physiol. Anim. Nutr.* **2009**, *93*, 366–372. [CrossRef] [PubMed]

Erratum

Erratum: Rufino-Moya, P.J., et al. Methane Production of Fresh Sainfoin, with or without PEG, and Fresh Alfalfa at Different Stages of Maturity is Similar, but the Fermentation End Products Vary. *Animals* 2019, 9, 197

Pablo José Rufino-Moya, Mireia Blanco, Juan Ramón Bertolín and Margalida Joy *

Centro de Investigación y Tecnología Agroalimentaria de Aragón (CITA), Instituto Agroalimentario de Aragón–IA2 (CITA-Universidad de Zaragoza), Avda, Montañana 930, 50059 Zaragoza, Spain; pjrufino@cita-aragon.es (P.J.R.-M.); mblanco@aragon.es (M.B.); jrbertolin@cita-aragon.es (J.R.B.)
* Correspondence: mjoy@aragon.es; Tel.: +34-976-716-442

Received: 28 June 2019; Accepted: 3 July 2019; Published: 5 July 2019

The authors wish to make the following correction to their paper [1].

In Table 2, the production of methane in alfalfa at the start-flowering should be 38 mL/g dOM and not 3 mL/g dOM.

Table 2. Effect of the substrate (S) and the stage of maturity [1] (SM) on gas and methane production (CH_4), potential gas production (A), rate of gas production (c), in vitro organic matter degradability (IVOMD), ammonia (NH_3-N), and volatile fatty acids (VFAs).

Item	Alfalfa			Sainfoin			Sainfoin+PEG			r.s.d.[2]	P-values		
	VEG	Start-F	End-F	VEG	Start-F	End-F	VEG	Start-F	End-F		S	SM	SxSM
pH	6.42 ^{a}	6.44 ^{a}	6.41 ^{a}	6.35 ^{bx}	6.32 ^{by}	6.33 ^{by}	6.37 ^{bx}	6.31 ^{bz}	6.34 ^{by}	0.032	<0.001	0.002	0.01
Gas production (mL/g dOM ^{3})	179	184	188	183	163	181	180	199	173	26.5	0.43	0.97	0.12
A (mL)	68 ^{x}	67 ^{bxy}	62 ^{y}	63 ^{y}	74 ^{bx}	70 ^{x}	65 ^{y}	78 ^{ax}	67 ^{y}	5.7	0.01	<0.001	<0.001
c (h^{-1})	0.2 ^{a}	0.22 ^{a}	0.19	0.14 ^{by}	0.16 ^{bxy}	0.19 ^{y}	0.15 ^{bx}	0.15 ^{bx}	0.19 ^{y}	0.037	<0.001	0.02	0.05
CH_4 production (mL/g dOM ^{3})	37	38	39	37	33	38	37	39	36	5.0	0.29	0.73	0.17
CH_4:gas (%)	14.5	14.2	14.1	14.1 ^{xy}	13.4 ^{y}	14.8 ^{x}	14.5	14.4	14.4	0.88	0.35	0.13	0.11
IVOMD (%)	85.13 ^{bx}	83.73 ^{bx}	76.97 ^{by}	84.97 ^{by}	92.59 ^{ax}	85.58 ^{ay}	86.81 ^{ay}	93.59 ^{ax}	86.41 ^{ay}	1.692	<0.001	<0.001	<0.001
NH_3-N (mg/L)	240 ^{a}	247 ^{a}	210	206 ^{b}	184 ^{b}	201	242 ^{a}	234 ^{a}	215	31.7	<0.001	0.06	0.20
Total VFAs (mmol/L)	102 ^{xy}	105 ^{x}	97 ^{y}	96 ^{y}	103 ^{x}	101 ^{xy}	99 ^{y}	105 ^{x}	102 ^{xy}	7.1	0.61	0.02	0.35
Acetic acid (C_2) (mol/100 mol)	63.1 ^{cy}	63.9 ^{bx}	64.0 ^{bx}	65.4 ^{ay}	66.5 ^{ax}	66.3 ^{ax}	63.9 ^{by}	64.6 ^{bx}	64.5 ^{bx}	0.76	<0.001	<0.001	0.95
Propionic acid (C_3) (mol/100 mol)	15.5 ^{a}	15.3 ^{a}	15.3 ^{a}	15 ^{bx}	14.5 ^{cy}	14.5 ^{by}	15.3 ^{ax}	14.9 ^{by}	15.2 ^{ax}	0.28	<0.001	<0.001	0.09
Butyric acid (mol/100 mol)	13.0 ^{a}	12.6 ^{a}	12.6 ^{a}	12.1 ^{c}	12.2 ^{b}	12 ^{b}	12.5 ^{b}	12.8 ^{a}	12.4 ^{a}	0.45	<0.001	0.20	0.16
Iso-butyric acid (mol/100 mol)	1.97 ^{ax}	1.91 ^{axy}	1.87 ^{ay}	1.82 ^{bx}	1.68 ^{by}	1.72 ^{by}	1.97 ^{ax}	1.83 ^{ay}	1.84 ^{ay}	0.100	<0.001	<0.001	0.63
Valeric acid (mol/100 mol)	2.14 ^{axy}	2.09 ^{ay}	2.22 ^{ax}	1.87 ^{bx}	1.7 ^{cy}	1.83 ^{cx}	2.12 ^{ay}	1.95 ^{bx}	2.04 ^{by}	0.107	<0.001	<0.001	0.11
Iso-valeric acid (mol/100 mol)	4.27 ^{ax}	4.09 ^{axy}	4.04 ^{ay}	3.87 ^{bx}	3.5 ^{by}	3.65 ^{by}	4.19 ^{ax}	3.91 ^{ay}	3.93 ^{ay}	0.223	<0.001	<0.001	0.69
C_2:C_3 (mol/mol)	4.1 ^{cy}	4.18 ^{cxy}	4.2 ^{bx}	4.39 ^{ay}	4.62 ^{ax}	4.59 ^{ax}	4.18 ^{by}	4.36 ^{bx}	4.25 ^{by}	0.109	<0.001	<0.001	0.13
CH_4:VFA (mL/mol)	2.41	2.24 ^{b}	2.35	2.53	2.22 ^{b}	2.48	2.62	2.73 ^{a}	2.4	0.313	0.01	0.26	0.10

[1] VEG: vegetative; Start-F: start of flowering; End-F: end of flowering; [2] residual standard deviation; [3] degraded organic matter. Within a parameter, means with different superscript (a,b,c) differ at $P < 0.05$ for the substrate effect in each stage of maturity; with different superscript (x,y,z) at $P < 0.05$ for the stage of maturity effect in each substrate.

The authors would like to apologize for any inconvenience caused.

Reference

1. Rufino-Moya, P.J.; Blanco, M.; Bertolín, J.R.; Joy, M. Methane Production of Fresh Sainfoin, with or without PEG, and Fresh Alfalfa at Different Stages of Maturity is Similar but the Fermentation End Products Vary. *Animals* **2019**, *9*, 197. [CrossRef] [PubMed]

MDPI

St. Alban-Anlage 66

4052 Basel

Switzerland

Tel. +41 61 683 77 34

Fax +41 61 302 89 18

www.mdpi.com

Animals Editorial Office

E-mail: animals@mdpi.com

www.mdpi.com/journal/animals

CPSIA information can be obtained
at www.ICGtesting.com
Printed in the USA
LVHW021632250820
664158LV00009B/287